INCOMPLETE NATURE

ALSO BY TERRENCE W. DEACON

The Symbolic Species

Incomplete Nature

HOW MIND EMERGED FROM MATTER

Terrence W. Deacon

W. W. NORTON & COMPANY   *New York • London*

Selection from *Number: The Language of Science* by Tobias Dantzig reprinted by permission of Simon & Schuster and the estate of Tobias Dantzig. Selection from *Brainstorms* by Daniel Dennett reprinted by permission of The MIT Press. Selection from "Behaviorism at Fifty" by B. F. Skinner reprinted by permission of University of Chicago Press. Selection from "Principles of the Self-Organizing System" by W. Ross Ashby reprinted by permission of the estate of W. Ross Ashby. Selection from "Some Remarks on the Notion of Order" by David Bohm reprinted by permission of Aldine Transaction Co. Selection from *Understanding Thermodynamics* by H. C. Van Ness reprinted by permission of Dover Publications. Selection from "Visions of Evolution: Self-organization Proposes What Natural Selection Disposes" reprinted by permission of The MIT Press. Selection from "The Origin of Selves" by Daniel Dennett reprinted by permission of the Philosophy Documentation Center. Selection from "The Big Idea: Can There Be a Science of Mind?" by Jerry Fodor reprinted by permission of the *Times Literary Supplement* / NI Syndication. Selection from "They're Made Out of Meat" by Terry Bisson reprinted by permission of Terry Bisson. Selection from *Reasoning and the Logic of Things: The Cambridge Conferences Lectures of 1898* by Charles Sanders Peirce, edited by Kenneth Laine Ketner, with an Introduction by Kenneth Laine Ketner and Hilary Putnam, reprinted by permission of the publisher. Copyright © 1992 by the President and Fellows of Harvard College.

For information about permission to reproduce selections from this book, write to Permissions, W. W. Norton & Company, Inc., 500 Fifth Avenue, New York, NY 10110

For information about special discounts for bulk purchases, please contact W. W. Norton Special Sales at specialsales@wwnorton.com or 800-233-4830

Manufacturing by Courier Westford
Book design by Dana Sloan
Production manager: Julia Druskin

Library of Congress Cataloging-in-Publication Data

Deacon, Terrence William.
Incomplete nature : how mind emerged from matter / Terrence W. Deacon. — 1st ed.
p. cm.
Includes bibliographical references and index.
ISBN 978-0-393-04991-6 (hbk.)
1. Consciousness. 2. Life—Origin. 3. Phenomenology. 4. Brain. I. Title.
QP411.D43 2012
612.8'2—dc23

2011030827

W. W. Norton & Company, Inc.
500 Fifth Avenue, New York, N.Y. 10110
www.wwnorton.com

W. W. Norton & Company Ltd.
Castle House, 75/76 Wells Street, London W1T 3QT

1 2 3 4 5 6 7 8 9 0

To my parents, Bill and JoAnne Deacon (and "the Pirates")

CONTENTS

ACKNOWLEDGMENTS

I t any of us can claim to see the world from a new vantage point it is because, as Isaac Newton once remarked, we are standing on the shoulders of giants. This is beyond question in my case, since each of the many threads that form the fabric of the theory presented in this book trace their origin to the life works of some of the greatest minds in history. But much more than this, it is often the case that it takes the convergence of like minds stubbornly questioning even the least dubious of our collective assumptions to shed the blinders that limit a particular conceptual paradigm. This is also true here. Few of the novel ideas explored here emerged from my personal reflections fully formed, and few even exhibited the rough-hewn form described in this book. These embryonic ideas were lucky to have been nourished by the truly dedicated and unselfish labors of a handful of brilliant, insightful, and questioning colleagues, who gathered in my living room, week after week, year after year, to question the assumptions, brainstorm about the right term for a new idea, or just struggle to grasp a concept that at every turn seemed to slip through our grasp. We were dubbed "Terry and the Pirates" (after the title of a postwar comic strip) because of our intense intellectual camaraderie and paradigm-challenging enterprise. This continuous thread of conversation has persisted over the course of nearly a decade and has helped to turn what was once nearly inconceivable into something merely counterintuitive.

The original Pirates include (in alphabetical order) Tyrone Cashman, Jamie Haag, Julie Hui, Eduardo Kohn, Jay Ogilvy, and Jeremy Sherman, with

Ursula Goodenough and Michael Silberstein playing the role of periodic drop-ins from across the country. Most recently Alok Srivastava, Hajime Yamauchi, Drew Halley, and Dillon Niederhut have joined, as Jamie and Eduardo moved away to careers elsewhere in the country. Most have read through major portions of this manuscript and have offered useful comments and editorial feedback. Portions of the chapter on "Self" were re-edited from a paper co-written with Jay Ogilvy and Jamie Haag, and fragments of a number of chapters were influenced by papers co-authored either with Tyrone Cashman or with Jeremy Sherman. The close involvement of these many colleagues was demonstrated also in the long days that Ty Cashman, Jeremey Sherman, Julie Hui, and Hajime Yamauchi spent with me struggling through more than 700 pages of copy-editing queries and doing the necessary error checking that led to the final manuscript. Without the collective intelligence and work of this extended mind and body, it is difficult for me to imagine how these ideas would ever have seen the light of day, at least in a semi-readable form. In addition, I owe considerable thanks to the many colleagues around the country who have read bits of the manuscript and have offered their wise counsel. Ursula Goodenough, Michael Silberstein, Ambrose Nahas, and Don Favereau, in particular, read through early versions of various parts and provided extensive editorial feedback. Also my editor on *The Symbolic Species*, Hilary Hinzman, did extensive editorial work on the first four chapters, which therefore will be the most readable. These efforts have helped to greatly improve the presentation by cleaning up some of the oversights and confusions, and simplifying my sometimes tortured prose. Indeed, I have received far more editorial feedback than I have been able to take advantage of, given time and space.

So, it is only by the great good fortune of being in the right place and time in the history of ideas—heir to countless works of genius—and surrounded by a loyal band of fellow travelers that these ideas can be shared in a form that is even barely comprehensible. This loyal band of "Pirates" has been critical, supportive, patient, and insistent in just the right mix to make it happen. The book is a testament to the marvelous synergy of these many converging contributions, which have made this effort both a most exciting journey and a source of enduring friendships. Thank you all.

Finally, and most important, has been the constant unfaltering encouragement of my wife and best friend, Ella Ray, who has understood more than anyone else the limitations of my easily distracted mind and how to help me focus on the challenge of putting these ideas on paper, month after month, year after year.

Berkeley, September 2010

INCOMPLETE NATURE

0

ABSENCE[1]

In the history of culture, the discovery of zero
will always stand out as one of the greatest single
achievements of the human race.

—TOBIAS DANTZIG, 1930

THE MISSING CIPHER

Science has advanced to the point where we can precisely arrange individual atoms on a metal surface or identify people's continents of ances-
try by analyzing the DNA contained in their hair. And yet ironically we
lack a scientific understanding of how sentences in a book refer to atoms,
DNA, or anything at all. This is a serious problem. Basically, it means that
our best science—that collection of theories that presumably come closest to explaining everything—does not include this one most fundamental
defining characteristic of being you and me. In effect, our current "Theory
of Everything" implies that we don't exist, except as collections of atoms.

So what's missing? Ironically and enigmatically, something missing is
missing.

Consider the following familiar facts. The meaning of a sentence is not
the squiggles used to represent letters on a piece of paper or a screen. It is
not the sounds these squiggles might prompt you to utter. It is not even the
buzz of neuronal events that take place in your brain as you read them. What
a sentence means, and what it refers to, lack the properties that something
typically needs in order to make a difference in the world. The information

1

conveyed by this sentence has no mass, no momentum, no electric charge, no solidity, and no clear extension in the space within you, around you, or anywhere. More troublesome than this, the sentences you are reading right now could be nonsense, in which case there isn't anything in the world that they could correspond to. But even this property of being a pretender to significance will make a physical difference in the world if it somehow influences how you might think or act.

Obviously, despite this something not-present that characterizes the contents of my thoughts and the meaning of these words, I wrote them because of the meanings that they might convey. And this is presumably why you are focusing your eyes on them, and what might prompt you to expend a bit of mental effort to make sense of them. In other words, the content of this or any sentence—a something-that-is-not-a-thing—has physical consequences. But how?

Meaning isn't the only thing that presents a problem of this sort. Several other everyday relationships share this problematic feature as well. The function of a shovel isn't the shovel and isn't a hole in the ground, but rather the potential it affords for making holes easier to create. The reference of the wave of a hand in greeting is not the hand movement, nor the physical convergence of friends, but the initiation of a possible sharing of thoughts and remembered experiences. The purpose of my writing this book is not the tapping of computer keys, nor the deposit of ink on paper, nor even the production and distribution of a great many replicas of a physical book, but to share something that isn't embodied by any of these physical processes and objects: ideas. And curiously, it is precisely because these ideas lack these physical attributes that they can be shared with tens of thousands of readers without ever being depleted. Even more enigmatically, ascertaining the value of this enterprise is nearly impossible to link with any specific physical consequence. It is something almost entirely virtual: maybe nothing more than making certain ideas easier to conceive, or, if my suspicions prove correct, increasing one's sense of belonging in the universe.

Each of these sorts of phenomena—a function, reference, purpose, or value—is in some way incomplete. There is something not-there there. Without this "something" missing, they would just be plain and simple physical objects or events, lacking these otherwise curious attributes. Longing, desire, passion, appetite, mourning, loss, aspiration—all are based on an analogous intrinsic incompleteness, an integral without-ness.

As I reflect on this odd state of things, I am struck by the fact that there is no single term that seems to refer to this elusive character of such things. So, at the risk of initiating this discussion with a clumsy neologism, I will refer to this as an *absential*[2] feature, to denote phenomena whose existence is determined with respect to an essential absence. This could be a state of things not yet realized, a specific separate object of a representation, a general type of property that may or may not exist, an abstract quality, an experience, and so forth—just not that which *is* actually present. This paradoxical intrinsic quality of existing with respect to something missing, separate, and possibly nonexistent is irrelevant when it comes to inanimate things, but *it is a defining property of life and mind*. A complete theory of the world that includes us, and our experience of the world, must make sense of the way that we are shaped by and emerge from such specific absences. What is absent matters, and yet our current understanding of the physical universe suggests that it should not. A causal role for absence seems to be absent from the natural sciences.

WHAT MATTERS?

Modern science is of course interested in explaining things that are materially and energetically present. We are interested in how physical objects behave under all manner of circumstances, what sorts of objects they are in turn composed of, and how the physical properties expressed in things at one moment influence what will happen at later moments. This includes even phenomena (I hesitate to call them objects or events) as strange and as hard to get a clear sense of as the quantum processes occurring at the unimaginably small subatomic scale. But even though quantum phenomena are often described in terms of possible physical properties not yet actualized, they are physically present in some as-yet-unspecified sense, and not absent, or represented. A purpose not yet actualized, a quality of feeling, a functional value just discovered—these are not just superimposed probable physical relationships. They are each an intrinsically absent aspect of something present.

The scientific focus on things present and actualized also helps to explain why, historically, scientific accounts have endured an uneasy co-existence with absential accounts of why things transpire as they do. This is exemplified by the relationship that each has with the notion of order. Left alone, an arrangement of a set of inanimate objects will naturally tend to fall into disorder, but we humans have a preference for certain arrangements, as do many species. Many functions and purposes are determined with respect to preferred arrangements, whether this is the arrangement of words in sentences or twigs in a bird's nest. But things tend not to be regularly organized (i.e., tend to be disordered). Both thermodynamics and common sense predict that things will only get less ordered on their own. So when we happen to encounter well-ordered phenomena, or observe changes that invert what should happen naturally, we tend to invoke the influence of absential influences, like human design or divine intervention, to explain them. From the dawn of recorded history the regularity of celestial processes, the apparently exquisite design of animal and plant bodies, and causes of apparently meaningful coincidences have been attributed to supernatural mentalistic causes, whether embodied by invisible demons, an all-powerful divine artificer, or some other transcendental purposiveness. Not surprisingly, these influences were imagined to originate from disembodied sources, lacking any physical form.

However, when mechanistic accounts of inorganic phenomena as mysterious as heat, chemical reactions, and magnetism began to ascend to the status of precisely formalized science in the late nineteenth century, absential accounts of all kinds came into question. So when in 1859 Charles Darwin provided an account of a process—natural selection—that could account for the remarkable functional correspondence of species' traits to the conditions of their existence, even the special order of living design seemed to succumb to a non-absential account. The success of mechanistically accounting for phenomena once considered only explainable in mentalistic terms reached a zenith in the latter half of the twentieth century with the study of so-called self-organizing inorganic processes. As processes as common as snow crystal formation and regularized heat convection began to be seen as natural parallels to such unexpected phenomena as superconductivity and laser light generation, it became even more common to hear of absential accounts described as historical anachronisms and illusions of a prescientific era. Many scholars now believe that developing a science capable of accurately characterizing complex self-organizing phenomena will be sufficient to finally describe organic and mental relationships in entirely non-absential terms.

I agree that a sophisticated understanding of Darwinian processes, coupled with insights from complex system dynamics, has led to enormous advances in our understanding of the orderliness observed in living, neuronal, and even social processes. The argument of this book will, indeed, rely heavily on this body of work to supply critical stepping stones on the way to a complete theory. However, I will argue that this approach can only provide intermediate steps in this multistep analysis. Dynamical systems theories are ultimately forced to explain away the end-directed and normative characteristics of organisms, because they implicitly assume that all causally relevant phenomena must be instantiated by some material substrate or energetic difference. Consequently, they are as limited in their power to deal with the representational and experiential features of mind as are simple mechanistic accounts. From either perspective, absential features must, by definition, be treated as epiphenomenal glosses that need to be reduced to specific physical substrates or else excluded from the analysis. The realm that includes what is merely represented, what-might-be, what-could-have-been, what-it-feels-like, or is-good-for, presumably can be of no physical relevance.

Beginning in the 1980s, it was becoming clear to some scholars that dynamical systems and evolutionary approaches to life and mind would fall short of this claim to universality. Because of their necessary grounding in what is physically here and now, they would not be able to escape this implicit dualism. Researchers who had been strongly influenced by systems thinking—like Gregory Bateson, Heinz von Foerster, Humberto Maturana, and Francisco Varela (to name only a few)—began to articulate this problem, and struggled with various attempts to augment systems thinking in ways that might be able to reintegrate the purposiveness of living processes and the experiential component of mental processes back into the theory. But the metaphysical problem of reintegrating purposiveness and subjectivity into theories of physical processes led many thinkers to propose a kind of forced marriage of convenience between mental and physical modes of explanation. For example, Heinz von Foerster in 1984 argued that a total theory would need to include, not exclude, the act of observation. From a related theoretical framework, Maturana and Varela in 1980 developed the concept of *autopoiesis* (literally, "self-creating") to describe the core self-referential dynamics of both life and mind that constitutes an observational perspective. But in their effort to make the autonomous observer-self a fundamental element of the natural sciences, the origin of this self-creative dynamic is merely taken for granted, taken as a fundamental axiom. The theory thereby avoids the challenges posed by phenomena whose existence is determined with respect to something displaced, absent, or not yet actualized, because these are defined in internalized self-referential form. Information, in this view, is not *about* something; it is a formal relationship that is co-created both inside and outside this autopoietic closure. Absential phenomena just don't seem to be compatible with the explanatory strictures of contemporary science, and so it is not surprising for many to conclude that only a sort of preestablished harmony between inside and outside perspectives, absential and physical accounts, can be achieved.

So, although the problem is ancient, and the weaknesses of contemporary methodologies have been acknowledged, there is no balanced resolution. For the most part, the mental half of any explanation is discounted as merely heuristic, and likely illusory, in the natural sciences. And even the most sophisticated efforts to integrate physical theories able to account for spontaneous order with theories of mental causality end up positing a sort of methodological dualism. Simply asserting this necessary unity—that an observing subject must be a physical system with a self-referential character—avoids the implicit absurdity of denying absential phenomena, and yet it defines them out of existence. We seem to still be living in the shadow of Descartes.

This persistent dualism is perhaps made most evident by the recent flurry of interest in the problem of consciousness, and the often extreme theoretical views concerning its nature and scientific status that have been proposed—everything from locating some hint of it in all material processes to denying that it exists at all. The problem with consciousness, like all other phenomena exhibiting an absential character, is that it doesn't appear to have clear physical correlates, even though it is quite unambiguously associated with having an awake, functioning brain. Materialism, the view that there are only material things and their interactions in the world, seems impotent here. Even major advances in neuroscience may leave the mystery untouched. As the philosopher David Chalmers sees it:

> For any physical process we specify there will be an unanswered question: Why should this process give rise to experience? Given any such process, it is conceptually coherent that it could be instantiated in the absence of experience. It follows that no mere account of the physical process will tell us why experience arises. The emergence of experience goes beyond what can be derived from physical theory.[3]

What could it mean that consciousness cannot be derived from any physical theory? Chalmers argues that we just need to face up to the fact that consciousness is non-physical and yet also not transcendent, in the sense of an ephemeral eternal soul. As one option, Chalmers champions the view that consciousness may be a property of the world that is as fundamental to the universe as electric charge or gravitational mass. He is willing to entertain this possibility because he believes that there is no way to reduce experiential qualities to physical processes. Consciousness is always a residual phenomenon remaining unaccounted for after all correlated physical processes are described. So, for example, although we can explain how a device might be built to distinguish red light from green light—and can even explain how retinal cells accomplish this—this account provides no purchase in explaining why red light *looks* red.

But does accepting this anti-materialist claim about consciousness require that there must be fundamental physical properties yet to be discovered? In this book I advocate a less dramatic, though perhaps more counterintuitive approach. It's not that the difficulty of locating consciousness among the neural signaling forces us to look for it in something else—that is, in some other sort of special substrate or ineffable ether or extra-physical realm. The anti-materialist claim is compatible with another, quite materially grounded approach. Like meanings and purposes, consciousness may not be something *there* in any typical sense of being materially or energetically embodied, and yet may still be materially causally relevant.

The unnoticed option is that, here too, we are dealing with a phenomenon that is defined by its absential character, though in a rather more all-encompassing and unavoidable form. Conscious experience confronts us with a variant of the same problem that we face with respect to function, meaning, or value. None of these phenomena are materially present either and yet they matter, so to speak. In each of these cases, there is something present that marks this curious intrinsic relation to something absent. In the case of consciousness, what is present is an awake, functioning brain, buzzing with trillions of signaling processes each second. But there is an additional issue with consciousness that makes it particularly insistent, in a way that these other absential relations aren't: *that which is explicitly absent is me.*

CALCULATING WITH ABSENCE

The difficulty we face when dealing with absences that matter has a striking historical parallel: the problems posed by the concept of zero. As the epigraph for this chapter proclaims, one of the greatest advances in the history of mathematics was the discovery of zero. A symbol designating the lack of quantity was not merely important because of the convenience it offered for notating large quantities. It transformed the very concept of number and revolutionized the process of calculation. In many ways, the discovery of the usefulness of zero marks the dawn of modern mathematics. But as many historians have noted, zero was at times feared, banned, shunned, and worshiped during the millennia-long history that preceded its acceptance in the West. And despite the fact that it is a cornerstone of mathematics and a critical building block of modern science, it remains problematic, as every child studying the operation of division soon learns.

A convention for marking the absence of numerical value was a late development in the number systems of the world. It appears to have originated as a way of notating the state of an abacus[4] when a given line of beads is left unmoved in a computation. But it literally took millennia for marking the null value to become a regular part of mathematics in the West. When it did, everything changed. Suddenly, representing very large numbers no longer required coming up with new symbols or writing unwieldy long strings of symbols. Regular procedures, algorithms, could be devised for adding, subtracting, multiplying, and dividing. Quantity could be understood in both positive and negative terms, thus defining a number line. Equations could represent geometric objects and vice versa—and much more. After centuries of denying the legitimacy of the concept—assuming that to incorporate it into reasoning about things would be a corrupting influence, and seeing its contrary properties as reasons for excluding it from quantitative analysis—European scholars eventually realized that these notions were unfortunate prejudices. In many respects, zero can be thought of as the midwife of modern science. Until Western scholars were able to make sense of the systematic properties of this non-quantity, understanding many of the most common properties of the physical world remained beyond their reach.

What zero shares in common with living and mental phenomena is that these natural processes also each owe their most fundamental character to what is specifically not present. They are also, in effect, the physical tokens of this absence. Functions and meanings are explicitly entangled with something that is not intrinsic to the artifacts or signs that constitute them. Experiences and values seem to inhere in physical relationships but are not there at the same time. This something-not-there permeates and organizes what is physically present in these phenomena. Its absent mode of existence, so to speak, is at most only a potentiality, a placeholder.

Zero is the paradigm exemplar of such a placeholder. It marks the columnar position where the quantities 1 through 9 can potentially be inserted in the recursive pattern that is our common decimal notation (e.g., the tens, hundreds, thousands columns), but it itself does not signify a quantity. Analogously, the hemoglobin molecules in my blood are also placeholders for something they are not: oxygen. Hemoglobin is exquisitely shaped in the negative image of this molecule's properties, like a mold in clay, and at the same time reflects the demands of the living system that gives rise to it. It only holds the oxygen molecule tightly enough to carry it through the circulation, where it gives it up to other tissues. It exists and exhibits these properties because it mediates a relationship between oxygen and the metabolism of an animal body. Similarly, a written word is also a placeholder. It is a pointer to a space in a network of meanings, each also pointing to one another and to potential features of the world. But a meaning is something virtual and potential. Though a meaning is more familiar to us than a hemoglobin molecule, the scientific account of concepts like function and meaning essentially lags centuries behind the sciences of these more tangible phenomena. We are, in this respect, a bit like our medieval forbears, who were quite familiar with the concepts of absence, emptiness, and so on, but could not imagine how the representation of absence could be incorporated into operations involving the quantities of things present. We take meanings and purposes for granted in our everyday lives, and yet we have been unable to incorporate these into the framework of the natural sciences. We seem only willing to admit that which is materially present into the sciences of things living and mental.

For medieval mathematicians, zero was the devil's number. The unnatural way it behaved with respect to other numbers when incorporated into calculations suggested that it could be dangerous. Even today schoolchildren are warned of the dangers of dividing by zero. Do this and you can show that 1 = 2 or that all numbers are equal.[5] In contemporary neuroscience, molecular biology, and dynamical systems theory approaches to life and mind, there is an analogous assumption about concepts like representation and purposiveness. Many of the most respected researchers in these fields have decided that these concepts are not even helpful heuristics. It is not uncommon to hear quite explicit injunctions against their use to describe organism properties or cognitive operations. The almost universal assumption is that modern computational and dynamical approaches to these subjects have made these concepts as anachronistic as phlogiston.[6]

So the idea of allowing the potentially achievable consequence characterizing a function, a reference, or an intended goal to play a causal role in our explanations of physical change has become anathema for science. A potential purpose or meaning must either be reducible to a merely physical parameter identified within the phenomenon in question, or else it must be treated as a useful fiction only allowed into discussion as a shorthand appeal to folk psychology for the sake of non-technical communication. Centuries of battling against explanations based on superstition, magic, supernatural beings, and divine purpose have trained us to be highly suspicious of any mention of such *intentional* and *teleological* properties, where things are explained as existing "for-the-sake-of" something else. These phenomena can't be what they seem. Besides, assuming that they are what they seem will almost certainly lead to absurdities as problematic as dividing by zero.

Nevertheless, learning how to operate with zero, despite the fact that it violated principles that hold for all other numbers, opened up a vast new repertoire of analytic possibilities. Mysteries that seemed logically necessary and yet obviously false not only became tractable but provided hints leading to powerful and currently indispensable tools of scientific analysis: in other words, calculus.

Consider the famous Zeno's paradox, which was framed in terms of a race between swift Achilles and a tortoise, which was given a slight head start. Zeno argued that moving any distance involved moving through an infinite series of fractions of that distance (1/2, 1/4, 1/8, 1/16 of the distance, and so on). Because of the infinite number of these fractions, Achilles could apparently never traverse them all and so would never reach the finish line. Worse yet, it appeared that Achilles could never even overtake the tortoise, because every time he reached that fraction of the distance to where the tortoise had just been, the tortoise would have moved just a bit further.

To resolve this paradox, mathematicians had to figure out how to deal with infinitely many divisions of space and time and infinitely small distances and durations. The link with calculus is that differentiation and integration (the two basic operations of calculus) represent and exploit the fact that many infinite series of mathematical operations converge to a finite solution. This is the case with Zeno's problem. Thus, running at constant speed, Achilles might cover half the distance to the finish line in 20 seconds, then the next quarter of the distance in 10 seconds, then the next smaller fraction of the distance in a correspondingly shorter span of time, and so forth, with each microscopically smaller fraction of the distance taking smaller and smaller fractions of a second to cover. The result is that the total distance can still be covered in a finite time. Taking this convergent feature into account, the operation of differentiation used in calculus allows us to measure instantaneous velocities, accelerations, and so forth, even though effectively the distance traveled in that instant is zero.

A ZENO'S PARADOX OF THE MIND

I believe that we have been under the spell of a sort of Zeno's paradox of the mind. Like the ancient mathematicians confused by the behavior of zero, and unwilling to countenance incorporating it into their calculations, we seem baffled by the fact that absent referents, unrealized ends, and abstract values have definite physical consequences, despite their apparently null physicality. As a result, we have excluded these relations from playing constitutive roles in the natural sciences. So, despite the obvious and unquestioned role played by functions, purposes, meanings, and values in the organization of our bodies and minds, and in the changes taking place in the world around us, our scientific theories still have to officially deny them anything but a sort of heuristic legitimacy. This has contributed to many tortured theoretical tricks and contorted rhetorical maneuvers in order either to obscure this deep inconsistency or else to claim that it must forever remain beyond the reach of science. We will explore some of the awkward responses to this dilemma in the chapters that follow.

More serious, however, is the way this has divided the natural sciences from the human sciences, and both from the humanities. In the process, it has also alienated the world of scientific knowledge from the world of human experience and values. If the most fundamental features of human experience are considered somehow illusory and irrelevant to the physical goings-on of the world, then we, along with our aspirations and values, are effectively rendered unreal as well. No wonder the all-pervasive success of the sciences in the last century has been paralleled by a rebirth of fundamentalist faith and a deep distrust of the secular determination of human values.

The inability to integrate these many species of absence-based causality into our scientific methodologies has not just seriously handicapped us, it has effectively left a vast fraction of the world orphaned from theories that are presumed to apply to everything. The very care that has been necessary to systematically exclude these sorts of explanations from undermining our causal analyses of physical, chemical, and biological phenomena has also stymied our efforts to penetrate beyond the descriptive surface of the phenomena of life and mind. Indeed, what might be described as the two most challenging scientific mysteries of the age—explaining the origin of life and explaining the nature of conscious experience—both are held hostage by this presumed incompatibility. Recognizing this contemporary parallel to the unwitting self-imposed handicap that limited the mathematics of the Middle Ages is, I believe, a first step toward removing this impasse. It is time that we learned how to integrate the phenomena that define our very existence into the realm of the physical and biological sciences.

Of course, it is not enough to merely recognize this analogous situation. Ultimately, we need to identify the principles by which these unruly absential phenomena can be successfully woven into the exacting warp and weft of the natural sciences. It took centuries and the lifetime efforts of some of the most brilliant minds in history to eventually tame the troublesome non-number: zero. But it wasn't until the rules for operating with zero were finally precisely articulated that the way was cleared for the development of the physical sciences. Likewise, as long as we remain unable to explain how these curious relationships between what-is-not-there and what-is-there make a difference in the world, we will remain blind to the possibilities of a vast new realm of knowledge. I envision a time in the near future when these blinders will finally be removed, a door will open between our currently incompatible cultures of knowledge, the physical and the meaningful, and a house divided will become one.

"AS SIMPLE AS POSSIBLE, BUT NOT TOO SIMPLE"

In this book I propose a modest first step toward the goal of unifying these long-isolated and apparently incompatible ways of conceptualizing the world, and our place within it. I am quite aware that in articulating these thoughts, I am risking a kind of scientific heresy. Almost certainly, many first reactions will be dismissive: "Haven't such ideas been long relegated to the trash heap of history?" . . . "Absence as a causal influence? Poetry, not science" . . . "Mystical nonsense." Even suggesting that there is such a gaping blind spot in our current vision of the world is like claiming that the emperor has no clothes. But worse than being labeled a heretic is being considered too uninformed and self-assured to recognize one's own blindness. Challenging something so basic, so well accepted, so seemingly obviously true is more often than not the mark of a crackpot or an uninformed and misguided dilettante. Proposing such a radical rethinking of these foundational assumptions of science, then, inevitably risks exposing one's hubris. Who can honestly claim to have a sufficient grasp of the many technical fields that are relevant to making such a claim? In pursuing this challenge, I expect to make more than a few technical gaffs and leave a few serious explanatory gaps.

But if the cracks in the foundation were obvious, if the intellectual issues posed were without risk to challenge, if the technical details could be easily mastered, then the attempt would long ago have been rendered trivial. That issues involving absential phenomena still mark uncrossable boundaries between disciplines, that the most enduring scientific mysteries appear to be centered around them, and that both academic and cultural upheavals still erupt over discussions of these issues, indicates that this is far from being an issue long ago settled and relegated to the dustbin of scientific history. Developing formal tools capable of integrating this missing cipher—absential influence—into the fabric of the natural sciences is an enterprise that should be at the center of scientific and philosophical debate. We should be prepared that it will take many decades of work by the most brilliant minds of the current century to turn these intuitions into precise scientific tools. But the process can't begin until we are willing to take the risk of stepping outside of the cul-de-sac of current assumptions, to try and determine where we took a slightly wrong turn.

The present exclusion of these absence-based relationships from playing any legitimate role in our theories of how the world works has implicitly denied our very existence. Is it any wonder, then, that scientific knowledge is viewed with distrust by many, as an enemy of human values, the handmaid of cynical secularism, and a harbinger of nihilism? The intolerability of this alienating worldview and the absurdity of its assumptions should be enough to warrant an exploration of the seemingly crazy idea that something not immediately present can actually be an important source of physical influence in the world. This means that if we are able to make sense of absential relationships, it won't merely illuminate certain everyday mysteries. If the example of zero is any hint, even just glimpsing the outlines of a systematic way to integrate these phenomena into the natural sciences could light the path to whole new fields of inquiry. And making scientific sense of these most personal of nature's properties, without trashing them, has the potential to transform the way we personally see ourselves within the scheme of things.

The title of this section is Albert Einstein's oft-quoted rule of thumb for scientific theorizing. "As simple as possible, but not too simple" characterizes my view of the problem. In our efforts to explain the workings of the world with the fewest possible basic assumptions, we have settled on a framework that is currently too simple to incorporate that part of the world that is sentient, conscious, and evaluative. The challenge is to determine in what way our foundational concepts are too simple and what minimally needs to be done to complicate them just enough to reincorporate us.

Einstein's admonition is also a recipe for clear thinking and good communication. This book is not just addressed to future physicists or biologists, or even philosophers of science. The subject it addresses has exceedingly wide relevance, and so this effort to probe into its mysteries deserves to be accessible to anyone with the willingness to entertain the challenging and counterintuitive ideas it explores. As a result, I have attempted to the best of my abilities to make it accessible to anyone whose intellectual curiosity has led them into this labyrinth of scientific and philosophical mysteries. I operate on the principle that if I can't explain an idea to any well-educated reader, with a minimum of technical paraphernalia, then I probably don't thoroughly understand it myself.

So, in order to reach a broad audience, and because it is the best guarantee of my own clarity of understanding, I have tried to present the entire argument in purely qualitative terms, even though this risks sacrificing rigor. I have minimized technical jargon and have not included any mathematical formalization of the principles and relationships that I describe. In most cases, I have tried to describe the relevant mechanisms and principles in ways that assume a minimum of prior knowledge on the part of the reader, and by employing examples that are as basic as I can imagine, but that still convey the critical logic behind the argument. This may make some accounts appear overly simplified and pedantic to technical readers, but I hope that the clarity gained will outweigh the time spent working cautiously step-by-step through familiar examples in order to get to concepts that are more counterintuitive and challenging. I will admit from the start that I have felt compelled to coin a few neologisms to designate some of the concepts for which I could find no commonly recognized terms. But wherever I felt that non-technical terminology would be adequate, even at the risk of dragging along irrelevant theoretical baggage, I have resisted the tendency to use specialized terminology. I have included a glossary that defines the few neologisms and technical terms that are sprinkled throughout the text.

This book is basically organized into three parts: articulating the problem, outlining an alternative theory, and exploring its implications. In chapters 1 to 5, I show that the conceptual riddles posed by absential phenomena have not been dealt with, despite claims that they have been overcome, but rather that they have been swept under the rug in various ingenious ways. By critiquing the history of efforts to explain them or explain them away, I argue that our various efforts have only served to insinuate these difficulties more cryptically into our current scientific and humanistic paradigms. In chapters 6 to 10, I outline an alternative approach—a theory of emergent dynamics that shows how dynamical process can become organized around and with respect to possibilities not realized. This is intended to provide the scaffolding for a conceptual bridge from mechanistic relationships to end-directed, informational, and normative relationships such as are found in simple life forms. In chapters 11 to 17, having laid the groundwork for an expanded science capable of encompassing these basic absential relations, I explore a few of the implications for reformulating theories of work, information, evolution, self, sentience, and value. This is a vast territory, and in these final chapters I only intend to hint at the ways that this figure/background shift in perspective necessarily reformulates how these fundamental concepts need to be reconsidered. Each of these chapters frames new theoretical and practical approaches to these presumably familiar topics, and each could easily be expanded into major books in order to do justice to these complex questions.

I consider this work to be only a first hesitant effort to map an unfamiliar and poorly explored domain. I can't even promise to have mastered the few steps I have taken into this strange territory. It is a realm in which unquestioned intuitions are seldom reliable guides, and where even the everyday terminology that we use to understand the world can insinuate misleading assumptions. Moreover, many of the scientific ideas that need to be addressed are outside my technical training and at the edge of my grasp. In those cases where I offer my best guesses, I hope that I have at least stated this still embryonic understanding with sufficient detail and clarity to enable those with better tools and training to clear away some of the ambiguities and confusions that I leave behind.

But although I cannot pretend to have fashioned a precise calculus of this physical causal analogue to operating with zero, I believe that I can demonstrate how a form of causality dependent on specifically absent features and unrealized potentials can be compatible with our best science. I believe that this can be done without compromising either the rigor of our scientific tools or the special character of these enigmatic phenomena. I hope that by revealing the glaring presence of this fundamental incompleteness of nature, it will become impossible to ignore it any longer. So, if I can coax you to consider this apparently crazy idea—even if at first only as an intellectual diversion—I feel confident that you too will begin to glimpse the qualitative outlines of a future science that is subtle enough to include us, and our enigmatically incomplete nature, as legitimate forms of knotting in the fabric of the universe.

1

(W)HOLES

Thirty spokes converge at the wheel's hub, to a hole
    that allows it to turn.
Clay is shaped into a vessel, to enclose an empti-
    ness that can be filled.
Doors and windows are cut into walls, to provide
    access to their protection.
Though we can only work with what is there, use
    comes from what is not there.

—LAO TSU[1]

A STONE'S THROW

A beach stone may be moved from place to place through the action of waves, currents, and tides. Or it may be moved thanks to the arm muscles of a boy trying to skip it over the surface of the water. In either case, physical forces acting on the mass of the stone cause it to move. Of course the waves that move beach stones owe their own movement to the winds that perturb them, the winds arise from convection processes induced by the energy of sunlight, and so on, in an endless series of forces affecting forces back through time. As for a child's muscle movements, they too can be analyzed in terms of forces affecting forces back through time. But where should that account be focused? With the oxidative metabolism of glucose and glycogen that provides the energy to contract muscles and move the arm that launches the otherwise inert stone? With the carbohy-

18

drates eaten a few hours earlier? Or with the sunlight that nourished the plants from which that food was made? One could even consider things at a more microscopic scale, such as the molecular processes causing the contraction of the child's muscles: for example, the energy imparted by adenosine triphosphate (ATP) molecules that causes the actin and myosin molecules in each muscle fiber to slide against one another and contract that fiber; the shifts in ionic potential propagated down millions of nerve fibers that cause acetylcholine release at neuromuscular synapses and initiate this chemical process; and the patterns of neuronal firing in the cerebral cortex that organize and initiate the propagation of these signals. We might even need to include the evolutionary history that gave rise to the hands and arms and brains capable of such interactions. Even so, the full causal history of this process leaves out what is arguably the most important fact.

What about the role of the child's *mental conception* of what this stone might look like skipped over the waves, his knowledge of how one should hold and throw it to achieve this intriguing result, or his fascination with the sight of a stone resisting its natural tendency to sink in water—if only for a few seconds? These certainly involve the physical-chemical mechanisms of the child's brain; but mental experience and agency are not exactly neural firing patterns, nor brain states. Neither are they phenomena occurring outside the child's brain, and clearly not other physical objects or events in any obvious sense. These neural activities are in some way *about* stone-skipping, and are crucial to the initiation of this activity. But they are not in themselves either past or future stones dancing across the water; they are more like the words on this page. Both these words and the mental images in the boy's mind provide access to something these things are not. They are both representations of something not-quite-realized and not-quite-actual. And yet these bits of virtual reality—the contents of these representations—surely are as critical to events that will likely follow as the energy that will be expended. Something critical will be missing from the explanation of the skipped stone's subsequent improbable trajectory if this absential feature is ignored.

Something very different from a stone's shifting position under the influence of the waves has become involved in a child's throwing it—something far more indirect, even if no less physical. Indeed, within a few minutes, this same boy might cause a dozen stones to dance across the surface of the water along this one stretch of beach. In contrast, prior to the evolution of humans, the probability that any stone on any beach on Earth might exhibit this behavior was astronomically minute. This difference exemplifies a wide chasm separating the domains in which two almost diametrically opposed modes of causality rule—two worlds that are nevertheless united in the hurtling of this small spinning projectile.

This exemplifies only one among billions of unprecedented and inconceivably large improbabilities associated with the presence of our species. We could just as easily have made the same point by describing a modern technological artifact, like the computer that I type on to write these sentences. This device was fashioned from materials gathered from all parts of the globe, each made unnaturally pure, and combined with other precisely purified and shaped materials in just the right way so that it could control the flow of electrons from region to region within its vast maze of metallic channels. No non-cognitive spontaneous physical process anywhere in the universe could have produced such a vastly improbable combination of materials, much less millions of nearly identical replicas in just a few short years of one another. These sorts of commonplace human examples typify the radical discontinuity separating the physics of the spontaneously probable from the deviant probabilities that organisms and minds introduce into the world.

Detailed knowledge about the general organization of the body and nervous system of the child would be usefully predictive of only the final few seconds of events as the child caught sight of an appropriately shaped stone. In comparison, knowing about the causal organization of the human body in only a very superficial way, but also knowing that a year before, during another walk on another beach, someone showed this child how to skip a stone, would confer far greater predictive power than would knowing a thousand times more physiological details. None of the incredibly many details about physical and physiological history would be as helpful as this bit of comparatively imprecise and very general information about an event that occurred months earlier and miles away. In many ways, this one past event on a different beach with a different stone in which the child may only have been an observer is even less physically linked to this present event than is the stone's periodic tumbling in the waves during the centuries that it lay on this beach. The shift from this incalculably minuscule probability to near certainty, despite ignoring an astronomically vast amount of physical data and only learning of a few nebulous macro details of human interaction, captures the essence of a very personal riddle.

The boy's idea that it might be possible to treat this stone like another that he'd once seen skipped is far more relevant to the *organization* of these causal probabilities than what he ate for breakfast, or how the stone came to be deposited in this place on this beach. Even though the force imparted to the stone is largely derived from the energy released from the chemical bonds in digested food, and its location at that spot is entirely due to the energetics of geology, wind, and waves, these facts are almost irrelevant. The predictive value of shifting attention to very general types of events, macroscopic global similarities, predictions of possible thought processes, and so forth, offers a critical hint that we are employing a very different and in some ways orthogonal logic of causality when we consider this mental analysis of the event rather than its physics. Indeed, this difference makes the two analyses appear counterintuitively incompatible. The thought is about a possibility, and a possibility is something that doesn't yet exist and may never exist. It is as though a possible future is somehow influencing the present.

The discontinuity of causality implicit in human action parallels a related discontinuity between living and non-living processes. Ultimately, both involve what amounts to a reversal of causal logic: order developing from disorder, the lack of a state of affairs bringing itself into existence, and a potential tending to realize itself. We describe this in general terms as "ends determining means." But compared to the way things work in the non-living, non-thinking world, it is as though a fundamental phase change has occurred in the dynamical fabric of the world. Crossing the border from machines to minds, or from chemical reactions to life, is leaving one universe of possibilities to enter another.

Ultimately, we need an account of these properties that does not make it absurd that they exist, or that *we* exist, with the phenomenology we have. Our brains evolved as one product of a 3-billion-year incremental elaboration of an explicitly end-directed process called life. Like the idle thoughts of the boy strolling along the beach, this process has no counterpart in the inorganic world. It would be absurd to argue that these differences are illusory. Whatever else we can say about life and mind, they have radically reorganized the causal fabric of events taking place on the surface of the Earth. Of course, life and mind are linked. The sentience that we experience as consciousness has its precursors in the adaptive processes of life in general. They are two extreme ends of a single thread of evolution. So it's not just mind that requires us to come to a scientific understanding of end-directed forms of causality; it's life itself.

WHAT'S MISSING?

This missing explanation exposes a large and gaping hole in our understanding of the world. It has been there for millennia. For most of this time it was merely a mysterious distraction. But as our scientific and technological powers have grown to the point that we are unwittingly interfering with the metabolism of our entire planet, our inability to fill in this explanatory hole has become more and more problematic.

In recent centuries, scientists' ceaseless probing into the secrets of nature has triumphed over mystery upon mystery. In just the past century, scientists have harnessed the power of the stars to produce electric power or to destroy whole cities. They have identified and catalogued the details of the molecular basis for biological inheritance. And they have designed machines that in fractions of a second can perform calculations that would have taken thousands of people thousands of years to complete. Yet this success has also unwittingly undercut the motivation that has been its driving force: faith in the intrinsic value for humankind that such knowledge would provide. Not only have some of the most elegant discoveries been turned to horrific uses, but applying our most sophisticated scientific tools to the analysis of human life and mind has apparently demoted rather than ennobled the human spirit. It's not just that we have failed to uncover the twists of physics and chemistry that set us apart from the non-living world. Our scientific theories have failed to explain what matters most to us: the place of meaning, purpose, and value in the physical world.

Our scientific theories haven't exactly failed. Rather, they have carefully excluded these phenomena from consideration, and treated them as irrelevant. This is because the content of a thought, the goal of an action, or the conscious appreciation of an experience all share a troublesome feature that appears to make them unsuited for scientific study. They aren't exactly anything physical, even though they depend on the material processes going on in brains.

Ask yourself: What does it mean to be reading these words? To provide an adequate answer, it is unnecessary to include information about the histories of the billions of atoms constituting the paper and ink or electronic book in front of you. A Laplacian demon[2] might know the origins of these atoms in ancient, burned-out stars and the individual trajectories that eventually caused them to end up as you see them, but this complete physical knowledge wouldn't provide any clue to their meaning. Of course, these physical details and vastly many more are a critical part of the story, but even this complete physical description wouldn't include the most crucial causal fact of the matter.

You are reading these words because of something that they and you are not: the ideas they convey. Whatever else we can say about these ideas, it is clear that they are not the stuff that you and this book are made of.

The problem is this: *Such concepts as information, function, purpose, meaning, intention, significance, consciousness, and value are intrinsically defined by their fundamental incompleteness.* They exist only in relation to something that they are not. Thus, the information provided by this book is not the paper and ink or electronic medium that conveys it, the cleaning function of soap is not merely its chemical interaction with water and oils, the regulatory function of a stop sign is not the wood, metal, and paint that it is composed of or even its distinctive shape, and the aesthetic value of a sculpture is not constituted by the chemistry of the marble, its weight, or its color. The "something" that each of these is not is precisely what matters most. But notice the paradox in this English turn of phrase. To "matter" is to be substantial, to resist modification, to be beyond creation or destruction— and yet what matters about an idea or purpose is dependent on something that is not substantial in any obvious sense.

So, what is shared in common between all these phenomena? In a word, nothing—or rather, something not present. In the last chapter, I introduced the somewhat clumsy term *absential* to refer to the essential absent feature of each of these kinds of phenomena. This doesn't quite tell the whole story, however, because as should already be obvious from the examples just discussed, the role that absence plays in these varied phenomena is different in each case. And it's not just something missing, but an orientation toward or existence conditional upon this something-not-there.

The most characteristic and developed exemplar of an absential relationship is *purpose*. Historically, the concept of purpose has been a persistent source of controversy for both philosophers and scientists. The philosophical term for the study of purposive phenomena is *teleology*, literally, the logic of end-directedness. The term has its roots in ancient Greek. Curiously, there are two different ancient Greek roots that get transcribed into English as *tele*, and each is relevant in its own way. The first, τελε (and also τελοσ or "end"), is the one from which teleology derives. It can variously mean completion, end, goal, or purpose. So that teleology refers to the study of purposeful phenomena. The second τηλε (roughly translated as "afar") forms the familiar prefix of such English terms as telescope, telephone, television, telemetry, and telepathy, all of which imply relationship to something occurring at a distance or despite a physical discontinuity.

Though the concept of teleology specifically derives from this first derivation of *tele*, by coincidence of transliteration (and possibly by some deeper etymological sound symbolism) the concept of separateness and physical displacement indicated by the second is also relevant. We recognize teleological phenomena by their development toward something they are not, but which they are implicitly determined with respect to. Without this intrinsic incompleteness, they would merely be objects or events. It is the *end* for the sake of which they exist—the possible state of things that they bring closer to existing—that characterizes them. Teleology also involves such relationships as representation, meaning, and relevance. A purpose is something represented. The missing something that characterizes each of these semiotic relationships can simply be physically or temporally dissociated from that which is present, or it can be more metaphorically distant in the sense of being abstract, potential, or only hypothetically existent. In these phenomena there is no disposition to bring the missing something into existence, only a marking of the essential relationship of what is present to something not present.

In an important sense, purpose is more complex than other absential relationships because we find all other forms of absential relationship implicit in the concept of purpose. It is most commonly associated with a psychological state of acting or intending to act so as to potentially bring about the realization of a mentally represented goal. This not only involves an *orientation* toward a currently non-existing state of affairs, it assumes an explicit *representation* of that end, with respect to which actions may be organized. Also, the various actions and processes typically employed to achieve that goal *function* for the sake of it. Finally, the success or failure to achieve that goal has *value* because it is in some way relevant to the *agency* for the sake of which it is pursued. And all these features are contributors to the sentience of simple organisms and the conscious experience of thinking beings like ourselves.

Because it is complex, the concept of mental purpose can be progressively decomposed, revealing weaker forms of this consequence-organized relationship that do not assume intrinsic mentality. Thus the concepts of function, information, and value have counterparts in living processes, which do not entail psychological states, and from which these more developed forms likely derive. The function of a designed mechanism or a biological organ is also constructed or organized (respectively) with respect to the end of promoting the production of some as-yet-unrealized state of things.

With respect to artifacts crafted by human users, there is no ambiguity. We recognize that an artifact or mechanical device is designed and constructed *for* a purpose. In other words, achieving this type of outcome guides the selection and modification of its physical characteristics. A bowl should be shaped to prevent its contents from draining onto the surface it sits on, and a nail should be made of a material that is more rigid than the material it needs to penetrate. The function that guides a tool's construction as well as its use is located extrinsically, and so a tool derives its end-directed features parasitically, from the teleology of the designer or user. It is not intrinsic.

In contrast, the function of a biological organ is not parasitic on any extrinsic teleology in the same sense.[3] An organ like the heart or a molecule like hemoglobin inherits its function via involvement in organism survival and reproduction. Unlike a mentally conceived purpose, a biological function lacks an explicit representation of the end with respect to which it operates. Nevertheless, it exists because of the consequences it tends to produce. Indeed, this is the essence of a natural selection explanation for certain of the properties that an organ exhibits.

Two similarly homonymous terms with different but related meanings are also associated with purpose, representation, and value. In common usage, the word "intention" typically refers to the predisposition of a person to act with respect to achieving a particular goal, as in intending to skip a stone across the water. Being intentional is in this sense essentially synonymous with being purposeful, and having an intention is having a purpose. In philosophical circles, however, the term *intention* is used differently and more technically. It is defined as the property of being *about* something. Ideas and beliefs are, in this sense, intentional phenomena. The etymology of the term admits to both uses, since its literal meaning is something like "inclined toward," as in tending or leaning toward something. Again, the common attribute is an involvement with respect to something extrinsic and absent: for example, a content, meaning, or ideal type. So both of these terminological juxtapositions suggest a deeper commonality amongst these different modes of intrinsic incompleteness.

Unfortunately, because it is not just mentality that exhibits intrinsic incompleteness and other-dependence, the terms *teleology* and *intentionality* (in both forms) are burdened with numerous mentalistic connotations. To include the more primitive biological counterparts to these mental relationships, such as function and adaptation, or the sort of information characteristic of genetic inheritance, we need a more inclusive terminology. I will argue that these more basic forms (or grades) of intrinsic incompleteness, characteristic of even the simplest organisms, are the evolutionary precursors of mentalistic relationships in an evolutionary sense. Organ functions exist *for the sake of* maintaining the life of the plant or animal. While this is not purpose in any usual sense, neither is it merely a chemical-mechanical relationship. Though subjective awareness is different from the simple functional responsiveness of organisms in general, both life and mind have crossed a threshold to a realm where more than just what is materially present matters.

Currently, we lack a single term in the English language (or others that I know of) that captures this more generic sense of existing with-respect-to, for-the-sake-of, or in-order-to-generate something that is absent that also includes function at one extreme and value at the other. The recognition that there is a common core property characterizing how each of these involves a necessary linkage with, and inclination toward, something absent argues for finding a term to refer to it. To avoid repeatedly listing these attributes, we need to introduce a more generic term for all such phenomena, irrespective of whether they are associated with minds or merely features of life.

To address this need, I propose that we use the term *ententional* as a generic adjective to describe all phenomena that are intrinsically incomplete in the sense of being in relationship to, constituted by, or organized to achieve something non-intrinsic. By combining the prefix *en-* (for "in" or "within") with the adjectival form meaning something like "inclined toward," I hope to signal this deep and typically ignored commonality that exists in all the various phenomena that include within them a fundamental relationship to something absent.

Ententional phenomena include functions that have satisfaction conditions, adaptations that have environmental correlates, thoughts that have contents, purposes that have goals, subjective experiences that have a self/other perspective, and values that have a self that benefits or is harmed. Although functions, adaptations, thoughts, purposes, subjective experiences, and values each have distinct attributes that distinguish them, they all also have an orientation to a specific constitutive absence, a particular and precise missing something that is their critical defining attribute. When talking about cognitive and semiotic topics (sign processes), I will, however, continue to use the colloquial terminology of teleology, purpose, meaning, intention, interpretation, and sentience. And when talking about living processes that do not involve any obvious mental features, I will likewise continue to use the standard terminology of function, information, receptor, regulation, and adaptation. But when referring to all such phenomena in general, I will use *entention* to characterize their internal relationship to a *telos*—an end, or otherwise displaced and thus non-present something, or possible something.

As an analysis of the concept of teleology demonstrates, different entential phenomena depend on, and are interrelated with, one another in an asymmetrical and hierarchical way. For example, purposive behaviors depend on represented ends, representations depend on information relationships, information depends on functional organization, and biological functions are organized with respect to their value in promoting survival, well-being, and an organism's reproductive potential. This hierarchic dependence becomes clear when we consider some of the simplest ententional processes, such as functional relationships in living organisms.

Thus, although we may be able to specify in considerable detail the physical chemical relationships constituting a particular molecular interaction, like the function of hemoglobin to bind free oxygen, none of these facts of chemistry provides the key news that hemoglobin molecules have these properties because of something that is both quite general and quite distinct from hemoglobin chemistry: the evolution of myriad molecular processes within living cells that are dependent on electrons captured from oxygen for energy, and the fact that they can't get it by direct diffusion in large bodies. Explaining exactly how the simplest entential phenomena arise, and how the more complex forms evolve from these simpler forms, is a major challenge of this book.

Scholars who study intentional phenomena generally tend to consider them as processes and relationships that can be characterized irrespective of any physical objects, material changes, or motive forces. But this is exactly what poses a fundamental problem for the natural sciences. Scientific explanation requires that in order to have causal consequences, something must be susceptible of being involved in material and energetic interactions with other physical objects and forces. Accepting this view confronts us with the challenge of explaining how these absential features of entential phenomena could possibly make a difference in the world. And yet assuming that mentalistic terms like *purpose* and *intention* merely provide descriptive glosses for constellations of material events does not resolve the problem. Arguing that the causal efficacy of mental content is illusory is equally pointless, given the fact that we are surrounded by the physical consequences of people's ideas and purposes.

Consider money. A dollar buys a cup of coffee no matter whether it takes the form of a paper bill, coins, an electronic debit on a bank account, or an interest-bearing charge on a credit card. A dollar has no essential specific physical substrate, but appears to be something of a higher order, autonomous from any particular physical realization. Since the unifying absential aspect of this economic process is only a similar type of outcome, and not any specific intrinsic attribute, one might be inclined to doubt that there is any common factor involved. Is the exchange of a paper dollar and a dollar's worth of coins really an equivalent action in some sense, or is it simply that we ignore the causal details in describing it? Obviously, there is a difference in many of the component physical activities involved, even though the end product—receiving a cup of coffee—may be superficially similar irrespective of these other details. But does having a common type of end have some kind of independent role to play in the causal process over and above all the specific physical events in between? More troublesome yet is the fact that this end is itself not any specific physical configuration or event, but rather a *general type* of outcome in which the vast majority of specific physical details are interchangeable.

Philosophers often describe independence from any specific material details as "multiple realizability." All ententional phenomena, such as a biological adaptation like flight, a mental experience like pain, an abstract convention like a grammatical function, a value assessment like a benefit, and so on, are multiply realizable. They can all be embodied in highly diverse kinds of physical-chemical processes and substrates.

Consideration of the not-quite-specific nature of the missing something that defines functions, intentions, and purposes will force us to address another conundrum that has worried philosophers since the time of the Greeks: the problem of the reality of *generals*. A "general," in this sense of the word, is a type of thing rather than a specific object or event. Plato's ideal forms were generals, as is the notion of "life" or the concept of "a dollar." The causal status of general types has been debated by philosophers for millennia, and is often described as the Realism/Nominalism debate.[4] It centers on the issue of whether types of phenomena or merely their specific individual exemplars are the source of causal influence. Ententional phenomena force the issue for this reason. And this further explains why they are problematic for the natural sciences. A representation may be about some specific thing but also about some general property, the same function may be realized by different mechanisms, and a purpose like skipping a stone may be diversely realized. So, accepting the possibility that ententional phenomena are causally important in the world appears to require that *types* of things have real physical consequences independent of any specific embodiment.

This requirement is well exemplified by the function of hemoglobin. This function is multiply realizable. Thus biologists were not too surprised to discover that there are other phyla of organisms that use an entirely different bloodborne molecule to capture and distribute oxygen. Clams and insects have evolved an independent molecular trick for capturing and delivering oxygen to their tissues. Instead of hemoglobin, they have evolved *hemocyanins*—so named for the blue-green color of their blood that results from the use of copper instead of iron in the blood protein that handles oxygen transfer. This is not considered biologically unusual because it is assumed that the detailed mechanism is unimportant so long as certain general physical attributes that are essential to the function are realized. The function would also be achieved if medical scientists devised an artificial hemoglobin substitute for use when blood transfusions were unavailable. Science fiction writers have regularly speculated that extraterrestrial life forms might also have blue or green blood. Perhaps this is behind the now colloquial reference to alien beings as "little green men." Indeed, copper-based blood was explicitly cited in the *Star Trek* TV series and movies as a characteristic of the Vulcan race, though the colorful characterization of alien beings appears to be much older than this.[5]

The idea that something can have real causal efficacy in the world independent of the specific components that constitute it is the defining claim of the metaphysical theory called "functionalism." Multiple realizability is one of its defining features. A common example used to illustrate this approach comes from computer science. Computer software comprises a set of instructions that are used to organize the circuits of the machine to shuttle signals around in a specific way that achieves some end, like adding numbers. To the extent that the same software produces the same effects on different computer hardware, we can say that the result is functionally equivalent, despite the entirely separate physical embodiment. Computational multiple realizability has been a crucial (though not always reliable) factor in composing this text on the different computers that I have used.[6]

It is the *interchangeability of consequences* that warrants classifying these processes as the same type of phenomenon despite the diversity of possible ways to achieve these consequences. But potential consequences are also generals, as well as being merely virtual (and thus absent) possibilities. Functions and purposes are only the same with respect to a type of outcome, not a specific fully described physical state. In contrast, mechanical conceptions of causality lead from specific antecedent physical conditions to specific consequent physical conditions.

Probably the most successful theory to be based on general consequences rather than specific physical chemical details is Darwin's theory of natural selection. For example, an adaptation like the oxygen-delivery

capability of blood may be embodied in each successful organism by a specific set of hemoglobin molecules produced in that body, but individuals
with evolved variants of hemoglobin molecules that don't produce any difference in oxygen binding are not distinguished from normal hemoglobin
by natural selection. Thus, this adaptation is a general type, not anything
physically particular.

DENYING THE MAGIC

Although we navigate our day-to-day existence in the domain of absential
relationships as much as in the domain of food, weather, houses, roads,
and other animals and plants, whenever scientists, philosophers, or theologians attempt to explain how these domains fit together, the result inevitably
seems either magical or absurd. Ententional causality appears like a sort of
magic, at least from a purely physical-chemical point of view, because it
assumes the immediate influence of something that is not present. Scientists
have tacitly entered into a polite agreement to ignore ententional properties and ban the use of them for scientific explanation. If asked, most would
deny that the idea of ententional causality makes any sense at all. Like children at a magic show, maybe we've simply been fooled into accepting this
trick by the greatest of all magicians, Mother Nature.

In this spirit, the widely read evolutionary biologist Richard Dawkins'
River Out of Eden challenges us to shed childish beliefs about meaning and
purpose in the universe. In his view, "The universe we observe has precisely
the properties we should expect if there is, at bottom, no design, no purpose,
no evil and no good, nothing but blind pitiless indifference."[7] And the Nobel
Prize–winning physicist Steven Weinberg's *The First Three Minutes* (. . . of
the universe!) similarly observes that "the more the universe seems comprehensible, the more it also seems pointless."[8]

These statements are concerned with the universe as a whole, and the
laws of the unfolding of cosmic events. It is not clear from this that either of
the two scientists would follow such implications out to the logical conclusion, that our own experience of meaning, purpose, and conscious agency is
likewise illusory. But the Nobel laureate and co-discoverer of the structure
and function of DNA Francis Crick's *The Astonishing Hypothesis* asserts:
"You, your joys and your sorrows, your memories and your ambitions, your

sense of personal identity and free will, are in fact *no more than* the behavior of a vast assembly of nerve cells and their associated molecules . . ." (emphasis added).[9]

These three scientists are certain that human conscious experience and agency are not magical or in any sense otherworldly. And as Crick suggests, it is taken for granted that they will ultimately be explained in terms of the same sorts of substances and processes as are found in the rest of the living and non-living world. However, as he struggles to explain in his book, the processes that constitute conscious experience are unusually complicated in their physical organization and difficult to study. Nevertheless, he says, they are just everyday physics and chemistry. And yet he hints at a slightly subtler version of this claim by arguing that these conscious experiences are the *emergent* consequences of these processes.

Exactly what is meant by *emergence* turns out to be subtly entangled with the problem of making sense of ententional phenomena. It is often invoked as a sort of stand-in for whatever is the special relationship that explains the difference between ententional and non-ententional physical phenomena that a complete theory of the world must ultimately provide. Addressing this challenge will occupy a good portion of this book. By invoking the concept of emergence, Crick is signaling that there is something special about conscious processes that distinguishes them from other physical processes. In the end, he finishes up telling us that this special quality is its complexity.

The difference between Crick's enterprise and that of this book is signaled by the words I have italicized in the above quote. With the phrase "no more than," Crick at the same time denies a mystical explanation and claims that these phenomena are completely understandable in terms of contemporary physics and chemistry.

As our sciences and technologies have grown ever more sophisticated, and as we have come to recognize the reliability of the tools they have provided, the fact that ententional phenomena appear invisible to these powerful methods leads inexorably to the possibility that such phenomena may involve nothing special after all. Of course, conscious experience is not magical, and I agree that the laws of physics and chemistry are not changed as they play constitutive roles in mental representation and conscious experience. I also agree with Dawkins and Weinberg that the part of the universe that our current physics and chemistry unambiguously describes *is* indeed devoid of function, meaning, and value. But denying that mind is magic does not explain the sleight-of-hand that appears to be magic with respect to the rest of the non-living universe. Something quite different happens with organisms and brains than with erupting volcanoes, turbulent rivers, or exploding stars.

As a scientist who has spent much of his career exploring the organization of brains and peering through microscopes at neurons, I am convinced that by studying such physical details we gain a better understanding of how this magnificent organ functions. At least as we understand them now, however, knowing many of the physical details of brains has not made it easier to explain how a pattern of neural activity can be *about* something that it is not. When the discussion turns to the interior experience of conscious awareness, the explanatory hole gapes even wider. Are we forever forced to use parallel non-translatable forms of explanation to discuss the sorts of phenomena that make up experience on the one hand and the physical processes that correlate with them on the other? Or will we, as some think, eventually be able to dispense with mentalistic talk altogether, at least in scientific circles, and finally get over the illusion that there is "someone home"? Are these the only choices?

This last question suggests that there is a deeper and more corrosive aspect to this impossible forced choice. It is a nightmarish thought, appropriate for *The Twilight Zone*, to imagine that everyone around you, every friend or loved one, has been replaced with robots or zombies, whose actions merely simulate the behavior of those persons. In such a world, there is no one to care and no one to care for, no reason for kindness, no one with whom to share the experience of beauty, sorrow, or discovery, and perhaps no value beyond one's own pleasures and pains. The nightmare implicitly becomes even more horrific, however, because in such a world even your own existence is a lie. It's hard enough to accept that your body is a chemical machine, or that the world as it appears to your senses is only the stuff that dreams are made on. But it is quite absurd to believe that your feelings, thoughts, and experiences, even in such a dreamworld, are also unreal. To whom or to what is each of us an illusion?

Conscious experience—that most commonplace phenomenon of everyday waking life—is the quintessence of an ententional phenomenon. It is intrinsically perspectival, representational, consequence-oriented, and normative. The possibility that it is illusory, that there really isn't anyone home, is just too absurd and devastating to contemplate. As the American philosopher and cognitive scientist Jerry Fodor puts it:

> [I]f it isn't literally true that my wanting is causally responsible for my reaching, and my itching is causally responsible for my scratching, and my believing is causally responsible for my saying . . . if none of that is literally true, then practically everything I believe about anything is false and it's the end of the world.[10]

In comparison, believing that conscious experience and other ententional phenomena are more like magic may seem at least more palatable, if not more plausible, than believing that such phenomena don't really exist at all. So it shouldn't surprise us that just when our scientific prowess appears on the verge of Godlike capacities to extend life, create new species, leave the Earth, manufacture new elements, and create machines able to perform tasks that once only people could perform, ancient fundamentalisms have swept across the globe in a rising tide of denial and defiance. Humanism is no sanctuary, if to be human is only to be a chemical computer running evolutionary programs. To navigate in a world without value is to be without rudder or destination, and yet without science, we navigate blind. To many, apparently, blindness is preferable.

TELOS EX MACHINA

The most sophisticated early recognition of a distinction between these very different modes of causality can be traced all the way back to Aristotle. In fact, he considered the problem even more complicated than I have summarized here. He distinguished four distinct modes of causality: material, efficient, formal, and final cause. In carpentry, for example, material cause is what determines the structural stability of a house, efficient cause is the carpenter's modifications of materials to create this structure, formal cause is the plan followed in this construction process, and final cause is the aim of the process, that is, producing a space protected from the elements. A final cause is that "for the sake of which" something is done. His classic example of a final cause is a man walking for his health. Health isn't a specific state, but a general condition defined by contrast to the many ways one can be unhealthy.

For Aristotle, these were different and complementary ways of understanding how and why change occurs. There has been an erosion of this plural understanding of causality since Aristotle that, although an important contribution to the unity of knowledge, has also contributed to our present intellectual (and spiritual) dilemma. Ententional processes and relationships can best be classed as expressions of final causality in Aristotle's terms. Their absential character is determined by the end toward which they tend, or the represented concept for the sake of which they were created. But with this erosion of the plurality of causal concepts, a place for a form of causal influence derived from something absent was no longer acceptable.

It was at the beginning of the Renaissance that the coherence of the concept of final causality came into question. Seminal thinkers like Francis Bacon, René Descartes, Baruch Spinoza, and others progressively chipped away at any role that final causality might play in physical processes. Bacon argued that teleological explanations were effectively redundant and thus superfluous additions to physical explanations of things. Descartes considered animate processes in animal bodies to be completely understandable in mechanical (i.e., efficient) terms, while mental processes composed a separate extensionless domain. Spinoza questioned the literal sense of final causality, since it was nonsensical to think of future states producing present states.

In their own ways, these thinkers each recognized that appealing to mental contents as physical causes accomplishes little more than pointing to an unopened black box. Positing motivations or purposes for actions still requires a physical account of their implementation. And inside that black box? Well, a further appeal to purposive agency only leads to vicious regress. Accordingly, it was held, purpose and meaning are intrinsically incomplete notions. They require replacement with something more substantial. A purpose, conceived as the "pull" of some future possibility, must be illusory, lacking the materiality to affect anything. Renaissance thinkers abandoned old forms of explanation, such as "nature abhors a vacuum" for the power of vacuums, or "being a substance lighter than water" for the buoyancy of materials, by progressively restricting the conception of causal influence to the immediate pushes and pulls of physical interaction. This heritage of modern science has guided a relentless effort to replace these black boxes and their end-directed explanations of function, design, or purposive action with mechanistic accounts. That effort has yielded astounding success.

Perhaps the greatest triumph of this enterprise came with the elucidation of the process of natural selection. It showed, in principle, that through accidentally produced variations, competition for resources, and selective reproduction or preservation of lineages, even something as complicated as a living organism with its apparent purposiveness and fittedness to local conditions could have evolved. Where the natural theology of William Paley in the early 1800s had concluded that observing functional organization, complexity, and perfection of design in an object (e.g., a watch) implied the operation of prior intelligence to fashion it, Charles Darwin and subsequent researchers could now demonstrate ways that living organization of even greater sophistication could be the result of preserved chance variation. So prior design and end-directedness became unnecessary assumptions even in the realm of life. From this perspective, organisms *were* mechanisms like watches, but their adaptive organization was conceived as arising serendipitously by matching accidentally formed mechanisms with conditions favoring their persistence.

The metaphor of the world as an immense machine full of smaller machines is, however, deeply infected with the special assumptions associated with human artifact design. When Richard Dawkins caricatures evolution as a Blind Watchmaker,[11] he still characterizes organisms as machines, and machines are assembled to do something for some end by some external process. Though we typically think of organisms as analogous to engineered artifacts performing some designed task, this analogy can set up quite misleading expectations. Design of engineered artifacts is a function of imposed order that derives from outside. The integration of parts in a machine results from the careful selection of materials, shaping of parts, and systematic assembly, all of which occur with respect to an anticipated set of physical behaviors and ends to be achieved. Although living processes have components that are at least as precisely integrated in their function as any man-made machine, little else makes them like anything engineered. Whole organisms do not result from bringing together disparate parts but by their parts' differentiating from one another. Organisms are not built or assembled; they grow by the multiplication of cells, a process of division and differentiation from prior, less differentiated precursors. Both in development and in phylogeny, wholes precede parts, integration is intrinsic, and design occurs spontaneously. The machine metaphor is a misleading oversimplification.

The tacit importation of a human artifact view of the world, with its implicit design logic, into a materialist metaphysics that restricts the introduction of anything like final causal relationships, creates the logical necessity of a *telos ex machina* universe, where design and purpose can only be imposed from the outside. In such a world, we appear as accidental robots blindly running randomly generated programs. But there is an implicit contradiction in this conception, though it is not due to the exclusion of absential properties as much as to the limitations of the machine metaphor. Machines are simplifications of the causal world. Abstractly conceived, a machine is finite, and all its features and future states are fully describable. They are essentially closed off from all physical variations except those that are consistent with a given externally determined function. Thus the whole notion of machine causality is predicated on a conception of causality that both excludes teleology from consideration and yet assumes it as the basis for distinguishing machines from generic mechanical interactions. The machine metaphor of the world implicitly begs for a watchmaker, even as it denies his or her existence.

EX NIHILO NIHIL FIT

The famous Latin pronouncement from the Roman philosopher-poet Lucretius—roughly translated as "From nothing, nothing comes"—is the precursor to the first law of thermodynamics, which states that energy (recognized as matter-energy after Einstein) can neither be created nor destroyed. Modern science has tested the conservation assumption expressed in this classic proposition to quite precise values. It is implicit in the conservation laws of physics as a cosmic bookkeeping rule. All of modern physics and chemistry stands on its reliable foundation.[12] It's also just common sense. Magic is trickery. Things don't simply vanish or appear from nowhere. It takes something to make something. And so on.

The *ex nihilo* postulate can also be interpreted to mean that existing structures, patterns, and forces are just shuffled versions of others that came before; nothing added, nothing removed. In a cosmic sense it assumes that the universe is closed, and it could also be taken to be a causal closure principle, which is to say that the basic causal laws of the universe also form a closed system—all changes come from within. This suggests that both materially and dynamically, absolute novelty is a fiction, and the causal properties present on Earth now are no more than shuffled versions of causal properties that have persisted since the Big Bang. With respect to the paradox of teleology, it can be taken as excluding the possibility that absential properties were ever fundamentally novel additions to the everyday physical-chemical causality that applies to rocks, flames, or star formation.

Yet this idea seems to lead to confusing consequences when we consider the nature of the sorts of processes and relationships that involve purposes, goals, meanings, values, and the like. It can hardly be doubted that there was a time when these phenomena were not in evidence on the surface of the Earth, although they are everywhere in evidence at present. Certainly, none were present when the Earth's surface was still hot enough to boil water, and many may only have arisen since the evolution of hominids. They probably didn't just drop in from elsewhere, though if they did, their origin would need to be explained there. Wherever and whenever they arose, however, they were unprecedented and "new," even if they are now mundanely common in our current experience. Before this, potential consequences did not play any role in what could happen. Now they do.

Doesn't such a claim violate the causal closure principle? If there truly is no antecedent set of conditions that has the production of this effect as a rule-governed consequence, then aren't we invoking an almost mystical claim about the appearance of a causal principle from nowhere? Actually, there is an even worse problem. It's not just the unprecedented appearance on the prehistoric Earth of these ententional phenomena that seems to violate the *ex nihilo* principle, it's the very nature of these phenomena themselves.

Since all ententional phenomena are intrinsically organized with respect to some property or state of affairs that does not currently exist, it is unclear from this simple materialistic perspective how these absent features could have causal influence. So from this perspective, although we describe function as a process that exists in order to produce an otherwise unlikely state of affairs, it would seem more appropriate to simply describe it as the causal antecedent to that end. Likewise, a representation is interpreted to be related to some displaced object or abstract property, which might not even exist, and so although the physical process of interpretation itself may have physical consequences, the represented content should be causally irrelevant. Finally, since purposes are determined with respect to some currently absent but conceivably achievable outcome, even if that potential is never realized, this end also appears causally irrelevant. These phenomena not only appear to arise without antecedents, they appear to be defined with respect to something nonexistent. It seems that we must explain the uncaused appearance of phenomena whose causal powers derive from something nonexistent! It should be no surprise that this most familiar and commonplace feature of our existence poses a conundrum for science.

The question of how nonexistent states can play a role in the physical goings-on of the world is hardly new. As Descartes pointed out four centuries ago, physical phenomena are extended in space and time and thus can directly affect other physical phenomena. But each of these various entential phenomena is determined with respect to something that is without local extension and in many cases can be entirely nonexistent. Shouldn't this render these absential phenomena unreal and physically irrelevant? If the referential content of ideas has no extension in the physical world, how can it make a physical difference? For Descartes, this problem sundered the mental from the physical world, mind from body, and forced him to seek an account that would bridge the gap. Since he first articulated this problem for philosophy, it has remained a major impediment to precisely articulating a natural science account of entential phenomena of all kinds.

As a result, scientific theories typically assume that concepts like information, representation, and function are merely shorthand proxies for otherwise well understood, but unfortunately long, clumsy, and arduous mechanistic accounts; that is, accounts involving only the extended physical properties of things. As we will soon discover when we dig a bit deeper, efforts to provide these accounts can quickly become circular or require the introduction of simpler and more primitive entential relationships, but fail to reduce out the critical absential quality. Attempts to explain away these phenomena just become progressively more cryptic, as ever more subtly camouflaged entential phenomena get introduced to remedy our inability to provide a thoroughly physical account. So, although naturalistic explanations can't countenance absential properties, they can't quite do without them either, unless we altogether abandon the effort to include the special sciences (e.g., psychology, sociology, economics) within the natural sciences. This incompatibility is particularly troublesome for the sciences that stand in between physics and psychology: biology and the neurosciences.

For some philosophers of mind, the difficulty of avoiding this dilemma leads them to argue that an ententional principle is an unanalyzably primitive attribute of all physical phenomena: a basic universal property of things like mass or electric charge. This evades the challenge of finding a reductionistic account for consciousness, but in its place it introduces a no less troubling problem. Making mind as ubiquitous as space and time is no more of an answer to this dilemma than was Aristotle's concept of final causality. If everything is ententional in some respect, then we are nevertheless required to specify why the absential properties of life and mind are so distinctive from the properties exhibited in the non-living world. We still need to draw a line and explain how it is crossed. Is experience a feature of the interactions of elementary particles? Is there a special form of absential property that only appears with the origins of life, or only with the origins of cognition, or perhaps only with the emergence of language and human minds? Wherever we place the threshold distinguishing ententional phenomena from the rest of natural processes, we need an account of what is involved in crossing that threshold. We don't escape this requirement by assuming that ententional properties are just universal givens. We merely shift the locus of the problem.

This is evident when we consider theories that are willing to ascribe absential properties even to fundamental physical processes. In this vein, a number of theorists have argued that absential properties must be numbered among the fundamental building blocks of the universe: often identified with the concept of information. In these theories, all physical processes involve a change in information. We will return to reanalyze the concept of information in subsequent chapters, and will discover that it is more complicated than is commonly recognized, but even without being explicit about exactly what information is, we can see that accepting this assumption does not help to resolve the dilemma. First of all, modern theories that for example argue that information is a critical feature of every quantum-level event are employing a special technical definition of information which explicitly excludes the ententional quality that defines our usual notion. Information in this technical sense is a measure of difference and order, irrespective of any referential function. So, although on the surface this appears to suggest that mind is implicit in the most basic processes of physics, on closer inspection it is a conception of mind devoid of any hint of meaning, experience, or even function.

But even if we were to accept the proposition that all physical processes are also information processes in a fuller sense, it would not eliminate the need to explain the difference between the causality that distinguishes flames and waterfalls from organisms and ideas. Nor does it resolve the mystery of how mental experience and end-directed behavior arose from the inorganic chemistry of the early Earth. It doesn't even explain why DNA molecules, alone among the wide range of complex polymers, have come to convey the information of heredity from generation to generation. All these should be sentient if this were true. What needs explaining is not how brains are like the weather, but how and why they are so different, despite the fact that both are highly complex physical processes. Only brains are organized with respect to the vast absential world of possible future events and abstract properties. Weather is organized only with respect to the immediate physical conditions. And although neural processes are also chemical processes, there are critical organizational differences distinguishing the chemistry occurring in brains from the chemistry occurring anywhere in the inorganic world. It is this difference that makes subjective experience possible. So, even if all physical processes involve some tiny hint of protomentality, we would still need an account that explains why two equally complex physical processes can differ so radically in this respect. Even if this were true, complex ententional phenomena like conscious experiences still wouldn't be explainable as a quantitative effect of some basic property of matter.

The response usually given to this criticism is that these systems differ in their *organization*. Indeed, I think that this *is* the appropriate kind of answer. But it blunts the point of the initial assumption. If the reason that human brains are a unique, intense, and persistent locus of absential properties (compared to rivers) is because they are differently organized, then *organization is effectively doing all the work* of explaining this fact, and the postulated *Ur*-mentality at the base of things—whatever else it explains— contributes nothing in addition. We still must ask, what difference in the organization of these phenomena constitutes the absential properties of the one that are not manifest in the other?

It's one thing to argue that we should avoid invoking teleology in our scientific explanations, but quite another to argue that the scientific world-view excludes the possibility of true teleology. And there is something particularly paradoxical to the claim that we have misrepresented our own merely mechanistic minds by describing them in mentalistic terms. Indeed, if this is an error, what sort of entity are we that we could make this error? I think that there is something disingenuous to the claim that teleological explanations are merely heuristics serving the purpose of standing in for those special mechanistic phenomena that are simply too complex in their organization for us to describe easily. Alternatively, to accept teleological explanations at face value, or to assume that they are primitive irreducible elements of the physical world, is tantamount to ceasing scientific inquiry prematurely and ceding explanation to magic and miracles. Both extremes deny us the possibility of an explanation. Left with these alternatives, we will neither expose nature's sleight-of-hand nor explain how the trick is done. Either way we find ourselves lacking an approach that offers a satisfactory solution to this very personal riddle.

WHEN LESS IS MORE

The secret to performing magic is to get the audience to pay attention to features of the performance that distract them from the important happenings. Continuing this metaphor we can ask, have we missed something that is right before our eyes? Have our own cognitive biases misdirected us with regard to where we train our scientific instruments? Though the full story involves many subtle and challenging conceptual twists and turns, we can at least begin with a hint concerning where we should be turning our attention, even if stating it so baldly may initially seem to make little sense. The core insight that guides this book can be grasped by taking the Taoist metaphor quoted at the beginning of this chapter seriously. We simply need to pay attention to the *holes*. As rhetorically ironic as this sounds, the thesis of this book is that the answer to the age-old riddle of teleology is not provided in terms of "nothing but . . ." or "something more . . ." but rather "something less . . ." This is the essence of what I am calling absentialism.

The roots of the problem derive from an ancient debate about the reality (i.e., physical causal efficacy) of general types (the more general analogues of Platonic forms). As part of our heritage from Enlightenment science, we tend to assume that something is only real and efficacious if it involves specific material objects and energies. But while entential phenomena are often particular in their embodiment, their causal consequences irreducibly depend on some general type of outcome or class of object. For example, the function of flight can be quite variably achieved by different means, the same idea can be expressed in different words and media, and the same end can oftentimes be achieved by very different means. Being a type of thing, however, is presumed to have no additional causal relevance over and above the specific details of its physical constitution in a specific place and time. Unrealized ends cannot be specific in every detail, because they do not yet exist, and represented things are as likely to be imaginary as real and are typically only represented with respect to a tiny fraction of their full detail. So for the content of a thought, the function of a device, or the value of a service performed, any physical influence they might have appears to depend upon unreal causes.

The fundamental shift in perspective that will lead to a way to make sense of these seemingly incompatible domains of reality can be expressed in a simple inversion of a common aphorism. Almost every text written about the way that complex systems like organisms come to exhibit end-directed organization has echoed the refrain that there must be something *more* that is involved than merely the basic physics and chemistry. In contrast, and not simply to be enigmatic, I will try to explain instead how the whole is *less* than the sum of its parts. To this we might add: and less than the interrelationships its parts can potentially realize.

My counterintuitive hypothesis is that whenever we recognize that a system exhibits entential properties, it is not because of something added to the physical processes involved, but rather quite literally because it depends on the physical fact of something specifically missing from that object or process. Commonsense English, for example, uses the metaphor of physical containment, as in the *content* of a thought, or transportation, as in a sign *vehicle*, to speak of the presumed basis for entential phenomena. Yet contrary to what such metaphors presuppose, the content of a thought or sign, the goal of an action, or the design purpose of something with function is not physically present. While it may seem empirically reasonable to ignore the content and simply look to the physical features of the sign vehicle or the associated neurological event to discover the link to the physical consequences of an idea, this hardly accords with our experience, which adamantly ascribes causality to what is represented rather than merely what does the representing.

In the chapters that follow, we will see that this simple inversion of perspective to pay attention to what is missing resolves a series of dilemmas assumed to have fatal consequences for theories of entential phenomena in all their many forms. A counterintuitive figure/background reversal, focusing on what is absent rather than present, offers a means to repair some of the serious inadequacies in our conceptions of matter, order, life, work, information, representation, and even consciousness and conceptions of value.

Showing how these apparently contradictory views can be reconciled requires that we rethink some very basic tacit assumptions about the nature of physical processes and relationships. It requires reframing the way we think about the physical world in thoroughly dynamical, that is to say, process, terms, and recasting our notions of causality in terms of something like the geometry of this dynamics, instead of thinking in terms of material objects in motion affected by contact and fields of force.

The current paradigm in the natural sciences conceives of causal properties on the analogy of some ultimate stuff, or as the closed and finite set of possible interactions between all the ultimate objects or particles of the universe. As we will see, this neat division of reality into objects and their interaction relationships, though intuitively reasonable from the perspective of human actions, is quite problematic. Curiously, however, modern physics has all but abandoned this billiard ball notion of causality in favor of a view of quantum processes, associated with something like waves of probability rather than discretely localizable stuff. Failure to overcome the view of the world as stuff rather than process has led us to think of entential phenomena as involving something added to normal physical processes. From this assumption it is only a small step to a dualistic conception of the contents of thoughts or the goals of functions as somehow non-physical stuff, leaving us with unexplainable physical consequences.

What we will discover is that entential processes have a distinctive and characteristic dynamical circularity, and that their causal power is not located in any ultimate stuff but in this dynamical organization itself. Our ultimate scientific challenge is to precisely characterize this geometry of dynamical forms which leads from thermodynamic processes to living and mental processes, and to explain their dependency relationships with respect to one another. It is a quest to naturalize teleology and its kin, and thereby demonstrate that we are the legitimate heirs of the physical universe. To do this, we must answer one persistent question. How can something not there be the cause of anything? Making sense of this "efficacy of absence" will be the central challenge of this book, and the key to embracing our ententional nature, rather than pretending to ignore or deny its existence.

2

HOMUNCULI

. . . wherever a theory relies on a formulation
bearing the logical marks of intentionality,
there a little man is concealed.

—DANIEL DENNETT, 1978[1]

THE LITTLE MAN IN MY HEAD

I remember the first time I ever thought seriously about what it means to have a brain. Although no doubt repeated recall and embellishment overwrote the details many times, the images and their emotional impression still come freshly to mind. I was seven or eight years old, watching a cartoon-illustrated science film about how bodies work. As I remember it now, the animated sequences provided a tour inside a person's body. Instead of the fearful undulating swamp of tubes and pumps I'd been led to expect lay hidden there, the animation depicted a little man engaged in managing something halfway between a large machine and a small factory. Busily shifting from task to task in his white coat, he monitored gauges, twiddled dials, and periodically erupted into panic when confronted with a surprising stimulus or a potential malfunction. Five devices that constituted the five senses produced signals that were routed to a single master control room in the head. In this room, the little man (who vaguely resembled Professor Einstein) sat at a control panel puzzling over his many instruments.

I was recently reminded of this childhood experience by a similar, though tongue-in-cheek, film called *Osmosis Jones*, released in 2001.[2] It told the half-filmed, half-animated story of a man, a disease, and a comic series of events from both his perspective and that of dozens of little men and women running the factory of his body. Thankfully, the animation was better, and it was a comedy. There were plots within plots (literally) about the physiological equivalents of criminality, pandemonium, and, yes, plenty of goo. The familiarity of the animated minions of the body helped make thinking about body functions fun; and as a way to reach kids, whose fears of knots of messy spaghetti and slimy guts might otherwise get in the way of understanding body functions, a vision of little men and women struggling to manage the body politic is both comprehensible and unthreatening.

Of course, even young children can see through these playful analogies. I knew the animation that I saw at seven or eight was really only somebody's whimsical way of explaining something that might be a whole lot more complicated. Despite this awareness, the animation both intrigued and disturbed me. If there was a little man in *my* head surveying the important information of my body and making all its important decisions to mobilize this or that mechanism, then where was I? Was I him? If I wasn't, who was I? Why didn't it seem like I was a little man trapped in a large machine? The idea twisted my imagination in knots. There was something weird here, that I couldn't quite put my finger on. And then it struck me—even as a kid: there would have to be a little man in *his* head, too!

The little man inside the man, responsible for analyzing sensory inputs and deciding on appropriate responses, is a *homunculus* (literally, a "little man"). In the sense that I will use the term here, it refers to a form of explanation that pretends to be offering a mechanistic account of some living or mental phenomenon, but instead only appeals to another cryptically equivalent process at some lower level. Although such an account appears to have the familiar form of explanation where a complicated mechanism at one level is analyzed into lower-level component mechanisms and their interactions, in this case these lower-level components exhibit properties that are no simpler than those they are purported to explain. Such an explanation is no explanation at all.

I discovered recently (ironically, from an essay written by the father of behaviorism, B. F. Skinner) that the animation I recalled seeing as a child was part of a television production called *Gateways to the Mind*. Skinner recounted this same fanciful exposition of sensation as the springboard for summarizing the contributions of behaviorism during the first fifty years of the field's existence. One of the principal motivations for behavioral psychology was the desire to avoid mental concepts in explaining why we act the way we do. To invoke mental states like beliefs and desires to explain our actions assigns the major parts of the explanation to unobservable and equally unanalyzed causes. At best it merely postpones the analysis; at worst it convinces us that no further analysis is required. Using the show's portrayal of the little man in the brain's control room to make this point, Skinner says: "The behavior of the homunculus was, of course, not explained. An explanation would, presumably require another film. And it in turn another."[3]

Skinner's underlying concern is that attributing behavior to mental constructs sneaks unanalyzed teleological relationships into the science of mind through the backdoor of common sense, when this is precisely what needs explanation. The result of ignoring Skinner's warning is a science of psychology erected on a foundation that assumes what it sets out to explain. Behaviorism aimed to remedy this tacit acceptance of mental faculties standing in for explanations. To say that a desire, wish, idea, scheme, and so on is the cause of some behavior merely redirects attention to a placeholder, which— although familiar to introspection and "folk psychology"—is an unanalyzed "black box" insofar as its causal mechanism is concerned.

Despite its overly minimalist dogma, behaviorism was one of the first serious efforts to explicitly design a methodology that would systematically avoid what has been termed the *homunculus fallacy*. The malady the behaviorists identified in the science of mind was real and quite serious. But their cure—pretending that mental experience is irrelevant for explaining behavior—was worse. Nothing is accomplished by attributing our capacity of perception to the perceptive abilities of a little man in the head; but attributing behavior to a simple mechanism for linking input stimuli to motor outputs simply sets the question of mental agency aside, it doesn't resolve it. Even if I am behaving only because of certain reward contingencies that made one action more probable than another in a given context, I am still unable to explain what constitutes my being aware of perceiving and behaving in this way, and in some cases deciding to act counter to what my immediate emotions dictate. Behaviorism merely pushes the homunculus off center stage. It doesn't hand him his walking papers.

Though the term is somewhat esoteric, the concept of a homunculus is quite familiar. In the movie version of the classic American fairy tale *The Wizard of Oz*, a little dog, Toto, exposes the all-powerful wizard as a fraud. He pulls back a curtain and reveals a man fiddling with the controls that project the frightful wizard's image in front of unwitting onlookers. Noticing his perplexed audience looking at him, the man causes the wizard image to command, "Pay no attention to the man behind the curtain!" Once the trick is exposed, however, it is too late to close the curtain and resume the masquerade. All trace of magical power and mystery evaporated, the great and powerful wizard shrinks into a meek little man twiddling the dials on an elaborate machine capable of nothing more than creating a special stage effect. This was a disappointment. Finding out that it was all just a trick meant that the presumed powers of this wizard were illusory as well. Exposing the homunculus, we find ourselves back where we started, though perhaps a bit wiser.

In this chapter, I will emulate Toto by trying to pull back the curtains from a number of slightly more cryptic homunculi. By doing so nothing gets explained except the trick itself, but seeing how this subterfuge gets perpetrated in different ways by different theories in different fields can provide some defense against falling into such traps as we begin to explore the rugged terrain of intentional phenomena.

HOMUNCULAR REPRESENTATIONS

Historically, the term *homunculus* has also been used to refer to any tiny or cryptic humanlike form or creature. The fairies, mischievous gremlins, or embodied spirits of the mythologies of most cultures are homunculi in this wider sense. The term has analogously been extended to describe forms that are only "pretenders" to humanness, such as the golem of Jewish mythology or Mary Shelley's fictional creature reanimated from dead tissues by Dr. Frankenstein. In all such cases the connotation is of being something slightly less than human, though exhibiting certain human attributes.

In the sciences, homunculus arguments are closely related to the concept of preformationism. One of the most widely recognized versions of a homuncular explanation is immortalized in a woodcut. It is a depiction of a minute, fetus-shaped "seed" crouched in the head of a sperm (see Figure 2.1). In the early days of biology, it was thought that the sperm might already

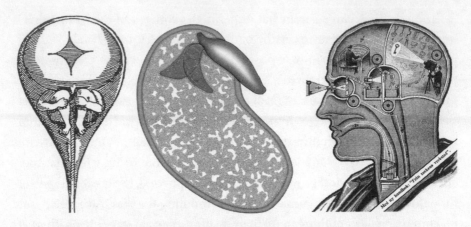

FIGURE 2.1: Three expressions of the homunculus concept. *Left:* an image
from a seventeenth-century woodcut exemplifying the belief that the head of
a sperm contained a preformed fetus. *Middle:* the presence of what might be
described as a "plantunculus" in a bean seed. *Right:* a fanciful depiction of a
homunculus constituting the functions of a person's mind and body. Draw-
ings from Nicolaas Hartsoeker's *Essai de dioptrique*, Paris, 1694 (left); the
author (center); cover illustration from *Allers Familj-Journal*, c. 1927 (right).

exhibit the form of a miniature human, and that maturation consisted of the
growth of this form, much the way that a plant might grow from the tiny
little shoots and leaves encased within a germinating seed (the cotyledons;
see also Figure 2.1). Of course, neither new plants nor new humans begin as
preformed bodies. Assuming that the human physique was preformed from
the beginning provided a material alternative to the then dominant idea that
some non-material essence or spiritual agency was responsible for shaping
the developing body from its formless beginnings. Rather than form being
imposed on the material of the developing body by an ineffable source, the
process could in this way be seen as entirely material, but only if the form
was already present in some minimal way from the beginning.

The assumption that in order to build a complex structure you need to
begin with a detailed plan or template made intuitive good sense. Like produces like. But, of course, there is an equally troublesome problem with this
view. Where did this homunculus come from? Did the parent organism
somehow impose its mirror image on the sperm or the fertilized egg? And
if so, where did that form ultimately originate?

The term *homunculus* has also made its way into contemporary science
in other, less problematic ways. For example, neurologists use the term to

describe the maplike patterns of tactile and motor responsive areas on the surface of the brain because they preserve a somewhat distorted topography of human body regions. Running vertically up and down the midline on either side of the cerebral cortex there are regions that topographically map tactile responsiveness of body surfaces and movement control of muscles. Anatomy and psychology texts often depict these regions as distorted body projections, with disproportionately enlarged mouth, face, hand, and foot for the opposite half of the body. Similar topographic projections are characteristic of the other major senses as well. Thus, the cortical area that receives retinal inputs is organized as a retinotopic map (a map that reflects position in the visual field), and the cortical area that receives inputs originating in the inner ear or cochlea is organized as a tonotopic map (a map organized from low- to high-frequency sounds).

This recapitulation of body organization in the brain is actually a bit more fragmentary and irregular than usually depicted. In tactile and motor areas, for example, the maps are not always continuous (for comparison, think of flat maps of the continents that are "cut" in ways that minimize distortion of continent size and shape). In the brain, the relative size distortions of certain homunculus features often reflect the density of innervation of the corresponding peripheral structure. So the center of gaze which settles on the most receptor-dense region of the retina (called the fovea) is proportionately much larger in its representation in the primary visual cortex than is the periphery of the visual field, which in actual dimensions is much larger though less dense in receptors. Similarly, those touch and movement systems with the most precise tactile and movement precision, respectively, are also represented in the cerebral cortex by disproportionately larger areas. So these distorted homunculi provide a useful heuristic for representing brain organization.

Unfortunately, the existence of these homuncular representations can also be an invitation to misleading shortcuts of explanation. This is because a map is a representation, and it can be tempting to mistake a map for the territory, in the sense of treating it as intrinsically interpreted. Although a map is similar in structure to what it represents, it is not intrinsically meaningful. It requires reading and comparison—but who or what does this in the brain? Using a map of a city, for example, requires knowing that certain lines correspond to roads, that distances are uniformly and systematically reduced to some tiny fraction of the real distances, and so forth. Tracing directions on the map is a reduced version of navigating through the city. In this sense, carrying the map is carrying a miniature version—a model—of the city. It is not surprising that people often feel as though finding a miniature map of the body in the brain somehow explains how sensation works. The correspondence is indeed important to its functionality, but this tells us nothing about how it is interpreted. Even mapping maps to other maps—for example, between the retinotopic map of vision and the somatotopic map of the body—won't get us far. It just passes the explanatory buck. We need to know something about the processes that constitute the interpretation. This homuncular form tells us only that the representations of different positions on a body surface are not jumbled or radically decoupled in the brain. It tells us nothing about the implicit homunculus that is required to interpret these maps.

In recent scientific literature, the concept of a homunculus has come to stand for something more subtle, and related to the misuse of teleological assumptions that Skinner criticizes. This is the sense of the concept I focus on here. Homunculi are the unacknowledged gap-fillers that stand behind, outside, or within processes that assume teleological properties such as those exhibited by life and mind, while pretending to be explanations of these properties. Treating aspects of brain function as if they are produced by little demonic helpers begs the main questions that need answering. Doing so cryptically, disguised as something else, actually impedes our efforts to understand these functions. Consider this analogy from Christof Koch and Francis Crick for the neural basis of perceptual consciousness:

> A good way to begin to consider the overall behavior of the cerebral cortex is to imagine that the front of the brain is "looking at" the sensory systems, most of which are at the back of the brain.[4]

Does the front of the brain "looking at" the back of the brain improve significantly on the following statement, written more than a century earlier by Samuel Butler in *Erewhon*:

> What is a man's eye but a machine for the little creature that sits behind in his brain to look through?

These sorts of blatantly homuncular rhetorical framings are often followed by caveats arguing that they are mere anthropomorphic heuristics, later to be replaced with actual mechanistic details; but even beginning this way opens the door for homuncular connotations to be attributed to whatever mechanism is offered. Are Koch and Crick arguing that the frontal cortex has its own form of internal visual perception to "see" what the visual system is merely registering? Of course, that can't be what they mean, but in even caricaturing the explanation in these terms, they invite their readers to frame the question in terms of looking for an ultimate homunculus.

The ancestors of today's scientific homunculi were gods, demigods, elves, fairies, demons, and gremlins that people held responsible for meaningful coincidences, human disasters, and unexpected deviations from the norm. Malevolent spirits, acts of sorcery, poltergeists, fates, and divine plans imbued natural events with both agency and meaningfulness. In past millennia and in technologically undeveloped societies, homunculus accounts were routinely invoked in cases where mechanical causes were obscure or where physical correlations appeared meaningfully linked. Although these agents from a netherworld have no legitimate place in contemporary science, they remain alive and well hiding at the frayed edges of our theories and popular culture, wearing various new guises. These modern theoretical gremlins often sit at their virtual control panels behind names for functions that carry out some teleologically glossed task, such as informing, signaling, adapting, recognizing, or regulating some biological or neurological process, and they lurk in the shadows of theoretical assumptions or unnoticed transitions in explanation where mechanistic descriptions get augmented by referring to functions and ends.

Homunculi figure more explicitly in the beginnings of science. Not surprisingly, considering the alchemical preoccupation with the purification of metals as both a goal of science and a route to the purification of the spirit, and possibly the secret of immortality, the spontaneous generation of life from non-life was a central concern of alchemy. This preoccupation was entangled in other confusions about the nature of the reproductive process, particularly the assumption of a privileged role for the male principle (in semen) in forming matter (provided by the female) into an organism. For example, alchemical notebooks contained recipes for the spontaneous production of simple animals and even animate human forms. The presumably not-quite-fully-human products of these imagined experiments were thus also referred to as homunculi.

The use of this concept that I wish to focus on is, however, more abstract than a miniature human or even a mental self. This broader conception might be usefully compared to Aristotle's notion of an *entelechy*. He argued that there must necessarily be some active principle present within each organism that guides growth and development to an adult form. This term modifies the Greek root *tele* with the prefix *en-* to indicate an internalized end. An entelechy can be understood both as a general template for the final result and the active principle that determines its development toward that end. A horse embryo develops into a horse and an acorn develops into an oak tree because, according to Aristotle, they are each animated by a distinctive end-directed tendency that is intrinsic to their material forms. Aristotle saw this as indisputable evidence that living processes are goal-directed and not merely passively reactive or accidental, as are many non-living processes. And since these target qualities are not present in the earliest stages of development, this entelechy must be present only as a potential at this stage. In this respect it is not a specific material attribute, but something like a purpose or agency embodied in some way within the matter of the undeveloped organism. An entelechy is, then, a general formative potential that tends to realize itself.

Aristotle's notion was probably in part influenced by his mentor Plato's notion of an ideal form that exists independent of any specific embodiment and that gets expressed in material form, though imperfectly. Thus, all physical spheres are imperfect spheres in some respect, deviating from the geometric ideal. Spherical objects are an expression of the influence of this ideal, to which they partially conform. Aristotle parted from his mentor on this point, arguing that all form is embodied form, and that the conception of disembodied ideals is a generalization from these exemplars, but not their cause. His notion of entelechy is often compared to the modern metaphor of a developmental program. This currently fashionable account of the causes of organism development likens the genome to a computer algorithm. A computer program is a set of instructions to be followed, and in this respect is in itself a passive artifact until embodied. In contrast, Aristotle considered the entelechy to be an active principle, intrinsic to the material substance of the organism. Millennia later, a disembodied conception of this self-actualizing living potential was also developed. This view was called "vital-ism," and although it retained the idea of an end-directed essence implicit in the organism, it also retained the Platonic focus on a disembodied source for this force.

Vitalists argued that living processes could not be understood in merely mechanistic terms. The early twentieth-century biologist Hans Driesch, for example, argued that the laws of chemistry and physics were incapa-ble of explaining this end-directedness of organisms. Thus something else, something non-material, a vital force or energy, was presumed necessary to endow merely mechanistic matter with the end-directedness that char-acterizes life. Indeed, given what was known at the time, the critique could not be easily dismissed; but the presumed alternative which vitalists offered, that living matter is distinguished from non-living matter by virtue of being infused with some ineffable and non-physical energy or essence—an *élan vital*—provided no basis for empirical study. In this respect, it too offered a homunculus argument in the broader sense of the term as I am using it. It is homuncular to the extent that the *élan vital* confers the property of living end-directedness, but simply posits the existence of this physically atypical complex property in this non-substance. So, curiously, both strict prefor-mationism and strict vitalism are at core homuncular theories, since each locates the entelechy of living organisms in some essence that pre exists its expression (whether in physical form or formative essence).

Whether a little man in my head, a demon who causes the crops to wither, the active principle that actualizes the adult body form, or an eternal and ineffable source of human subjectivity, there is an intuitive tendency to model processes and relationships that seem to violate mechanistic explana-tions on human agency. Because of this dichotomy, it is tempting to reify this distinction in the image of the difference that we are hoping to account for: the difference between a person and inanimate matter, between the apparent freedom of thought and the rigid predictability of clockwork. This is a difference that we directly encounter in our own experience, and see reflected in other people. It has the phenomenology of emerging as if with-out antecedent cause. So it is natural that this mirror image should appear intuitively acceptable to account for these kinds of phenomena. For this reason, I think it is appropriate that the homunculus should symbolize this critical essence of a human life and of mind. It is the model on which the *élan vital*, the life principle, the cryptic source of animacy and agency, and the source of meaning and value are all based on. In Enlightenment terms, it is Descartes' *cogito*, the rational principle, or one's conscience and "inner voice." In spiritual traditions, it is the eternal soul, the Hindu *atman*, or the ghost that persists after the material body succumbs. These many images all point to something more fundamental that all share in common with some of the most basic phenomena in our lives: meaning, mental representation, agency, inclination toward a purpose or goal, and the normative value of that goal. Some of these attributes are characteristically human, but many are also characteristic of animate life in general. What they share in common is something more basic than mind. It is a property that distinguishes the living from the non-living.

THE VESSEL OF TELEOLOGY

For all of these reasons, I see the homunculus as the enigmatic symbol of the most troubling and recalcitrant challenge to science: the marker for the indubitable existence of ententional phenomena in all their various forms, from the simplest organic functions to the most subtle subjective assessments of value. Their existence has posed an unresolved paradox to scientists and philosophers since the beginnings of Western thought, and the tendrils of this paradox remain entangled in some of the most technologically developed branches of modern science, where a failure to successfully integrate these properties within current theoretical systems continuously interferes with their growth. The homunculus symbolizes ententional phenomena as they are contrasted against the background of mechanistic theories describing otherwise typical physical causes and effects. Homuncular concepts and principles are the residue left behind, placeholders for something that is both efficacious and yet missing at the same time.

Historically, human beings invoke homunculi or their abstract surrogates whenever it is necessary to explain how material bodies or natural events exhibit active ententional principles. For the most part, the fact that they are in "man-analogues" often goes unnoticed. The tendency not to recognize homuncular explanations as placeholders probably derives from the familiarity of interacting with other persons whose thoughts and motivations are hidden from us. But a willingness to leave the causality of these sorts of phenomena only partially explained probably also stems from a deep intuition that a radical discontinuity distinguishes the behavior of material systems influenced by purposes and meanings from those that are not. Homunculi mark this boundary. What makes entential processes so cryptic is that they exhibit properties that appear merely superimposed on materials or physical events, as though they are something in addition to and separate from their material-physical embodiment.

This feature epitomizes human behaviors and human-designed artifacts and machines, especially when these involve meaningful relationships or actions designed to achieve a specific consequence. Human beliefs and purposes can shape events in ways that often have little direct relationship to current physical conditions and are produced in opposition to physical tendencies that are intrinsically most likely. So it is only natural that we should consider that a humanlike guiding principle is the most obvious candidate for events that appear to exhibit these discontinuities. Of course, eventually we need to explain how such a discontinuity of causality could arise in the first place, without any appeal to prior homunculi, or else just give up and decide that it is beyond any possible explanation.

Faith that these properties are not forever beyond the reach of scientific explanation is drawn from the fact that the range of phenomena where homuncular attributions/explanations are still common has been steadily shrinking. With the rise in prominence of scientific explanations, more and more of these processes have succumbed to non-teleological explanation. So it is legitimate to speculate about an extrapolated future in which all teleological explanations have been replaced by complex mechanical explanations.

Why should entential explanations tend to get eliminated with the advance of technological sophistication? The simple answer is that they are necessarily incomplete accounts. They are more like promissory notes standing in for currently inaccessible explanations, or suggestive shortcuts for cases that at present elude complete analysis. It has sometimes been remarked that teleological explanations are more like accusations or assignments of responsibility rather than accounts of causal processes. Teleological explanations point to a locus or origin but leave the mechanism of causal efficacy incompletely described. Even with respect to persons, explaining their actions in terms of motives or purposes is effectively independent of explaining these same events in terms of neurological or physiological processes and irrespective of any physical objects or forces.

Like an inscrutable person, an ententional process presents us with a point at which causal inquiry is forced to stop and completely change its terms of analysis. At this point, the inquiry is forced to abandon the mechanistic logic of masses in motion and can proceed only in terms of functions and adaptations, purposes and intentions, motives and meanings, desires and beliefs. The problem with these sorts of explanatory principles is not just that they are incomplete, but that they are incomplete in a particularly troubling way. It is difficult to ascribe energy, materiality, or even physical extension to them.

In this age of hard-nosed materialism, there seems to be little official doubt that life is "just chemistry" and mind is "just computation." But the origins of life and the explanation of conscious experience remain troublingly difficult problems, despite the availability of what should be more than adequate biochemical and neuroscientific tools to expose the details. So, although scientific theories of physical causality are expected to rigorously avoid all hints of homuncular explanations, the assumption that our current theories have fully succeeded at this task is premature.

We are taught that Galileo and Newton slew the Aristotelean homunculus of a prime mover, Darwin slew the homunculus of a divine watchmaker, Alan Turing slew the homunculus of disembodied thought, and that Watson and Crick slew the homunculus of the *élan vital*, the invisible essence of life. Still, the specter of homunculus assumptions casts its shadow over even the most technologically sophisticated and materialistically framed scientific enterprises. It waits at the door of the cosmological Big Bang. It lurks behind the biological concepts of information, signal, design, and function. And it bars access to the workings of consciousness.

So the image of the homunculus also symbolizes the central problem of contemporary science and philosophy. It is an emblem for any abstract principle in a scientific or philosophical explanation that imports an unanalyzed attribution of information, sentience, reference, meaning, purpose, design, self, subjective experience, value, and so on—attributes often associated with mental states—into scientific explanations. I will call such theories homuncular insofar as these attributes are treated as primitives or unanalyzed "black boxes," even if this usage is explicitly designated as a placeholder for an assumed, yet-to-be-articulated mechanism. The points in our theories where we must shift into homuncular terminology can serve as buoys marking the shoals where current theories founder, and where seemingly incompatible kinds of explanation must trade roles. Homunculi both indicate the misapplication of teleological principles where they don't apply and offer clues to loci of causal phase transitions where simple physical accounts fail to capture the most significant features of living processes and mental events.

HIDING FINAL CAUSE

As suggested in the previous chapter, homunculi are placeholders for sources of what Aristotle called "final causality": that for the sake of which something occurs. Aristotle described this form of causality as "final" because it is causality with respect to ends; it is something that occurs or is produced because of the consequences that are likely to result. Superficially, this applies to all sorts of human actions, indeed, everything that is done for some purpose. All designed objects, all purposeful performances, all functions, all things produced for communicating, and ultimately all thoughts and ideas are phenomena exhibiting final causality in some form. They are produced in order to bring about a consequence that contributes to something else. As a result, they exist with respect to something that does not exist yet, and for something that they are not. They exist *in futuro*, so to speak, and in this sense they are incomplete in a crucial respect.

As we have seen, Aristotle compared final causality to three other ways of understanding the concept of cause: material cause, efficient cause, and formal cause. Since the Renaissance, the concept of efficient cause has become the paradigm exemplar for all fully described conceptions of cause in the natural sciences, and Aristotle's other modes of causality have fallen into comparative neglect. However, in human affairs, from psychology to anthropology to law, some version of final causality is still the presumed fundamental influence. One's personal experience of causal agency directed to achieve some end is the archetypical expression of final causality. Although some philosophers dispute the veracity of this experience, it is hard to argue with the very strong intuition that much of our behavior is directed to change things with respect to achieving some desired state, or to keep them from changing from such a state. And where there is some argument against it, it is not based upon a denial of this experience, but rather on the argument that it is in some way logically incoherent, and therefore that our experience is in some sense false.

In its literal sense, final causality *is* logically incoherent. The material future does not determine what goes on in the present, but a conceivable future *as represented* can function to influence present circumstances with respect to that possibility. But we are not on any more solid footing attributing this causal influence to a represented future. That which does the representing is intrinsically homuncular as well, and the represented content is neither this homunculus nor the signs that do the representing. The content of a thought has no more of a material existence than a possible future.

There is one important difference between ancient and modern scientific homunculi: those that show up in modern scientific theories are by definition not supposed to exist. They are understood as markers of something yet to be explained. So this denial, paired with the inclusion of cryptically homuncular terminology, indicates the locus of a particular type of conceptual problem. As a result, *homunculus* has become a pejorative term that is applied to errors of reasoning in the natural sciences which invoke intentional properties where otherwise physical-chemical relationships should be brought to bear. Appealing to the schemes, plans, and whims of invisible demons allowed our ancestors to make sense of obscure natural processes and explain meaningful coincidences and bad luck. But appealing to an agency that is just beyond detection to explain why something happened is an intellectual dead end in science. It locates cause or responsibility without analyzing it. When this unexplained cause is treated as though it has the attributes of an intentional agent, it can masquerade as a power to explain what even mechanical explanations cannot. This is because an intentional agent can bring together diverse mechanisms to correspond with its interpretations of possible outcomes or meanings. But in the natural sciences, whose purpose is to leave the least number of gaps in an explanation, the introduction of such a wild card is for the most part counterproductive.

There is one exception: when it is done with the full disclosure that it is merely a marker for unfinished analysis. If investigators are unable to fill in all the pieces of a multipart scientific puzzle, it is considered a legitimate tactic to temporarily set aside one aspect of the problem, by assuming that it somehow "just gets done," in order to be able to work on another part of the puzzle. Here, an explicitly posited man-analogue stands in for the yet-to-be-completed analysis. The unexplained aspects are simply treated as though they are given as "accomplished," and are hopefully not critical to explaining those aspects of the puzzle that are the focus of explanatory effort. These are provisional homunculi, recognized for what they are, and slated for replacement as soon as possible. What is not accepted scientific practice is to allow a homuncular concept to play a central explanatory role. When the surrogates for little men in the works handle all the important jobs, suspension of disbelief has allowed what is most in need of explanation to be bracketed out of consideration.

GODS OF THE GAPS

A politically contentious contemporary example of the explicit use of homunculus thinking is the so-called Intelligent Design (ID) explanation for evolution. Politically, ID is a thinly veiled battle by Christian religious fundamentalists to sneak vestiges of biblical creationism or supernaturalism into the educational system. As a cover story for this infiltration effort, ID inherits homunculus assumptions distilled from a long history of beliefs in gods and goddesses thought to hold the strings of fate. But although I see ID as a troublesome mix of proselytization, ideological propaganda, and disingenuous politics, I will not comment further on these judgments, and only focus on its rhetorical use of a homunculus argument. Ultimately, it is not the religious implications nor the criticisms of theory that trouble scientists most about ID claims, but rather the deep incompatibility of this use of a homunculus argument with the very foundations of science. This threatens not just evolutionary theory but also the very logic and ethic of the scientific enterprise.

The theory of Intelligent Design almost entirely comprises critiques of contemporary explanations of evolution by natural selection. The claim is that explanatory weaknesses in current Darwinian-based theory are so serious that they demand another sort of explanation as well, or in its place. What ID advocates consider to be the most devastating criticism is that Darwinian approaches appear to them insufficient to account for certain very complex structures and functions of higher organisms, which they assert are irreducibly complex. I will return to the irreducible complexity claim in later chapters. What is at stake, however, is not whether the current form of evolutionary theory is entirely adequate to account for every complex aspect of life. It is whether there is a point at which it is legitimate to argue that the (presumed) incompleteness of a mechanistic theory requires that explicit homunculus arguments be included.

Evolutionary biology is still a work in progress, and so one should not necessarily expect that it is sufficiently developed to account for every complicated phenomenon. But evolutionary biologists and intelligent design (ID) advocates treat this incomplete state of the art in radically different ways. From the point of view of ID, evolutionary theory suffers from a kind of incompleteness that amounts to an unsalvageable inadequacy. Although it is common for theoretical proponents of one scientific theory to prosecute their case by criticizing their major competitors, and thus by indirect implication add support to their own favored theory, there is something unusual about ID's particular variant of that strategy. It is in effect a metacriticism of scientific theory in general, because it attempts to define a point at which the ban against homuncular explanations in science must be lifted.

The Intelligent Designer is a permanently unopenable black box. The work of scientific explanation cannot, by assumption, penetrate beyond this to peer into its (His) mechanism and origins. So this argument is an implicit injunction to stop work when it comes to certain phenomena: to shift from an analytical logic of causes and effects to an entential logic of reasons and purposes. In this respect, it has the same effect as would a scientist's claim that demonic interference caused a lab experiment to fail. This would hardly be considered a useful suggestion, much less a legitimate possibility. Taken literally, it undermines a basic assumption of the whole enterprise of experimental scientific research. Demons can, by definition, act on whim, interfering with things in such a way as to modify physical conditions without detection. Were demonic influence to be even considered, it would render experimentation pointless. What obtained one day would not necessarily obtain on another day, despite all other things being equal. There may indeed be unnoticed influences at work that cause currently unpredictable variations in experimental outcome, but that is taken as evidence that control of the experimental conditions is insufficient or consistency of analysis is wanting. Such problems motivate an effort to discover these additional intervening variables and thus bring them under control. If the scientists were to instead believe in demonic intervention, their response would be quite different. They might try to placate the demon with gifts or sacrifices, or ask a sorcerer or priest to perform an exorcism. Of course, I am deliberately creating an absurd juxtaposition. The culture of science and the culture of belief in this sort of supernatural being do not easily occupy the same intellectual space. As this fanciful example shows, engaging in the one interpretive stance requires abandoning the other.

This does not mean that they aren't ever juxtaposed in everyday life. People tend to be masters at believing incompatible things and acting from mutually exclusive motivations and points of view. Human cognition is fragmented, our concepts are often vague and fuzzy, and our use of logical inference seldom extends beyond the steps necessary to serve an immediate need. This provides an ample mental ecology in which incompatible ideas, emotions, and reasons can long co-exist, each in its own relatively isolated niche. Such a mix of causal paradigms may be invoked in myths and fairy tales, but even here such an extreme discontinuity is seldom tolerated. Science and philosophy compulsively avoid such discontinuities. More precisely, there is an implicit injunction woven into the very fabric of these enterprises to discover and resolve explanatory incompatibilities wherever possible, and otherwise to mark them as unfinished business. Making do with placeholders creates uneasiness, however, and the longer this is necessary, the more urgent theoretical debate or scientific exploration is likely to be.

So where homunculi show up in science and philosophy they are almost always smuggled in unnoticed—even by the smuggler—and they are seldom proposed outright. In the cases where homunculi sneak in unnoticed, they aren't always full man-analogues either, but rather more general teleological faculties and dispositions that are otherwise left unanalyzed. Invoking only fragments of human capacity (rather than all features of intentionality and agency) makes it easier for homuncular assumptions to enter unnoticed. This may even contribute to the misleading impression that it achieves explanatory advance. But even fractional homuncular capacities can end up carrying most of the explanatory load when invoked to identify intentional and teleological features of a system. Such minimalistic homunculi can still serve as cryptic stand-ins for quite complex functions.

This possibility of fractionating homuncular assumption requires that we come up with a more general definition of what constitutes a homunculus than merely a "man-analogue." To serve this purpose, I will define a homunculus argument as one in which *an ententional property is presumed to be "explained" by postulating the existence of a faculty, disposition, or module that produces it, and in which this property is not also fully understood in terms of non-ententional processes and relationships.* This does not require that ententional properties must be *reduced* to non-ententional properties, only that there can be no residual incompatibility between them. Describing exactly how this differs from reductive explanation will be a topic for later chapters, but suffice it to say that this requires ententional properties to be constructible from (and thus emergent from) relationships among non-ententional properties.

One of my favorite parodies of this trick of providing a pseudoexplanation by simply postulating the existence of a locus for what needs explaining was provided by the great French playwright Molière in his *Le Malade Imaginaire*. One scene depicts a medical student being examined by his mentors. One examiner asks if the student can explain why opium induces sleep. He replies that it is because opium contains a "soporific factor," and all the examiners praise his answer. Molière is, of course, lampooning the unscientific and pretentious physicians of his day. This is not an explanation but an evasion. Within the play, no one notices that the answer merely rephrases the point of the question as an unanalyzed essence. It just shifts the question one step back. Opium is a substance that causes sleep because it contains a substance that causes sleep. One cannot fail to notice the ironic double entendre in this example. This sort of answer exemplifies critical reason having fallen asleep!

PREFORMATION AND EPIGENESIS

Historically, homuncular explanations have not always been associated with immaterial causality. In the early days of biology, a homuncular interpretation of organism development, on the model of Aristotle's entelechy but framed in explicitly material terms, was proposed as an alternative to a disembodied developmental principle. The preformationist theory of biological development was the paradigm materialist alternative to the then competing "epigenetic" view. Epigenetics was criticized for its presumably mystical assumptions. Epigeneticists argued that the form of a developing plant or animal was not materially present at conception, and that an organism's form was acquired and shaped by some intangible vital influence as it grew and matured. Epigenesis was criticized for being comparatively unscientific because it postulated a non-physical source of form, often described in terms of a vital essence. Both the preformationists and the epigeneticists recognized that organisms followed definite trajectories of development whose form must in some sense be present from the beginning. But whereas preformationists argued that this form was present in reduced form in the fertilized egg (or just in the sperm), the epigeneticists argued that it was only present in some virtual or potential form, and that this potential was progressively imposed upon the developing embryo, like a lump of clay spinning on a potter's wheel slowly taking shape from the potter's guiding touch. So the idea of a pre-existing form guiding development was not so much the distinguishing character dividing preformation from epigenesis as it was a difference of opinion as to its locus. Preformationists imagined the locus to be a physical feature of the organism, just in need of expansion; epigeneticists imagined it to be non-material.

Two linked problems posed by the hypothesis of epigenesis made it harder to defend than preformationism. First, if the controlling form were not located within the gametes, the fertilized egg, or the developing embryo (presumably at all stages), where was it, and how did it impose itself on the matter of the developing body? If not physically present in material form, how could it shape a physical process? Indeed, what sense could be made of a Platonic disembodied form? Second, if the mature form is not intrinsic to its matter in some way, how is it that organisms of the same species always give rise to offspring with the same form? Though the analogy to an artifact shaped by human hands and human intentions was intuitively appealing and preserved a role for a designer, intelligent life principle, or vital force (*élan vital*), it inevitably appealed to non-material influences, and by the nineteenth century the appeal to mystical sources of influence was a perspective that many natural philosophers of the era were eager to avoid. Ultimately, preformationism became difficult to support in view of growing evidence from the new fields of microscopy and embryology showing that gametes, zygotes, and even the early stages of developing embryos were essentially without any trace of the form of the resulting organism.

This classic version of preformationism had another Achilles' heel. Even had embryology delivered different results, this would still have posed a serious logical problem: a version of the same problem that Skinner pointed out for mental homunculi. The need for a preformed antecedent to explain the generation of a new organism requires a prior preformed antecedent to that antecedent, and so on, and so on, homuncular sperm in the body of each homuncular sperm. Or else the antecedent must be always already present, in which case it could not be in material form, because ultimately it would be reduced to invisibility. In light of this, it is not an explanation of how the form arose so much as a way of avoiding dealing with the origins question altogether.

A related theory called "sensationalism" also flourished at the beginning of the nineteenth century. Charles Darwin's grandfather, Erasmus Darwin, enthusiastically promoted this view. Like epigenesis, sensationalism argued that developing embryos incrementally acquired their distinctive forms, beginning from an undifferentiated fertilized egg. But whereas epigeneticists appealed to an intrinsic life force to explain the drive toward adult form, sensationalists argued that bodies were actively molded by fetal experience during development. In this way, analogous to empiricist theories of mind, the developing embryo began as a sort of undifferentiated lump of clay on which this experience would sculpt its influence. Sensationalists could explain the initial similarity of child to parent as a consequence of experiences imposed by the parent body, for example, on the embryo in the womb or on the gamete prior to fertilization, and then later as a consequence of being cast in the same ecological role in the world, with the same distinctive demands imposed by the material, spatial, and energetic demands of that niche. In other words, the form of an organism was thought to be the result of its sensitivity and responses to the world. Form was imposed from without, though the material of the organism played a constraining and biasing role. Unfortunately, this view also appeared contrary to biological observation. Diverse alterations of incubation and developmental conditions had little effect on the basic body plan. There was unquestionably something intrinsic to organisms that dictated body form.

Though theories of evolution were not directly invoked as deciding influences in these debates, the rise of evolutionary theories and the decline of the preformationist/epigeneticist debate ran in parallel, and influenced each other. Probably the first evolutionist to propose a comprehensive solution to this problem was Jean-Baptiste de Lamarck. He glimpsed a middle path with his theory of evolution in which body plans were intrinsically inherited, but could be subtly modified from generation to generation as organisms interacted with their environment. The evolutionist approach thus allowed materially embodied, preformed, inherited tendencies (even if not expressed in the early embryo) to be the determinant of body form without implicating an infinite regress of preformed antecedents. This, for the first time, offered a means to integrate a materialist-preformationist theory with an environmental source of structural influence. It replaced the creative contribution of the epigeneticists' immaterial influence with the concrete influence of organism-environment interactions.

Charles Darwin's theory of natural selection provided a further tweak to this middle path by potentially eliminating even the strivings and action of the organism. Lamarck's view implicitly held that strivings to seek pleasurable experiences, avoid pain, and so forth were critical means by which the organism-environment interaction could produce novel forms. But these strivings too had the scent of homuncular influence; wouldn't these predispositions need to precede evolution in some way? Darwin's insight into these matters was to recognize that the strivings of organisms were not necessary to explain the origins of their adaptive forms. Even if these traits were the consequence of chance mechanisms produced irrespective of any function potential, if they coincidentally just happened to aid reproduction they would be preferentially inherited in future generations.

In this way, Darwinism also dealt a blow to strict preformationism. Living forms were not always as they had been, and so nothing was ultimately preformed. What was preformed in one generation was a product of the interaction of something preformed in previous generations, modified by interaction with the environment. Thus, the form-generating principles of organism development could have arisen incrementally, and even if only cryptically embodied in the embryo, they could be understood as physically present. This completed the synthesis, preserving the materialist principles of preformationism while introducing a creative dynamic in the form of natural selection acting on accidental variation. The battle over the origins of living form therefore shifted from development to phylogenetic history. This appeared to remove the last hint of prior intelligence, so that no designing influence or even organism striving need be considered. Darwinism thus appeared to finally offer a homunculus-free theory of life.

Darwin's contribution did not, however, finally settle the debate over theories of the genesis of organism form. Indeed, the theory of natural selection made no specific predictions concerning the process of embryogenesis and was agnostic about the mechanisms of reproduction and development. The debate was reignited at the end of the nineteenth century by two developments. First was the proposal by the German evolutionist Ernst Haeckel that phylogenetic evolution was driven by the progressive addition of developmental stages to the end of a prior generation's developmental sequence.[5] Second was the rediscovery of the genetic basis of biological inheritance. Haeckel's theory had an intrinsic directionality implicit in its additive logic, and suggested that evolution might be driven by a sort of developmental drive. The genetic theory of inheritance additionally suggested that the maturation of organic form might be embodied in a sort of chemical blueprint. Both ideas were influential throughout much of the early twentieth century. Only in recent decades have they been shown to be fallacious and oversimplified, respectively.

In hindsight, however, both preformation and vitalism have contributed something to our modern understanding of embryogenesis, even though their classic interpretations have been relegated to historical anachronism. The unfertilized egg begins as a relatively undifferentiated cell. The process of fertilization initiates a slight reorganization that polarizes the location of various molecular structures within that cell. Then this cell progressively divides until it becomes a ball of cells, each with slightly different molecular content depending on what part of the initial zygote gave rise to them. As they continue to divide, and depending on their relative locations and interactions with nearby cells, their cellular progeny progressively differentiate into the many thousands of cell types that characterize the different organs of the body. Although the features that distinguish these different, specialized cell types depend on gene expression, the geometry of the developing embryo plays a critical role in determining which genes in which cells will be expressed in which ways. This interdependency is best exemplified by the potential of embryonic stem cells at the blastula (initial ball-like) stage of development to become any cell type of the body. Taken from one blastula and inserted anywhere else in another, these cells will develop in accordance with their insertion point in the second new embryo. So at early stages of development, individual cells do not have preformed developmental fates. Their ultimate fates are determined collectively, so to speak. Thus, although this developmental potential reflects a kind of preformed organism plan attributable to the genetic code contained within each cell, this alone is insufficient to account for the details of development. The term *epigenesis* is retained to describe the contribution of such extragenomic influences as the cell-cell interactions that alter which genes are or are not active in a given cell line and the geometric relations between cells that determine which cells can interact.

So, although many popular accounts of DNA treat it as though it contains the equivalent of a blueprint for building the body or else a program or set of construction instructions, this modern variant of preformationism is a considerable oversimplification. It ignores the information ultimately embodied in the elaborate patterns of interaction among cells and how these affect which genes get expressed. The vast majority of this structural information is generated anew in each organism as an emergent consequence of a kind of ecology of developing cells. The structural information passed on by the DNA nevertheless contributes sufficient constraints and biases embodied in protein structure to guarantee that in this information generation process, humans always give rise to humans and flies to flies. But it also guarantees that every fly and every human is unique in myriad ways—even identical twins. This is because genes aren't like assembly instructions or computer programs. There is no extrinsic interpreter or assembly mechanism. Patterns of gene expression depend on patterns of embryo geometry, and changes of gene expression influence embryo geometry in cycle upon cycle of interactions.

MENTALESE

A more seriously entertained, widely accepted, and hotly debated modern variant of a homuncular explanation is the concept of a language faculty, conceived as the repository of the principles that predetermine what sort of forms any natural language is able to assume. This set of principles has been given the name "universal grammar," because it is postulated that they must be expressed in any naturally arising human language. Although contemporary proponents of this view will bristle at the comparison, I believe that it is not merely a caricature to call this a preformationist and indeed a homuncular theory.

The first modern form of this sort of theory was developed by the linguist Noam Chomsky in the late 1950s and early 1960s. His intention was to develop a systematic formal description of the principles necessary to predict all and only the class of sentences that a normal speaker of the language would recognize as well formed, or grammatically correct. This is complicated by the fact that language is generative and open-ended. There is no clear upper limit to the variety of possible grammatical sentences. So it was a finite faculty, capable of infinite productivity. The remarkable success of this modeling strategy, though tweaked again and again over the intervening decades, demonstrates that to a great extent languages are able to be modeled as axiomatic systems derived from a modest set of core design principles. But although this approach has provided a very precise representation of the kind of rule-governed system that speakers would need to have available if they were in fact to use a rule-governed approach to producing and interpreting speech, it does not guarantee that this is the only way to produce and interpret speech.

One reason *not* to expect a neat mapping between this formalism and the implementation of linguistic processes in the brain is that human brain architecture is the product of dozens of millions of years of vertebrate evolution, adapting to physiological and behavioral challenges whose attributes had little or nothing in common with the otherwise anomalous structure of language. This neural architecture has changed only subtly in human evolution (producing no unprecedented human-specific anatomical structures). This suggests that the otherwise anomalous logic of linguistic communication is implemented by a system organized according to very ancient biological design principles, long preceding the evolution of language.

This mismatch with the evolutionary logic of neural function is not an insurmountable problem. But an effort to reify grammatical logic as existing in this algorithmic form in the mind ceases to use it as a placeholder for a neural-behavioral-intentional process and instead treats it as a preformed homunculus. A set of algorithms that has been explicitly devised by testing its predictions against the massive corpus of utterances judged to be grammatical should by design be predictive of these utterances. It is in this sense merely a representation of this behavioral predisposition in a more compact form, that is, a methodological stand-in for a human speaker's competence. Postulating that this algorithmic system is *in the mind* in this sense is a redescription relocated: a homuncular move that mistakes a map for the territory.

If such a rule system comes preformed in each human infant, no linguistic, social-psychological, or developmental explanation is required. The rules of sentence formation are known before any word is learned. In Steven Pinker's widely acclaimed book *The Language Instinct*, for example, he begins his analogy for explaining the relationship between thought and language by pointing out that

> People don't think in English or Chinese or Apache; they think in a language of thought. This language of thought probably looks a bit like all of these language. . . . But compared with any given language, mentalese [this hypothetical language of thought] must be richer in some ways and simpler in others.

Here the problem to be explained has found its way into the explanation. With an elaboration of this analogy, he makes this clear.

> Knowing a language, then, is knowing how to translate mentalese into strings of words and vice versa. . . . Indeed, if babies did not have a mentalese to translate to and from English, it is not clear how learning English could take place, or even what learning English would mean.[6]

If thought equals language, then all that needs to be explained is the translation. But wait a minute! Exactly who or what is producing and comprehending mentalese? Modeling thought on language thus implicitly just pushes the problem back a level. In this sense, it is analogous to the caricature of perception offered by Koch and Crick earlier in the chapter. For a language to be "spoken" in the head, an unacknowledged homunculus must be interpreting it. Does this internal speaker of mentalese have to translate mentalese into homunculese? Is mentalese self-interpreted? If so, it is not like a language, and it is unclear what this would mean anyway.

The notion that words of a spoken language correspond to and are derived from something like words in a language of thought is not new. A similar idea was critical to the development of the logic of the late Middle Ages at the hands of such scholars as Roger Bacon and William of Occam. By dispensing with the messiness of reference to things in the world, inference could be modeled as formal operations involving discrete signs. This was a critical first step toward what today we understand as the theory of computation. In its modern reincarnation, pioneered by philosophers like Jerry Fodor, the mental language hypothesis turns this around and models cognition on computation. And by similarly ignoring problems of explaining representation and interpretation, it effectively shifts the problem of explaining the mysteries of language from psychology and linguistics to evolutionary biology and neuroscience. Of course, the analytical tools of evolutionary biology and neuroscience do not include propositions, meanings, or symbolic reference. Indeed, nearly all of the fundamental principles that must be included in this preformed language of thought have intentional properties, whereas brain structures do not. So it effectively asserts that the grammar homunculus is crucial to language, but irrelevant to explaining it.

MIND ALL THE WAY DOWN?

In *A Brief History of Time*, the physicist Stephen Hawking recounts an old anecdote (attributed to many different sources from William James to Bertrand Russell) about an exchange that supposedly took place at the end of a lecture on cosmology (or a related subject). Hawking tells it like this:

> At the end of the lecture, a little old lady at the back of the room got up and said: "What you have told us is rubbish. The world is really a flat plate supported on the back of a giant tortoise." The scientist gave a superior smile before replying, "What is the tortoise standing on?" "You're very clever, young man, very clever," said the old lady. "But it's turtles all the way down!"[7]

One way to avoid the challenge of explaining the intentional phenomena that characterize mental processes is to assume that they intrinsically constitute part of the physical substrate of the world. The suspicion that attenuated mental attributes inhere in every grain of sand, drop of water, or blade of grass has an ancient origin. It is probably as old as the oldest spiritual belief systems, tracing its roots to animistic beliefs present in nearly all the world's oldest spiritual traditions. As a systematic metaphysical perspective, it is known as *panpsychism*.

Panpsychism assumes that a vestige of mental phenomenology is present in every physical event, and therefore suffused throughout the cosmos. The brilliant Enlightenment thinkers Spinoza and Leibniz, among others, espoused this belief, and in nineteenth-century philosophy Gustav Fechner, Arthur Schopenhauer, William Clifford, Wilhelm Wundt, Ernst Haeckel, Josiah Royce, and William James, just to name a few, all promoted some form of it. Although panpsychism is not as influential today, and effectively plays no role in modern cognitive neuroscience, it still attracts a wide following, mostly because of a serendipitous compatibility with certain interpretations of quantum physics.

Although it is common for panpsychism to be framed in terms of a form of mentality that permeates the universe, panpsychist assumptions can be framed in more subtle and general ways. Using the broader conception of ententional relationships that we have been developing here—in which conscious states are only one relatively complex form of a more general class of phenomena constituted by their relationship to something explicitly absent and other—we can broaden the conception of panpsychism to include any perspective which assumes that ententional properties *of some kind* are ultimate constituents of the universe. This leaves open the possibility that most physical events have no conscious aspect, and yet may nonetheless exhibit properties like end-directedness, information, value, and so forth. In this view, physical events, taking place outside of organisms with brains, are treated as having a kind of unconscious mentality.

Many past panpsychic theories have espoused something less than the claim that all physical processes are also mental processes. For example, the evolutionary biologist Ernst Haeckel attributed variously attenuated forms of mentality to even the most simple life forms, but he apparently stopped short of attributing it to non-living processes, and William James, in a perspective dubbed "neutral monism," argued that a middle ground between full panpsychism and simple materialism offered a way out of Descartes' dualist dilemma. Neutral monism suggests that there is an intrinsic potential for mentality in all things, but that it is only expressed in certain special physical contexts, like the functioning of a brain. Of course, this begs the question of whether this property inheres in some ultimate substance of things or is merely an expression of these relationships alone—a question we will turn to below.

In quantum physics, many scientists are willing to argue that the quantum to classical transition—in which quantum probabilities become classical actualities—necessarily involves both information and observation, in the form of measurement. One famous interpretation of quantum mechanics, formulated in Copenhagen in the early twentieth century by Niels Bohr and Werner Heisenberg, specifically argued that events lack determinate existence until and unless they are observed. This has led to the popularity of many unfortunate overinterpretations that give human minds the status of determining what exists and what does not exist. Of course, a literal interpretation of this would lead to vicious regress, because if things aren't real until we observe them, then our own existence as things in the world is held hostage to our first being able to observe ourselves. My world may have come into existence with my birth and first mental awakening, but I feel pretty sure that no physicist imagines that the world only came into existence with the first observers. But what could it mean for the universe to be its own observer irrespective of us?

A less exotic homuncularization of quantum theory simply treats quantum events as intrinsically informational. Overinterpreted, this could lead to the proposal that the entire universe and its physical interactions are mindlike. Taking one step shy of this view, the quantum computation theorist Seth Lloyd describes the universe as an immense quantum computer. He argues that this should be an uncontroversial claim:

> The fact that the universe is at bottom computing, or is processing information, was actually established in the scientific sense back in the late 19th century by Maxwell, Boltzmann, and Gibbs, who showed that all atoms register bits of information. When they bounce off each other, these bits flip. That's actually where the first

measures of information came up, because Maxwell, Boltzmann, and Gibbs were trying to define entropy, which is the real measure of information.[8]

Although James Clerk Maxwell, Ludwig Boltzmann, and Josiah Gibbs were collectively responsible for developing thermodynamic theory in the nineteenth century before the modern conceptions of computing and information were formulated, they did think of thermodynamic processes in terms of order and the information necessary to describe system states. But Lloyd writes almost as though they had modern conceptions of computing and information, and conceived of thermodynamic processes as intrinsically a form of information processing. Intended as a heuristic caricature, this way of telling the story nonetheless misrepresents the history of the development of these ideas. Only after the mid-twentieth-century work of Alonzo Church and Alan Turing, among others, showed that most physical processes can be assigned an interpretation that treats them as performing a computation (understood in its most general sense) did it become common to describe thermodynamic processes as equivalent to information processing (though, importantly, it is equivalence *under an interpretation*). But from this assumption it can be a slippery slope to a more radical claim. We get a hint of this when Lloyd says that "here is where life shows up. Because the universe is already computing from the very beginning when it starts, starting from the Big Bang, as soon as elementary particles show up."[9]

On the assumption that organisms and brains are also merely computers, it may seem at first glance that Lloyd views teleological processes as mere complications of the general information processing of the universe. But here Lloyd walks a tightrope between panpsychism and its opposite, the eliminative view that computation is only physics—a physics that has no room for entential properties. This is because he deploys a special technical variant of the concept of information (to be discussed in detail in later chapters), which might better be described as the potential to convey information, or simply as order or difference. Being able to utilize one physical attribute or event as information about another requires some definite physical distinctions, and to the extent that the quantum-classical transition involves the production of definite physical distinctions, it also involves the necessary foundation for the possibility of conveying information. But this notion of information excludes any implication of these distinctions being *about* anything. Of course, this is the ententional property that distinguishes information from mere physical distinction.

Minds are, indeed, interpreters of information *about* other things. They are not merely manipulators of physical tokens that might or might not be assigned an interpretation. But the bits of information being manipulated in a computing process are only potentially interpretable, and in the absence of an interpreting context (e.g., a humanly defined use), they are only physical electrical processes. Making sense of this added requirement without invoking homuncular assumptions will require considerable unpacking in later chapters of this book. For now, suffice it to say that potential information is not intrinsically about anything. So whether or not the conception of the universe as a computer is truly a homuncular theory turns on whether one thinks that minds are doing more than computing, and whether in making this assumption, mind is projected down into the quantum fabric of things or explained away as an illusory epiphenomenon of the merely physical process of computing.

Physicists struggling to make sense of the Big Bang theory of the origins of the universe have also flirted with a homuncular interpretation of the orderliness of things. It turns out that a handful of physical constants—for example, the fine-structure constant, the value of the strong nuclear force, the gravitational constant—must be in exquisite balance with one another in order to have produced a universe like ours, with its longevity, heavy atoms, complex molecules, and other features that lead to the possibility of life. In other words, because we are here to attest to this fact, precisely balanced fine-tuning of these constants is also a necessary fact. This is generally referred to as the *anthropic principle*.

Assuming that things could be otherwise and that no fact excludes the possibility of an infinity of "settings" of these constants, however, makes it either highly providential or extraordinarily lucky that the universe favors life as we know it. Some, who see it as providential, espouse a cosmic variant of the Intelligent Design theme, arguing that the astronomically improbable conditions of our existence must result from some ultimate design principle or designer, for which or for whom the possibility of complex forms of matter (such as that responsible for humanlike consciousness) was a projected outcome.

The anthropic principle admits to two alternative, non homuncular interpretations, however. First, it could be that all other settings have been randomly generated in other universes, most of which did not yield the balance of properties that could lead to us. In this respect, our universe would be only one special case among the infinitude of failed exemplars. This is sometimes called the weak anthropic principle. Alternatively, the fine-tuning of these values might be the outcome of a cosmic process in which a self-organizing or evolutionlike process culminating in the Big Bang spontaneously converged toward these well-balanced values. It is not clear that we can ever evaluate these alternatives empirically.

The mathematician-philosopher Alfred North Whitehead produced what is probably the most sophisticated effort to make twentieth-century physics compatible with teleological principles in his process metaphysics.[10] Though describing and critiquing Whitehead's rethinking of metaphysics is far beyond the scope of this essay, his sophisticated effort to reconceptualize fundamental physics in a way that addresses the paradox of ententional causality has had a powerful influence, even if not in the sciences, and has often been characterized as a variant of panpsychism. His novel account of the nature of physical processes argues that physical change necessarily follows from an intrinsic incompleteness in every physical "occasion," by which he means something like a localized physical event. In this view, process is more fundamental than the matter and energy by which we track such physical changes.

In a rough sense, Whitehead conceived of specific physical occasions as defined by the incessant assimilation of adjacent occasions into a new descendant occasion, via a process he termed *prehension* (roughly, "grasping"). In this process, each physical occasion assimilates features of adjacent occasions, and changes accordingly. In this transition new potential relations become concrete, moment by moment, through an active extending and linking that incessantly occurs between just previously "concreted" loci in space and time. His technical term for this is *concrescence*, and in many ways it appears to be the analogue to the quantum physicist's notion of the "collapse of the (probability-) wave function" as marking the transition from quantum indeterminacy to classical determination. If no object, event, or interaction—down to the most fundamental physical interactions, such as between elementary particles—is complete in itself, then all aspects of physical causality implicitly depend on something extrinsic that is not physically present "there." Perhaps the most important aspect of this intrinsic incompleteness of every occasion is an intrinsic sense of animacy. Whitehead conceived of prehension in active terms, as an intrinsic tendency of previously isolated occasions to connect to and be completed by some adjacent occasion. In this sense, he understood it in intentional terms, though at the most minimal level. He thus envisioned prehension as the simplest physical exemplar of an interpretation.

Taking a micro-panpsychist approach to mind (or what might likewise be characterized as a micro-panvitalist approach to life) is, however, not as helpful as we might hope. In the ascent in scale and complexity from micro to macro—for example, from particle interactions to living creatures with minds—Whitehead's theory explains the special characteristics of living and mental phenomena in terms of the special organization of the "societies" (systems?) of prehensions that compose these complex phenomena. A mind thus becomes a particularly complex and synergistic society of prehensions, in which vastly many micro-interpretive actions together account for the emergence of subjective experience.

Unfortunately, this compounding of micro homunculi to make a macro homunculus ignores an important difference. If simply being constituted by myriads of prehensional events was sufficient to produce sentience, then raging rivers and storm clouds should exhibit at least as much mentality and intentional causal power as human brains, and vastly more than mouse brains. Presumably, for Whitehead this can be explained by virtue of the vast differences in structural-functional organization that make all brains like each other and radically different from non-living systems. The system of interactions in even the simplest brain is far more complex, flexible, and diverse than that composing a river or storm cloud. Indeed, Whitehead appeals to distinctive forms of organization to explain why some complexes of prehensions are associated with overt subjectivity and others are not. Yet, if specific organizational complexity is what matters, then little explanatory significance is added by the assumption that some level of micro intentionality was also suffused throughout all the component processes. The difference in causal powers and subjectivity, distinguishing the processes taking place in brains from those in rivers, lies in their organizational differences. Unless we are willing to imagine that rivers have subjective states that are merely causally impotent, the organizational features that distinguish brains from rivers are providing all the explanatory power that distinguishes sentient from insentient matter, and ententional phenomena from merely mechanical processes. We are still left needing to explain the distinctive features of mental attributes in terms of this organization.

So, while Whitehead's conception of physical process offers a physical precursor to end-directedness (in its active other-assimilating tendency) and intentionality (in the fundamental incompleteness of actual occasions), it mostly leaves the distinctive features of life and mind to be explained by their special forms of organization.

Something about this Heraclitean vision of the world must be right. Process is indeed fundamental both to physics and to all conceptions of ententional properties. There seems to be an aspect of all physical phenomena that has a self-undermining and intrinsically incomplete character to it, as is also true for ententional phenomena. But even if the fundamental incompleteness of physical entities is a necessary precondition for a universe including teleological relationships, panpsychic assumptions do not explain why the character of physical processes associated with life and mind differs so radically from those associated with the rest of physics and chemistry—even the weird physics of the quantum.

3

GOLEMS

Science does not dehumanize man,
it de-homunculizes him

—B. F. SKINNER, 1971[1]

ELIMINATION SCHEMES

The habit of reflecting on oneself in homuncular terms comes naturally, even though this leaves us caught in a mental hall of mirrors where endless virtual tunnels of self-reflections make it impossible to navigate our way out. There is little reason to expect that evolution would have equipped us with a special set of corrective glasses for dealing with the problems posed by such self-reflection. It is in the realm of social interaction with other creatures like ourselves that we need tools for navigating the challenges created by ententional processes. We don't have to worry very often why it is that we ourselves do the things we do, and we are seldom caught entirely by surprise by our own actions. But this is often the case with others. It is one of the claimed benefits of years of psychoanalysis or meditation that they can provide a modest capacity for intervention in our otherwise unanalyzed habits and predispositions. But social life constantly demands that we guess at, anticipate, and plan for the actions of others. Many students of mental evolution argue that this is a capacity that is only well developed in *Homo sapiens*. The ability to develop a mental model of another's experiences and intentions is often given the tongue-in-cheek name "mind reading" by behavioral researchers. However, despite our comparatively better evolved capacity, we are still notoriously bad at it. Because it is both difficult and fraught with error, we spend a considerable fraction of our days anticipating others, and from an early age appear predisposed to project the expectation of another mind into inanimate toys.

So one reason we may tend to think in homuncular terms, even in contexts where rationally we know it makes no sense (as in superstitious behaviors, or wondering at the "meaning" of some apropos coincidence), is that it just comes naturally. This psychological habit should not, however, absolve us as philosophers and scientists from the requirement that we think more clearly about these issues. And yet, as we will see in the examples discussed below, avoiding thinking this way, without throwing out the baby with the bathwater, is both remarkably difficult to achieve and deeply counterintuitive. Not only do we homuncularize without thinking, even careful theorists who systematically try to avoid these errors often fall easy prey to only slightly more subtle and cryptic versions of this fallacy. For the same reason that the homunculus fallacy is so seductive, it is also a slippery target. Thinking that we have finally destroyed all vestige of this crafty adversary, we often become complacent and miss its reappearance in more subtle and cryptic forms.

As noted in the last chapter, most considerations of ententional phenomena implicitly treat the critical details of their causal dynamics as though they are hidden in a black box, and worse, invoke the causal influence of explicitly absent entities. Because of this, researchers in the natural sciences have little choice but to make every effort to avoid assigning explanatory roles to ententional processes in their theories. And wherever a field such as cellular biology or cognitive neuroscience encounters issues of information or functional organization, it treats them as heuristic placeholders and endeavors to eventually replace them with explicit physical mechanisms. Indeed, biologists and cognitive neuroscientists treat this as an imperative, and rigorously scour their theories to make sure they are free of any hint of teleology. In philosophical circles, this methodological presumption has come to be known as *eliminative materialism* because it presumes that all reference to ententional phenomena can and must be eliminated from our scientific theories and replaced by accounts of material mechanisms.

But can this eliminative strategy be carried out exhaustively such that all hint of ententional explanation is replaced by mechanistic explanation?

And even if it can, will the result be a complete account of the properties that make life and mind so different from energy and matter? I have little doubt that a universe devoid of ententional phenomena is possible. In fact, I believe that at one point in the distant past, the entire universe was in just such a state. But that is not the universe in which we now live. For this reason, I believe that this eliminative enterprise is forced to sacrifice completeness for consistency, leaving important unfinished business in its wake. We need to explain the ententional nature of our own existence, not explain it away.

HEADS OF THE HYDRA

Since the Enlightenment, the natural sciences have progressively dealt with the homunculus problem by trying to kill it. The presumption is that it is illegitimate as an explanatory principle, and worse, that accepting it risks readmitting gods, demons, and other such supernatural surrogates and argument-stoppers back into the discourse of science. As the philosopher Daniel Dennett rightly warns, accepting these sorts of explanations incurs a serious explanatory debt.

> Any time a theory builder proposes to call any event, state, struc-
> ture, etc., in any system (say the brain of an organism) a signal or
> message or command or otherwise endow it with content, he takes
> out a loan of intelligence. . . . This loan must be repaid eventually
> by finding and analysing away these readers or comprehenders;
> for failing this, the theory will have among its elements unana-
> lysed man-analogues endowed with enough intelligence to read
> the signals, etc.[2]

Homunculi are stand-ins for incomplete explanations, and since, accord-ing to current scientific canon, good explanations should take the form of mechanistic analysis, all references to the teleological properties attributed to homuncular loci need to be replaced by mechanisms. But often, efforts to explain away blatant homunculi lead to their unwitting replacement with many less obvious homunculi. Instead of a little man in the head, there are sensory maps; instead of an *élan vital* animating our bodies, there are genes containing information, signaling molecules, receptor sites, and so on, to do the teleological work.

This reminds me of the classic tale of Hercules' battle with the Hydra. The Hydra was a monster with many heads. The very effort to cut off one of the Hydra's heads caused two more to grow where the one had been before. In the story, Hercules' helper Iolaus prevented the regrowth of new heads by searing each neck with flame as each head was chopped off, until eventually the one immortal head was also removed and buried under a boulder. The head might rightly be described as the organ of intention: the source of meaning and agency. As in the story, the effort to remove this organ of intention only compounds the challenge, progressively ceding power to the foe one is trying to subdue, because it reappears in other places. In the modern theoretical analogues of this myth, positing a locus of ententional cause merely passes the explanatory buck to an as-yet-to-be-explained process occurring at that locus or elsewhere, that serves this same role—only less obviously. The process is thus implicitly incomplete, and possibly not even able to be completed, requiring a similar effort at some later point in the analysis to deal with these new creations. The effort to deny the special character of ententional processes at that locus, and thus eliminate altogether any reference to teleological phenomena, only serves to displace these functional roles onto other loci in the explanatory apparatus. Instead of one homunculus problem, we end up creating many. And in the end the original homunculus problem remains. Removed from the body of scientific theory but unable to be silenced, it can only be hidden away, not finally eliminated. The analogy is striking, and cautionary.

An explicit homunculus-slaying proposal for dealing with mental causality was articulated by the artificial intelligence pioneer Marvin Minsky. In his book *Society of Mind*, he argues that although intelligence appears to be a unitary phenomenon, its functional organization can be understood as the combined behavior of vast numbers of very stupid mindless homunculi, by which he ultimately means robots; simple computers running simple algorithms. This is a tried-and-true problem-solving method: break the problem down into smaller and smaller pieces until none appears too daunting. Mind, in this view, is to be understood as made up of innumerable mindless robots, each doing some tiny fraction of a homuncular task. This is also the approach Dan Dennett has in mind. Of course, everything depends on mental processes being a cumulative effect of the interactions of tiny mind-less robots. Though the homunculus problem is in this way subdivided and distributed, it is not clear that the reduction of complex intentionality to many tiny intentions has done any more than give the impression that it can be simplified and simplified until it just disappears. But it is not clear where this vanishing point will occur. Though intuitively one can imagine simpler and simpler agents with stupider and stupider intentional capacities, at what point does it stop being intentional and just become mechanism?

So long as each apparently reduced agent must be said to be generating its mindless responses on the basis of information, adaptation, functional organization, and so on, it includes within it explanatory problems every bit as troubling as those posed by the little man in the head—only multiplied and hidden.

What are presumed to be eliminable are folk psychology concepts of such mental phenomena as beliefs, desires, intentions, meanings, and so forth. A number of well-known contemporary philosophers of mind, such as Richard Rorty, Stephen Stich, Paul and Patricia Churchland, and Daniel Dennett, have argued for versions of an eliminative strategy.[3] Although they each hold variant interpretations of this view, they all share in common the assumption that these folk psychology concepts will eventually go the way of archaic concepts in the physical sciences like phlogiston, replaced by more precise physical, neurological, or computational accounts. Presum-ably the teleological framing of these concepts provides only a temporary stand-in for a future psychology that will eventually redescribe these mental attributes in purely neurological terms, without residue. Stronger versions of this perspective go further, however, arguing that these mentalistic con-cepts are vacuous: fictitious entities like demons and magic spells.

While these eliminative efforts superficially appear as explanatory vic-tories, and briefly create the impression that one can re-describe a sentient process in terms of mechanism, this ultimately ends up creating a more difficult problem than before. In the examples below, we will see how the presumption that entential properties can be eliminated by just excising all reference to them from accounts of living or mental processes inadver-tently reintroduces them in ever more cryptic form.

Fractionation of a complex phenomenon into simpler component features is a common strategy in almost all fields of science. Complicated problems that can be decomposed into a number of simpler problems, which can each be solved independently, are not hard to find. The greatest successes of science are almost all the result of discovering how best to break things down to manageable chunks for which our tools of analysis are adequate. This is because most physical phenomena we encounter are componential at many levels of scale: stones composed of microscopic crystal grains, crystal grains composed of regular atomic "unit cells," crystal cells composed of atoms, atoms composed of particles, and some of these composed of yet smaller particles. Wherever we look, we discover this sort of compositional hierarchy of scale. However, this does not mean that it is universal, or that the part/whole relationship is as simple as it appears. Mathematicians recognize that some problems do not allow solution by operating independently on separate parts, and computational problems that do not decompose into chunks that can be computed independently are not aided by the power of parallel processing. So, although it may be nearly tautological to claim that all complex phenomena have parts and can be described in terms of these parts, it is not necessarily the case that the same complex whole can be understood one part at a time. What appear to be "proper parts" from the point of view of description may not have properties that can be described without reference to other features of the whole they compose.

For this reason I prefer to distinguish between *reducible* systems and *decomposable* systems. Reduction only depends on the ability to identify graininess in complex phenomena and the capacity to study the properties of these subdivisions as distinct from the collective phenomenon that they compose. Decomposition additionally requires that the subdivisions in question exhibit the properties that they exhibit in the whole, even if entirely isolated and independent of it. For example, a clock is both reducible and decomposable to its parts, whereas a living organism may be analytically reducible, but it is not decomposable. The various "parts" of an organism require one another because they are generated reciprocally in the whole functioning organism. Thus, a decomposable system is by definition reducible, but a reducible system may not be decomposable. Just because something is complicated and constituted by distinguishable subdivisions doesn't mean that these subdivisions provide sufficient information about how it functions, or how it is formed, or why as a complex whole it exhibits certain distinctive properties.

The question before us is whether entential phenomena are merely reducible or are also decomposable. I contend that while entential phenomena are dependent on physical substrate relationships, they are *not* decomposable to them, only to lower-order entential phenomena. This is because although entential phenomena are necessarily physical, their proper parts are not physical parts.

At this point, the distinction may sound like an overly subtle and cryptic semantic quibble, but without going into detail, we can at least gain some idea of why it might lead to the Hydra problem. If complex entential phenomena are reducible but not decomposable into merely physical processes, it is because components are in some sense "infected" with properties that arise extrinsic to their physical properties (e.g., in the organization of the larger complex context from which they are being analytically isolated). So, analyzed in isolation, the locus of these properties is ignored, while their expression is nevertheless recognized. Since this is the case for each analyzed part, what was once a unitary entential feature of the whole system is now treated as innumerable, even more cryptic micro-entential phenomena.

Consider, for example, the complex DNA molecules that, by reductionistic analysis, we recognize as components of an organism. Each nucleotide sequence that codes for a protein on a DNA molecule has features that can be understood to function as an adaptation that evolved in response to certain demands posed by the conditions of that species' existence. But to attribute to these sequences such properties as being adaptive, or serving a function, or storing information, is to borrow a property from the whole and attribute it to the part. These properties only exist for the nucleotide sequence in question in a particular systemic context, and may even change if that context changes over the course of a lifetime or across generations. What may be functional at one point may become dysfunctional at another point, and vice versa.

As we will see below, whether we ascribe cryptically simplified entential properties to computer algorithms or biological molecules, if we treat these properties as though they are intrinsic physical properties, we only compound the mystery. In the effort to dissolve the problem by fractionation, we end up multiplying the mysteries we have to solve.

DEAD TRUTH

Jewish folklore of the late Middle Ages tells of a creature called a golem. A golem is formed from clay to look like a man and is animated by a powerful rabbi using magical incantations. Whereas the Almighty Jehovah had the capacity to both form a man from clay and also imbue him with a soul, the mystic could only animate his figure, leaving it soulless. Like a sophisticated robot of contemporary science fiction, the golem could behave in ways similar to a person, but unlike a normal person there would be no one home. The golem would perceive without feeling, interact without understanding, and act without discernment. It is just an animated clay statue following the explicit commands of its creator, like a robot just running programs.

If we take the homunculus as an avatar of cryptic entential properties smuggled into our theories, we can take the golem as the avatar of its opposite: apparently mindlike processes that are nonetheless devoid of their own entential properties. If a homunculus is a little man in my head, then the golem is a hollow-headed man, a zombie.

Zombies are closely related mythical creatures that have recently been invoked in the debates about whether mental phenomena are real or not. The popular concept of a zombie traces its origins to voodoo mythology, in which people are "killed" and then reanimated, but without a mind or soul, and thus enslaved to their voodoo master. They are, to use a somewhat enigmatic word, "undead." A zombie in the philosophical sense is in every physical respect just like a person—able to walk, talk, drive an automobile in traffic, and give reasonable answers to complicated questions about life—but completely lacking any subjective experience associated with these behaviors. The plausibility of zombies of this sort, truly indistinguishable from a normal person in every other respect but this, is often proposed as a *reductio ad absurdum* implication of a thoroughly eliminative view. If subjective mental phenomena, such as the sense of personal agency, can be entirely explained by the physics and neurochemistry of brain processes, then the conscious aspect of this is playing no additional explanatory role. It shouldn't matter whether it is present or not.

Stepping back from these extremes, we can recognize familiar examples of what might be termed near zombiehood. We often discover that minutes have passed while we engaged in a complicated skill, like driving a car, and yet have no recollection of making the decisions involved or being alerted by changes of scenery or details of the roadway. It's like being on autopilot. In fact, probably the vast majority of brain processes contributing to our moment-to-moment behavior and subjective experience are never associated with consciousness. In this respect, we at least have a personal sense of our own partial zombie nature (which probably makes the concept intuitively conceivable). And many of these behaviors involve beliefs, desires, and purposes.

The real problem posed by golems and zombies is that what appears superficially to be intrinsic purposiveness in these entities is actually dead-cold mechanism. In the classic medieval golem story, a golem was animated to protect the discriminated Jewish population of Prague, but in the process it ended up producing more harm than good because of its relentless, mindless, insensate behavior. According to one version of this story, the power of movement was "breathed" into the clay figure by inscribing the Hebrew word for "truth"—אמת—on its forehead.[4] Thus animated, the golem was able to autonomously follow the commands of its creator. But precisely because the golem carried out its creator's missions relentlessly and exactly, this very literalness led to unanticipated calamity. When it became clear that the golem's behavior could not be channeled only for good, it had to be stopped. To do so, one of the letters had to be erased from its forehead, leaving the word for death—מת. With this the golem would again become an inanimate lump of clay.

This myth exemplifies one of many such stories about the inevitable ruin that comes of man trying to emulate a god, with Mary Shelley's story of Dr. Frankenstein's monster as the modern prototype. It is no surprise that variants on the golem and Frankenstein themes have become the source of numerous contemporary morality tales. Contemporary science fiction writers have found this theme of scientific hubris to be a rich source of compelling narratives. The mystic's assumption that the golem's behavior would be fully controllable and the scientists' assumption that the processes of life and mind can be understood like clockwork have much in common. Like homunculi, golems can be seen as symbolic of a much more general phenomenon.

The golem myth holds a subtler implication embodied in its truth/death pun. Besides being a soulless being, following commands with mechanical dispassion, the golem lacks discernment. It is this that ultimately leads to ruin, not any malevolence on either the golem's or its creator's part. Truth is heartless and mechanical, and by itself it cannot be trusted to lead only to the good. The "truth" that can be stated is also finite and fixed, whereas the world is infinite and changeable. So, charged with carrying out the implications that follow from a given command, the golem quickly becomes further and further out of step with its context.

Golems can thus be seen as the very real consequence of investing relentless logic with animate power. The true golems of today are not artificial living beings, but rather bureaucracies, legal systems, and computers. In their design as well as their role as unerringly literal slaves, digital computers are the epitome of a creation that embodies truth maintenance made animate. Like the golems of mythology, they are selfless servants, but they are also mindless. Because of this, they share the golem's lack of discernment and potential for disaster.

Computers are logic embodied in mechanism. The development of logic was the result of reflection on the organization of human reasoning processes. It is not itself thought, however, nor does it capture the essence of thought, namely, its quality of being about something. Logic is only the skeleton of thought: syntax without semantics. Like a living skeleton, this supportive framework develops as an integral part of a whole organism and is neither present before life, nor of use without life.

The golem nature of logic comes from its fixity and closure. Logic is ultimately a structure out of time. It works to assure valid inference because there are no choices or alternatives. So the very fabric of valid deductive inference is by necessity preformed. Consider the nature of a deductive argument, like that embodied in the classic syllogism:

1. All men are mortal
2. Socrates is a man

Therefore

3. Socrates is mortal

Obviously, 2 and 3 are already contained in 1. They are *implied* (folded in) by it, though not explicitly present in those exact words. The redundancy between the mention of "men" in 1 and "man" in 2 requires that if 1 and 2 are true, then 3 must follow necessarily. Another way one could have said this is that (1) the collection of all men is contained within the collection of all mortals; and (2) Socrates is contained within the collection of all men; so inevitably, (3) Socrates is also contained within the collection of all mortals. Put this way, it can be seen that logic can also be conceived as a necessary attribute of the notion of containment, whether in physical space or in the abstract space of categories and classes of things. Containment is one of the most basic of spatial concepts. So there is good reason to expect that the physical world should work this way, too.

It should not surprise us, then, that logic and mathematics are powerful tools for modeling natural processes, and that they should even provide prescient anticipations of physical mechanisms. Deductive inference allows only one and always the same one consequence from the same antecedent. The mechanical world appears to share the same non-divergent connectivity of events, hence its predictability. Mathematics is thus a symbol manipulation strategy that is governed by the same limitations as physical causality, at least in Newtonian terms. Mechanical processes can for this reason be constructed to precisely parallel logico-mathematical inferences, and vice versa. But mathematical symbolization is finite and each side of an equation specifying a possible transformation is complete and limited.

Machines, such as we humans construct or imagine with our modeling tools, are also in some sense physical abstractions. We build machines to be largely unaffected by all variety of micro perturbations, allowing us to use them as though they are fully determined and predictable with respect to certain selected outcomes. In this sense, we construct them so that we can mostly ignore such influences as expansions and contractions due to heat or the wearing effects of friction, although these can eventually pose a problem if not regularly attended to. That machines are idealizations becomes obvious precisely when these sorts of perturbing effects become apparent—at which point we say that the machine is no longer working.

This logico-mathematical-machine equivalence was formalized in reverse when Alan Turing showed how, in principle, every valid mathematical operation that could be precisely defined and carried out in a finite number of steps could also be modeled by the actions of a machine. This is the essence of computing. A "Turing machine" is an abstraction that effectively models the manipulations of symbols in an inferential process (such as solving a mathematical equation) as the actions of a machine. Turing's "universal machine" included a physical recording medium (he imagined a paper tape); a physical reading-writing device so that symbol marks could be written on, read from, or erased from the medium; and a mechanism that could control where on the medium this action would next take place (e.g., by moving the paper tape). The critical constraint on this principle is complete specification. Turing recognized that there were a variety of problems that could not be computed by his universal machine approach. Besides anything that cannot be completely specified initially, there are many kinds of problems for which completion of the computation cannot be determined Both exemplify limits to idealization.

Consider, however, that to the extent that we map physical processes onto logic, mathematics, and machine operation, the world is being modeled as though it is preformed, with every outcome implied in the initial state. But as we just noted, even Turing recognized that this mapping between computing and the world was not symmetrical. Gregory Bateson explains this well:

> In a computer, which works by cause and effect, with one transistor triggering another, the sequences of cause and effect are used to simulate logic. Thirty years ago, we used to ask: Can a computer simulate all the processes of logic? The answer was "yes," but the question was surely wrong. We should have asked: Can logic simulate all sequences of cause and effect? The answer would have been: "no."[5]

When extrapolated to the physical world in general, this abstract parallelism has some unsettling implications. It suggests notions of predestination and fate: the vision of a timeless, crystalline, four-dimensional world that includes no surprises. This figures into problems of explaining intentional relationships such as purposiveness, aboutness, and consciousness, because as theologians and philosophers have pointed out for centuries, it denies all spontaneity, all agency, all creativity, and makes every event a passive necessity already prefigured in prior conditions. It leads inexorably to a sort of universal preformationism. Paradoxically, this ultimate homunculus move eliminates all homunculi, and in the process provides no assistance in understanding our own homuncular character. But we should be wary of mistaking the map for the territory here. Logical syntax is constituted by the necessities that follow when meanings are assumed to be discrete, fixed, and unambiguous. A mechanism is a similar abstraction. Certain properties of things must be held constant so that their combinations and interactions can be entirely predictable and consistent. In both cases, we must pretend that the world exhibits precision and finiteness, by ignoring certain real-world details.

Curiously, even assuming this sort of total ideal separability of the syntax from the semantics of logic, complete predictability is not guaranteed. Kurt Gödel's famous 1931 incompleteness proof is widely recognized as demonstrating that we inevitably must accept either incompleteness or inconsistency in such an idealization. So long as the syntactic system is as powerful as elementary algebra and allows mapping of expressions to values, it must always admit to this loophole. The significance of this limitation for both computation and mental processes has been extensively explored, but the deliberations remain inconclusive. In any case, it warns that such an idealization is not without its problems. A complete and consistent golem is, for this reason, unobtainable.

To simplify a bit, the problem lies in the very assumption that syntax and semantics, logic and representation, are independent of one another. A golem is syntax without semantics and logic without representation. There is no one at home in the golem because there is no representation possible—no meaning, no significance, no value, just physical mechanism, one thing after another with terrible inflexible consistency. This is the whole point. The real question for us is whether golems are the only game in town that doesn't smuggle in little man-analogues to do the work of cognition. If we eliminate all the homunculi, are we only left with golems?

As we've seen, golems are idealizations. Formal logic already assumes that the variables of its expressions are *potential* representations. It only brackets these from consideration to explore the purely relational constraints that must follow. We might suspect, then, that whenever we encounter a golem, there is a hidden homunculus, a man behind the curtain, or a rabbi and his magical incantations pulling the nearly invisible strings.

THE GHOST IN THE COMPUTER

Behaviorism was conceived of as a remedy to the tacit acceptance of homuncular faculties standing in for psychological explanations. To say that a desire, wish, idea, scheme, and so on, is the cause of some behavior merely redirects attention to a placeholder, which—although familiar to introspection and folk psychology—is an unanalyzed black box insofar as its causal mechanism is concerned. B. F. Skinner and his colleagues believed that a scientific psychology should replace these homuncular notions with observable facts, such as the stimuli presented and the behaviors emitted by organisms. The result would be a natural science of behavior that was solidly grounded on entirely unambiguous empirical matters of fact.

Unfortunately, in hindsight, the behaviorist remedy was almost deadly for psychology. In an effort to avoid these sorts of circular explanations, behaviorism ignored the role of thoughts and experiences, treating them as taboo subjects, thus effectively pretending that the most notable aspects of having a working brain were mere metaphysical fantasies. Behavioral researchers also had to ignore the "behavior" going on inside of brains as well, because of technical limitations of the period. In recent decades, however, though classic behaviorism has faded in influence, neuroscientists have married behaviorism's methodology with precise measurements of neurological "behavior," such as are obtained by electrode recordings of neuronal activity or *in vivo* imaging of the metabolic correlates of brain activity. But even with this extension inwards, this logic still treats the contents of thoughts or experiences as though they play no part in what goes on. There is only behavior   of whole organisms and their parts.

With its minimalist dogma, behaviorism was one of the first serious efforts to explicitly design a methodology for systematically avoiding the homunculus fallacy in psychology and thus to base the field firmly in physical evidence alone. But perhaps what it illustrated most pointedly was that the homunculus problem won't go away by ignoring it. Although remarkable insights about brain function and sensory-motor control have come from the more subtle and neurologically based uses of the behaviorist logic, it has ultimately only further postponed the problem of explaining the relationship between mental experience and brain function.

During the 1960s, the dominance of behaviorism faded. There were a few identifiable battles that marked this turning point,[6] but mostly it was the austerity and absurdity of discounting the hidden realm of cognition that was the downfall of behaviorism. Still, the homunculus allergy to behaviorism was not abandoned in this transition. Probably one of the keys to the success of the new cognitive sciences, which grew up during the 1970s, was that they found a way to incorporate an empirical study of this hidden dimension of psychology while still appearing to avoid the dreaded homunculus fallacy. The solution was to conceive of mental processes on the analogy of computations, or algorithms.

Slowly, in the decades that followed, computing became a more and more commonplace fact of life. Today, even some cars, GPS systems, and kitchen devices have been given synthesized voices to speak to us about their state of operation. And whenever I attempt to place a phone call to some company to troubleshoot an appliance gone wrong or argue over a computer-generated bill for something I didn't purchase, I mostly begin by answering questions posed by computer software in a soothing but not quite human voice. None of these are, of course, anything more than automated electronic-switching devices. With this new metaphor of mind, it seemed like the homunculi that Skinner was trying to avoid had finally been exorcised. The problem of identifying which of the circuits or suborgans of the brain might constitute the "self" was widely acknowledged to be a non-question. Most people, both laymen and scientists, could imagine the brain to be a computer of vast proportions, exquisite in precision of design, and run by control programs of near-foolproof perfection. In this computer there was no place for a little man, and nothing to take his place. Finally, it seemed, we could dispense with this crutch to imagination.

In the 1980s and 1990s, this metaphor was sharpened and extended. Despite the consciously acknowledged vast differences between brain and computer architectures, the computer analogy became the common working assumption of the new field. Many university psychology departments found themselves subsets of larger cognitive science programs in which computer scientists and philosophers were also included. Taking this analogy seriously, cognition often came to be treated as software running on the brain's hardware. There were input systems, output systems, and vast stores of evolved databases and acquired algorithms to link them up and keep our bodies alive long enough to do a decent job of reproducing. No one and nothing outside of this system of embodied algorithms needed to watch over it, to initiate changes in its operation, or to register that anything at all is happening. Computing just happens.

In many ways, however, this was a move out of the behaviorist frying pan and into the computational fire. Like behaviorism before it, the strict adherence to a mechanistic analogy that was required to avoid blatant homuncular assumptions came at the cost of leaving no space for explaining the experience of consciousness or the sense of mental agency, and even collapsed notions of representation and meaning to something like physical pattern. So, like a secret reincarnation of behaviorism, cognitive scientists found themselves seriously discussing the likelihood that such mental experiences do not actually contribute any explanatory power beyond the immediate material activities of neurons. What additional function needs to be provided if an algorithm can be postulated to explain any behavior? Indeed, why should consciousness ever have evolved?

Though it seemed like a radical shift from behaviorism with its exclusively external focus, to cognitive science with its many approaches to internal mental activities, there is a deeper kinship between these views. Both attempt to dispose of mental homunculi and replace them with physical correspondence relationships between physical phenomena. But where the behaviorists assumed that it would be possible to discover all relevant rules of psychology in the relationships of input states to output responses, the computationalists took this same logic inside.

A computation, as defined in computer science, is a description of a regularly patterned machine operation. We call a specific machine process a computation if, for example, we can assign a set of interpretations to its states and operations such that they are in correspondence with the sequence of steps one might go through in performing some action such as calculating a sum, organizing files, recognizing type characters, or opening a door when someone walks toward it. In this respect, the workings of my desktop computer, or any other computing machine, are just the electronic equivalents of levers, pendulums, springs, and gears, interacting to enable changes in one part of the mechanism to produce changes in another part. Their implementation by electronic circuitry is not essential. This only provides the convenience of easy configurability, compact size, and lightning speed of operation. The machines that we use as computers today can be described as "universal machines" because they are built to be able to be configured into a nearly infinite number of possible arrangements of causes and effects. Any specific configuration can be provided by software instructions, which are essentially a list of switch settings necessary to determine the sequence of operations that the machine will be made to realize.

Software is more than an automated process for setting switches only insofar as the arrangement of physical operations it describes also corresponds to a physical or mental operation performed according to some

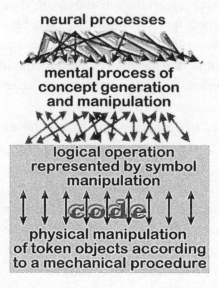

FIGURE 3.1: The relationship of computational logic to cognition. Computation is an idealization made possible when certain forms of inference can be represented by a systematic set of operations for writing, arranging, replacing, and erasing a set of physical markers (e.g., alphanumeric characters) because it is then possible to arrange a specific set of mechanical manipulations (e.g., patterns of moving beads on an abacus or of electrical potentials within a computer circuit) that can substitute for this symbol manipulation. It is an idealization because there is no such simple codelike mapping between typographical symbolic operations and thought processes, or between mental concepts and neurological events.

meaningful or pragmatic logic organized to achieve some specified type of end. So, for example, the description of the steps involved in solving an equation using paper and pencil has been formalized to ensure that these manipulations of characters will lead to reliable results when the characters are mapped back to actual numerical quantities. Since these physical operations can be precisely and unambiguously described, it follows that any way that a comparable manipulation can be accomplished will lead to equivalent results. The implications of being able to specify such correspondence relationships between mechanical operations and meaningful manipulations of symbols was the insight that gave rise to the computer age that we now enjoy. An action that a person might perform to achieve a given end could be performed by a machine organized so that its physical operations match one-to-one to these human actions. By implication it seemed that if the same movements, substitutions, and rearrangements of physical tokens could be specified by virtue of either meaningful principles or purely physical principles, then for any mental operation one should be able to devise a corresponding mechanical operation. The mystery of how an idea could have a determinate physical consequence seemed solved. If teleologically defined operations can be embodied in merely physical operations, then can't we just dispense with the teleology and focus on the physicality?

THE BALLAD OF DEEP BLUE

The classic version of this argument is called the computer theory of mind, and in some form or other it has been around since the dawn of the computer age. Essentially it can be summarized by the claim that the manipulation of the tokens of thoughts—the neural equivalents of character strings, or patterns of electric potential in a computer—completely describes the process of thinking. The input from sensors involves replacing physical interactions with certain of these "bits" of information; thinking about them involves rearranging them and replacing them with others; and acting with respect to them involves transducing the resultant neural activity pattern into patterns of muscle movement. How these various translations of pattern-to-pattern take place need not require anyone acting as controller, only a set of inherited or acquired algorithms and circuit structures.

But is computation fully described by this physical process alone, or is there something more required to distinguish computation from the physical shuffling of neural molecules or voltage potentials?

A weak link in this chain of assumptions is lurking within the concept of an algorithm. An *algorithm* is effectively a precise and complete set of instructions for generating some process and achieving a specific consequence. Instructions are descriptions of what must occur, but descriptions are not physical processes in themselves. So an algorithm occupies a sort of middle position. It is neither a physical operation nor a representation of meanings and purposes. Nor is an algorithm some extra aspect of the mechanism. It is, rather, a mapping relationship between something mechanical and something meaningful. Each interpretable character or character string in the programming language corresponds to some physical operation of the machine, and learning to use this code means learning how a given machine operation will map to a corresponding manipulation of tokens that can be assigned a meaningful interpretation. Over the many decades that computation has been developing, computer scientists have developed ever more sophisticated ways of devising such mappings. And so long as an appropriate mapping is created for each differently organized machine, the same abstract algorithm can be implemented on quite different physical mechanisms and produce descriptively equivalent results. This is because the level of the description comprising the algorithm depends only on a mapping to certain superficial macroscopic properties of the machine. These have been organized so that we can ignore most physical details except for those that correspond to certain symbol manipulations. Additional steps of translation (algorithms for interpreting algorithms) allow an algorithm or its data to be encoded in diverse physical forms (e.g., patterns of optically reflective pits on a disk or magnetically oriented particles on a tape), enabling them to be stored and transferred independent of any particular mechanism. Of course, this interpretive equivalence depends on guaranteeing that this correspondence relationship gets reestablished every time it is mapped to a specific mechanism (which poses a problem as computer technology changes).

In this respect, algorithms—or more colloquially, software—share a useful attribute with all forms of description: they don't specify all the causal details (e.g., all the way down to electrons and atomic nuclei). This is not even provided by the translation process.[7] This ability to ignore many micro causal details is possible because a well-designed computing device strictly limits subtle variations of its states, so that it tends to assume only unambiguously distinguishable discrete states. Many different physical implementations of the same computing process are thus possible. It only needs to be sufficiently constrained so that its replicable macro states and its possible state transitions match the descriptive level of detail required by the algorithm.

Like words printed in ink on a page (which is often one way to encode and store software for access by humans), any particular embodiment of this code is just a physical pattern. Its potential to organize the operations of a specially designed machine is what leads us to consider it to be something more than this. But this "something more" is not intrinsic to the software, nor to the machine, nor to the possibility of mapping software to machine operations. It is something more only because *we* recognize this potential. In the same way that the printed text of a novel is just paper and ink without a human mind to interpret it, the software is just pattern except for being interpretable by an appropriate machine for an appropriate user. The question left unanswered is whether the existence of some determinate correspondence between this pattern and some pattern of machine dynamics constitutes an *interpretation* in the sense that a human reader provides. Is the reader of a novel herself also merely using the pattern of ink marks on the page to specify certain "machine operations" of her brain? Is the implementation of that correspondence sufficient to constitute the meaning of the text? Or is there something more than these physical operations? Something that they are both about?

One of the signal events of the last decade of the twentieth century with regard to intelligence and computation was the defeat of the world's chess master, Garry Kasparov, by a computer chess program: Deep Blue. In many ways this is the modern counterpart to one of the great American folk tales: the ballad of John Henry.

John Henry was, so the ballad says, a "steel-drivin' man," the epitome of a nearly superhuman rail worker, whose job was to swing a massive hammer used to drive steel spikes into the ties (the larger wooden cross members) that held the tracks in place. He was a massive man, who was by reputation the most powerful and efficient of all steelmen. The introduction of a steam-driven spike driver in the mid-nineteenth century threatened to make this job irrelevant.[8] The tale centers on a contest between John Henry and the machine. In the end, John Henry succeeds at keeping up with the machine, but at the cost of his own life.

In the modern counterpart, Kasparov could usually play Deep Blue to a draw, but for the man this was exhausting while the machine felt nothing. In the end of both contests, the machines essentially outlasted the men that were pitted against them. Deep Blue's victory over Kasparov in 1997 signaled the end of an era. At least for the game of chess, machine intelligence was able to overcome the best that biology had to offer. But in the midst of the celebrations by computer scientists and the laments of commentators marking the supremacy of silicon intelligence, there were a few who weren't quite so sure this was what they had just witnessed. Was the world chess master playing against an intelligent machine, or was he playing against hundreds of chess-savvy programmers? Moreover, these many dozens of chess programmers could build in nearly unlimited libraries of past games, and could take advantage of the lightning-fast capacities of the computer to trace the patterns of the innumerable possible future moves that each player might make following a given move. In this way the computer could compare vastly many more alternative consequences of each possible move than could even the most brilliant chess master. So, unlike John Henry's steel competitor that matched force against force, with steel and steam against muscle and bone, Deep Blue's victory was more like one man against an army in which the army also had vastly more time and library resources at its disposal. Garry Kasparov was not, in this sense, playing against a machine, but against a machine in which dozens of homunculi were cleverly smuggled in, by proxy.

Like the machinery employed by the Wizard of Oz to dazzle his unwitting subjects, today's computers are conduits through which people (programmers) express themselves. Software is a surrogate model of what some anthropomorphic gremlin might do to move, store, and transform symbols. Ultimately, then, software functions are human intentions to configure a machine to accomplish some specified task. What does this mean for the computer model of cognition? Well, to the extent that a significant part of cognition is merely manipulating signals according to strict instructions, then the analogy is a good one. As Irving J. Good explained at the dawn of the computer age: "The parts of thinking that we have analyzed completely could be done on the computer. The division would correspond roughly to the division between the conscious and unconscious minds."[9]

But what parts of thinking can be analyzed completely? Basically, only that small part that involves thoroughly regular, habitual, and thus mechanical thought processes, such as fully memorized bits of addition or multiplication, or pat phrases that we use unconsciously or that are required by bureaucratic expediency. Thus, on this interpretation of Good's assessment, computation is like thoroughly *unconscious* mental processes. In those kinds of activities, no one is at the controls. Throw the right switches and the program just runs like a ball rolling downhill. But who throws the switches? Usually some homunculus; either a conscious organism like you or me, or someone now out of the picture who connected sensors or other kinds of transducers to these switches such that selected conditions could stand in for flipping the switches directly.

In all these examples, the apparent agency of the computer is effectively just the displaced agency of some human designer, and the representational function of the software is effectively just a predesigned correspondence between marks of some kind and a machine that a human programmer has envisioned. Is there any sense in which computers doing these kinds of operations could be said to have their *own* agency, purpose, meaning, or perhaps even experience, irrespective of or despite these human origins and reference points? If not, and yet we are willing to accept the computer metaphor as an adequate model for a mind, then we would be hard-pressed to be able to attribute agency to these designers and programmers either, except by reference to some yet further outside interpreter. At the end of this line we either find a grand black box, an ultimate homunculus, infinite regress, or simply nothing—no one home all the way up and down.

Many critics of the computer theory of mind argue that ultimately the algorithms—and by implication the cognitive level of description with its reference to intentions, agency, and conscious experience—provide no more than a descriptive gloss of a mapping from operation to operation. What matters is the correspondence to something that the algorithm and the computer operations are taken to refer to by human users. If we presume that what is being represented is some actual physical state of affairs in the world, then what we have described is ultimately an encoded parallelism between two physical processes that its users can recognize, but this human-mediated correspondence is otherwise not intrinsic to the computer operation and what it is taken to refer to. This problem is inherited by the computational theory of mind, rendering even this human mediation step teleologically impotent. A computer transducing the inputs and outputs from another computer is effectively just one larger computer.

But notice that there may be innumerable mappings possible between some level of description of a machine state and some level of description of any other physical state. There is nothing to exclude the same computer operation from being assigned vastly many mappings to other kinds of meaningful activities, limited only by the detail required and the sophistication of the mapping process. This multiple interpretability is an important property from the point of view of computer simulation research. Yet, if innumerable alternative correspondence relationships are possible, are we forced to conclude that the same physically implemented operation is more than one form of computation? Is this operation only a computation with respect to one of these particular mappings? Does this mean that given any precisely determined mechanical process, we are justified in calling it a *potential* computation, on the assumption that some interpretable symbolic operation could ultimately be devised to map onto it?

This last point demonstrates that treating computation as intrinsic to the machine operations essentially forces us to understand every physical process as a potential computation. Alternatively, if we define something as a computation only when actualized by an additional process that determines the relationship between a given machine operation and a specific corresponding process in some other domain, such as mathematical calculation, then this additional process is what ultimately distinguishes computation from mere mechanism. Of course, in the realm of today's electronic computers, this process is the activity of a human mind; so, if human minds are just computers, then indeed there is no one home anywhere, no interpretation. It's all just machine operations mapped to machine operations mapped to machine operations.

There is another equally troubling way that the logic of computation undermines the explanation of the special features of mental processes. It also seems to suggest that it may be impossible for minds to expand themselves, to develop new meanings, to acquire new knowledge or capacities.

The *reductio ad absurdum* of this argument was probably best articulated by the philosopher Jerry Fodor. He poses the problem this way:

> There literally isn't such a thing as the notion of learning a conceptual system richer than the one that one already has; we simply have no idea of what it would be like to get from a conceptually impoverished to a conceptually richer system by anything like a process of learning.[10]

In his characteristically blunt and enigmatic manner, Fodor appears to be saying that we are permanently boxed in by the conceptual system that we must rely on for all our knowledge. According to this conception of mental representation, what a mind can know must be grounded on a fixed and finite set of primitives and operations, a bit like the initial givens in Euclid's geometry, or the set of possible machine operations of a computer. Although there may be a vast range of concepts implicitly reachable through inference from this starting point, what is conceivable is essentially fixed before our first thoughts. At face value, this appears to be an argument claiming that all knowledge is ultimately preformed, even though its full extent may be beyond the reach of any finite human mind. This interpretation is a bit disingenuous, however, or at least in need of considerable further qualification. At some point the buck must stop. We humans each *do* possess a remarkably complex conceptual system, and almost certainly many aspects of this system arose during our evolution, designed into our neural computers by natural selection (as evolutionary psychologists like to say). But unless we are prepared to say that our entire set of mental axioms magically appeared suddenly with humans, or that all creatures possess the same conceptual system, our inherited conceptual system must have been preceded at some point in our ancestors' dim past by a less powerful system, and so on, back to simpler and simpler systems. This is the escape route that Fodor allows. The argument is saved from absurdity so long as we can claim that evolution is not subject to any such stricture.

Presumably, we arrived at the computational system we now possess step by step, as a vast succession of ancestors gave birth generation upon generation to progressively more powerful conceptual systems. By disanalogy then, evolution *could be* a process that generates a richer (i.e., functionally more complex) system from a simpler one. But could cognition be similar to evolution in this respect? Or is there some fundamental difference between evolution and learning that allows biology to violate this principle but not with respect to cognition? One fundamental difference is obvious. The principles governing physical-chemical processes are different from those governing computation. If evolution is merely a physical-chemical process, then what can occur biologically is not constrained to computational strictures. There will always be many more kinds of physical transformations occurring in any given biological or mechanical process than are assigned computational interpretations. So a change in the structure of the computational hardware (of whatever sort) can alter the basis for the mapping between the mechanism and the interpretation, irrespective of the strictures of any prior computational system. In other words, precisely because computation is not itself mechanism, but rather just an interpretive gloss applied to a mechanism, changes of the mechanism outside of certain prefigured constraints will not be "defined" within the prior computational system. Computation is fragile to change of its physical instantiation, but not vice versa. With respect to the computational theory of mind, this means that changes of brain structure (e.g., by evolution, maturation, learning, or damage) are liable to undermine a given computational mapping.

This might at the same time preserve Fodor's claim and yet overcome the restriction it would otherwise impose, though undermining the force of it in the process. Thus, on the analogy of a function that is not computable on a Turing machine, an intrinsically interminable machine process could, for example, be halted by damaging its mechanism at some point to make it stop (indeed, simply shutting off my computer is how I must sometimes stop a process caught in an interminable loop). This outside interference is not part of the computation, of course, any more than is pulling the plug on my computer to interrupt a software malfunction. If interference from outside the system (i.e., outside the mechanistic idealization that has been assigned a given computational interpretation) is capable of changing the very ground of computation, then computation cannot be a property that is intrinsic to anything. Computation is an idealization about cognition based on an idealization about physical processes.

Turning this logic around, however, offers another way to look at this loophole. The fact that computation can be altered by mechanistic failure imposed from without is a clue that both the mechanism and the correspondence relationship that links the mechanism to something else in the world are idealizations. Certain mechanistic processes must be prevented from occurring and certain properties must be ignored in establishing the representational correspondence that defines the computation. Computation is therefore derived from these extrinsic, simplifying constraints on both mechanical operations and the assignment of correspondences. These constraints are not only imposed from the outside, they are embodied in relations that are determined with respect to operations that are prevented or otherwise not realized. Paying attention to what is not occurring is the key to a way out of this conceptual prison. We must now ask: What establishes these mechanical and mapping constraints? It cannot be another computing relationship, because this merely passes the explanatory buck. Indeed, this implies that computation is parasitic on a conception of causality that is not just different from computation, it is its necessary complement.

So whatever processes impose these constraints on the physical properties of the mechanism and on the mapping from mechanism to use, these are what determine that something is a computation, and not merely a mechanical process. Of course, for digital computers, human designers-users supply these constraints. But what about our own brains? Although we can pass some part of the explanatory buck to evolution, this is insufficient. Even for evolution there is a similar mapping problem to contend with. Like computation, only certain abstract features of the physiological processes of a body are relevant to survival and reproduction. These are selectively favored to be preserved generation to generation. But although natural selection preserves them with respect to this function, natural selection does not produce the physical mechanisms from which it selects. Although natural selection is not a teleological process itself, it depends on the intrinsic end-directed and informational properties of organisms: their self-maintenant, self-generative, and self-reproducing dynamics. Without these generative processes that are the prerequisites for evolution, and require certain supports from their environment, there is no basis for selection. Like computation, the determination of what needs to be mapped to what and why is necessarily determined by prior ententional factors.

The ententional properties that make something a computation or an adaptation must ultimately be inherited from outside this mapping relationship, and as a result their ententional properties are parasitic on these more general physical properties. This is why, unlike computation, the messiness of the real-world physics and chemistry is not a bug, it's an essential feature of both life and mind. It matters that human thought is not a product of precise circuits, discrete voltages, and distinctively located memory slots. The imprecision and noisiness of the process are what crucially matters. Could it be our good fortune to be computers made of meat, rather than metal and silicon?

This brief survey of the homunculi and golems that have haunted our efforts to explain the properties of life and mind has led to the conclusion that neither preformation nor elimination approaches can resolve the scientific dilemmas posed by their ententional phenomena. The former merely assumes the existence of homunculi to supply the missing teleology; the latter denies their existence while at the same time smuggling in golems disguised as physical principles. To accept ententional properties as fundamental and unanalyzable is merely to halt inquiry and to rest our explanations of these phenomena on a strategy of simply renaming the problem in question. To deny the existence of these properties in the phenomena we study inevitably causes them to reappear, often cryptically, elsewhere in our theories and assumptions, because we can't simply simplify them down and down until they become purely chemical or mechanical processes. To pretend that merely fractionating and redistributing homunculi will eventually dissolve them into chemistry and physics uses the appearance of compositionality to distract our attention from the fact that there is something about their absential organization and their embeddedness in context that makes all the difference.

4

TELEONOMY

Teleology is like a mistress to a biologist:
he cannot live without her but he's
unwilling to be seen with her in public.

—J. B. S. HALDANE[1]

BACK TO THE FUTURE

Living and mental processes appear to work against an otherwise universal trend of nature. When it comes to processes that produce new structures, the more rare, complicated, or well designed something appears, the less likely that it could have happened spontaneously. It tends to take more effort and care to construct something that doesn't tend to form on its own, especially if it is composed of many complicated parts. In general, when things work well despite the many ways they could potentially fail, when they exhibit sophisticated functional matching to their context, especially where this matching is highly contingent, then such fittedness tends to be a product of both extensive effort and intelligent planning, at least in human experience. Good luck is rare, bad luck is the norm, and most problems left unattended don't improve spontaneously.

This tendency of things to fall effortlessly into messiness is the essence of the second law of thermodynamics: the relentless, inevitable increase in entropy. The term *entropy* was originally introduced into physics by the grandfather of thermodynamics, Rudolf Clausius, in 1865. In a general sense, it can be understood as the relative state of "mixed-upness" of a system, with its maximum value characterizing the state in which no additional change could make it more thoroughly mixed. The second law of thermodynamics tells us that, all other things being equal and without outside interference or loss (or more specifically, in a hypothetically isolated physical system in which energy neither enters nor leaves), entropy will inevitably tend to increase until it reaches this maximum. In simple terms, things just tend to get as mixed up as they possibly can. At this point we say that equilibrium has been reached, by which we mean that any future change is as likely to occur in one direction as the other, and over time directional changes will tend to cancel out, resulting in no net change overall.

Apparently, however, living and mental phenomena violate this presumably universal law. Expending effort to adjust conditions and to mobilize resources in order to alter future conditions toward an otherwise unlikely state is one of the most common attributes of living systems. In this way, the complex adaptive functions of organism bodies tend to increase orderliness, stabilize correspondences, and generate complex patterns of molecular interactions that are precisely complementary to one another and well suited to the contexts in which they occur. Yet in the non-living world, future possibilities do not directly influence present processes, and most physical and chemical processes are unlikely to have any supportive relationship to one another. Though non-living chemicals can become more organized when they freeze and solidify into crystals, even in this case the total surrounding environment ultimately gets even less organized as heat energy is dissipated to the surroundings in the process. In the non-living world, we only observe a reversal of this tendency in videos of events played in reverse. But of course time isn't running backwards in living and mental processes.

On the surface, then, the second law of thermodynamics appears to all but exclude the possibility of ententional processes, such as are characteristic of living and mental phenomena. There is the dead, pointless, uncaring world and its rules, and the living, striving, feeling world and its rules, and the two seem to be working in quite contradictory ways. Because the spontaneous order generation that is so characteristic of life and mind runs counter to this otherwise exceptionless current of nature, it demands that we take seriously the possibility that our usual forms of explanation might be inadequate. When unrealized future possibilities appear to be the organizers of antecedent processes that tend to bring them into existence, it forces us to look more deeply into the ways we conceive of causality and worry that we might be missing something important.

These enigmatic features of the living world are in fact hints that there must be a loophole in the second law. A first step toward discovering it is to recognize that the second law of thermodynamics is actually only a description of a tendency, not a determinate necessity. This makes it different from the inverse-square law that defines the distribution of an electric or gravitational field in space or Newton's law of force ($F = ma$). For all practical purposes, however, this thermodynamic tendency is so nearly inviolate that it can be treated like a law. It is a tendency that is so astronomically biased as to be almost universally the case. But "almost" is an important qualification. The second law of thermodynamics is only a probabilistic tendency, not a necessity, and that offers some wiggle room.

A superficial appearance of time reversal is implicit in describing functional and purposive processes in terms of the ends that they appear organized to produce. As we saw earlier, being organized for the sake of achieving a specific end is implicit in Aristotle's phrase "final cause." Of course, there cannot be a literal ends-causing-the-means process involved, nor did Aristotle imply that there was. And yet there is something about the organization of living systems that makes it appear this way. This time-reversed appearance is a common attribute of living processes, albeit in slightly different forms, at all levels of function. The highly ordered interactions of cellular chemistry succeed in maintaining living systems in far-from-equilibrium states. They locally counter the increase of internal entropy, as though thermodynamic time were stopped or even reversed within the organism. The increasing complexity of organism structures and processes that characterizes the grand sweep of evolution also appears as though it is running counter to the trend of increasing messiness and decreasing intercorrelations. Furthermore, mental processes recruit energy and organize physical work with respect to the potential of achieving some future general state of things that does not yet exist, thus seeming to use even thermodynamics against itself.

Of course, time is neither stopped, nor running backwards in any of these processes. Thermodynamic processes are proceeding uninterrupted. Future possible states of affairs are not directly causing present events to occur. So, what is responsible for these appearances?

THE LAW OF EFFECT

Order generation is a necessary property of life, but it is the production of orderliness in the service of perpetuating this same capacity that distinguishes life from any inorganic processes. Thus the localized opposition to thermodynamic degeneration that is exemplified in organisms has a kind of reflexive character. It is this reflexivity that warrants describing organism structures and processes as *functions* as opposed to mere tendencies. We are familiar with functions that are a product of design, but only since Darwin have we been able to understand function as emerging spontaneously—with its end-directed character emerging irrespective of any anticipated end.

The appeal to natural selection logic has long been considered to be the one proven way around the homunculus problem. Though there are still periodic uprisings, mostly from fundamentalist religious interests, as in the case of creationism or Intelligent Design proponents, it can fairly be said that the logic of natural selection has dealt a serious blow to any claim that teleological processes are necessary to explain complex structure and function in the biological world. Daniel Dennett in his book *Darwin's Dangerous Idea* considers the logic of natural selection to be the intellectual equivalent to a universal acid. And what it universally dissolves is teleology.

The theory of natural selection is ultimately just statistics at work. As Darwin argued, all that is required to explain how organisms become suited to their particular environments is (1) reproduction, with offspring inheriting traits from parents; (2) some degree of spontaneous variation from strict inheritance; and (3) reproduction in excess of the potential support that can be supplied by the local environment. This limitation will inevitably result in reproductive inequality among variant individuals. Those lineages with individual variants that are better suited to utilize available resources in order to support reproduction will gradually replace those lineages that are less well suited. Though Darwinism is often caricatured in contrast to Lamarck's theory of the inheritance of acquired characters, this was not a necessary requirement for Darwin, and he himself entertained the possibility of Lamarckian-like mechanisms in some editions of *The Descent of Man* in what he called a "gemule theory." This modern myth of intellectual history, which focuses on inheritance issues, actually obscures the fundamental insight that made Darwin's analysis so revolutionary and challenging to nineteenth-century sentiments.

The core distinguishing feature of Darwin's explanation—and what made it so revolutionary—was, instead, the *after-the fact* logic of this mechanism. Where previously it seemed natural to assume that the processes responsible for orderly function and adaptive design necessarily preceded their effects—as in Lamarck's view of functional use and disuse determining the evolution of future functions—Darwin showed that, in principle, antecedent "testing" of functional responses by trial and error was unnecessary to achieve adaptive outcomes. The process that generated variant forms could be completely uncorrelated with the process that determined which variant forms were functionally superior to the others in a given environment. So long as the options with favorable outcomes were preferentially reproduced or retained and re-expressed in future contexts, it did not matter why or how they were generated.

This after-the-fact logic for producing adaptive "design" is not just applicable to evolution. The early American behaviorist Edward Thorndike realized that one could even describe the way organisms adapted their behaviors to the environment in response to reinforcement (e.g., reward or punishment delivered after the fact) in these same terms. He called it the "Law of Effect."

One might wonder why such a simple idea had to wait for the mid-nineteenth century to come to the attention of scientists and natural philosophers. One need look no further than the current stridency of anti-Darwinian sentiment that continues to complicate the politics of American education, and even persists among scholars outside biology, to recognize that this form of causal logic remains counterintuitive to this day. On the one hand, the counterintuitive logic of design-without-designing strains credulity. On the other hand, we are entirely familiar with the human ability to conceive and design mechanisms suitable to serve some purpose. So it is only a modest conceptual leap to imagine that a far greater intelligence and technology might be capable of designing mechanisms like organisms. The reverse logic is not so congenial to common sense. It goes against common sense to imagine that something so complex as an organism, which is also extremely well fitted to its environment, could arise spontaneously from something less complex and less in sync with surrounding conditions. Experience relentlessly reminds us of the opposite tendency. For this reason alone, it should not surprise us that systematically describing the logic of such a process took millennia to be achieved.

But Darwin was not the first to think in these terms. In fact, we can probably credit a near contemporary of Aristotle with the first version of such an explanation. The man who first articulated this notion was a Greek philosopher-poet named Empedocles. We know about him because he was both famous and notorious in his own right, and because Aristotle took his ideas to be worthy exemplars of a way of thinking about the natural world that was deeply flawed.

Empedocles' cosmology was based on the four "elements" earth, water, air, and fire (roughly analogized to the four phases of matter: solid, liquid, gas, and plasma) and on forces of attraction and repulsion that he believed ruled their interactions. But what makes his analysis modern—and what made it seem incoherent to Aristotle—was his claim that even the orderliness of living bodies might be able to arise spontaneously, simply by preserved accident. From these sparse beginnings all the more complex features of the world, like living bodies, could arise by combinations of more basic parts which themselves arose and combined by accident.

Empedocles argued that even aimless mechanical processes could produce functionally useful complexity. Initially, all that would be produced would be incongruous combinations and monstrosities. For example, he invokes the possibility of chimeric creatures, such as a man-faced ox, that might willy-nilly be products of such a wild natural shuffling of materials and forms. But, he notes, most such combinations would be awkward and inappropriate to the world, and would quickly perish. Only those combinations that exhibited both an appropriate balance of component features and fittedness to the surroundings would be likely to persist in the world (and presumably, if living, would produce progeny). Further, he suggested that these creatures would sort themselves out into different environments, with those suited to water more likely to spend time there and those more suited to air more likely to spend time there, where each would then, by the same logic, get honed by elimination into progressively more specialized and distinguishable types. So, he claimed, without prior design, a blind and promiscuous process of mixing of elements could by dint of the elimination of the less coherent, less elegant, less appropriate variations, after the fact, result in forms that were both internally and relationally well designed and well suited to their contexts.

Aristotle, however, found this proposal to be incongruous with the extraordinary design elegance and behavioral sophistication of living things. He therefore devoted considerable effort to refuting this view. As a careful observer of nature, he noted that the elaborate process of organism development—which inevitably produced horses from horses and oaks from acorns from oaks—could not be understood without postulating the presence of an intrinsic end-directed principle (an entelechy). This active principle was inherent in these organisms from the earliest stages of development and was responsible for guiding their maturation toward the adult form. And it wasn't merely a mechanical principle, because it would function to bring about this end form despite the vicissitudes of an unpredictable environment. Thus a tree might grow its limbs through a fence or grow its roots around buried boulders in its striving toward its adult form. If such a principle is necessarily present in each organism, how could it be denied as an intrinsic contributor to the origination of living forms as well?

Darwin, too, apparently struggled with the counterintuitive logic that eventually became enshrined in his theory of natural selection. Most historians locate the point at which the idea of this reversal of logic came to him after reading Thomas Malthus' book on the statistics of population growth. Although Lamarck had previously developed the idea that useful traits should naturally replace less useful traits over the generations, Darwin realized that their usefulness did not need to be involved in their initial generation. There was no need for prior experience or prescient design. Because organisms reproduce, their traits would inevitably be differentially preserved and transmitted irrespective of how they were generated, so long as there was competition among variants for the means of reproduction that favored the preservation of some lineages over others. The mechanism responsible for the origin of a given trait would therefore be irrelevant, so long as that trait got successfully preserved in the future population. In the world of reproducing organisms, an effect could in this way be the explanation for the current existence of functional organization, even if that effect had no role in producing it. The mechanisms for producing forms and for preserving them were entirely separable.

This provided a way to get around both the necessity of prior design and the role of an organism's striving to adapt to local conditions. Adaptation and functional correspondence could even be a consequence of preserved accident, so long as there was a favorable reproductive outcome. Darwin could thus explain the presence of a living form bearing the marks of design, without invoking any hint of intelligence, representation, agency, or pre-science. Nothing more than blind chance and being consistent with the necessities of reproduction (*Chance and Necessity*, as Jacques Monod was to much later paraphrase it in the title of his celebrated book) might provide a sufficient basis for replacing *all* appeals to teleological processes in natural science.

PSEUDOPURPOSE

This causal dichotomy separating living from non-living nature is real, but the appearance of causal incompatibility is partly an unfortunate accident of conceiving of organisms as though they are machines. Although they are indeed functionally organized, living organisms aren't just complicated chemical machines. In many ways, living systems are the inverses of man-made mechanisms. As designers and users, we determine the form of a machine to be suited or not to a particular task, but this task otherwise has no relation to the machine's existence. Organism forms evolve in the process of accomplishing a task critical to maintaining the capacity to produce this form, so the task space and the form of the organism are essentially insepa-rable. Machines require assembly from parts that are produced separately and in advance. Organisms spontaneously develop. Their parts differentiate from an undifferentiated starting point. There are almost never additions, and parts almost never need to be brought together for assembly. The func-tional integration of the components of a machine must be anticipated in advance. Organisms' components are integrated and interdependent from the beginning and, as noted above, they exist as a consequence of having at some point been relevant for already fulfilling some function.

Despite these almost exactly inverted modes of construction, the comparison between organisms and complex machines is almost irresistible. Indeed, I think that it is accurate to say that biologists on the whole accept the machine analogy, at least in a vague sense of the concept, and rely on it to make sense of the obvious end-directedness of living processes. In large part, this satisfaction with the machine analogy is the result of the fact that there is now a straightforward way to explain the end-directedness of both organic and mechanical processes without any of the trappings of teleology in the classic sense, which still retains a vocabulary rich in teleological connotations. Functional organization can be treated as a variation on a theme common to simpler mechanistic processes lacking any whiff of meaning, purpose, or value. In an influential paper published in 1974, the Harvard evolutionary biologist Ernst Mayr argued that we can avoid confusing the different sorts of asymmetrically directed activities exhibited by intelligent agents, living organisms, and even some machines by distinguishing a number of variations on the theme of asymmetric change, respectively. Thus, for example, there is the asymmetric change exhibited by thermodynamic processes, the apparent goal-directed behavior of thermostats and guidance systems, the asymmetric development and evolution exhibited by biological processes, and the purposeful design and conscious goal-directed activity of human agents.

As with many of the prominent evolutionary biologists of the late twentieth century, Mayr was eager to complete the metaphysical rout of teleology that Darwin had initiated. He recognized the need to use end-directed and information-based terminology for describing biological phenomena, but wanted a way to do this without also accepting its metaphysical baggage. Mayr, in fact, believed that the problem could be solved by incorporating insights from cybernetics and computational theories into the evolutionary paradigm. He quotes his Harvard colleague, the philosopher-logician Willard Van Ormand Quine, as agreeing:

> The great American philosopher Van Quine, in a conversation I had with him about a year before his death, told me that he considered Darwin's greatest philosophical achievement to consist in having refuted Aristotle's final cause. . . . At the present we still recognize four teleological phenomena or processes in nature, but

they can all be explained by the laws of chemistry and physics, while a cosmic teleology, such as that adopted by Kant, does not exist."[2]

Whether or not we can credit Darwin with such a monumental philosophical achievement, this assertion nonetheless captures the belief held by most biologists that even the action of organisms can be understood as thoroughly mechanistic and on a par with the other physical processes of the inorganic world, except for being incorporated into a living body.

But recognizing that organisms behave in ways that differ from inorganic processes, Mayr sought a terminological distinction that would exemplify a middle ground between mere mechanism and purpose. He turned to a term coined by Colin Pittendrigh in a paper in 1958: *teleonomy*. The Greek combining form -*nomy* merely implies lawlike behavior, and so one could use it to describe behavior that was asymmetrically oriented toward a particular target state, even in systems where there was no explicit representation of that state (much less an intention to achieve it) but only a regular predictable orientation toward an end state. By only specifying the way a mechanical process can be organized so that it converges toward a specific state, rather than behavior *for the sake of* an end, this conception avoided sneaking mentalistic assumptions about teleology back into biological discourse. Mayr agrees with Pittendrigh that this makes it possible to feel no discomfort in saying that "A turtle came ashore to lay her eggs," instead of just, "She came ashore and laid her eggs." According to Mayr, "There is now complete consensus among biologists that the teleological phrasing of such a statement does not imply any conflict with physico-chemical causality."[3]

By coining a term which implied target-directedness but was agnostic about how this behavior was produced or came to exist, biologists and engineers could continue to use teleologically loaded descriptions, but without the metaphysical baggage that tended to come with them. Moreover, to the extent that the process of natural selection could be conceived as consequence-blind and accident-driven, this target-directed tendency could also be treated as accidental. It seemed as though teleology had been fully reduced to mere mechanism.

A classic example of a purely mechanistic teleonomic effect is provided

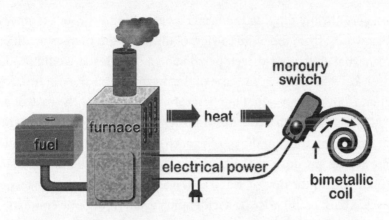

FIGURE 4.1: A diagram of a classic thermostat circuit that uses a bimetal coil
and a mercury switch to turn a furnace for heating a room on and off and
to maintain it at a constant temperature. Although changes of different sorts
propagate around this causal circle, the circular iterations of these influences
compound with themselves again and again incessantly. This causal analogue
to self-reference is what is responsible for the deviation-reduction dynamics
of the whole. Reversing the orientation of the mercury switch would alter the
circuit from deviation reduction (negative feedback) to deviation amplification (positive feedback).

by the cycle of causes and effects constituting an old-style thermostat circuit
used to maintain the temperature of a room (see Figure 4.1). We can trace
a continuous loop of differences that make a difference in this system (following Gregory Bateson's way of describing this) as follows: The difference
in the measured temperature of the room (1), makes a difference in the
winding or unwinding of a bimetallic strip (2), which makes a difference
in the tilt of a mercury switch (3), which makes a difference in the flow of
current through the electric circuit (4), which makes a difference in the fuel
introduced into the furnace (5), which makes a difference in the temperature of the room (1), and so on. The result is that the whole open system,
consisting of the room and heating system, embedded in a larger context
which introduces uncorrelated variations, exhibits a pattern of behavior that
is specifically convergent toward a given end state. So, although there is no
explicit internal representation of this target state, the structure of the circuit constraining the relationships between room temperature and furnace-
energy use produces a pattern of activity that would also be produced by a

person periodically flipping the heater switch to adjust room temperature. In other words, from the point of view of the behavior, there is no difference between what a purpose-driven person and a thermostat would produce.

The *telos*—the end—in the case of the thermostat is the minimization of variation from a given target temperature. Of course, in the case of a thermostat, that end, as well as the means to achieve it, is the product of a human mind. Being embodied in the purely mechanical organization of heater and thermostat circuit does not change the fact that its ententional character arose extrinsic to the device. But it is at least conceivable that an analogous feedback circuit could arise by accident in certain inorganic contexts.

Consider, for example, the famous Old Faithful geyser, located in Yellowstone National Park. It erupts on a highly regular basis because of the way that the temperature and pressure of the subsurface water is self-regulated around a mean value by the interplay of geothermal heating, the pressure of steam, and the weight of the column of water in the underground vent. As the water is heated to the point of boiling, the pressure of the released steam reaches a value sufficient to drive the overlaying column of water upward; but in so doing it "resets" the whole system, allowing new water to accumulate, boil, and reach this same threshold, again and again. So the water temperature and stem pressure oscillates around a constant value and the geyser erupts at regular intervals.

This demonstrates that such a teleonomic mechanism can arise by accident. If evolution can also be understood as nothing more than retained and reproduced accidental organization, then doesn't this suggest that all ententional processes might be reducible to accidental mechanisms? What are the implications for the source of the end-directed processes in an organism? Or a mind? Can't the purposiveness of a person turning a heater on and off to regulate room temperature also be ultimately traced to an accidental origin? Didn't our capacity to do this evolve by natural selection, and isn't that process reducible to physical-chemical accident? This is indeed the implication that many philosophers and biologists have drawn from this analysis.

Consider a typical biological analogue to the thermostat: the temperature regulating "circuit" of the human body. As in the case of the thermostat, although the material details are considerably different, body temperature regulation also depends on a mechanism that merely embodies this tendency in its overall organization. It isn't governed by an explicit representation of the goal state. Indeed, body temperature regulation is accomplished by a thermostatlike feedback mechanism. But, of course, the optimal temperature for body functions is not an accidental value. This mechanism evolved because this temperature was optimal for the many metabolic processes that ensured health and reproductive success of a long lineage of warm-blooded animals. So, like the electrical-mechanical thermostat, the target state that it tends to converge toward is derived extrinsic to the mechanism itself, even though this target is implicitly embodied in the structure of the circuit. This is not the case for Old Faithful. There the balance of factors that determines its regularity *is* determined solely by the accidental contingencies of the local geology.

There is, however, an extended sense in which an analogous claim could be made for mammalian temperature regulation. It is the contingency of best temperatures for the relevant metabolic biochemical processes that contributed to the preservation of this value rather than some other. This, too, is due to a merely physical-chemical contingency. A biochemical mechanism converging to this value was merely the most successful surviving variant from among many that were its probable competitors in evolution. Each was the product of a genetic accident, which became subject to the culling process of natural selection. In this respect, following a standard neo-Darwinian line of reasoning, we can say that this end wasn't the product of design. Its end-oriented tendency was produced by accident, and was retained merely because of its serendipitous contribution to the reproduction of organisms happening to exhibit this particular variant.

Writing in the early years of cybernetics and systems theory, Pittendrigh and Mayr were aware that target-directed behavior could be embodied in simple mechanisms, such as automatic guidance systems, and that these mechanisms could be designed to alter their output so that they could even track erratic targets or irregularly changing conditions. Analogous organic processes (such as temperature regulation) were already well known, as described in the work of physiologists like Walter Cannon and others. To illustrate the parallelism between goal-directed artifact behavior and organism behavior, Mayr cites the example of a guidance system: "A torpedo that has been shot off and moves toward its target is a machine showing teleonomic behavior." Mayr goes further than Pittendrigh in specifying how he believes this concept can be more precisely specified in biology. At the time of Mayr's appropriation of the term *teleonomy*, the informational function of DNA had been firmly established and the development of the science of computing had begun to reach maturity (though laptop computing was still a decade away). This led Mayr to generalize the concept one step further: conceiving biological goal-directedness in software terms. "A *teleonomic process or behavior is one which owes its goal-directedness to the operation of a program.*"[4]

This latter analogy was to prove highly influential because the computer program analogy can be extrapolated to any given level of complexity. Thus, in principle, there was no biological phenomenon that couldn't be modeled as the output of an algorithm. And like the cybernetic analogy, it could be understood in completely mechanistic terms. But also like the thermostat, when stripped of reference to human users and their interpretations and goals, a computer is simply a machine. Attributing the complex determinate behavior of a computer program to goal-directedness is thus an implicit reference to a potential human user/designer; otherwise (as we've seen), computation reduces to machine operation. Although computer behavior can be highly complicated, and thus can parallel highly organized organically produced behaviors—including those that converge toward specific end states—the appeal to the computer metaphor introduces considerable baggage in the form of non-trivial assumptions about what distinguishes computation from any other machine operation. As we argued in the last chapter, it is a mistake to confuse the abstract description applied to the machine's operations as something that is intrinsic to it. And yet without these extrinsic interpretive glosses of its machine operations, its apparent target convergence reduces to mere asymmetric mechanical change.

Teleonomy is, however, more than merely asymmetric physical change. To indicate the distinctiveness of teleonomic processes in both biology and human mechanisms, Mayr contrasted teleonomic processes with processes that converge toward a *terminal state* that is no longer characterized by *asymmetric* change. These are natural processes occurring spontaneously in the non-living physical world. In this respect, the fall of an object in a gravitational field or the development of a chemical reaction toward thermodynamic equilibrium have asymmetric tendencies that develop toward a specific stable end stage. He described these sorts of processes as *teleomatic*, literally, automatically achieving an end:

Many movements of inanimate objects as well as physico-chemical processes are the simple consequences of natural laws. For instance, gravity provides the end-state for a rock which I drop into a well. It will reach its end/state when it has come to rest on the bottom. A red-hot piece of iron reaches its "end-state" when its temperature and that of its environment are equal. All objects of the physical world are endowed with the capacity to change their state and these changes follow natural laws. They are "end-directed" only in a passive, automatic way, regulated by external forces or conditions. Since the end/state of such inanimate objects is automatically achieved, such changes might be designated as *teleomatic*.[5]

Mayr thus suggests that we should classify together all spontaneous physical processes that have a terminus at which change stops. This is probably a bit too broad. Others who have used the term *teleomatic* seem more content to restrict its reference to asymmetric causal processes that converge toward a more or less stable end state. This characterizes thermodynamic processes approaching equilibrium. Although mechanically moved objects may come to rest due to friction and falling objects may be specifically oriented in their movement and cease upon reaching the ground, these terminal states are relational and to some extent artificial, since they are just the result of being extrinsically impeded from continuing. In contrast, a chemical reaction or convection process developing toward thermodynamic equilibrium is defined by precisely excluding outside relational interactions. These are effectively intrinsically determined asymmetries and ends. Time-reversible processes, such as the interactions described by Newtonian mechanics, do not have intrinsic terminal states in the same sense, though field effects like the pull of gravity fall less clearly into one or the other category. This tripartite division of teleologic, teleonomic, and teleomatic is widely accepted (if not always framed in these terms). Many, including Quine and Mayr, would thus even deny that thought processes are anything more than highly complex forms of teleonomy.

The major problem with the term *teleonomy* is its implicit agnosticism with respect to the nature of the mechanism that exhibits this property. This is, of course, considered to be one of its main advantages, according to Mayr and others. On the one hand, if what is implied is organization on the anal-ogy of a thermostat or computer, there seems little that is intrinsic to these processes to distinguish their principles of operations from those of other mechanisms. With respect to its users, a thermostat is useful because of its highly reliable convergence toward a specific end state from a wide variety of initial conditions and perturbations. While this description might also apply to the second law of thermodynamics, the convergence mechanisms are quite different. Specifically, the thermostat achieves a degree of near sta-bility by doing work to perturb a thermodynamic process in a way that is intended to counter the second law effect as closely as possible. More use-fully, a thermostat can be set to many different arbitrary convergence values. Certain of these values will be consistent with the interests of its designers or users.

The analogy to the programmed operation of a digital computer is also an entirely mechanistic process that can be reduced to a collection of teleo-matic processes, which may be even more linearly deterministic and para-sitic on extrinsic description than is a feedback circuit, such as is embodied in a thermostat. Ultimately, in both of these interpretations, once we remove the implicit reference to a use, teleonomic processes reduce to teleomatic processes.

If these were presumed to be the only candidates, no wonder Mayr was willing to entertain the idea that teleology doesn't exist. It remains unclear whether other defenders of the concept also assume that teleonomic pro-cesses are entirely reducible to teleomatic processes. Justification for denying this reduction comes from those who focus on the special organization of teleomatic processes that constitutes feedback processes. In the case of feed-back, one teleomatic process is arranged so that it undermines the activity of another, and vice versa. Again, this is easily demonstrated by the action of a thermostat in which a change in room temperature activates a circuit, which has the effect of activating a furnace to change the room temperature, which ultimately has the effect of inactivating that circuit.

Both the thermostat and the computer analogies leave us with an unan-swered question, which is the issue at stake. What is the difference, if any, between attributing the construction of these sorts of mechanisms to human design versus natural selection versus serendipitous accident? As products of human design, we feel justified in locating the source of the *telos* outside the artifact. As products of pure accident, we feel justified in assuming that there is no *telos*, only the appearance of it under some description. But as products of evolution, it depends. If we conceive of this process as merely preserved accident, then it follows that the presumed ententional character of the mechanism is likewise merely an imposed descriptive gloss; and if we conceive of evolution as itself somehow intrinsically ententional, we appear to have invoked a variant of the human designer analogy. But of course, that merely begs the question, since human minds themselves evolved.

From the perspective of a kind of evolutionary hindsight, the end-directedness of an organism's adaptive features, even if they arose accidentally, currently exists *because of their contribution to something else*, a larger system that they happen to be a part of. Importantly, this system has the special and unusual property of being able to reproduce the many details of its organization, as well as this reproductive capacity. This embeddedness is relevant, in somewhat the same way that embeddedness in a human intentional context is relevant to the end-directedness of a thermostat or a computer program. The question is whether this makes the organism adaptation more than just teleonomic. Is a teleonomic mechanism that is preserved because it aids the persistence of a larger self-reproducing system equivalent to one that is entirely generated by accident? The answer depends on whether the larger system can also be adequately explained in this way. This means that in order to determine whether these processes are appropriately considered teleological, teleonomic, or just teleomatic (in the terms of this debate) requires that we carefully identify what comprises the organization of this larger system from which living processes derive their distinctive features.

Within evolutionary biology, it is generally assumed that the ultimate source for the special end-directed appearance of living processes is natural selection. So, examining evolution more closely can offer critical clues for deciding whether life's apparently purposive and representational processes are or are not reducible to teleomatic logic, and thus ententional in appearance only.

BLOOD, BRAINS, AND SILICON

During the twentieth century, the abstract logic of natural selection was realized to be relevant to other domains as well in which adaptive complexity requires explanation or where unguided "design" might be important. Darwinian processes were identified as major factors in the organization of the humoral immune response and in the development of the nervous system, and Darwinian mechanisms have been employed as computational tools and as molecular genetic tools. Even though Darwinism was instrumental in challenging teleological claims in nineteenth-century biology, the power of Darwin's proposal was not in his critique of prior views of external agency forming the myriad creatures, but in the elegant simplicity of the mechanism he proposed. It was a recipe for replacing purposeful design strategies with mindless selection mechanisms wherever they might arise. And as Darwin had managed to eliminate the need to appeal to an intelligent creator—a grand homunculus—in order to explain the functional consequences of biological evolution, others subsequently invoked the Law of Effect to eliminate homunculi from quite different domains.

One of the most resounding success stories for the Law of Effect was its application to the explanation of the humoral immune response—the basis for disease resistance in us and our close animal relatives. By the late nineteenth century it was known that there were factors in the blood of people who had once contracted a disease and had overcome it that made them immune to a subsequent infection. In addition, it was discovered that a defensive immune response could be built up by exposure (e.g., via inoculation) to a dead or less virulent strain of the disease agent, such as a bacterium or virus. By being exposed to the disease agent in some form, immune molecules were produced that were matched to that agent. Common sense suggested that this immune response was therefore something analogous to learning. Perhaps, it was thought, some molecular mechanism made a sort of mold of the disease agent and then caused many copies of it—antibodies—to be made. This mechanism was analogous to a Lockean blank slate, which was ready to accept whatever form was presented to it. Despite its intuitive appeal, the mechanism turned out to be almost opposite in its organization. Instead, during prenatal life, a vast array of highly variant antibody forms are produced by virtue of a sort of super fragility and shuffling of the bits of

DNA that code for the binding ends of the antibody molecules. This is called *hypermutation*, and it occurs only in the line of somatic cells destined to become antibody-producing cells (thus having no transmission to progeny). Later, during exposure to disease agents, invaders with molecular signatures different from one's own cells bind with those antibodies from among this vast array of preformed variants which just happen to be approximate mirror-image fits to the disease agent's molecular signature's. By fitting, it triggers rapid reproduction of the associated antibody-producing cells, and thus floods the circulation with appropriate antibodies.

All the potential variety of adapted forms is therefore preformed at random before any function is established, and as a result, a huge fraction of this variety will never become activated. Moreover, the fit between antibody and antigen (the molecular shape on the invader that the antibody attaches to, which is the *anti*-body *gen*-erator) is only approximate. But the fact that there are already antibodies present for any possible infectious agent gives the body a head start.

The immune response does not have information that anticipates its enemy, nor does it acquire information from this enemy. Yet in many ways it behaves as though both are the case. Because any structural variant is functional only after the fact of a serendipitous fit between randomly generated antibody and unpredictable antigen, there is no intentional end-directedness involved. And if we add to this account the fact that the immune mechanism itself evolved due to the Darwinian process—both of which exemplify a blind generation of variants and after-the-fact fitting to the environment—it would appear to have avoided any homunculus fallacy.

Probably the first serious application of Darwin's strategy to a quite different domain was in the psychological dogma of behaviorism, discussed in chapter 3. B. F. Skinner quite consciously modeled his operant conditioning paradigm on Darwin's after-the-fact logic and appealed to it to justify the anti-mentalist doctrine that was its base. The undirected emission of spontaneous exploratory behaviors was the analogue to spontaneous genetic variation, the rewarding or aversive stimuli that served as reinforcers were the analogues to selection pressures, and the biasing of behavior leading to habit formation and habit extinction that resulted was the analogue of differential reproduction. So, on the analogy to Darwin's natural selection, Skinner believed that by considering nothing else but these factors and their various schedules and contingencies, it might be possible to explain the formation of complex behavioral dispositions without ever invoking mental or other teleological influences. Just as Darwin had seemingly banished the teleology of a divine mind from biology, Skinner believed he had exorcised mind from psychology, leaving only mechanism.

Throughout the latter part of the twentieth century, other writers have also found the Darwinian model to offer a way out of the strictures of both simple mechanistic and homuncular accounts of ententional phenomena. The philosopher of science Karl Popper, for example, suggested that natural selection offered a useful analogy for explaining the accumulation of reliable knowledge which characterized the development of scientific theories. Although individual researchers may be motivated by personal insights and pet hypotheses to explore and promote certain ideas, the collective enterprise could also be understood as having a kind of inevitable Darwinian quality. Comparing mutational variants to conjectures and the culling effect of natural selection to refutations, Popper offered a similarly nonteleological scenario for explaining the progressive "adaptation" of scientific knowledge to its empirical subject matter. Over his career, he generalized this logic to apply to the evolution of knowledge as it is acquired in any domain, and suggested that this might be a way to define knowledge in general, a view that he aptly described as "evolutionary epistemology." This approach was later embraced and generalized further by the American psychologist Donald T. Campbell, who articulated what is probably the most generalized characterization of selection processes as distinct from other forms of knowledge creation. He characterized the essence of selection (as opposed to instruction) theories of adaptation and knowledge generation with the simple aphorism "blind variation and selective retention." This catchphrase highlights two distinguishing features of the selection logic: first, that the source of variation is unrelated to the conditions that favor preservation of a given variant; and second, that there is differential persistence of some variants with respect to others. As we will discuss in greater detail in later chapters, the defining criterion is not the absence of ententional processes, but the absence of any specific control of the variants produced by the process that determines their differential persistence.

A selection approach to global brain function has been explored by

neuroscientists as an alternative to computational approaches. Most notably, Gerald Edelman, who received the 1972 Nobel Prize in medicine for
his work on the structure of immunoglobulins, has compared the way the
nervous system is able to register and adapt to an unpredictable world of
stimuli to the strategy that the immune system employs to respond to the
unpredictable diversity of disease agents in the world (as described above).
He argued that an initial burst of variation in the production of neural connections is generated independent of later function, and that certain of these
variant circuits are subsequently selectively recruited with respect to their
relative "fit" with input from the senses, while others are eliminated. This
was not just an extrapolation from immune function, but also was influenced by evidence of an intermediate-level selection process taking place
during the embryonic development of the nervous system. During the
1970s and 1980s, numerous studies demonstrated that in developing mammal brains there is initially an overabundance of neurons and connections
produced, and that subsequently many are culled by activity-dependent
processes which favor neurons and connections that best reflect converging
signal correlations. In all these contexts—immune function, the adaptation
of neural circuits to inputs, and development of neural circuits—minimally
constrained ("blind") variations are generated before having a significant
portion of this variation sculpted away to meet specific functional requirements of the context in which they are embedded. Edelman additionally
argued that this could provide a neural computational logic as well.

The combination of the computer theory of mind with the Darwinian
paradigm seemed poised to offer the ultimate argument for a homunculus-
free mind. In biology, the analogue of the computer designer or software
engineer is this mindless, purposeless physical process: natural selection.
Edelman argued, in effect, that design and function of brains was selection
all the way up. Evolved computations do not require any external homunculus to determine their specific correspondence relationships to the world.
Correspondence results from a process of blind variation and selective preservation of those variant circuits and algorithms that best "fit" within their
context, both within the brain and within the world, and not in response
to any reflection, purpose, or design foresight. Given that natural selection
theory appeared to offer a non-teleological mechanism for improving the

correspondence between the organic processes and specific features of the environment, it would seem we might finally have found a way to evade the homunculus trap.

Variations on this theme have been explored extensively in many other fields as well. For example, in a field often described as artificial life and in a related field called evolutionary computing, algorithms are created that randomly generate populations of other algorithms, which are then pitted against one another with respect to their success at achieving some criterion of operation. The relative success at achieving this result is then used to pre- serve some and discard others, so that the preserved variants can be used as the starting forms from which new variants are generated for a new cycle. Because natural selection is taken to be teleology-free—merely remembered accident—generating these computational systems presumably requires no additional teleological aspect. Of course, in an artificial life simulation, the computation seen as an adaptation, the grounds of competition between algorithms, the reasons that some variants are preserved and others elimi- nated, and the environmental context providing the limited resource over which there is competition, are all specified extrinsically. They are models of selection for the purpose of pursuing a human-conceived result, even if only to explore this sort of mechanism. And this is not selection based on any physical parameters of the machine operations and their environment— these are bracketed from consideration and not allowed to interfere. Only the logical idealization of selection is being embodied in these determinate machine operations.

FRACTIONS OF LIFE

Organisms are not organized like thermostats. To the extent that they do self-regulate, it is in the service of some other outcome: reproduction. But can't this too be reduced to chemistry? Isn't the "secret of life" the replica- tion of DNA? Indeed, according to the so-called central dogma of molecu- lar biology, all forms of life depend on the ability of the DNA molecule to serve as a template for two essential functions: (1) determining the amino acid sequences constituting proteins and thus influencing their shapes and reactive surfaces; and (2) serving as a model for making duplicate copies of itself. Even viruses that contain RNA rather than DNA depend on the cellular DNA-based molecular machinery of their hosts to get their parts synthesized and their RNA replicated. With the discovery of this molecular basis for genetic inheritance, the mechanistic conception of life also was transformed. Life could be understood in information-processing terms. Mayr's conception of teleonomy on the analogy of "control by an algorithm" had an unambiguous molecular interpretation:

- DNA contains algorithms in the form of a triplet "code" of bases (purines and pyrimidines) constituting codons.
- A sequence of codons on a DNA molecule "codes for" a sequence of amino acids that constitute a protein.
- Proteins control the chemistry that does the work of constructing our cells and running our bodies, including the process of reproduction.
- During reproduction, bodies make copies of the DNA molecules that control them and pass them on to offspring, thereby passing on the algorithms that make the whole process repeatable in a new body.

The use of algorithms to control mechanisms that are capable of replicating those algorithms is a pretty reasonable caricature of the capability of a digital computer. And if we expand this vision to include a computer controlled factory for manufacturing computers, also controlled by algorithms that it can copy and transfer to newly manufactured computers, it seems as though we have pretty thoroughly described the logic of life in purely mechanistic and information-processing terms.

From here, it is a small step to expand the account further to include a computational account of natural selection. Errors in the code that alter the algorithm, that alter the function of the computer, that alter the manufacturing process, and so on, if they produce more efficient variants of this process will tend to replace the less efficient forms. Violà—evolution! This vision leads to an interesting speculation: If the computer-manufacturing machinery is entirely controlled by the algorithms that are being copied, can't we ignore the implementation mechanisms and focus only on the information embodied in the algorithms?

Probably the most famous and popular modern account of the process of natural selection is provided by Richard Dawkins in *The Selfish Gene* (1976). It is relevant to this discussion because of the way it both appears to invoke ententional properties in evolution and denies them at the same time. With the discovery of the molecular nature of genes—how they replicate and determine the structure of proteins—it became possible to recast evolutionary theory in terms of information and computing, as well as chemistry. The non-repetitive and linear structure of the DNA molecule is in many respects the perfect biological analogue to the classic computer tape, as envisioned in the original Turing machine, which was still in use for data transfer in the early 1970s. So the recognition that genetic inheritance is accomplished by virtue of duplicating the sequence structure of one DNA to make two copies could realistically be understood in information-processing terms. Dawkins' genius was the recognition that it should therefore be possible to think of evolution in informational terms.

A DNA molecule is capable of embodying an arbitrary linear pattern consisting of alternating "coding" elements (its four bases). This pattern indirectly organizes the functional dynamics of the chemistry occurring within living cells by providing a template for specifying the structure of life's workhorse macromolecules: proteins. In this respect, the base sequences embodied in DNA are often treated as the analogue to software for the organism as computer. To make the analogy more complete, one must think of DNA as carrying the code for actually building the computer on which it runs. Because a DNA molecule can serve as a template for the assembly of new DNA molecules with the same base sequence, it can further be claimed that it is analogous to self-writing software. Although the gene concept has become progressively complicated, and to some extent deconstructed, over the past couple of decades as additional variant functions of DNA sequences have been identified, the commonly understood use of the term is applied to that length of a DNA molecule that corresponds to the structure of a given protein product. This turns out not to be the only feature of DNA that is functionally relevant, yet it still serves as a useful approximate way to partition DNA into distinct informational chunks relevant to evolution.

Dawkins describes genes as *replicators*. The suffix "-or" suggests that genes are in some sense the locus of this replication process (as in a machine designed for a purpose like a refrigerator or an applicator), or else an agent accomplishing some function (such as an investigator or an actor). This connotation is a bit misleading. DNA molecules only get replicated with the aid of quite elaborate molecular machinery, within living cells or specially designed laboratory devices. But there is a sense in which they contribute indirectly to this process: if there is a functional consequence for the organism to which a given DNA nucleotide sequence contributes, it will improve the probability that that sequence will be replicated in future generations. Dawkins describes genes as *active* replicators for this reason, though the word "active" is being used rhetorically here to indicate this contribution, because they are otherwise dynamically passive players in this process. So, although Dawkins explicitly employs a terminology that connotes agency in order to avoid what he argues would be tedious circumlocutions, it is largely for heuristic reasons.

Replicator theory thus treats the pattern embodied in the sequence of bases along a strand of DNA as information, analogous to the bit strings entered into digital computers to control their operation. Like the bit strings stored in the various media embodying this manuscript, this genetic information can be precisely copied again and again with minimal loss because of its discrete digital organization. This genetic data is transcribed into chemical operations of a body analogous to the way that computer bit strings can be transcribed into electrical operations of computer circuits. In this sense, genes are a bit like organism software.

Replicators are, then, patterns that contribute to getting themselves copied. Where do they get this function? According to the standard interpretation, they get it simply by virtue of the fact that they do get replicated. The qualifier "active" introduces an interesting sort of self referential loop, but one that seems to impute this capacity to the pattern itself, despite the fact that any such influence is entirely context-dependent. Indeed, both sources of action—work done to change things in some way—are located outside the reputed replicator. DNA replication depends on an extensive array of cellular molecular mechanisms, and the influence that a given DNA base sequence has on its own probability of replication is mediated by the physiological and behavioral consequences it contributes to in a body, and most importantly how these affect how well that body reproduces in its given environmental context. DNA does not autonomously replicate itself; nor does a given DNA sequence have the intrinsic property of aiding its own replication—indeed, if it did, this would be a serious impediment to its biological usefulness. In fact, there is a curious irony in treating the only two totally passive contributors to natural selection—the genome and the selection environment—as though they were active principles of change.

But where is the organism in this explanation? For Dawkins, the organism is the medium through which genes influence their probability of being replicated. But as many critics have pointed out, this inverts the location of agency and dynamics. Genes are passively involved in the process while the chemistry of organism bodies does the work of acquiring resources and reproducing. The biosemiotician Jesper Hoffmeyer notes that, "As opposed to the organism, selection is a purely external force while mutation is an internal force, engendering variation. And yet mutations are considered to be random phenomena and hence independent of both the organism and its functions."[6]

By this token the organism becomes, as Claus Emmeche says, "the passive meeting place of forces that are alien to itself."[7] So the difficulty is not that replicator theory is in error—indeed, highly accurate replication is necessary for evolution by natural selection—it's that replicators, in the way this concept has generally been used, are inanimate artifacts. Although genetic information is embodied in the sequence of bases along DNA molecules and its replication is fundamental to biological evolution, this is only relevant if this molecular structure is embedded within a dynamical system with certain very special characteristics. DNA molecules are just long, stringy, relatively inert molecules otherwise. The question that is begged by replicator theory, then, is this: What kind of system properties are required to transform a mere physical pattern embedded within that system into information that is both able to play a constitutive role in determining the organization of this system and constraining it to be capable of self-generation, maintenance, and reproduction in its local environment? These properties are external to the patterned artifact being described as a replicator, and are far from trivial. As we will see in subsequent chapters, it can't be assumed that a molecule that, under certain very special conditions, can serve as a template for the formation of a replica of itself exhibits these properties. Even if this were to be a trivially possible molecular process, it would still lack the means to maintain the far-from-equilibrium dynamical organization that is required to persistently generate and preserve this capacity. It would be little more than a special case of crystallization.

Indeed, the example of *prions* both qualifies as a form of replicator dynamics and demonstrates the fundamental systemic dependency of both the replication process and information properties. Prions are the cause of some quite troublesome diseases, such as so-called mad cow disease (bovine spongiform encephalopathy, BSE) and Kuru, which are acquired by eating body parts (nervous tissue) that contain prions. A prion is not a virus or a bacterium, but a protein molecule with a distinctive three-dimensional shape. It can impose this shape on other protein molecules formed by the same amino acid sequence structure, but which are folded into a different three-dimensional structure. Only when the protein is in the prion shape-configuration can it exert this shape-templating effect. As a result, adding some prions to a context where there are other non-transformed (pre-prion) molecules will induce differential refolding of more and more of the pre-prion forms as time goes on and the concentration of prions increases. Pre-prion molecules are generated as a consequence of normal brain function and are routinely metabolized; but if transformed into the prion form, they are not only difficult to get rid of but their accumulation is damaging, eventually producing severe loss of function and death.

In this sense, prions are active replicators because they both get their form copied, and they also contribute to perpetuating this process. Indeed, distinctive lineages of prions have been identified that are specifically associated with different animal species (e.g., cows, sheep, cats and dogs, and humans), having evolved variants that are "adapted" to these species. But prions are not parasitic on animal brains, as are viruses or bacteria, because prions don't cause other pre-prion or prion molecules to be synthesized— pre-prion molecules are synthesized by cells within the brain—rather, they just catalyze a change in shape of the pre-prion molecules that are already present. If it weren't for the production of this protein by a host brain, there would be no shape replication. In fact, prion proteins are only able to produce this effect because the prion shape is more energetically stable than the pre-prion shape. They do no chemical work. They just influence the less stable forms to more easily transform into a more "relaxed" state. If it were not for the far-from-equilibrium metabolism of the nervous system, prions would be impossible, because there would be no substrates to embody the form. Something physical must be generated and multiplied for evolution to be possible, and this process is necessarily dependent on a special kind of dynamical system: an organism. The form that gets replicated must be embodied, and because generating embodied form is a process that runs counter to the second law of thermodynamics, work must be done to accomplish this. Work can't be done by an inert form. It takes a special kind of dynamical system, as we will develop in later chapters.

So evolution is not merely differential replication of pattern, and information is not an intrinsic property of physical pattern (even though we may be able to measure the information-bearing potential of a given pattern). Were this so, then crystal growth would count as evolution. Even if we treat DNA base sequences as program code (data plus instructions), there still needs to be something serving the role played by computers and their users for this pattern to count as information. So when Dawkins suggests that evolution is the result of the differential replication of information, he is not incorrect, but equating "information" with pattern smuggles unspecified systemic properties of its context into the account, while pretending they can be provided by a passive artifact.

The dependency of gene replicator theory on organism systems and organism-environment relationships in order to account for their informational, adaptive, and functional features of biological evolution suggests that something critical is ignored in the analytic effort to collapse a systemic relationship down to one of its apparent component parts. In doing so, the material processes that create the possibility of creating this part get ignored.

The embryologist Paul Weiss, writing in the late 1960s, posed this conceptual problem clearly in his description of the effect of uncritical reductionistic interpretations of biological processes:

> In trying to restore the loss of information suffered by thus lifting isolated fragments out of context, we have assigned the job of reintegration to a corps of anthropomorphic gremlins. As a result, we are now plagued—or blessed, depending on one's party view—with countless demigods, like those in antiquity, doing the jobs we do not understand: the organizers, operators, inductors, repressors, promoters, regulators, etc.—all prosthetic devices to make up for the amputations which we have allowed to be perpetrated on the organic wholeness, or to put it more innocuously, the "systems" character, of nature and of our thinking about nature.[8]

Weiss argues that the analytic dissection of living organization into independent parts, which presumably reduces an organism to a mere machine,
and thereby exorcises the homunculus of an *élan vital*, only serves to shift its
locus to more cryptic contributors of teleological functions: genetic codes,
translation, regulation, signaling, and so forth. Even though analytically dissecting the organic wholeness of a living system doesn't remove anything
from the material components of life, it nevertheless segregates the whole
into parts. This provides a powerful tool for breaking up the work involved
in the exploration of the complex system that is an organism, yet it also
precisely brackets from analysis what is most relevant: the "organic wholeness." The life of an organism is not resident in its parts. It is embodied in
the global organization of the living processes. Moreover, the so-called parts
that analysis produces—the individual molecules, organelles, cells, tissue
types, and organs—are not parts in the sense that machine parts are.

Organisms aren't composed by assembling independently produced
and grouped parts. What we interpret as parts are in most cases the consequence of differentiation processes in which structural discontinuities and
functional modularization emerged from a prior, less-differentiated state,
whether in evolution or development. And even those that biologists believe
to have originated independently, like the mitochondria of eukaryotic cells,
are no longer fully separable cell components. Though mitochondria carry
their own chromosome of genes involved in their assembly and function,
much of the genetic information necessary for their maintenance has been
relocated outside of the mitochondrion, in the cell's nucleus. Actually, however, the teleological properties that we identify in the functioning of a cell
or whole organism are a function of the global organization of the many processes that have given rise to these superficial boundaries. The boundaries
and divisions do not correspond to relationships between functional units,
because the functional units are synergistically integrated processes, not
distinct parts. So the *élan vital* has not really been banished by these analyses; it has just been transformed into simpler and more insidious homunculi
a level or two down, assigned to parts presumed to be simply physical but
which cryptically provide the missing consequence organization.

THE ROAD NOT TAKEN

Missing from this way of conceiving of the end-directed processes involved in life is a consideration of what sort of physical process is minimally necessary for evolution to take place. Evolution not only requires reproduction of information in the form of pattern but also reproduction of a system capable of utilizing and copying that pattern and, not incidentally, building a replica of itself. In other words, self-reproduction necessarily involves the production of a system that is capable of countering the ubiquitous teleomatic tendency of thermodynamic degradation by generating new forms, in order to replicate these forms to at least keep pace with the inexorable breakdown of each. As some mechanism to achieve this thermodynamic feat is the prerequisite for any form of natural selection, it seems that the appeal to Darwinism for a complete constructive definition of teleonomy cannot resolve the problem either. Natural selection could not have produced the conditions that made natural selection possible. These conditions that enable an individuated dynamical system—an organism—to defeat the second law of thermodynamics locally, by repairing and replicating its parts and by producing replicas of itself, are prerequisites to natural selection. Attributing the entential properties of organisms to natural selection is not then equivalent to attributing it to "remembered" accident. These properties must be additionally attributed to whatever constitutes these conditions—conditions which themselves appear to exhibit entential features—conditions that predate natural selection.

This brings attention to what is bracketed from consideration in the contemporary conception of evolution in informational terms: the device itself—the computer that performs the operations and the physical properties that make it capable of these operations. Do these details matter? Indeed, materiality *is* the critical difference between biological evolution and simulated evolution. The specific molecular substrates, the energy required to synthesize them and animate their interactions, and the work required to generate the structures and process that get reproduced with variation *are* what determines why a given organization persists or doesn't. In the computer world this part of the story is taken for granted, and even if these factors are part of what is being simulated, the actual physics of computing is still a given. Patterns of signals are produced, transformed, and erased by physical processes whose details are irrelevant to the algorithmic game being played. In life, however, it is precisely this physical embodiment that matters. It is not just pattern and correspondence that matters; the properties of the substrates that things are made of and the thermodynamics of the work being performed on them determine what persists and what perishes over time. Could the apparent ability to eliminate teleological assumptions from simulated evolution be an artifact of having ignored the physicality of its implementation? Can the view of biological evolution as thoroughly non-teleological be an artifact of bracketing out consideration of the details of molecular structure, chemical reaction probabilities, thermodynamics, and so forth?

The logic of the Law of Effect showed how the appearance of final causality could be understood in terms of blind mechanism. No future consequence need be considered to explain its presence. But the notion of mechanism that we need to invoke in this analysis is not as simple as it might first appear. As we will explore in greater depth later, the power of Darwin's insight to challenge the necessity of teleology depends on some rather complex assumptions about the nature of the reproductive process. But the assumptions in question are not the usual ones that have been repeatedly raised by critics and addressed by supporters, for example, about genetic inheritance, sources of variation, and the concept of species. Darwin's mechanism is curiously agnostic about mechanism. But from the point of view of the physics and chemistry of life, explaining its most critical process—reproduction—confronts us with some challenging questions.

Reproduction and the generation of organism forms are necessary requirements for Darwin's logic, even though *how* this is accomplished is largely irrelevant to the existence of natural selection consequences. Precisely because of this logical disconnect, natural selection is indeed a thoroughly non-teleological process. Yet the specific organic processes which this account ignores, and on which it depends, are inextricably bound up with teleological concepts, such as adaptation, function, information, and so forth. The differential reproduction of organisms with respect to their fittedness to local ecological conditions is also ultimately determined with respect to the thermodynamics of these form-generating processes and their correspondence with these conditions. In other words, the successful overthrow of large-scale teleological conceptions of the evolution of species appears dependent on the existence of lower-level entential properties. And because the possibility of natural selection depends on them, ultimately they cannot be the products of natural selection, even if they have been fine-tuned by it.

The fact of evolution and the logic of natural selection may not in themselves hold the solution to the problem posed by entential phenomena, but they offer a tantalizing hint. This is because entential phenomena evolve. Their forms have become more sophisticated and complex since life began on Earth. And although the process of evolution itself is not end-directed, it has produced all the forms of entential phenomena we know of. So far, efforts to explain these enigmatic properties of life and mind have proceeded from the top down, so to speak. Whether beginning from an effort to make sense of human consciousness, or else appealing to the design of regulatory devices designed to produce end-directed mechanical tendencies, we are already assuming what we need to explain, and then trying to find the best way to disassemble these phenomena to understand their composition. Maybe, however, something blocks this approach. What if, for some reason, these phenomena can't be analyzed this way? What if, in the process of the emergence of these phenomena, their most important foundational features somehow get obscured or even lost in some way? It may be possible to discover how entential phenomena come into existence, and yet not possible to trace the process in reverse.

So, instead of trying to eliminate entential properties from science, I propose we try to understand how they could have come into existence where none existed before. In other words, we need to start without any hint of *telos* and end up with it, not the other way around. In some ways, this is a far more difficult enterprise because we are not availing ourselves of the entential phenomena most familiar to us: living bodies and conscious minds. Even worse, it forces us to explore domains in which there may be little science to support our efforts. It has the advantage, however, of protecting us from our own familiarity with teleology, and from the appeal of allowing homuncular assumptions to do the work that explanation should be doing.

For no other problem in the sciences are there so many potential pitfalls, false leads, and possibilities for self-deception. Besides the extremes of eliminative reductionism, panpsychism, and idealism in all their various forms, which attempt to avoid resolving the matter-mind transition by defining it away, there have been at least as many failed efforts to explain how the properties of mind are not in conflict with those of physics and chemistry. These include many unnoticed sleight-of-hand moves that merely rename these homunculi in terms of familiar human artifactual creations, such as feedback circuits or computations. These approaches also halt the analysis prematurely. Whether beginning by ignoring one or the other aspect, or else beginning with the assumption that there is no ultimate distinction between mind and matter, one can pretend to ignore what it is that determines the difference between simple mechanism and teleological organization. To assume incompatibility belies the constant entanglement of physical and ententional processes and treats the difference between them as merely a matter of degree. But this denies the obvious inversion of causal logic that distinguishes them.

No fractionation of ententional functions into modules of even smaller scope and proportion allows the apparent arrow of causality to reverse. There is no point where ententional dynamics just fades smoothly into thermo dynamics. Minds are not just made of minds of simpler form made of minds of yet simpler form that eventually become so "stupid" as to be modeled by simple mechanisms. Nowhere down this ever smaller rabbit hole can we expect that the normal laws of causality will imperceptibly transform into their mirror image. Ententional processes organized with respect to represented ends, and mechanical processes organized with respect to immediate determinate physical interactions and nothing else, stand on opposite sides of the looking glass.

But if all efforts to show that there is no fundamental distinction between ententional processes and mechanistic processes turn out to be ways to avoid the critical issue, or to define it away, and all efforts to show that they are fundamentally incompatible lead to absurd consequences, what's left? Is there a path through this maze that, in the end, avoids the pitfalls of the homunculus fallacy, and yet does not deny the existence of the ententional properties that we attribute to the homunculi we invoke in our common-sense talk and in our scientific theories? Despite the long history of failures to make progress in this effort, there are good reasons for believing that it is not an intractable mystery. We human beings are, after all, finite physical phenomena that evolved spontaneously. Why shouldn't we be able to explain how it is that the material processes of our bodies and brains constitute conscious experiences of teleological processes? Mostly, it appears that the major problem is not so much that the details are too complex, but rather that we are unwilling to give up dualism, while at the same time being intolerant of the seeming paradoxes that show up because of it.

Though most people who have thought deeply about it intuit that it must ultimately be an incoherent idea, dualism is nonetheless deeply compelling. Why? The answer is simple: the causal logic of spontaneous physical processes is radically different than the ententional logic of life and mind. Efforts to deny this difference are as doomed to failure as are efforts to pretend that one or the other aspect is illusory. This difference is an essential fact to be explained, not explained away. But we can't stop at this dualistic appearance any more than at a dualistic metaphysics, because in all its cryptic forms, dualism is a halting move. It leaves all the critical questions unasked.

There is no use denying that there is a fundamental causal difference between these domains that must be bridged in any comprehensive theory of causality. The challenge of explaining why such a seeming reversal takes place, and exactly how it does so, must ultimately be faced. At some point in this hierarchy, the causal dynamics of teleological processes do indeed emerge from simpler blind mechanistic dynamics, but we are merely restating this bald fact unless we can identify exactly how this causal about-face is accomplished. We need to stop trying to eliminate homunculi, and to face up to the challenge of constructing teleological properties—information, function, aboutness, end-directedness, self, and even conscious experience—from unambiguously non-teleological starting points.

In one sense, what follows is an effort to save homunculi rather than to bury them. They may indeed be empty placeholders in our present theories, but they are neither standing in for ineffable forces nor for incompletely analyzed mechanical relationships. They are the representatives of gaps of a more serious sort. The properties they supply for our theories do not merely pose an impediment to progress in cognitive science and philosophy. They ultimately raise troublesome questions about the very nature of physical causality. These homunculi disguise a major bit of unfinished business at the very foundations of science: until we can explain the transformation by which this one mode of causality becomes the other, our sciences will remain dualistically divided, with natural science in one realm and the human sciences in the other.

So now that we know where to look, and what to avoid, how should we proceed? Hints that there must be a constructive way to approach this problem come initially from two observations about the ententional processes of which we are most familiar:

1. Life had its origins in non-living matter. It was not always present on Earth. Life first originated, then evolved. Following the dawn of life, biological evolution eventually gave rise to autonomously mobile creatures—animals—and they needed brains to model the environments into and out of which they moved. They became progressively more responsive, more sentient, and eventually conscious, in the sense that we are. Minds are an evolved property of complex animals.

2. Human thoughts and experiences emerge out of the seeming cacophony of simple electrochemical communications incessantly shuttled between cells (neurons) in the brain. Conscious experience varies with these material processes, and can be altered in precise and distinctive ways by chemical and physical alteration of neural processes. In other words, minds originated and constantly emerge from simpler living processes, both in evolutionary time and at each moment, just as living processes originally emerged from and continue to emerge from non-living processes and materials.

Ultimately, I will contest the claim that what is often called "causal power" can be vested in a thing's materiality alone or even in the energetic interactions between things. One might well ask: So what else is there? As we'll learn in later chapters, there is a "what else," but not exactly a "something else." And it is not just causality that is at issue in this analysis, but something more akin to the concept of work. Work is a more complex concept than mere cause, because it is a function of relational features. And relations, both actual and potential, are precisely where the action is. More important, work is only possible because of limitation.

To abuse an old metaphor: the fabric of mind is not merely the thread that composes it. Indeed, these physical threads can be traced, and give the fabric its solidity and resistance. But the properties that constitute the fabric vanish if we insist on unraveling the intricate weave until only thread is left. By analogy, the fabric of *telos* is woven, if you will, from the same material and energetic thread that constitutes all unfeeling, unthinking, inanimate phenomena. In this respect, there is complete continuity. It's the distinctive pattern of the self-entanglement of these threads that contributes the critical and fundamental difference in properties distinguishing the ententional world from the world described by contemporary physics and chemistry. The intertwined threads of a fabric systematically limit each other's possibilities for change. There is of course no new kind of stuff necessary to transform the one-dimensional thread into a two-dimensional sheet; just a special systematic form of reciprocal limitation. By analogy, to really understand how the additional dimension of ententional properties can emerge from a substrate that is dimensionally simpler and devoid of these properties, it is necessary to understand how the material and energetic threads of the physical universe became entangled with one another in just the right way so as to produce the additional dimension that is the fabric of both life and mind. This is the problem of emergence: understanding how a new, higher dimension of causal influence can be woven from the interrelationships among component processes and properties of a lower dimension.

5

EMERGENCE

. . . we need an account of the material world in
which it isn't absurd to claim that it produced us.

—TRANSLATED AND REPHRASED BY THE AUTHOR FROM
ILYA PRIGOGINE AND ISABELLE STENGERS, 1979[1]

NOVELTY

Whereas the formation of the first stars, the formation of the nuclei of the first heavier-than-hydrogen elements, the coalescence of atoms into the first molecules, and so on, have each had a profound effect on the interactions and distributions of substances in the universe, a common set of physical equations is able to give a reasonably complete description of the properties that resulted. The same cannot be said of the transitions that led to life and to mind. Though the laws of physics and chemistry were not violated or superseded in these transitions, their appearance led to a fundamental reorganization in how these properties were expressed and distributed in the world and how they were organized with respect to each other. The reliable reversal of typical thermodynamic tendencies, and the insinuation of informational relationships into the causal dynamics of the world, make the transitions from inorganic chemistry to life, and from mechanism to thought, seem like radical contraventions of causality-as-usual. So, although I am convinced that no tinkering with basic physics is required to make sense of these phenomena, we still need to account for the radical about-face in how these principles apply to everyday entential phenom-

143

ena. At some point, the reliable one-dimensional lockstep of just one thing after another that had exclusively characterized physical events throughout the universe took an abrupt turn and headed the other way.

The appearance of the first particles, the first atoms, the first stars, the first planets, and so on, marked fundamental new epochs in the 13-billion-year history of the universe, yet none of these cosmic transitions contorted the causal fabric of things as radically as did the appearance of life or of mind. Even though these living transitions only took place on a comparatively insignificant scale compared to other cosmic transitions, and even though no new kind of matter or energy came into existence with them, what they lack in scale and cosmic effect they make up in their organizational divergence from the universal norm. Consider the following:

- There were no ententional properties in the universe for most of its 13-billion-year history (i.e., before significant amounts of heavier elements were produced by dying stars).
- There was nothing resembling a function on the Earth until just over 3 billion years ago, when life first appeared.
- There was no hint of mental awareness on Earth until just a few hundred million years ago, when animals with brains first evolved.
- There was nothing that was considered right or wrong, valuable or worthless, good or evil on our planet until the first human ancestors began thinking in symbols.

All these innovative ways of organizing matter and energy, producing unique forms of influence over the events of the world, popped into existence from antecedent forms of organization that entirely lacked such properties. Physics and chemistry continued as they had before, but radical and unprecedented changes in the way materials and events could become organized followed these transitions wherever and whenever they occurred.

Such major transitions in the organization of things are often described as *emergent*, because they have the appearance of spontaneous novelty, as though they are poking their noses into our world from out of a cave of non-existence. And while they are not exactly something coming from nothing, they have a quality of unprecedented discontinuity about them—an almost magical aspect, like a rabbit pulled from an apparently empty hat.

In the way the term is often used, there is a close kinship between the concept of emergence and ideas of novelty and newness, as well as an implication that predictability is somehow thwarted. Although I hope to show how these are misleading attributions, this view is both common and attractive. For example, one of the more widely read books on emergence, written by the eminent physical chemist Harold Morowitz, itemizes and describes over twenty "emergences" in the history of the cosmos, including everything from the formation of stars to the appearance of language. At each of these transitions in the history of the cosmos and of the Earth, new organizations of matter appeared that were not present previously, at least at those locations. From this perspective, the transition that separates living processes from other physical-chemical processes is only one among many emergent transitions, such as the formation of the first stars or the production of heavy elements in dying stars.

But just being the first instance of something, or being novel or unpredictable, are not particularly helpful distinctions. Newness is in one sense the very nature of all physical change. However, a consideration of the sorts of transitions that characterize emergences for Morowitz indicates that there is a hierarchic aspect to this conception of emergence. Each transition involves the formation of a higher-order structure or process out of the interrelationships of smaller, simpler components. Emergence in this sense involves the formation of novel, higher-order, composite phenomena with coherence and autonomy at this larger scale.

What about predictability? Being unpredictable, even in some ultimate sense, is only a claim about the limits of representation—or of human intellect. Even if certain phenomena are "in principle" unpredictable, unexplainable, or unknowable, this doesn't necessarily imply a causal discontinuity in how the world works. There may be a determinate path from past to future, even to a radically divergent form of future organization, even if this causal influence is beyond precise representation. Indeed, all representations of worldly events are simplifications, so we should expect to find many physical transitions that exceed our capacity to represent the basis of this transition. And it is often the case that what was once beyond predication becomes more tractable with better representational tools and more precise measuring instruments. The history of science has provided many examples of what once were apparently mysterious phenomena, assumed to be intrinsically intractable, that eventually succumbed to unproblematic scientific explanation. Such was the case with the once mysterious forces constraining the possible transmutations of substances, as explored by alchemists, which were eventually explained by chemistry, and to an even greater depth with quantum physics.

Without question, phenomena such as life and mind owe some of their mysterious character to limitations in our present state of science. I'm confident that these limitations of our current theoretical tools can be overcome and that these phenomena can also become integrated into the larger fabric of natural science. The question is whether in accomplishing this, their distinctive ententional characteristics (function, representation, end-directedness, self, and so on) will be explained rather than merely explained away.

What is interesting and challenging about ententional phenomena is that they appear to involve a global reorganization of their component dynamical interactions and interdependencies that only makes sense with respect to non-intrinsic relationships. So proving ultimate unpredictability isn't critical, nor do we need to demonstrate a kind of radical discontinuity of causal influence. But we do need to explain how such phenomena can exhibit causal organization that is (superficially at least) the inverse of the pattern of causal organization which is otherwise ubiquitously present throughout the inanimate world. Instead of postulating discontinuous jumps, in which novel physical properties appear like rabbits out of empty hats as we cross certain thresholds of compositionality, we might better focus on making sense of the apparent causal reversals that motivate these special accounts. As in a magic act, there can be subtle effects that are difficult to detect, and there can be cognitive biases that cause us to look in the wrong place at the wrong time, thus missing the important features. We just need to know what to look for, and what we can ignore.

THE EVOLUTION OF EMERGENCE

The concept of emergence is fairly new in the history of science. This is because it was originally formulated as a contrast to a causal paradigm that only reached its full elaboration in the nineteenth century. By the mid-nineteenth century, a more thoroughly mechanistic and statistical approach to natural philosophy had begun to coalesce and displace the previously more teleological and platonic forms of explanation. At midcentury, electricity and magnetism were being tamed, the general concept of energy was becoming formalized, thermodynamic principles were being derived from micro-Newtonian dynamics, and alchemy was being replaced by an atomic theory of chemical interactions. Even the complex logic of organism design and adaptation appeared subject to entirely material processes, as outlined in Charles Darwin's theory of natural selection. Herbert Spencer had even suggested that these same principles might be applicable to human psychology and social processes. Mystical powers, intangible forces, immaterial essences, and divine intervention in the goings-on of nature were seen as prescientific anachronisms. Nature's designs, including living and mental processes, were now viewed through the lens of materialism: reducible to matter in motion. Antecedent teleology and metaphysical claims concerning the directionality of evolutionary change were no longer legitimate scientific assumptions.

This posed some troubling questions. Given their superficially inverted form of causal influence—with current processes structured with respect to future possibilities—how can the teleological appearance of living and mental processes be accounted for in these same terms? And if these ententional properties are not already prefigured in the inorganic world, how can their novel features be derived from those non-ententional processes alone? In response to these challenges, a number of scientists and philosophers of science realized the necessity of reconciling the logic of physical science with the logic of living and mental teleology. A true reconciliation would need to accept both the unity of material and living/mental processes and the radical differences in their causal organization. Investigators could neither accept ententional properties as foundational nor deny their reality, despite this apparent incompatibility. The key concept that came to characterize an intermediate position was that of *emergence*.

This use of the term was introduced by the English philosopher and critic George Henry Lewes, in his *Problems of Life and Mind* (1874–79), where he struggles with the problem of making scientific sense of living and mental processes. He defines emergence theory as follows.

Every resultant is either a sum or a difference of the co-operant forces; their sum, when their directions are the same—their difference, when their directions are contrary. Further, every resultant is clearly traceable in its components, because these are homogeneous and commensurable. It is otherwise with emergents, when, instead of adding measurable motion to measurable motion, or things of one kind to other individuals of their kind, there is a co-operation of things of unlike kinds. The emergent is unlike its components insofar as these are incommensurable, and it cannot be reduced to their sum or their difference.[2]

Lewes appears to have been influenced by John Stuart Mill's effort to unite logic with the methodologies of such new sciences as chemistry and biology. Mill was particularly struck by discontinuities of properties that could be produced by combinatorial interactions such as chemical reactions. Thus two toxic and dangerous substances, chlorine gas and sodium metal, when combined together, produced common table salt—which is both ubiquitous and an essential nutrient in the living world. The chemical reaction that links these two elements neutralizes their highly reactive natures and in their place yields very different ionic properties on which all life depends. Mill viewed this radical change in properties due to chemical combination to be analogous to the combinatorial logic that produced life from mere chemistry. He argues in *A System of Logic* that

All organised bodies are composed of parts, similar to those composing inorganic nature, and which have even themselves existed in an inorganic state; but the phenomena of life, which result from the juxtaposition of those parts in a certain manner, bear no analogy to any of the effects which would be produced by the action of the component substances considered as mere physical agents. To whatever degree we might imagine our knowledge of the properties of the several ingredients of a living body to be extended and perfected, it is certain that no mere summing up of the separate actions of those elements will ever amount to the action of the living body itself.[3]

Over the course of the previous centuries, the idea that living organisms might be machinelike and constructed of the same basic chemicals as inorganic objects had gained credence, but an account of how organism structure and function arose was still mysterious. Before Darwin, their exquisite construction was generally assumed to be the work of a "divine intelligence." But even centuries before Darwin, the successes of materialist science had convinced many that a spontaneous mechanistic approach had to be possible. This notion was given added support by the success of the empiricist theory of mind described by John Locke in the late 1600s. Locke's account of the gradual and spontaneous development of complex ideas out of the association of sense data acquired via interaction with the world provided an atomistic notion of knowledge as the compound resulting from the associative bonds between these "atoms" of experience. That the complex interrelations of ideas could be the result of the environmental impressions on a receptive medium suggested to Locke that a material analogue might also be conceivable. In the century and a half that followed, thinkers like Erasmus Darwin (Charles' grandfather) and later Jean-Baptiste de Lamarck struggled to articulate how an analogous process might also explain the way that the adaptively associated combinations of parts which constituted plant and animal bodies might have similarly arisen. As Locke had conceived of mind forming from an unformed merely impressionable beginning, couldn't the association of living processes arise from the living equivalent of a blank slate? If such functional associations could arise by spontaneous mechanism, then antecedent design would be redundant.

This turned the classic "Chain of Being" logic on its head. As Gregory Bateson has described it,

> Before Lamarck, the organic world, the living world, was believed
> to be hierarchic in structure, with Mind at the top. The chain, or
> ladder, went down through the angels, through men, through
> the apes, down to the infusoria or protozoa, and below that to
> the plants and stones. What Lamarck did was to turn that chain
> upside down. When he turned the ladder upside down, what had
> been the explanation, namely: the Mind at the top, now became
> that which had to be explained.[4]

Previously, starting with the assumption of the infinite intelligence of a designer God, organisms—including those capable of flexible intelligent behavior—could be seen as progressive subtractions and simplifications from Godlike perfection. In the "Great Chain of Being," mental phenomena were primary. The mind of God was the engine of creation, the designer of living forms, and the ultimate source of value. Mind was not in need of explanation. It was a given. The living world was derived from it and originally folded within it, preformed, as a definite potential. The evolutionary reconception proposed by the elder and younger Darwins, Lamarck, and others therefore posed a much more counterintuitive possibility. Mind and its teleological mode of interacting with the world could be described as an end product—not an initiating cause—of the history of life. Lamarck's vision of evolution was in this sense ultimately emergent. He conceived of life in purely materialistic terms, and his evolutionary argument proposed a means by which mind might have emerged in a continuous process from otherwise mindless processes. Mill's analysis opened the door to thinking of this association process as potentially resulting in something quite unlike the "atoms" (whether experiential or chemical) from which the process began.

The event that probably played the key role in precipitating the articulation of an emergentist approach to life and mind was the publication in 1859 of Darwin's *On the Origin of Species*. Although Darwin was not particularly interested in the more metaphysical sorts of questions surrounding the origins of mind, and did not think of his theory in emergence terms, he *was* intent on addressing the teleological issue. He, like Lamarck, was seeking a mechanistic solution to the problem of the apparent functional design of organisms. But where Lamarck assumed that the active role of organism striving to adapt to its world was necessary to acquire "instruction" from the environment—a cryptic homunculus—Darwin's theory of natural selection required no such assumption. He reasoned that the same consequence could be reached due to the differential reproduction of blindly generated variant forms of organisms in competition for limited environmental resources. This eliminated even the tacit teleological assumption of a goal-seeking organism. Even this attribute could be achieved mechanistically. Thus, it appeared that teleology could be dispensed with altogether.

The theory of natural selection is not exactly a mechanistic theory, however. It can best be described as a form of statistical inference that is largely agnostic about the mechanisms it depends on. As is well known, Darwin didn't understand the mechanism of heredity or the mechanisms of reproduction. He didn't have a way to explain how forms are produced during development. And he didn't have an account of the origins of spontaneous variations of organism form. But he didn't need them. As a keen observer of nature, he saw the regular consequences of these mechanisms (e.g., inheritance of traits, competition for the resources needed to develop and reproduce, and individual variation) and drew the inevitable statistical implications. Fittedness of organisms to their environment was a logical consequence of these conditions.

What Galileo and Newton had done for physics, Lavoisier and Mendeleyev had done for chemistry (and alchemy), and Carnot and Clausius had done for heat, Darwin had now done for the functional design of organisms. Functional design could be subsumed under a lawlike spontaneous process, without need of spirits, miracles, or extrinsic guidance.

Of course, there were many (including Alfred Russel Wallace, the co-discoverer of natural selection) who felt that the argument could not be extended to the domain of mental agency, at least as is implicit in human experience. Even though Wallace agreed that the teleological influence of a divine intelligence was unnecessary to explain organism design in general, there were features of human intelligence that seemed to be discordantly different from what would have been merely advantageous for survival and reproduction. Despite these misgivings, as the nineteenth century waned and the twentieth century began, it became easier to accept the possibility that the mechanistic view of life could also be imagined to hold for mind. Teleology appeared to be tameable, so long as it was evolvable. If functional design could arise as an after effect of accidental variation, reproduction, and resource competition, then why not mental function as well?

Yet it wasn't quite this simple. Mental processes are defined in teleological terms. Although the evolutionary paradigm offered a powerful unifying framework that promised to answer vast numbers of puzzling questions about biological function and design, it also drew attention to what was previously an irrelevant question.

If a teleological account of organism design is actually superfluous, then couldn't it also be superfluous to an account of the teleological features of thought as well? To put it more enigmatically, couldn't Darwinian logic allow science to dispense with teleological accounts at all levels and in all processes? This is of course the motivation behind the various "golem" arguments we critiqued in chapter 3. If something as complex as the fitted functional organization of a body and brain can be generated without the assistance of teleology, then why should we assume that complex adaptive behavior, including even human cognition, requires a teleological account to make sense of it? The rise of emergentism in the late nineteenth and early twentieth century can be seen as an effort to rescue teleological phenomena from this ultimate elimination by conceiving of them as derived from a sort of cosmic evolution. With the Great Chain of Being inverted, teleology must be constructed.

REDUCTIONISM

Emergentism was also a response to another achievement of nineteenth-century science: a new methodology for analyzing nature known as *reductionism*. If complicated phenomena can be analyzed to component parts, and the properties of the parts analyzed separately, then it is often possible to understand the properties of the whole in terms of the properties and the interactions of these parts. This approach was highly successful. The atom theory of chemistry had made it possible to understand the properties of different substances and their interconvertability in terms of combinatorial interactions between more basic elemental units, atoms, and molecules. The understanding of heat and its convertability into motive power was explained in terms of the movements of molecules. Even organisms could be understood in terms of cellular interactions. Reductionistic analysis was a natural extension of the atomism that began with ancient scholars like Empedocles and Democritus. Atomism was just the extreme form of a tried-and-true explanatory strategy: break complex problems into simpler parts, then if necessary break these into even smaller parts, and so on, until you can go no further or else encounter ultimately simple and indivisible parts. Such indivisible parts—called atoms (literally, "not cut-able")—should, by assumption, exhibit less complicated properties and thereby reduce complicated collective properties to combinations of simpler properties. Of course, the quest to find nature's smallest units has led to the discovery that the

atoms of matter are not indivisible, and even to the realization that the particles that constitute them may be further divisible as well.

All macroscopic objects are indeed composed of smaller components,
and these of yet smaller components. So the assumption that the properties
of any material object can be understood in terms of the properties of its
component parts is quite reasonable. Thomas Hobbes, for example, argued
that all phenomena, including human activity, could ultimately be reduced
to bodies in motion and their interactions. This was an assumption that
birthed most of modern science and set it on a quest to dissect the world,
and to favor explanations framed at the lowest possible level of scale.

Reflecting on this assumption that smaller is more fundamental, the
Canadian philosopher Robert Wilson dubbed it "smallism."[5] It is not obvious, however, that things do get simpler with descent in scale, or that there
is some ultimate smallest unit of matter, rather than merely a level of scale
below which it is not possible to discern differences. Nevertheless, it is
often the case that it is possible to cleanly distinguish the contributions of
component objects from their interactions in explaining the properties of
composite entities. Unfortunately, there are many cases where this neat segregation of objects and relationships is not possible. This does not mean that
complex things lack discernable parts, only that what exactly constitutes
a part is not always clear. Nevertheless, phenomena susceptible of simple
decomposition led to many of the greatest success stories of Western science.

By the middle of the nineteenth century, it was becoming obvious that
the chemistry of life was continuous with the chemistry that applied to
all matter. Living metabolism *is* in this sense just a special case of linked
chemical reactions. The discovery of the structure of DNA at the middle of
the twentieth century marked the culmination of a century-long effort to
identify something like the philosopher's stone of living processes, and was
widely heralded as the "secret of life." In a parallel science, it was becoming
clear that the smallest units of chemical reactions—atoms—were themselves
composed of even smaller components—electrons, protons, and neutrons—
and these were eventually found to be further dissectible. The study of brain
function likewise progressed from bumps on skulls to the ultrastructure of
neurons and synapses. Contemporary neuroscience can boast a remarkably detailed and nearly complete theory of synaptic function, and only a
slightly less complete understanding of neurons. In all these sciences, the

push to understand the properties of ever smaller and presumably more basic components has led to a profoundly thorough map of the microverse of elementary physical particles, and yet has made comparatively less dramatic progress toward a theory of ordinary-scale compositional dynamics and systemic functions, especially when it comes to entential phenomena.

A critical shortcoming of methodological smallism, despite its obvious successes, is that it implicitly focuses attention away from the contributions of interaction complexity. This gives the false impression that investigating the organizational features of things is less informative than investigating component properties. This bias has only quite recently been counterbalanced by intense efforts to study the properties of systems in their own right. The sciences of the twenty-first century have at last begun to turn their attention away from micro properties and toward problems of complex interaction dynamics. As the astrophysicist Stephen Hawking has said: "I think the next century will be the century of complexity."[6]

THE EMERGENTISTS

In many ways, the first hints of a scientific reaction to the seductive influence of reductionism, and its minimization of the special entential characteristics of life and mind, date to a group of late nineteenth- and early twentieth-century philosophers of science. They struggled to articulate an alternative middle path between reductionism and vitalism, believing that it must be possible to avoid the reductionist tendency to explain away the special features of entential phenomena, and yet also avoid invoking enigmatic non-material causes or special forms of substance to account for them. It took some decades for the subtle implications and complications of this emergence perspective to become sufficiently explored to expose both its promise and its weaknesses. Indeed, its promise has yet to be fully realized. The first systematic efforts to explore these implications are associated with a small circle of theorists who came to be known as "the British emergentists." Perhaps the most prominent were Samuel Alexander, C. D. Broad, and Conwy Lloyd Morgan, each of whom was influenced by the ideas of Lewes and Mill.

Mill argued that the principles governing the properties of higher-level entities are often quite distinct and unrelated to those of the components that constitute them. This is particularly the case with organisms, about which he says: "To whatever degree we might imagine our knowledge of the properties of the several ingredients of a living body to be extended and perfected, it is certain that no mere summing up of the separate actions of those elements will ever amount to the action of the living body itself."[7] By "summing up," Mill is invoking an analogy to Newtonian dynamics where different force vectors literally sum their component dimensional values. In contrast, he draws attention to chemical composition, where the properties of elements like sodium or chlorine gas are quite distinct from those of simple table salt that is composed by the ionic bonding of these two elements. His notion of levels is based on this distinction. Within a level, he says, interactions are compositional and summative while adjacent levels are distinguished by the appearance of radically different properties resulting from a non-summative interaction among components. Although subsequent emergence theorists differ in the way levels are defined and how properties are related to one another across levels, this conception of higher-level properties distinguished from lower-level properties by virtue of non-additive forms of interaction has been a consistent characteristic of the concept of emergence invoked by all subsequent theorists—a discontinuity of properties despite compositional continuity.

Mill distinguished laws for interactions taking place within a level from those determining relationships across levels, by calling the former *homopathic* laws and the latter *heteropathic* laws. Homopathic laws were expected to have an additive character, producing highly predictable patterns of causal interactions. Heteropathic laws, however, were presumed to be somewhat idiosyncratic *bridging* laws, linking quite different classes of homeopathic properties across levels. Thus, the advent of organisms and minds involved the appearance of novel causal laws but not any new laws of chemistry or physics. These higher-level properties were not discontinuous from lower-level properties, just not predictable using the lower-level laws alone.

Though written before the appearance of Darwin's *On the Origin of Species*, Mill's argument for the appearance of unprecedented heteropathic laws, which were due to special combinatorial processes occurring at a lower level, left open the possibility that the evolution of novel functions might not merely involve simple mechanical combination. Mill's analysis only went so far as to provide a framework for dealing with the apparent discontinuities between causal laws and properties at different compositional levels; but the general logic of this approach made it reasonable to consider the possibility of the evolution of progressively more complex causal domains from simple physics to social psychology, an idea that Herbert Spencer was later to advocate (though largely remaining within a mechanistic paradigm).

An important bridge between evolutionism and emergentism was provided by the British comparative psychologist and evolutionary theorist Conwy Lloyd Morgan. In the 1890s, he co-discovered an evolutionary mechanism (independently with James Mark Baldwin, for whom the effect is now named) by which strict natural selection could produce Lamarckian-like effects. He argued that behavioral (and thus goal-driven) responses to environmental challenges, including those that could be learned, might become assimilated into inherited adaptations in future generations because of the way they would alter the conditions of natural selection. Where the strict neo-Darwinians of the period argued that functionality could only arise in a *post hoc* fashion from blind chance variations,[8] Morgan suggested that prior functional and even goal-directed behavior (whether or not arising *post hoc* in evolution) could create conditions whereby these tendencies were more likely to become innate within a lineage. It could do so because it would partially spare those lineages able to acquire this adaptation experientially; and if by chance members of such a spared lineage inherited a slightly more efficient means to acquire this adaptation, this variant would likely replace any less efficient means.

In more general terms, this scenario implied that evolutionary change might not be limited to the generation of merely incremental adjustments of past mechanisms, but could at times generate truly unprecedented transitions that took advantage of what Mill would have called heteropathic effects at the organism level. There is no evidence that Morgan developed this idea following Mill's notion, but decades later Morgan enlarged upon the idea in a book titled *Emergent Evolution*,[9] where he argued that the evolutionary process regularly produced emergent properties. Unfortunately, partly because of the theological implications he drew from this insight, his prescient anticipation of the challenge of explaining major evolutionary transitions was not widely influential. The problem posed to natural selection theory by such major evolutionary transitions as the origins of multicellular organisms was not recognized again until the late 1980s and 1990s,[10] and remains a topic of debate.

In many respects, the motivation behind the emergence theories of the early twentieth century was to bridge an explanatory gap between the physical sciences and the so-called special sciences of psychology and the social sciences (to which I would add biology). As we have seen, the critical fault line between these sciences can be traced to their approach to ententional processes and properties. The natural sciences must exclude ententional explanations, whereas the so-called special sciences cannot. Biology is in the awkward position of at the same time excluding ententional accounts and yet requiring ententional properties such as representation and function. The question at issue was whether some new kind of causality was at work in the phenomena considered by these special sciences (i e , the life sciences), or whether their apparently unusual forms of causal relationships could be entirely reduced to forms of causal relationship that are found in the physical sciences more generally.

Two prominent British philosophers brought the topic of emergence into mainstream discourse during the first part of the twentieth century. Samuel Alexander and C. D. Broad each had slightly different ideas about how to frame the emergence concept. Broad's view was close to that of Mill in arguing that emergent properties changed the causal landscape by introducing properties that were fundamentally discontinuous from any that characterized the component interactions from which they emerged. Alexander was closer to Lewes' argument, proposing that the principal characteristic of emergent properties is an intrinsic inability to predict these properties from the properties and laws at the lower level. This distinction in approach to the characterization of emergent phenomena has been made more explicit over the intervening decades and is generally identified with ontological versus epistemological conceptions of emergence, respectively. Despite this difference in characterizing the nature of the discontinuity and novelty of higher-level emergent phenomena, both men echoed the common theme that low-level component mechanistic processes could be the ground for non-mechanistic emergent properties at a higher level.

Broad, following closely on Mill's ideas, conceived of processes at different levels as obeying distinct level-specific laws that, although incompatible from level to level, were nevertheless related to one another by bridging laws (which he calls "trans-ordinal laws"). Broad further claimed that it is, in principle, impossible to deduce the higher-level properties even from complete knowledge of lower-level laws. Only after observing an actual instance of the emergence of a higher-level process can we retrospectively reconstruct the lawful relationships that hold between these levels. In other words, even given a complete description of all the intrinsic properties of the parts plus their arrangements and interactions, one could not predict certain properties of the whole higher-level phenomenon.[11]

Invoking predictability is tricky in this account, since it is not clear whether Broad means that this is primarily a problem of not being able to extrapolate from knowledge of these relationships, or something stronger: there not being any definite physical determination of the higher- from the lower-level laws.

Broad clarifies this somewhat by arguing that this unpredictability is the result of the ultimate irreducibility of the higher-level phenomena to properties of their lower-level components. For him, the unpredictability is a symptom of a deep incompatibility of causal properties distinguishing levels. By this he must mean that the trans-ordinal laws do not directly map the higher-level properties onto more basic level properties, but only describe and itemize the lower-level conditions from which the higher-level property emerged in the past and will likely do so in the future. The incompatibility between levels, and the non-deducibility it entails, are for Broad evidence that there may be no fixed and definite unity to the fabric of causal laws.

Alexander, who was influenced by Morgan, also argued that higher-level emergent entities and their properties cannot be predicted from component properties and interactions at a lower level.[12] But for him this is not because of some metaphysical incompatibility of the entities and laws at different levels. It is the result of problems intrinsic to the way this transitional relation must be represented. We are forced to recognize emergent transitions merely as matters of "brute empirical fact," which cannot be anticipated or subsumed under a determinate explanation because of *our* limitations, not any fundamental incompatibility. He admits that a Laplacian demon with complete knowledge of the universe's laws and its state prior to life could in fact predict all future distributions of matter and energy in full detail, but he still would fail to be able to describe such qualities as life or consciousness.

In this respect, Alexander is adamant that higher levels exhibit fundamental new qualities that are not predictable even from a full knowledge of all
relevant lower-level facts. Living and mental processes are fundamentally
new, created afresh, and not merely the "resultant" of chemical and neural
interactions. But this doesn't mean that they aren't determined from the
previous state of things, only that the logic of this determination is some
how intrinsically beyond computation. Here Alexander's proposal parallels
contemporary theories based on notions of determinate chaos (see below),
which argue that the non-linear features of certain composite systems makes
prediction of their future behaviors incalculable beyond a short period. But
this fact could justify his claim that despite emerging from lower-level roots,
emergent laws could be special to the higher level.

Unfortunately, these early attempts to base the notion of emergence on a
solid philosophical foundation raised more questions than they answered.
In their effort to make sense of the relationship between the physical sciences
and the special sciences, these philosophers mixed issues of predictability
and incompatibility, novelty and discontinuity. Like Mill, they worked more
toward justifying apparent descriptive discontinuities observed in nature,
rather than deriving the emergence concept from first principles. This left
both future emergentists and their critics to argue over the candidate principles. The result has been a wide divergence of emergentists committed
to either a mostly epistemological definition (i.e., based on predictability
and representational issues) or committed to a mostly ontological definition
(i.e., based on assuming fundamental discontinuity of physical laws).

Many other ways of defining emergence have grown up around the different ways that levels, discontinuity, novelty, and predictability are used to
define the concept of emergence. So, for example, theorists are often distinguished as either being "weak" or "strong" emergentists, referring to their
stance on the question of causal discontinuity and whether emergence is
compatible or incompatible with reductionism. Strong emergentism argues
that emergent transitions involve a fundamental discontinuity of physical
laws; weak emergentism argues that although there may be a superficially
radical reorganization, the properties of the higher and lower levels form a
continuum, with no new laws of causality emerging. However, this distinction does not capture many more subtle differences, and the perspective
developed in this book is not easily categorized in these terms.[13]

The conceptions of emergence offered by the British philosophers were subsequently treated to harsh criticism in the early part of the twentieth century, though the concept of emergence was also often invoked by prominent scientists to explain otherwise surprising natural phenomena. For example, the developmental biologist Paul Weiss in his explorations of spontaneous pattern formation in cellular, molecular, and embryological context regularly pointed to the importance of the spontaneous emergence of higher-order organization as a critical contributor to biological form. And Ludwig von Bertalanffy's effort to establish a general systems theory repeatedly stressed that complex systems almost always exhibited properties that could not be understood solely in terms of their components and their interactive properties, but also were characterized by novel system-level properties.

The Nobel Prize–winning neuroscientist Roger Sperry also updated classic emergentist arguments about the nature of consciousness. In an article published in 1980, Sperry argued that although there is no change in basic physics with the evolution of consciousness, the property of the whole system of interacting molecules constituting brains that we call "consciousness" is fundamentally different from any collective property they would exhibit outside of brains. In this way, he offers a configurational view of emergence. He illustrates this with the example of a wheel. Although the component particles, atoms, and molecules forming the substance of the wheel are not changed individually or interactively by being in a wheel, because of the constraints on their relative mobility with respect to one another, they collectively have the property of being able to move across the ground in a very different pattern and subject to very different conditions than would be exhibited in any other configuration. The capacity to roll is only exhibited as a macroscopic collective property. It nevertheless has consequences for the component parts. It provides a means of displacement in space that would be unavailable otherwise. In this sense, Sperry argues that being part of this whole indirectly changes some of the properties of the parts. Specifically, it creates some new possibilities by restricting others. This trade-off between restriction and constraint on the one hand and unprecedented collective properties on the other will be explored more fully in the next few chapters.

More generally, Sperry uses this example to exemplify another common theme that recurs in discussions of emergence, and increasingly so in more recent discussions. This is the concept of downward causation. For

Sperry, the wheel example provides a case of downward causation, because being a part in a particular kind of whole alters possible movement options available to the parts. In Sperry's example, we would not describe this as exactly changing any specific properties intrinsic to the parts, but rather altering how these properties (e.g., the possibility of translation in space) are changed in their probability of being realized. Outside of their inclusion in a wheel, individual atoms are exceedingly unlikely to move by spiraling in the plane of forward movement; but inclusion in a wheel makes this highly likely. As we will discuss below, this configurational effect that the whole has on its parts might more accurately be described in terms of constraints

Molecules forming the solid structure of a wheel are constrained in their ability to move in space by their tight attachment to neighbors. The key factor is that this constraint on movement is additionally influenced by the geometric structure of the whole. Although still constrained by being within a solid aggregate, if the aggregate structure is round, this constraint is potentially overcome at the macroscopic level. Thus, if one wants to move a large mass of some substance, say a ton of hay, it is easier if the whole aggregate is shaped into a large cylinder and rolled. Sperry uses this analogy to argue that the configuration of brain processes similarly changes what can be done with the parts—the brain's neurons, molecules, and ionic potentials. But it is a bit of a misnomer to call this a form of causation, at least in modern parlance. The downward (in levels) causation (from whole to part) is in this sense *not* causation in the sense of being induced to change (e.g., due to colliding or chemically interacting with neighboring molecules), but is rather an alteration in causal probabilities.

This "downward" sort of causality might better be framed in Aristotelean terms as a species of formal cause (see chapter 1), whereas the notion of being induced to change (e.g., in position or configuration) might be analogized to Aristotle's efficient cause. The constraint of being incorporated into a wheel makes an atom's mobility more subject to formal (geometric) features of the whole, and mostly irrespective of its individual properties. It's not that a given atom of the wheel could not be moved in some other way, if, say, the wheel was broken and its parts scattered; it's just that rotational movement has become far more probable.

The notion of downward or top-down causation has, however, been subject to considerable debate. There are almost as many interpretations of this concept as there are emergence theories. It is considered by many to be the most important determinate of emergence and it is also one of the most criticized concepts. To some degree, these confusions reflect an unfortunately simplified conception of causality inherited from our Enlightenment forbears, but it also implicitly incorporates assumptions of smallism in the sense that it is treated as the reciprocal to a bottom-up notion of causal influence. These assumptions about the nature of causality will be extensively reexamined in the following chapters, but for now it is sufficient just to recognize that emergence theories are generally defined in terms that counter bottom-upism. They differ in how they do this.

It is not essential, however, to argue for a form of top-down causality in order to define emergence. For example, an alternative clarification of the part/whole issue has been provided by the American philosopher Paul Humphreys. Rather than arguing that the interactions of parts of an emergent whole produce new properties, inherit new properties by virtue of their involvement in the whole, or exhibit new properties imposed by the whole configuration, he argues that in many cases parts are significantly transformed as a result of being merged with one another in some larger configuration. Humphreys maintains that in some cases the very constitution of parts is changed by inclusion in some larger unity. He calls this modification *fusion*. By virtue of their systemic involvement with each other, they are no longer entirely distinguishable. As a result, reductionist decomposition cannot be completed because what were once independently identifiable parts no longer exist.[14]

Humphreys' argument has its roots in quantum physics, where the individuation of events and objects is ambiguous (as in quantum entanglement, etc.). It is intended to circumvent the problem of double-counting causal influences at two levels. It solves this problem by arguing that fusion results in a loss of some constituent properties. They literally cease to exist as parts are incorporated into a larger configuration. To slightly oversimplify an example from quantum physics, consider the way that quantum properties such as quantum uncertainty (not being able to simultaneously specify the position *and* momentum of an elementary particle) effectively disappear in interaction with macroscopic instruments, resulting in retrospective certainty. This is the so-called quantum-classical transition, often described as the "collapse" of the Schrödinger wave function, which defines a "probability wave." These strange quantum properties are effectively swallowed up in the transition to an event at a higher scale.

An argument that is loosely related to both Sperry's and Humphreys' accounts of higher-order properties can be applied to the presumed part/whole analysis of organisms. Because organism components (e.g., macromolecules) are reciprocally produced across time, the very nature of what constitutes a part can only be determined with respect to its involvement in the whole organism. Indeed, the vast majority of molecules constituting an organism are enmeshed in a continual process of reciprocal synthesis, in which each is the product of the interactions among many others in the system. They exist and their specific properties are created by one another as a result of this higher-order systemic synergy. With cessation of the life of the organism—i.e., catastrophic dissolution of critical reciprocal interrelationships—the components rapidly degrade as well. Thus their structures and resultant properties were in large part derived from this systemically organized dynamic. Even though these macromolecular properties were also constrained by possibilities offered by the atomic properties of the constituent atoms, these combinatorial possibilities are effectively infinite for molecules composed of hundreds or thousands of atoms. As part of a functioning organism, however, the range of their possible interactions and combinatorial configurations is vastly constrained, and the molecules that they form are themselves highly restricted in their interactions and distributions.

Notice that this is related to Sperry's notion of altering certain constraints affecting the parts by virtue of being incorporated into the emergent configuration. But while Sperry's analogy of being affected by the geometric configuration of the whole does not fundamentally affect any local properties of the parts, Humphreys' notion of fusion (interpreted more broadly) does. In both the wheel and the organism, there is a change in the *constraints* affecting lower-level interactions; but in the wheel this affects relational properties and in the organism it additionally affects intrinsic properties. In the wheel, independent mobility of the contained atoms is lost, but intrinsic properties, like mass and charge, are unaffected. In the organism, the properties of the molecules are a consequence of their incorporation into this system. This apparent difference may in part reflect a complexity difference. The closer analogy is between atoms in a wheel and atoms in a molecule.

As we will see in the following two chapters, this distinction exemplifies two hierarchically distinct *orders* of emergent transition. The organism example assumes the emergence of entential organization; the wheel shape does not. There is also a clear sense in which the wheel provides us with an emergence notion that is straightforwardly compatible with a reductionistic account of the higher-order property. The wheel can be dissected into its parts and reconstructed without loss, but a living organism taken apart suffers the Humpty-Dumpty problem. Most of its parts are themselves unstable entities outside of the context of a living organism, so it becomes a problem to decide what exactly are the "proper parts" of an organism that in interaction determine its emergent character. One of the crucial differences is that the emergent relationship in the wheel example is synchronic; but whereas there is a synchronic analysis possible for the organism, the property that is crucial to the constitution of its proper parts is the dynamics of reciprocal synthetic processes that is intrinsically diachronic, as well as the larger diachrony of the evolutionary process that resulted in this metabolic system. Thus, at least for higher-order forms of emergence, the part/whole distinction and the synchrony/diachrony distinction are intertwined.[15]

A HOUSE OF CARDS?

The most influential critiques of ontological emergence theories target these notions of downward causality and the role that the emergent whole plays with respect to its parts. To the extent that the emergence of a supposedly novel higher-level phenomenon is thought to exert causal influence on the component processes that gave rise to it, we might worry that we risk double-counting the same causal influence, or even falling into a vicious regress error—with properties of parts explaining properties of wholes explaining properties of parts. Probably the most devastating critique of the emergentist enterprise explores these logical problems. This critique was provided by the contemporary American philosopher Jaegwon Kim in a series of articles and monographs in the 1980s and 1990s, and is often considered to be a refutation of ontological (or strong) emergence theories in general, that is, theories that argue that the causal properties of higher-order phenomena cannot be attributed to lower-level components and their interactions. However, as Kim himself points out, it is rather only a challenge to emergence theories that are based on the particular metaphysical assumptions of substance metaphysics (roughly, that the properties of things inhere in their material constitution), and as such it forces us to find another footing for a coherent conception of emergence.

The critique is subtle and complicated, and I would agree that it is devastating for the conception of emergence that it targets. It can be simplified and boiled down to something like this: Assuming that we live in a world without magic (i.e., the causal closure principle, discussed in chapter 1), and that all composite entities like organisms are made of simpler components without residue, down to some ultimate elementary particles, and assuming that physical interactions ultimately require that these constituents and their causal powers (i.e., physical properties) are the necessary substrate for any physical interaction, then whatever causal powers we ascribe to higher-order composite entities must ultimately be realized by these most basic physical interactions. If this is true, then to claim that the cause of some state or event arises at an emergent higher-order level is redundant. If all higher-order causal interactions are between objects constituted by relationships among these ultimate building blocks of matter, then assigning causal power to various higher-order relations is to do redundant bookkeeping. It's all just quarks and gluons—or pick your favorite ultimate smallest unit—and everything else is a gloss or descriptive simplification of what goes on at that level. As Jerry Fodor describes it, Kim's challenge to emergentists is: "why is there anything except physics?"[16]

The concept at the center of this critique has been a core issue for emergentism since the British emergentists' first efforts to precisely articulate it. This is the concept of supervenience. *Supervenience* is in many respects the defining property of emergence, but also the source of many of its conceptual problems. The term was first used philosophically by Lloyd Morgan to describe the relationship that emergent properties have to the base properties that give rise to them.[17] A more precise technical definition was provided by the contemporary philosopher Donald Davidson, who defines it in the context of the mind/body problem as follows: "there cannot be two events exactly alike in all physical respects but differing in some mental respects, or that an object cannot alter in some mental respects without altering in some physical respects."[18]

This defines an asymmetric dependency in this hierarchic relationship, which is sometimes stated in aphoristic form as: there cannot be changes in mental (aka emergent) properties without a change in neurophysiological (aka physical substrate) properties. It is an emergentist no-free-lunch restriction. The fundamental challenge of classical emergentism is to make good on the claim that higher-order (supervenient) properties can in some critical sense not be reduced to the properties of their component lower-level (subvenient) base, while at the same time being entirely dependent on them. So, if one agrees that there can be no difference in the whole without a difference in the parts, how can it be possible that there is something about the whole that is not reducible to combinations of properties of the parts?

Looking at the world in part/whole terms seems like an unimpeachable methodology. It is as old as the pre-Socratic Greek thinkers. It figures prominently in the thinking of Plato and Aristotle, and remains almost an axiom of modern science. Philosophically, the study of compositionality relationships and their related hierarchic properties is called *mereology*. The term quite literally means "the study of partness." The concept of emergence, which has its roots in efforts to challenge the notion that the whole is just the sum of its parts, is thereby also predicated on mereological assumptions. This critique, if coherent, suggests that this foundational assumption of emergentism renders it internally inconsistent. But mereological assumptions may prejudice the case.

Effectively, Kim's critique utilizes one of the principal guidelines for mereological analysis: defining parts and wholes in such a way as to exclude the possibility of double-counting. Carefully mapping all causal powers to distinctive non-overlapping parts of things leaves no room to find them uniquely emergent in aggregates of these parts, no matter how they are organized. Humphreys' concept of fusion appears on the surface to undermine this stricture by challenging the basis for whole/part decomposition. The example of macromolecules in an organism also suggests that at least simple decomposition is problematic. But this does not entirely escape the critique if we recognize that these system-modified parts still analyze into smaller ultimate parts. Shuffling the part/whole relationship at one level does not alter it all the way down.

Kim's critique is most troublesome for emergence theories that have been developed to support functionalism (see chapter 1). Functionalism claims that the same *form* of operation can be physically embodied in different media (e.g., the same program run on two different computers) and that the different physical implementations will nonetheless have the same causal powers, despite having entirely different ultimate components embodying this operation. As we saw in chapter 1, this claim for multiple realizability is also a feature of natural-kind concepts like solid, liquid, and gas. Thus water and alcohol at room temperature both exhibit surface tension, viscosity, the capacity to propagate transverse waves, and so on. It's only when we ask for an accounting of *all* the causal powers that different liquids can be discriminated. Indeed, there is a subset of behaviors in each case that are expressed despite the difference in physical substrates (sometimes called universality classes); but because there is some level of detail at which differences can be discriminated, we must conclude that these differences as well as the similarities are attributable to the different physical constituents of each.

There have been many challenges and responses to this line of criticism, which traces back to the time of the British emergentists,[19] but here I will focus on one that will serve as the starting point for considering a very different sort of alternative. This is the fact that the substance metaphysics that supports this mereological analysis does not accurately reflect what we currently know of the physics of the very small: quantum physics.

The scientific problem is that there aren't ultimate particles or simple "atoms" devoid of lower-level compositional organization on which to ground unambiguous higher-level distinctions of causal power. Quantum theory has dissolved this base at the bottom, and the dissolution of this foundation ramifies upward to undermine any simple bottom-up view of the causal power. At the lowest level of scale there are only quantum fields, not indivisible point particles, or distinguishable stable extended configurations. Quantum fields have ambiguous spatiotemporal origins, have extended properties that are only statistical and dynamically definable, and are defined by a dynamical quality, a wave function, and not any discrete extensional boundary. At this level, the distinction between dynamics and the substrate of dynamics dissolves. Quantum interactions become "classical" in the Newtonian sense in interactions involving macroscopic "measurement," and only then do they exhibit mereologically identifiable properties (recall Humphreys' notion of fusion). But at this presumed lowest level, discrete parts do not exist. The particulate features of matter are statistical regularities of this dynamical instability, due to the existence of quasi-stable, resonantlike properties of quantum field processes. This is why there can be such strange unparticlelike properties at the quantum level. Only with the smoothing of statistical scale effects do we get well-behaved mereological features.

This is not meant to suggest that we should appeal to quantum strangeness in order to explain emergent properties, nor would I suggest that we draw quantum implications for processes at human scales. However, it does reflect a problem with simple mereological accounts of matter and causality that is relevant to the problem of emergence. A straightforward framing of this challenge to a mereological conception of emergence is provided by the cognitive scientist and philosopher Mark Bickhard. His response to this critique of emergence is that the substance metaphysics assumption requires that at base, "particles participate in organization, but do not themselves have organization." But, he argues, point particles without organization do not exist (and in any case would lead to other absurd consequences) because real particles are the somewhat indeterminate loci of inherently oscillatory quantum fields. These are irreducibly processlike and thus are by definition organized. But if process organization is the irreducible source of the causal properties at this level, then it "cannot be delegitimated as a potential locus of causal power without eliminating causality from the world."[20] It follows that if the organization of a process is the fundamental source of its causal power, then fundamental reorganizations of process, at whatever level this occurs, should be associated with a reorganization of causal power as well.

This shift in emphasis away from mereological interpretations of emergence evades Kim's critique, but it requires significant rethinking of the concept of emergence as well. In many respects, supervenience was one of the defining features of classic emergence theories. But it shouldn't have surprised us to find that a synchronic understanding of this relationship is an insufficient basis for the concept of emergence. For one reason, emergence itself is a temporal conception. Most if not all of the higher-order properties considered to be emergent were not always present in the universe, and not always present in the local contexts in which they currently exist. Certainly in the case of life and mind, their emergent characteristics are relatively recent phenomena in the history of the universe and of this tiny sector of our galaxy and our solar system. These emergent phenomena and their ententional properties emerged as new forms of physical-chemical process organization developed among the existing atoms and their energetic interactions. As they did so, new kinds of components also came into existence. Static notions of part and whole are for this reason suspect since the wholes we are interested in are dynamical and the parts are constantly in flux, being constantly synthesized, damaged, and replaced, while the whole persists. Rethinking the concept of emergence in dynamical terms has, additionally, been aided by the discovery of an array of new exemplars of emergence, understood somewhat differently than in philosophical circles, as well as new descriptive and experimental tools for studying them. We turn to these next.

COMPLEXITY AND "CHAOS"

The concept of emergence began to take on a new meaning in the last years of the twentieth century. This reemergence of emergence as a major topic of study arose as a consequence of two related developments: the study of a variety of inorganic phenomena with distinctive and discontinuous changes in aggregate behavior due to lower-level dynamics; and the development of computational simulation techniques able to model complex iterative processes and highly non-linear dynamics.

Compared to life and mind, inorganic processes that generate scale-related discontinuities of aggregate behavior are simpler, easier to study in detail, and metaphysically unproblematic. Specifically, so-called self-organizing processes—such as the growth of snow crystals, the spontaneous "assembly" of lipids into membranes, and the transition to superconductivity at extremely low temperatures, to name a few very diverse examples—offer a highly suggestive set of phenomena that satisfy the general criteria characterizing emergence: higher-order aggregate properties arising from the interactions of components lacking these properties.

It is unlikely that the study of naturally occurring self-organization would have received the attention it did were it not for the insights gained from computation. This is because computational model systems allowed complete control and manipulation of the initial variables of an interactive process involving large numbers of component interactions and which could be set running for hundreds of thousands of iterations to see where these interactions led. Study of these sorts of phenomena was also aided by the fact that analytically insoluble relationships could nevertheless be modeled as step-by-step interactions so that their trajectories of change could be traced. The result of such simulations even with simple operations, iterated thousands or millions of times, often produced highly complex results that were both difficult to predict and surprising. Quasi-regularities could even be discerned in largely chaotic processes that would not have been detected by analytic techniques. But they could be directly perceived in graphic presentations of the process in action. So the importance of these tools was not merely, or even mostly, due to the way they could help model large-scale physical interactions, but primarily to the ability to create "toy" worlds where critical interaction variables could be manipulated at will.

Probably the most widely studied and influential computational exemplars of this new way of thinking about emergence involved algorithms called *cellular automata*. They were so named because they consisted of algorithms for computing some operation whose result would be entered into one of an array of cells in a matrix, such as the pixels on a display screen. In a typical example, the operation involves repeated computations in which some value assigned to one cell is computed by virtue of an operation involving the values present in a select number of adjacent cells. If the operation is iterated for each cell over the entire matrix and the whole process is repeated incessantly, it can superficially simulate local physical interactions in a spatially distributed medium. As researchers explored the behavior of these iterative algorithms, it became clear that simple algorithms could often produce quite complex dynamically regular patterns.

A famous simple example is a simulation dubbed the *Game of Life* by its creator John Conway in the late 1960s. He called it "life" because the algorithm determined whether cells were on (alive) or off (dead) by virtue of being activated by proximity to live neighbors or inactivated due to lack of live neighbors. What resulted from different starting configurations and algorithmic rules were repeating configurations that could persist and move about the screen as they were reproduced as a whole configuration. So by analogy a cell-by-cell level property was emerging at a configuration level, and new properties such as "mobility" of such a configuration were also emerging. In a simple sense, this was an exemplification of at least the spirit of the concept of emergence.

In a related early variant of this logic, the Santa Fe Institute complexity theorist Stuart Kauffman explored the behavior of randomly interconnected cells controlled by similar sorts of interaction rules, but randomly assigned. Employing binary on/off rules and low connectivity between cells (nodes in the network), these randomly connected networks nevertheless exhibited remarkably consistent activation patterns across time, rapidly falling into cycles of global states of roughly the same number of steps, comprising only a fraction of states possible, and they would do so with predictable consistency from any random configuration. Noise in— order out.

This logic of generating complex global regularities and identifiable "entities" from simple iterated operations became a new model for emergent processes. The emergent character appeared even more distinctive because such regularities would arise spontaneously from randomly distributed starting conditions. The relevance to many naturally occurring physical processes is that most processes involve local interactions and are in this respect analogous to interactions between "neighboring" cells or pixels. This analogy has spawned claims that this general logic offers a new paradigm for understanding complex physical processes in general (an idea promoted by Steven Wolfram in his *A New Kind of Science* in 2002).

Previously, mathematicians had also explored the behavior of non-equilibrium dynamical systems using the tools of thermodynamics. The equations for representing these complex physical processes are non-linear, where solutions-out are also values-in, thus producing a recursive computational logic. Because such recurrent calculations can change values with each iteration, they generally do not converge to simple solutions, but they can nevertheless be mapped into a "phase space"[21] as a trajectory of changing values with each step of calculation. By graphing the values of tens of thousands of such calculations, it is often found that the results are not entirely regular, and yet not entirely chaotic either. This quasi-regularity of repeated operations means that the product of an unspecified number of iterations will be quite unpredictable, with similar values often producing widely divergent results, and yet can also produce values that are likely to be confined to distinct subregions of the phase space. The biasing of trajectories to repeatedly traverse these zones within phase space is sometimes termed *itinerance*, by analogy to a person frequenting the same general region, even though never in exactly the same way twice. Trajectories of iterated solutions thus tend to "orbit" through these zones in the graphic space of sequential solutions. To the extent that such itinerant zones can be described with respect to their proximity to a central tendency trajectory (e.g., a loop or complex curve), even though no one trajectory matches it, this structure is called an *attractor*. This is a useful term, though it has somewhat misleading connotations. The "attraction" is only figurative, since there is no attractive force involved, only a statistical bias.

One of the first, and probably the most well known, attractors was discovered by Edward Lorenz, an American meteorologist, in 1963 while he was graphically modeling the solutions of equations intended to represent fluid flow behavior in the atmosphere. The configuration of this attractor produced a sort of twisted butterfly shape in 3D (see Figure 5.1). Since attractors tend to have distinctive geometry, they are often given names. This one has come to be known as the Lorenz attractor and is one of the most widely reproduced exemplars of what has come to be known as deterministic chaos.

The convergence of these various lines of research in the 1980s began to

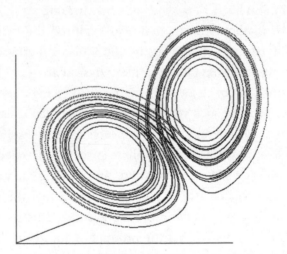

FIGURE 5.1: A plot of the Lorenz attractor produced by computing the values in three dimensions of the progressively iterated solutions of an equation that Edward Lorenz studied as a way of describing simple meteorological processes. Values from each prior solution are fed back into the equation to compute the next point on a continuous trajectory. By computing vast numbers of results a continuous curve is produced which, although it never exactly repeats the same values, tends to produce values that remain within a definite domain and trace out a sort of twisted butterfly pattern. This most probable region within which the solutions to the equation tend to lie has come to be called an attractor.

suggest that it might be possible to discern general principles that applied to a wide range of complex, multicomponent, iterative processes. In different hands this general topic area became known as complexity theory, chaos theory, synergetics, or just complex systems theory (among other names). And among the various whole fields that have grown from these initial insights, two of the more prominent have been the study of so-called genetic algorithms (because they utilize a variation and selection logic) and artificial life. Not only have these computational approaches provided important new tools for studying complex phenomena from economic change to ant colony foraging, they have also significantly reframed the problem of emergence.

These approaches have provided well-understood physical and mathematical exemplars of processes that produce complex regularities of global organization from unorganized and even randomly distributed interactions among components. Such processes are thoroughly non-mysterious and mechanistic. They generate spontaneous regularities and synergies at a global level that depend solely on cumulative interaction effects at a lower level. They are not consequences of extrinsic imposed regularity (such as containment) or the simple summation of intrinsic component properties (as is, say, mass or electric charge) or component dynamical properties (as is total heat).

Many of the common descriptors used to characterize emergent phenomena also applied to these dynamical processes. The regularities observed at a higher level result from otherwise unorganized dynamical interactions of lower-level components. These regularities are not mere summations or superpositions of these lower-level properties, and exhibit properties that have no lower-level counterparts. And these higher level properties are often incredibly difficult to predict, even knowing in detail the lower-level details. They are more often recognized simply by observation of consequences. Moreover, although these simple physical and computational processes do not exhibit end-directed organization, their attractor structure can be loosely compared to end-directedness in the sense of being teleonomic (see chapter 4). In many respects, these fairly well understood physical and computational processes are legitimately described as emergent, at least in a limited sense.

So, despite not providing an account of the ententional properties that motivated the early emergentists, these phenomena both offered numerous exemplars of processes where lower-order interactions gave rise to distinctive, unprecedented, higher-order consequences and offered powerful heuristic tools for developing an empirically grounded account of a whole class of emergent processes. Besides providing potentially informative exemplars, and the space necessary for a rethinking of the problem of emergence from a viewpoint unburdened by the baggage of metaphysical justification, there is another less obvious but more important contribution of this new conception of emergent process: it initiated a shift in perspective from part/whole to process conceptions of emergence.

Following these developments in the physical and computational sciences, the concept of emergence has begun to take on a new and primarily descriptive meaning in many fields. This has done much to legitimate its usage, free of philosophical baggage. It is now often applied to any case where there is spontaneous production of complex dynamical patterns from uncorrelated interactions of component parts. As a result, the term has come to be routinely used within the social sciences, economics, and even in business application, as well as in the natural sciences.

This shift from a largely philosophical to a more descriptive usage of the term *emergence* has been both salutary and misleading. One consequence is that the meanings of the term have diversified and proliferated so that what it refers to in one field may be quite different than in another, with each of the associated concepts of novelty, unpredictability, ascent in scale, synergy, and so on, serving as the principal feature being highlighted. But a more troublesome consequence of identifying exemplars of emergent relationships in physical and computational processes is that it suggests that all examples of phenomena might be accounted for as members of this (albeit diverse) class of processes. This motivates a tendency to presume that although these specific forms do not exhibit obvious entential properties, such properties can nevertheless be explained, or else explained away, in these terms. As we will see in the following chapters, this presumption is premature, and in any case it still evades questions about the origins and physical nature of entential phenomena. Nevertheless, by unambiguously modeling the possibility of the emergence of global order from lower-level chaos, this work provides an important "intuition pump"[22] to help conceive of how this additional emergent step might be possible. As we will see, although this conception of emergent processes does not directly address ententional issues, it may provide a critical missing link from mechanism to teleology.

PROCESSES AND PARTS

One of the most significant consequences of the growing interest in self-organizing physical processes and complex computational systems is that it has shifted attention to dynamical rather than structural characterizations of emergence. However, process approaches don't easily fit into the classic categories of emergence theories. This is because they require rethinking two of the central concepts typically used to define emergence: the part/whole (mereological) distinction and the supervenience relation. If we dispense with these, will we still have something we can call emergence? Actually, this may be the only way to save the concept.

At the end of a paper discussing his process approach to emergence, Mark Bickhard boldly asserts: "Mental states do not exist, any more than do flame states—both are processes."[23] This may be a bit too extreme, but it drives home a crucial point: these phenomena consist in the special character of the transformations *between* states, not in the constitution of things at any slice in time. So, trying to build a theory of the emergence of living or mental processes based on a synchronic conception of emergence theory is bound to fail from the beginning. The phenomena we are interested in explaining are intrinsically historical and dynamic. Being alive does not merely consist in being composed in a particular way. It consists in *changing* in a particular way. If this process of change stops, life stops, and unless all molecular processes are stopped as well (say by quick-freezing), the cells and molecules distinctive to it immediately begin to degrade. We can of course dissect organisms and cells, and isolate and study the molecules, molecular complexes, and chemical reactions that they consist of; but as important as this is to the understanding of the living process, it is the organization of that process—involving all those components and component processes—that is what we ultimately hope to be able to reconstruct once we've understood a significant fraction of these details. Yes, the process involves those component parts and subordinate processes; but as was pointed out in the last chapter, these tens of thousands of molecules in vast interaction networks reciprocally synthesize and organize each other. None are parts, as we would understand the concept from an engineering perspective. In an organism, the very notion of a part is process-dependent.

This should be even more obvious with respect to mind. No one would seriously suggest that the process of thought could be explained solely in terms of neural structures or neurotransmitter molecules. Of course, these are critical to the account, but without understanding the processes that they are enmeshed within, we are merely doing anatomy. What matters is the process, and this is hierarchically complex, though not mereological in any simple sense. Nevertheless, we find it hard to resist part/whole simplifications.

Consider functional brain imaging, such as positron emission tomography (PET) or functional magnetic resonance imaging (fMRI). Although these are some of the most exciting new techniques available, and have provided an unparalleled window into brain function, they also come with an intuitive bias: they translate the correlates of cognitive processes into stationary patterns. These techniques identify regions with particularly elevated metabolism, compared to control states, during a given mental task. There are good reasons to believe that local oxygen and glucose consumption are correlated with local signal-processing demands, and that this is a clue to the degree to which different brain areas are involved in the handling of different kinds of cognitive tasks. But it would be a serious mistake to imagine that the function in question is in any sense "located" in the identified "hot spots," or to believe that a metabolic "snapshot"[24] would be in any sense a simple correlate to what is involved even at a gross level in the performance of that activity.

Even the simplest conscious cognitive act inevitably involves rapidly shifting, fleeting, and low-level metabolic changes across dozens of diverse brain regions—some increasing and some decreasing in metabolism at different moments in the process—in order to be completed. Thus an averaged scan, collapsed to a single image, inevitably erases information about this pattern of change. And as two excerpts from pieces of music can analogously demonstrate, regions whose metabolism is constant in different tasks (e.g., control and test) can have the same energy and yet strikingly different "melodies." Knowing the neuronal and molecular substrates, their potential interactions, and the locations of intense activities (indicated by metabolic change) are essential bits of evidence, just far from sufficient for an explanation. Ultimately, we will at least need to develop methods to track the second-by-second regional changes, and eventually even the systemic activities of thousands of interacting neurons, in order to have a methodology appropriate to the challenge. In other words, what we need to know are the relationships between brain system *processes* and cognitive processes.

A process conception of emergence both resolves certain conceptual dilemmas associated with classic synchronic conceptions of emergence and poses new conceptual problems that were previously irrelevant. As noted earlier, Bickhard undermined the force of some of the most devastating criticisms of emergence theories by pointing out that the particulate assumptions of substance metaphysics are incompatible with the ultimate inseparability of process and substance. If we recast the problem in terms of process and organization instead of parts and wholes, the concept of emergence no longer suffers from the criticisms of causal redundancy and explanatory circularity. Because there are no material entities that are not also processes, and because processes are defined by their organization, we must acknowledge the possibility that organization itself is a fundamental determinant of physical causality. At different levels of scale and compositionality, different organizational possibilities exist. And although there are material properties that are directly inherited from lower-order component properties, it is clear that the production of some forms of process organization is only expressed by dynamical regularities at that level. So the emergence of such level-specific forms of dynamical regularity creates the foundation for level-specific forms of physical influence.

Of course, this does not in itself solve the problem of explaining the nature of the transitions we are interested in. It only argues for their plausibility. As Bickhard himself notes, the shift to a process perspective merely clears some serious roadblocks standing in the way of understanding emergent phenomena. We still need to explain what kind of change in process organization actually constitutes any given emergent transition.

This also signals a paradigm shift in what we take to be an emergent phenomenon. In fact, a process approach forces us to abandon the two pillars of classic emergence theory: supervenience and mereology. For example, the philosopher Timothy O'Connor argues that composition is not well defined for processes.[25] This is because components may be continually changing (as in, say, a vortex or an organism) while the emergent attributes are persistent. Although the component elements in such processes are critical, the dynamics of their transitions are more so, and it is this pattern of change that is critical.

Classically, particles are conceived as specific individuated things, with specific location in space and time, specific extension into these dimensions, and specific properties. As we have seen, quantum physics appears to undermine the lowest level of particulateness, and thus denies us the ability to ultimately ground emergence on this concept, even if at higher levels we feel pretty confident that individual objects and events can be disambiguated from one another. Thus individual atoms, molecules, cells, and bodies seem fairly unambiguously distinguishable and can be assigned unambiguous causal properties with respect to each other. And yet, even classic conceptions of emergence have an irreducible temporal aspect, because in most cases there was a time before and after an emergent transition. Many specific emergent features of the world have not always been around. Depending on what emergences we consider—for example, Morowitz's list of cosmic emergence events, or maybe just the origins of life and mind—pretty much all of them are associated with a before and after.

The role of temporal transition is relevant for the analogue of fusion as it applies to the molecules constituting an organism. The existence of these molecules is not explained without at least implicitly describing their origins in a prior phase of the process. The majority of biomolecules exhibit process-dependent properties in the sense that they are reciprocally producers and products, means and ends, in a network of synthetic pathways. This means that although there appears to be a supervenience relationship between the molecules and certain organism-level traits (e.g., collagen molecules contributing elasticity to skin), this appearance is misleading. Supervenience is defined by the ontological priority of the components compared to the higher-level features they generate. But in this case, this hierarchic ontological dependency is tangled in what Douglas Hofstadter has called "strange loops."[26] Over the course of evolution, it is likely that there were modifications to essentially every molecular component found in a living organism. These modifications were produced by changes in the synthetic processes, which in turn were produced by changes in the molecules involved, which in their turn were produced by changes in the synthetic processes, and so on. Even though biomolecules are synthesized anew in each organism, the synthetic processes that produced them owe their existence to being the object of natural selection in long-past epochs. Processes dependent on processes dependent on processes.

Bickhard's and O'Connor's defenses of a process conception of emergence only clear away assumptions that have been blocking progress. To accept this critique is to be enjoined to provide an alternative to these more classic conceptions of emergence that nonetheless achieves what they did not. So we must begin to consider how one would construct a theory of emergent processes that articulates the dynamical alternatives to the classic defining properties of mereological organization, supervenience, and causal inflection. These criteria define what the emergence of a new level of process entails. It specifically requires developing new criteria for defining hierarchically distinct *dynamic regimes* and their relationship to lower-level dynamic regimes. Although the relationship is in this sense loosely analogous to a supervenience relationship, in that higher-level dynamical regimes require lower-level dynamical regimes, it cannot be understood in terms of simple part/whole compositionality. Nor is the existence of lower-level properties explainable irrespective of higher-level properties. Higher-order processes are more accurately "nested" within lower-level dynamics as distinctive special cases. Thus, for example, the circular flow of a whirlpool is a special case of laminar flow of liquid and occurs as a local disturbance of this general flow pattern.

What must be explained, then, is how the organization of a process can be a locus of a distinctive mode of causality that is independent of the patterns of causal interaction among individual components of that process. Beneath this question lies another that is even more basic: Isn't organization merely a *description* of some relationships between components? Indeed, isn't it inevitably a simplification and an abstraction? We must still ask whether there is something about the organization of a process that is causally relevant over and above lower-level processes and their causal influences. If organization is not an intrinsic property, but rather merely a comparative assessment or idealization, it is rendered physically epiphenomenal.

This is relevant to the adequacy of reductionism for analyzing processes whose components are ultimately processes as well. Classic conceptions of emergence were specifically formulated as critiques of reductionism and eliminativism, and in particular they contested the priority of the very small. But framing emergence in dynamical terms does *not* necessarily favor either bottom-up or top-down priority of causal influence. As in the case of the interdependence of organization and compositionality of the biomolecules that constitute an organism, reductionistic analysis is both informative and yet only fully explanatory in the context of an evolved organism and its relevant environments. Reductive analysis that assumes this larger systemic context as a kind of unanalyzed boundary condition is in fact the typical case in the biological sciences. This is not simple reductionism, since the emergent features are already assumed.

In conclusion, when it comes to the two classes of phenomena that we would be most willing to call emergent in a robust sense—life and mentality—the dependency between the dynamical regularities that bring them into being at different levels of scale is neither supervenient nor mereological. The nested and interdependent forms of process organization that characterize these ententional phenomena are not compositional in any clear sense, and do not yield to simple decompositional analysis. Processes are not decomposable to other simpler processes. Or, to put this in simpler terms: processes don't have other processes as their parts. This may explain the failure of classic emergence theories to provide a coherent account of the phenomena they were intended to explain. Although the exploration of non-linear and far-from-equilibrium dynamical systems has recently provided the impetus to reframe emergence concepts in process terms, it has not provided an account of the ententional properties that originally inspired the development of emergence theories.

This is seen by many as their strength, since the elimination of ententional assumptions is often thought to be a fundamental methodological imperative in the natural sciences. If this is in fact a complete and comprehensive paradigm within which emergent phenomena can be explained, then indeed the classic project will have failed. For although it is possible to explain much of the complexity of spontaneous form production and hierarchical discontinuities of dynamical levels within natural and living systems in this way, their ententional properties must be treated as mere descriptive heuristics.

There is, however, another way to view the contribution of these dynamical systems insights. The shift to a dynamical perspective provides a way to explain the spontaneous generation of organization, despite the relentless dissolution of orderliness that is implicit in the second law of thermodynamics. Although dynamical order is not in itself ententional in any sense, all ententional phenomena are defined with respect to order, and so depend on the presence of organized processes for their existence. Perhaps the emergence of dynamical regularity from a context of far-from-equilibrium thermodynamics can serve both as a stepping stone contributing to the emergence of ententional processes and as a model of the logic of emergent transitions in general.

6

CONSTRAINT

... while, in the past, biologists have tended to
think of organization as something extra, something
added to the elementary variables, the modern
theory, based on the logic of communication,
regards organization as a restriction or constraint.

—W. ROSS ASHBY, 1962[1]

HABITS

What emerges? New laws of physics? New forms of matter? Clearly, even if there is considerable diversity of opinion about what should be considered emergent, most if not all of the candidate examples of emergent transitions involve some radical and all-but-discontinuous reorganization of physical processes, but without a departure from what we might describe as physics-as-usual. Consider the two transitions that will most occupy us in the remainder of this book: the transitions from non-life to life and from insentient mechanism to mind. Each involves an unprecedented alteration in the pattern of the way things change—one that inverts some previous nearly exceptionless pattern of causal influences. Of course, a causal pattern isn't any specific individual change. It is something like a general tendency that is exhibited by a vast many events. Though it needn't be some new "law" of physics, or even a violation of existing physical laws, in these cases at least there is a consistent and predictable expression of what would be exceedingly unlikely prior to this transition. As in the case of the second law of thermodynamics, a tendency to change in a distinctive asymmetric way can be nearly ubiquitous, and yet at the same time not be universal. As a first pass at a general definition, then, we can say an emergent transition is a change in conditions that results in a fundamental shift in the global symmetry of some general causal tendency, despite involving the same matter, energy, and physical laws. But what sort of thing is a global causal tendency that is not a law?

The American philosopher Charles Sanders Peirce, writing at the end of the nineteenth century, argued that, at base, change is spontaneous and singular, and thus intrinsically uncorrelated: a metaphysical assumption he termed *tychism*.[2] And yet, he also argued that there is a tendency of things to fall into habits. *Habit* was Peirce's general term referring to regularities of behavior arising in both physical and organic contexts. The term could thus apply equally to the tendency for hurricanes to circle counterclockwise in the northern hemisphere and the yearly migration of Monarch butterflies between the Yucatán and New England. Peirce did not offer an explicit account of why otherwise uncorrelated events might ever tend to become correlated with one another, and thus habitual, but he postulated the existence of a universal habit of habit formation, considering it to be fundamental to the nature of things. In an extreme framing of this principle, he even suggested that what we think of as fundamental physical laws might themselves have evolved from less regular, less habitual tendencies to change. He characterized this most fundamental property of nature with the phrase, "habits tend to beget habits." One way to characterize the point of this chapter is that it is an attempt to give this notion a more precise physical characterization.

This is a somewhat archaic use of the term *habit*, as compared to our modern colloquial use, which typically characterizes repetitive human or animal behaviors. But even this connotation is relevant, since we often consider something to be a habit even though it only periodically exhibits a precisely repetitive characteristic. Thus a person's smoking habit or a habit of clearing one's throat when talking may only periodically, and somewhat irregularly, be expressed. A "habit" in this sense is a behavioral bias or predisposition that may or may not be overtly expressed, but is always a tendency. This is an important caveat, because it will allow us to consider a much more general class of only quasi-regular phenomena that don't easily correspond to notions of order, pattern, or organization. Peirce, too, had a statistical notion of habit in mind. For example, he metaphorically described the invariant regularities of inorganic matter as processes that are "hidebound with habit" in comparison to organic matter.

Peirce's concept of habit was the foundation for his somewhat icono- clastic defense of metaphysical realism. Realism is a classic metaphysi- cal viewpoint which argues that ideal forms, regular tendencies, types of things, and general properties of phenomena—not just specific singular events, objects, or effects—are important in determining what happens in the world. Realism is generally traced to the philosophy of Plato, who famously argued that the material world only imperfectly reflects the influ- ence of the ideal universal forms. These forms were at once both perfect and unchanging, while their physical manifestations were inevitably flawed and impermanent. Though Peirce's version of realism did not postulate the existence of a realm of ideal forms, he nevertheless argued that physical regularities of a general type—habits—can influence the creation of other physical regularities of the same or different type, despite being embodied in diverse substrates. He argued that regularity itself, rather than any specific materially regular feature, is what is causally most relevant.

The problem of the physical influence of general types dates back to the late Middle Ages and the question of whether general terms referring to types of things rather than individual things (sometimes referred to as "tokens" of a type) actually refer to anything "real," since no two examples of things of the same type are truly identical. The issue was whether types could have any real influence in the world except as a consequence of the specific features of individual exemplars of the type. If being a type of thing only mattered to an observer interested in clustering phenomenal details into classes or collections for cognitive or communicative convenience, then being of that type would effectively be epiphenomenal, and of no physical consequence. But more to the point, can we really assume that the laws of physics and chemistry are any more than idealized descriptions, since they apply across individual instances? And what about properties of things like hardness or opacity?

Ironically, realism claimed that general properties, laws, and physical dispositions to change in such-and-such a way are fundamental facts about reality irrespective of anyone's having considered them, whereas the view that these generalizations are merely conveniences of thought, abstracted from observation and otherwise epiphenomenal in the world of physical cause and effects, was called *nominalism*.

Nominalism arose as a critique of realism. In its most extreme form, nominalism claims that general types of physical properties, like being hard, liquid, reflective, living, or general dispositions of behavior, are all ultimately epiphenomenal. Accordingly, they contribute nothing to explain causal properties and merely provide a way of mentally clustering individually diverse physical interactions. They are according to this view mere projections of our descriptive simplifications. However, each utterance of such a term, that is, the name of a type, is itself an instance of a discrete physical phenomenon capable of having a discrete effect on the world, even if its general content corresponds to no real thing. This is the source of the term describing this view, where *nom-* refers to "name." In effect, then, only the name of the general type is a real part of the world. Peirce characterized the core issue in terms of "whether *laws* and general *types* are figments of the mind or are real." A general type, in Peirce's metaphysics, is not merely an aspect of mind, but is "the most important element of being." It is the ultimate locus of physical causality so discovering these general features should be "the end and aim of knowledge."[3]

The concept of emergence assumes that new general types of properties and physical dispositions—new causal habits of nature—can arise *de novo*. Hence, this classic unresolved debate is fundamental to questions of emergence.

Realism and nominalism were serious issues for Scholastic thinkers struggling to make sense of theological problems, because ideals and norms, such as characterize the true, the good, and the beautiful, were not specific instances but general properties. The debate is nearly as old as Western philosophy. The existence of Plato's ideal forms and properties and their presumed physical influence was challenged even by Aristotle, who argued that there is a necessary link between matter and its formal properties. While not denying the efficacy of formal influences (implicit in his notions of formal cause), Aristotle nevertheless argued that the forms, regular tendencies, and general properties were always embodied materially and specifically. Even the end-directed tendencies (final causes) of living organisms were literally embodied in their material constitution. His concept of an *entelechy* was for him inseparable from the material of that body. This compromise resisted serious challenges until the late medieval period.

The beginning of a serious challenge to realism can probably be traced to William of Occam. Occam's interest in developing principles of precise logic led him to argue that although different material objects exhibit characteristic forms and properties, the similarities among classes of these are solely the result of how the human mind ("soul" in his terminology) applies the same mental sign (i.e., distinct concept) to aspects of these objects, due to limitations in mental capacity. Like the words themselves, he argues, these mental signs are distinct and individual, but the general aspects and properties of things to which they refer do not exist.

Basically, Occam considered it mistaken to assume that universal properties exist independent of specific individual instances. Mental signs are themselves singular individual entities, and every instance to which these are applied is a singular and distinct individual case. There is no general feature that can be understood without being singularly embodied. There are only things to which these signs apply. So, while mental signs apply plurally, there is no unambiguous general property in the world that corresponds to a given mental sign. Most scholars point out that Occam did not deny the possibility that there might be actual universal properties in the world, but nevertheless felt that the existence of words referring to them did not guarantee their existence. The words may only reflect the finiteness and limitations of the mind. Although this does not directly follow from his famous exhortation to eliminate redundant hypotheses in favor of the simplest or fewest—memorialized in what came to be called Occam's razor—it is consistent with it.

The reductionistic assumptions characteristic of modern science exemplify an analogous denial of any simple notion of general realism by treating all properties of things as reducible to properties of their parts and their interactions. Curiously, however, modern science remains agnostic about this assumption in other respects. For example, the acceptance of general physical laws as determining the way things can change and physically affect each other can be taken to imply that at least these general laws have some kind of real efficacy, independent of the physical events that exhibit them.

It would take us too far afield to review and critique the many subtleties of argument that have been proposed on both sides of this long and complex philosophical debate, and yet it should be obvious that some form of realism must hold for an emergentist view of the physical world to be accurate. For an emergence theory to offer anything more than a descriptive reframing of events, it must explain how general regularities and dispositions of causal interaction can be responsible for which particular causal interactions occur and which don't, over and above any particular physical antecedent state of things. This is an issue that ultimately cannot be ignored. As we've seen above, the force of the critique of the efficacy of general types derives from a particular set of Platonic assumptions about what constitutes something general, and so this apparent dilemma may dissolve if these assumptions can be dispensed with.

REDUNDANCY

To reexamine these assumptions of realism, let's consider what is implied by Peirce's concept of habit. Habits and patterns are an expression of redundancy. Redundancy is a defining attribute of dynamic organization, such as the spiraling of galaxies, the symmetries defining geometric forms such as polygons, and the global regularities of physical processes such as are expressed by the second law of thermodynamics. Each is a form of organization in which some process or object is characterized by the repetition of similar attributes. But this characterization of organization in terms of repetition or similarity brings up another long-standing philosophical debate. Consider an individual whirlpool in a stream. What distinguishes its individuality and continuity?

Obviously, the individuality of the whirlpool is not determined by its material constitution. The whirlpool is composed of different water molecules, moment to moment, and none remain more than a few seconds. What about the pattern of the movement of water molecules through the whirlpool? An individual whirlpool is easily distinguished from the surrounding flowing water because of its roughly circular symmetry and the general movement of water around a center of rotation. And it is the persistence of both this circular symmetry and the locus of the rotational center that we implicitly refer to when describing this flow pattern as the same whirlpool moment to moment. But of course, on closer inspection, an actual whirlpool is slightly irregular, and this irregularity of shape is constantly changing. Moreover, if we were to explore the flow of water through an individual whirlpool (e.g., by injecting a stream of dye just upstream or tracking the movement of individual molecules within the flow), we would probably discover that some of the water even passes more or less directly through the whirlpool and does not follow a circular or spiraling path at all. And if we were able to examine it in even more precise submicroscopic molecular detail, it is likely that we would find that no individual water molecule ever actually follows a precise, circularly symmetric path. Each will be differently deviant in its trajectory from this ideal pattern.

Is there anything that is constant—anything that we can use to define something as the same whirlpool moment to moment, other than a general tendency that we notice only because we ignore these many details? Indeed, it might be argued that defining a whirlpool as something stable and persistent is merely an artifact of having failed to attend to these many differences. There really is nothing quite the same, and nothing exactly regular in the whirlpool, from one moment to the next. So, is the regularity on which we base its individuality merely a simplification due to descriptive abstraction?

This example is of course generalizable. No two instances of a general category of dynamical organization—whether two whirlpools in water, two members of the same species, or even "identical" twins—are ever truly identical. Furthermore, similarities of an even more general sort, such as characterize the commonalities between whirlpools in water and in spiral galaxies, may only have very abstract characteristics in common. In all cases, similarity is assessed by ignoring differences at some level of detail, and in some cases by ignoring almost all differences, including vast differences of scale (as in the spiraling of water and of stars). So describing both of these dynamical objects as exemplars of the same general type of phenomenon is an account that only makes sense at a very high level of abstraction. Does this mean that there is nothing in common about the causes of these similar-appearing processes? Is the rough circular symmetry of these two processes only a commonality in the mind of an observer, but otherwise of no causal consequence? Does the effect of the symmetry of global angular momentum in each have any independent causal relevance over and above the molecular collisions and cohesion in the one and the center of gravitational attraction in the other? Or is this apparent commonality also merely a descriptive feature imposed by external analysis and descriptive simplification? In other words, shouldn't we stop thinking of the spiraling as being "out there" when it is just a case of observational simplification? Isn't this just a result of what we ignore?

Of course, a description is not a physical attribute of the thing being described. Comparison to some ideal type of geometric object, such as a regular spiral, is merely something that goes on in the mind. If these similarities are only descriptive glosses, then indeed invoking them as causes of anything is physically irrelevant because it is redundant. What matters are all the many individual physical interactions, not the vague similarities we notice. Analogously, then, if emergence is defined in terms of descriptive abstractions alone—regularities and patterns that are merely projections of our ideal types, mental categories, and formal models—it would seem as though we are identifying emergence in merely descriptive (and thus epistemological) terms, but not causal (ontological) terms. Emergence in this sense would only exist in the mind of the beholder.

When we attribute some general property or regularity to some natural phenomenon, like spiral motion, we have certainly made an assessment based on our experience. The regularities that we focus on to distinguish different types of things are indeed constructions of our perceptual and cognitive apparatus. Our abilities are only sensitive to very few of its physical characteristics, and usually we are only attentive to a fraction of the details that we can discern. Does this necessarily imply that these similarities and regularities are somehow mere figments of mind? Can they be mental abstractions *and* physically relevant causal features of the world at the same time? One reason to think that mental abstraction does not imply physical epiphenomenalism is that mental processes are themselves also dependent on physical processes. To even perceive regularity or pattern this act of observation must itself be grounded in a *habit of mind*, so to speak. In other words, to attribute physical regularity to some perceived or measured phenomenon presumes a prior mental regularity or habit with respect to which the physical regularity is assessed. So, unless we grant mentality some other undefined and mysterious means of observational abstraction, claiming that regularities, similarities, and general types are only in the mind but not in the world simply passes the buck, so to speak. Some general tendency must already exist in order to attribute a general tendency to something else.

MORE SIMILAR = LESS DIFFERENT

Even if we grant that general tendencies of mind must already exist in order to posit the existence of general tendencies outside the mind, we still haven't made any progress toward escaping this conceptual cul-de-sac. This is because comparison and abstraction are not physical processes. To make physical sense of ententional phenomena, we must shift our focus from what is similar or regularly present to focus on those attributes that are *not* expressed and those states that are *not* realized. This may at first seem both unnecessary and a mere semantic trick. In fact, despite the considerable clumsiness of this way of thinking about dynamical organization, it will turn out to allow us to dispense with the problem of comparison and similarity, and will help us articulate a physical analogue to the concept of mental abstraction.

The general logic is as follows: If not all possible states are realized, variety in the ways things can differ is reduced. Difference is the opposite of similarity. So, for a finite constellation of events or objects, any reduction of difference is an increase in similarity. Similarity understood in this negative sense—as simply fewer total differences—can be defined irrespective of any form or model and *without even specifying which differences are reduced*. A comparison of specifically delineated differences is not necessary, only the fact of some reduction. It is in this respect merely a quantitative rather than a qualitative determination of similarity, and consequently it lacks the formal and aesthetic aspects of our everyday conception of similarity.

To illustrate, consider this list of negative attributes of two distinct objects: neither fits through the hole in a doughnut; neither floats on water; neither dissolves in water; neither moves itself spontaneously; neither lets light pass through it; neither melts ice when placed in contact with it; neither can be penetrated by a toothpick; and neither makes an impression when placed on a wet clay surface. Now, ask yourself, could a child throw both? Most likely. They don't have to exhibit these causal incapacities for the same reasons, but because of what they don't do, there are also things that both can likely do or can have done to them.

Does assessing each of these differences involve observation? Are these just ways of assessing similarity? In the trivial example above, notice that each negative attribute could be the result of a distinct individual physical interaction. Each consequence would thus be something that fails to occur in that physical interaction. This means that a machine could be devised in which each of these causal interactions was applied to randomly selected objects. The objects that fail all tests could then get sorted into a container. The highly probable result is that any of these objects could be thrown by a child. No observer is necessary to create this collection of objects of "throwable" type. And having the general property of throwability would only be one of an innumerable number of properties these objects would share in common. All would be determined by what *didn't* happen in this selection process.

As this example demonstrates, being of a similar general type need not be a property imposed by extrinsic observation, description, or comparison to some ideal model or exemplar. It can simply be the result of what doesn't result from individual physical interactions. And yet what doesn't occur can be highly predictive of what can occur. An observational abstraction isn't necessary to discern that all these objects possess this same property of throwability, because this commonality does not require that these objects have been assessed by any positive attributes. Only what didn't occur. The collection of throwable objects is just everything that is left over. They need have nothing else in common than that they were not eliminated. Their physical differences didn't make a difference in these interactions.

Had we started this thought experiment with a large collection of randomly chosen objects and subjected them to this series of interactions, the resulting objects would have been a small subset of the original collection, and the differences in properties among the objects in that subset would have been only a fraction of the differences exhibited by the initial collection. To say this in other terms: the variety of properties exhibited was significantly *constrained* in the resultant set of objects, compared to this initial variety. This concept of constraint can provide a negative approach to realism,[4] though it might also be paradoxically described as a *nominalism of absences*, since it is determined by discrete interaction effects that don't occur.

Along these lines, the introductory quote from W. Ross Ashby—a pioneering figure in the history of cybernetics—offers a negative way to define concepts like order, organization, and habit. The concept of *constraint* is, in effect, a complementary concept to order, habit, and organization, because it determines a similarity class by exclusion. Paying attention to the critical role played by constraints in the determination of causal processes offers us a figure/background reversal that will turn out to be critical to addressing some of the more problematic issues standing in the way of developing a scientific theory of emergence. In this way, we avoid assuming that abstract properties have physical potency, and yet do not altogether abandon the notion that certain general properties can produce other general properties as causal consequences. This is because the concept of constraint does not treat organization as though it is something added to a process or to an ensemble of elements. It is not something over and above these constituents and their relationships to one another. And yet it neither demotes organization to mere descriptive status nor does it confuse organization with the specifics of the components and their particular singular relationships to one another. Constraints are what is not there but could have been, irrespective of whether this is registered by any act of observation.

In statistical mechanics, *constraint* is a technical term used to describe some reduction of degrees of freedom of change or a restriction on the variation of properties that is less than what is possible. In colloquial terms, constraint tends to be understood as an external limitation, reflecting some extrinsically imposed factor that reduces possibilities or options. So, for example, railcars are constrained in their movement by the location of tracks, farmers are constrained in what can be grown by the local climate, and citizens are constrained in their behaviors by laws. In each case, there are reduced degrees of freedom that could potentially be realized in the absence of these constraints.

In a 1968 paper entitled "Life's Irreducible Structure," the philosopher-scientist Michael Polanyi argued that the crucial difference between life and chemistry is the constraint that DNA imposes on the range of chemical reactions that tend to occur in an organism. Genetic information introduces these constraints by virtue of the ways that it generates molecules which either act as catalysts to facilitate certain chemical reactions or serve as structural elements which minimize certain other molecular interactions. Polanyi argues that this formal influence is what separates life from non-living chemistry, and distinguishes both man-made machines and living organisms from other physical systems. For this reason, he argues that this formal aspect cannot be reduced to chemistry. This is an important insight, but it falls short of resolving the issue of life's ententional features in one important respect: it treats the source of constraint as extrinsic. Indeed, this is critical to his argument for irreducibility; but as the analogy to machines indicates, treating this influence as external to the chemistry passes the explanatory buck to DNA as an instruction manual and ultimately to natural selection as its author. If this constraint is the defining feature of life, but must be created and imposed extrinsically, then life's ententional features are inherited from an extrinsic homunculus, which in this case doesn't exist.

Constraints can also be intrinsic. The etymology of the term—which derives from the Latin, and roughly means "that which binds together"—does not imply anything about the cause of that restriction or reduction of variety; for example, whether it is a property that is intrinsic or extrinsic to whatever it is that is thereby constrained. The term *constraint* thus denotes the property of being restricted or being less variable than possible, all other things being equal, and irrespective of why it is restricted. This latter usage has become common in statistical and engineering applications, such as when describing the range of values for some measured feature of a collection of objects, a set of observations, or a sample of informant answers. Used in an intrinsic sense, we might say that the errors in a set of measurements fall within certain constraints or tolerances. Or we might comment that the limitations of a given measuring technique impose constraints on the accuracy of the results. In this sense, constraint is a property of a collection or ensemble of some sort, but a negative property. It is a way of referring to what is not exhibited, but could have been, at least under some circumstances. As we will see, distinguishing between an imposed restriction and a restriction that develops because of intrinsic properties and processes is critically important to the project of making sense of ententional properties, and avoiding any implicit realist assumptions that compromise the concepts of order and organization.

As critics of emergence have long argued, we tend to categorize different forms of pattern and organization in terms of possessing abstract formal properties. Since such properties are not intrinsic to the collection of interacting components involved but are instead based on comparisons to some abstract ideal forms, they are indeed causally epiphenomenal. We can now counter that general formal properties do exist independent of external observed comparison, but they are not positive attributes in this traditional sense. They are, so to speak, correlates of the physical simplification that follows from constraint. Thinking in terms of constraint in this way changes how we must understand the concepts of order, organization, and pattern—and thus emergence.

Organization is a general property that can be more or less specifically described, but it is typically defined with respect to some model. When we describe the organization of stars in a galaxy, it is with respect to being globular or spiral in shape, and when we characterize a human organization, such as the United Nations, it is with respect to a charter and purpose. Similarly, describing something as exhibiting order typically appeals to some abstract ideal, such as arranging a deck of playing cards according to numerical values and suits. Even just saying that events follow a pattern implies that we observe regularly repeated similar attributes. Modern science and philosophy have long rejected the Platonic notion that the objects and events in the world are merely flawed instantiations of ideal forms, in favor of the recognition that form and organization refer to idealizations abstracted from observation. Thus it is accurate to assume that scientific descriptions framed in terms of order and form are to be understood heuristically. This is even often applied to that ubiquitous asymmetry of change, the second law of thermodynamics, which characterizes the predictable pattern of change toward increasing disorder. When we describe an organized room as "ordered" and a messy room as "disordered," we are in effect privileging one arrangement of objects over another, and characterizing the tendency to spontaneously fall into messiness as a loss of something. Statistical mechanics simply recognizes that there are a great many more arrangements considered messy, according to human preferences, than there are orderly ones, and that random change tends to inevitably produce the more probable results. So describing a high-entropy state as disordered implicitly suggests that we cannot recognize any pattern.

By abandoning descriptive notions of form, however, and thinking of regularity and organization in terms of possible features being excluded—real possibilities not actualized—two things are gained. First, individuation: possible features *not* expressed and degrees of freedom *not* exhibited are specific individual facts. They are precisely measurable in space and time, in just the same way as are specific trajectories and velocities of moving atoms. Second, extension: constraints can have definite extension across both space and time, and across whole ensembles of elements and interactions. But although the specific absences that constitute a constraint do not suffer the epiphenomenality of descriptive notions of organization, they are nevertheless explicitly not anything that is present. This requires that we show how what is absent is responsible for the causal power of organization and the asymmetric dynamics of a physical or living process. Beginning to explain how this might be possible is the major task of this chapter. But a full explanation (undertaken in subsequent chapters) will require that we eventually reconceptualize even something as basic as physical work, not to mention concepts like self-organization and evolution—concepts that are typically defined in terms of explicit model relationships.

There is one other advantage of this approach. Employing the concept of constraint instead of the concept of organization (as Ashby proposes in the epigraph) not only avoids observer-dependent criteria for distinguishing patterns, it also undermines value-laden notions of what is orderly and not. As in the case of the messiness of a room, order is commonly defined relative to the expectations and aesthetics of an observer. In contrast, constraint can be objectively and unambiguously assessed. That said, order and constraint are intrinsically related concepts. Irrespective of specific observer preferences, something will tend to be assessed as being more orderly if it reflects more constraint. We tend to describe things as more ordered if they are more predictable, more symmetric, more correlated, and thus more redundant in some features. To the extent that constraint is reduced variety, there will be more redundancy in attributes. This is the case with respect to any change: when some process is more constrained in some finite variety of values of its parameters or in the number of dimensions in which it can vary, its configurations, states, and paths of change will more often be "near" previous ones in the space of possibilities, even if there is never exact repetition.

The advantage of this negative way of assessing order is that it does not imply any model-based conception of order, regularity, or predictability. It is only something less than varying without limit. But this assessment encompasses a wide range of phenomena between those that we normally describe as orderly and those that we describe as entirely chaotic. As we saw earlier, chaos theory provides an important context for demonstrating the usefulness of this figure/background shift in the analysis of order and organization. Calling a pattern "chaotic" in this theoretical context does not necessarily mean that there is a complete absence of predictability, but rather that there is very little redundancy with which to simplify the description.

This way of characterizing disorder is exemplified by an information-theoretic means of measuring complexity termed *Kolmogorov complexity*, after the theoretician who first promoted its use, the Russian mathematician Andrey Nikolaevich Kolmogorov (1903–1987). It can most easily be understood in terms of a method for analyzing or generating a string of numbers. If the same string can be generated by an algorithm that is shorter than that string, it is said to be compressible to that extent. Such an algorithm effectively captures a form of redundancy that is not superficially exemplified in its product. So, for example, even an apparently non-repeating sequence, such as the decimal digits of π, can be generated by an algorithm that progressively calculates an approximation of the relationship between the diameter and circumference of a circle. This is how a short and simple computer algorithm can generate the numbers of this infinite sequence to any length desired. Similarly, a superficially simple non-linear equation[5] may produce an irregular non-repeating trajectory when plotted on a graph. This is the case with the Lorenz attractor graph, discussed in the last chapter, and yet iterating successive values of the equation shows that the non-repetition of values nevertheless is considerably less variable than fully chaotic. But if a sequence of values (such as in a string of numbers or coordinates of a trajectory) cannot be generated by an algorithm that is shorter in length than the sequence itself, that sequence is described as maximally complex. Since there is no algorithm able to extract some regularity from such a sequence, it is in this sense also maximally disordered.

Calling this theoretical field "chaos theory" was in this sense somewhat ironic, since the focus is not on maximal disorder, but on what order could be discovered amidst the apparent disorder. As a result, the field is more commonly and more appropriately described as "complexity theory." But note that the term *complexity* is given contrasting connotations here. The word "complex" etymologically derives from roots that together mean something like woven, braided, or plaited together, and not merely involving many details. So what may appear chaotic may be merely the result of a simple operation that tends to entwine with itself to the point that any regularity is obscured. Alternatively, being impossible to simplify means that there are only details to contend with, nothing simpler. Irrespective of whether all real phenomena are maximally algorithmically complex—as extreme nominalism would suggest—or are algorithmically compressible without residue—as extreme realism would suggest—a constraint view of orderliness still applies. It is tantamount to allowing sloppy compression. For example, if all we require of the prediction algorithm is that it specifies a numerical sequence in which each number is within ±2 of the predicted value for some specified finite sequence length, then many more sequences will be approximately describable by the same algorithm. In some sense, this is analogous to the way we psychologically identify similarity, and thus characterize order.

Thinking of Peirce's notion of habit as the expression of constraint is implicit in his doctrine of *tychism*, which he considers to be the most primitive feature of his ontology. This is critical to his conception of "law," which, although it can be framed in Platonic terms, was for Peirce merely a habit of nature that "evolved" in some sense. From a Peircean point of view, a natural law should be described as an invariant tendency that requires us to explain why it is not tychistic. By translating Peirce's notion of habit into the terms of constraint, his unique reframing of metaphysical realism can be made more precise. Recasting the Realism/Nominalism debate in terms of dynamics and constraints eliminates the need to refer to both abstract generals, like organization, and simple particular objects or events lacking in organization. Both are simplifications due to our representation of things, not things in themselves. What exist are processes of change, constraints exhibited by those processes, and the statistical smoothing and the attractors (dynamical regularities that form due to self-organizing processes) that embody the options left by these constraints.

CONCRETE ABSTRACTION

To return to the conundrum that began this exploration: Are constraints in the head or also in the world? Do they exist irrespective of observation, and if so, how do they make a difference in the world? To answer this, we need to know if causal interactions in the physical world also, in effect, tend to "ignore" certain details during physical interactions. If so, this negative characterization of regularity and habit would be both a more accurate representation of the underlying physics and ultimately the relevant causal factor. Recall that the nominalist critique of Platonic realism argues that the regularities and similarities that we discern in the world are merely descriptive simplifications—abstractions away from the details—and thus causally irrelevant. However, if certain physical interactions also tend to drop out selective details, then there can also be a purely physical analogue to the abstraction process. Constraint is precisely this: the elimination of certain features that could have been present.

But can the concept of constraint account for something like concrete abstraction: a physical analogue to mental abstraction? One important clue is that *what something doesn't exhibit, it can't impose on something else via interaction.* In other words, where there is no difference, it can cause no difference. To the extent that differences in one system can produce differences in the other, those variations not expressed in one will not be transferred to the other during their interaction. Although this may initially sound like there is no causality involved, it's not that simple. As we will see later (in chapter 11), this is because the presence of constraint—the absence of certain potential states—is a critical factor in the capacity to perform work. Thus, it is only because of a restriction or constraint imposed on the release of energy (e.g., the one-directional expansion of an exploding gas in a cylinder) that a change of state can be imposed by one system on another. It is precisely by virtue of what is not enabled, but could otherwise have occurred, that a change can be forced.

So the nature of the constraint (and therefore the absent options) indirectly determines which differences can and cannot make a difference in any interaction. This has two complementary consequences. Whenever existing variations are suppressed or otherwise prevented from making a difference in any interaction, they cannot be a source of causal influence; but whenever new constraints are generated, a specific capacity to do work is also generated.

In massively componential processes involving linearly interacting components, such as in a simple near-equilibrium thermodynamic system, low-probability fluctuations will tend to be statistically "washed out" with increase in scale. Microscopic fluctuations will have minimal macroscopic influence. But this means that macroscopic interactions between such systems will largely be insensitive to these microscopic fluctuations, and the system's capacity to do work that alters another system will be determined by this constraint. Alternatively, in massively componential dynamical systems where there is the possibility of extensive non-linear interactions, such as in systems that are persistently maintained far from equilibrium, constraints can sometimes amplify to become macroscopically expressed. In these cases, the emergence of these new constraints at the macroscopic level can be a source for new forms of work. We will focus on the first of these dynamical conditions—near-equilibrium thermodynamic processes—in the remainder of this and the next chapter, and will take up the second of these conditions—persistent far-from-equilibrium processes—in chapter 8.

Ascent in scale plays a critical role, because it tends to reduce the contributions of micro fluctuations at the macroscopic level by virtue of the way that the distribution of fluctuations tends to even out. Time-scale differences also play a role. Component micro interactions tend to occur at rates that are many orders of magnitude more rapid than interactions between systems. Perturbed micro components will return to stable near-equilibrium conditions at rates that are orders of magnitude more rapid than the time involved in macro interactions. As a result, the highest-probability micro conditions will tend to correlate with the most redundant micro-interaction effects. To say this in different terms: the detailed micro fluctuations of the processes underlying higher-order system properties mostly cancel each other out over modest distances, so that only the most generalized attractor features end up having causal influence at the higher level. It is as though the details of what is happening at the lower level are filtered out, only allowing the most redundant features to be manifest at the higher level.

This is a property related to Humphreys' concept of fusion—a loss of properties with ascent of scale—but is much more general, and less exotic. A decrease in the diversity of variations able to exert influence on systemic variation with ascent in scale means that some lower-level details are lost. There is nothing surprising about this. Consider again one of the first attempts to describe emergence: Mill's account of chemical properties distinguishing chemical compounds and those of their component elements (his example of sodium, chlorine, and their ionically bonded compound, table salt). In the compound, the properties of the individual atoms are masked as each is in some sense modified by the bond with the other (in simplified terms, due to transfer of an electron). As a result, elemental properties are effectively suppressed (reduced to extremely low probability), whereas the properties associated with their stable (i.e., highly constrained) interactions are vastly more likely to be expressed in any interactions with other molecules.

This feature of physical hierarchies was highlighted in a 1972 *Science* review article by the Nobel Prize–winning physicist Philip Anderson, appropriately titled "More Is Different," and has been reiterated in various forms by a number of theorists since. The basic argument is that material properties (especially those exhibited in solid-state forms of matter) inevitably exhibit discretely segregated levels of organization, with vast differences in mean levels of energy of interaction and intervals between interaction effects, correlated with vast differences in spatial dimensions. This is particularly striking for solid states of matter, because the atomic-scale interactions have become highly localized in their dynamics compared to atomic-scale interactions in other phases (e.g., liquid or gas). The bonds between atoms are, however, in constant flux, often flip-flopping between alternative orientations in proportion to their relative probabilities, but at rates vastly more rapid than these effects could possibly propagate to neighboring regions of the solid. This allows statistical scale effects to produce a kind of mutual insulation of the dynamical interaction possibilities between adjacent levels. For example, the specific heat of a solid is a reflection of the level of constant jostling of component molecules; but these constant micro fluctuations cycle through millions of states in the time that a solid-to-solid or solid-to-liquid interaction takes place, and so are almost entirely causally irrelevant. So, although we tend to think of solids as made up of unchanging components and liquids as uniform continua, this is an illusion of statistical smoothing.

Following in this tradition, a focus on what gets lost with ascent in scale in physical systems has become a key notion in the Nobel Prize–winning physicist Robert Laughlin's conception of emergence. We can employ a rough caricature of Laughlin's concept of *protected states* to describe this sort of insulation between levels of physical dynamics.[6] Thus the reason that the atoms or molecules in a solid, liquid, or gas can largely be treated as simple billiard ball–like objects is that each tends to be in a dynamically regular stable ground state (i.e., constrained to an attractor) with a very high probability. And it will immediately return to this state shortly after being perturbed, at a rate that is effectively instantaneous compared to the interaction rates at the intermolecular scale. The electrons comprising these molecules and atoms are continually in quantum dynamical fluctuation within the confines of their orbitals. But the time frame of these fluctuations is orders of magnitude more rapid than the time frame of the exchange of energy in molecular collisions. Thus, in the time it takes colliding molecules to rebound, there has been time for a vast number of intraatomic fluctuations to occur. With respect to the intermolecular scale events, the intraatomic system has had plenty of time for vast epochs of stochastic fluctuation of electron interactions to reach quasi-stable quantum values. It is as though the micro fluctuations weren't occurring.

This is somewhat analogous to the macroscopic thermodynamic properties of systems near equilibrium. Even at equilibrium, there are incessant deviations of local micro-level fluctuations away from mean values of molecular momentum, and so forth. These are reflected in the so-called Brownian motion of dust particles in stable air or pollen grains suspended in liquid. But although the individual particle movements contribute to the transfer of heat between two bodies, these micro deviations have little effect on the exchange of properties at the macro level. Those lower-level fluctuations that are deviant from equilibrium are, in effect, *micro differences that don't make a macro difference*. Voting in a simple two party election provides a crude analogy to this. Within a voting district there may be quite complicated patterns of voting preferences, but after an election it is a winner-take-all situation; and even if the differential is only a fraction of a percent, the result can be that political decisions are applied as though the results were highly skewed. Political negotiation between representatives from different districts can thus appear as though their populations differed radically in their opinions, even though in both the votes were extremely close. In this way, subtleties of difference can be entirely obscured.

Across levels of scale, then, higher-level dynamical properties can exhibit global properties that effectively hide any lower-level fluctuations by statistical smoothing. Thus detailed particle-by-particle dynamics will seldom propagate effects up level, while "average" values of dynamical interactions will be robustly represented. This limits bottom-up notions of causality. It is as though specific individual influences get erased in the causal bookkeeping. It also means that reductionistic analysis cannot recover these differences after the fact. Often, the best that a reductionistic analysis can do—and all that is necessary, in most cases—is to accurately characterize the constraints (e.g., the average values) converged on by the lower level of dynamics.

Translating this into Peircean terms, if not all states of a process are realized, or if there is a bias in the probability of their occurrence, there is a habit. There will necessarily be some degree of redundancy simply by virtue of the fact that changes of state will typically traverse only a fraction of the space of possibilities and extremes will be rare. A habit can thus be described as the expression of a constraint. It is an attractor disposition—whether dynamically or statistically generated—that can arise irrespective of any outside mental assessment.

A constraint is relational—not relative to a representation or mentally conceived type, but rather with respect to other dynamical options. Both the determination of the specific features of a physical disposition (or habit) and the causal efficacy of that general property are realized in relation to some other dispositions of some other phenomenon: habits determined with respect to other habits; constraints realized with respect to other constraints. Peirce's vision of mental categorization and representation was a special case of this general principle. The general properties by which we assess the similarities and differences among the diverse phenomena of our experience (e.g., pattern and order) are a consequence of the interaction between habits of our brains and habits of the world. So, putting this in constraint terminology, we can dispense with the notion that habit, pattern, and organization presuppose mental representation and comparison. The causal efficacy of these redundant features of the world is presupposed in the very notion of a mental category, not the other way around.

But then Peirce must also be right about the physical efficacy of habits. Habits must be real causes in their own right. They are, according to him, the ultimate locus of causality. In the terms used here, habits begetting habits can be translated as *constraint propagation*. In these terms, states that are not realized or that occur only improbably in a given process, can play the critical causal role in the formation of further constraints at higher levels. Although what is never or seldom realized does not play the kind of causal role we usually think of in Newtonian terms, once we shed the assumption of the priority of substance, we can begin to see how to reframe this in terms of the interaction of constraints. As we will see in the next chapter, the spontaneous nature of constraint reduction is the principal source of resistant properties of things. And it is ultimately the production and propagation of constraints that make physical work possible. For example, containing the expanding gases in an internal combustion engine, and thus constraining expansion to occur in only one direction, allows this release of energy to be harnessed to do work on other systems, such as propelling the vehicle which contains the engine up a steep incline. So to argue that constraint is critical to causal explanation does not in any way advocate some mystical notion of causality. We can restate this causal logic as follows: reduction of options for change in one process can lead to even greater reduction of options in another process that in some way depends on the first.

By recasting our understanding of habit and order in negative terms, we can begin to disentangle ourselves from the "something more" fallacy of traditional emergence theories, and give new meaning to Peirce's "law of habits." Emergent properties are not something added, but rather a reflection of something restricted and hidden via ascent in scale due to constraints propagated from lower-level dynamical processes. In terms of the Realist/Nominalist debate, generals determined by constraint relationships are not types or classes, but neither are they individuals. Both concepts positively frame what I have been arguing must be understood negatively.

NOTHING IS IRREDUCIBLE

Thinking in terms of constraint can help resolve one of the core questions raised by emergence theories: Are there emergent phenomena that cannot be fully understood by reductionistic analysis? The concept of emergence is part of a general critique of the completeness of reductionism. This critique has been particularly motivated by the assumption that if reductionism provides a complete and sufficient account of physical causality, it would appear necessary to treat ententional processes and relationships as epiphenomenal heuristics. There can be little doubt that reductionistic science is fundamentally sound. It has provided unparalleled predictive power for explaining physical-chemical processes across unimaginable ranges of scale and diversity of phenomena. It would be pointless to even imagine that it is somehow misguided. Moreover, the most devastating critiques of emergence have identified its fatal flaw in attempting to both retain a mereological and a supervenient conception of levels of physical causation, and yet at the same time claiming that some causal relationships invert this logic. Shifting to an emergence conception framed in terms of constraint on dynamics, however, undermines both of these criteria and thus escapes this critique; and yet it does not lead to anti-reductionism in the traditional sense. All the material and energetic features of a given system are subject to mereological analysis without residue. There is nothing left out. Or rather, "what is not exemplified" is exactly what *is* left out in reductionistic analyses. What is *not* there or not exemplified is not anything that is reducible, because there are no components to what is absent.

If the constraints rather than the properties of parts are what determine the causal power of a given phenomenon, then a reductive analysis that decomposes a complexly constrained phenomenon into its observable parts and relationships can end up dropping out what may be the critical feature. If the constraints that are most prominent are intrinsic and emergent, all trace of these constraints will be "erased" by simple decomposition, unless the history of this higher-order constraint generation is already assumed. In other words, while it may often be possible to model constraint propagation and thus make general predictions about the emergence of higher-order dynamical constraints, it may not be possible to work backwards. Since, in complex dynamical systems, attractors may be converged on from many quite diverse initial conditions, this convergence history is an irreducible factor. While a general attractor form may itself be predictable by simulation, being at some specific locus within the phase space of that attractor dynamic is not highly informative about initial conditions, nor is its specific emergent history of particular relevance to the causal properties that result. This of course is just another way to conceive of the multiple-realizability relationship, but in the negative.

We can summarize the source of irreducibility in such cases in a simple slogan: Absence has no components, and so it can't be reduced or eliminated. Or, to be a bit less cryptic: Constraint is the fact of possible states not being realized, and what is not realized is not subject to componential analysis. Reductive analysis can thus irretrievably throw away information about the basis of higher-order causal power. In cases where constraints at a higher level are linear extrapolations of those at a lower level, there is no loss due to componential analysis. Thus, in these cases (e.g., simple-equilibrium thermodynamics), there is no loss due to reductive analysis. But in systems where there is non-linear constraint, propagation with increase in scale (as in processes discussed in the next two chapters), and thus double protection of higher-order attractor features from lower-order fluctuations, both physical and analytical decomposition eliminate the source of this constraint, and hence the source of its causal power. Such cases should therefore be paradigm examples of emergent transitions.

7

HOMEODYNAMICS

Nothing endures but change.

—HERACLITUS

WHY THINGS CHANGE

What causes things to change? Is there any difference between things that are forced to change and things that change on their own? Do we need to think of causality differently in these two conditions? Since ancient times, people have assumed that when things change, they must have been induced to do so, and that lack of change, that is, stasis or stability, requires no explanation. Of course, the reality is not quite this simple. For the most part it does require intervention to change things, but this is because it is far more common to find things in a stable low-energy state or at rest than in an unstable changing state. So we normally think of causes as disturbances of otherwise stable states of affairs. When asked: "What caused X to happen?" we naturally assume that the state prior to X would have remained unchanged were it not for some perturbation, that we understand as the cause. Nevertheless, some changes happen spontaneously and may need to be actively or passively prevented from occurring. For example, objects suspended above the surface of the Earth will tend to fall unless supported, and even if propelled away from the Earth by the force of muscle or chemical explosion, unless they are accelerated to escape velocity—just under seven miles per second off the surface of the Earth—they will eventu-

206

ally reverse direction, fall back, and finally come to rest when stopped by the Earth's surface. We can prevent this only by constantly providing propulsion or erecting some support to prevent movement all the way to the Earth's surface.

Not all spontaneous changes are movements. Most organic material that is no longer part of a living body, such as food left out in warm open air, will spontaneously decay. Of course, in the case of organic decay, the process is actively supported by the influence of bacteria and mold. But even without this active decomposition, aided by living organic processes, molecular structure and chemical composition will eventually degrade spontaneously due to the breakdown of unstable chemical bonds in warm, wet conditions. Spontaneous chemical breakdown of physical and molecular structure is nevertheless a function of micro-scale movement—the incessant jostling of molecules in air and water that allows chemical bonds to be broken and rearranged. This is why freezing can slow this process, as well as halting the actions of microorganisms.

Reflecting on the movement of physical bodies, Aristotle came to the commonsense conclusion that a moving object will persist in movement only so long as it is constantly pushed or pulled; otherwise, it will eventually stop moving of its own accord. So, from what he could see, it appeared that the natural state of things was to be at rest. The problem with this view, as was subsequently demonstrated, is that not all forms of change require continuous intervention and not all forms of stability are changeless. While objects flying through the air or pushed along the ground do tend to come to rest if not continually pushed, as every modern schoolchild soon learns, this is because they are slowed by the friction of the comparatively stable medium they are in contact with and through or over which they must move. Although even shortly after Aristotle some of his own students began to question his theory of persistent movement, the final refutation of this view can probably be traced to the medieval scholar John Buridan, who conceived of the concept of *impetus*—that attribute of a moving mass that intrinsically tends to perpetuate its movement unless resisted. Following the further refinement of this idea at the hands of such later geniuses as Galileo and Newton, modern textbooks inform us that if an object is not in motion, it takes a push to get it moving, but once set in motion it will continue until it meets some resistance. Does this mean that simple movement should *not* be considered a form of change? Or does this mean that some forms of change are not caused, at least in the colloquial sense of that word?

There is a partial analogue of this property of resistance to change in thermodynamics as well: the resistance of a thermodynamic system (a solid, liquid, or gas) to a change in state. Of course, the resistance of a physical object to any change in trajectory or velocity is not a statistical phenomenon. It is a simple single-value property, defined as its inertial mass—and precisely correlated with the pull that gravity has on it. To be inert is to be unchanging, and so in one sense this is merely a way of naming the extent to which it resists being moved or altered in movement. The analogue in a thermodynamic system is also a single global property, which also is a function of motion, and likewise can be assessed in terms of resistance to change, or inertness. Thus what we might call thermodynamic inertia is exemplified by how difficult it is to induce modification of a spontaneous trajectory of thermodynamic change toward equilibrium, or to induce change away from thermodynamic equilibrium.

A thermodynamic system in its equilibrium state is not at rest microscopically, only in terms of its global distribution of macroscopic features (e.g., temperature); but to cause this collective motion state to diverge from equi-distribution requires that this system interact with another that is in a different thermodynamic state (e.g., a hotter or colder system). Thus the interaction of two systems with different collective motion values modifies these values. This is the case whether both, one, or none are in thermodynamic equilibrium. If one or both are in the process of spontaneous change toward equilibrium, interaction will alter these global rates and/or initiate change with respect to equilibrium. The case of perturbation from equilibrium can thus be analogized to the initiation of movement of an object from a state of rest, and the case of change in the rate at which a system changes toward equilibrium can be analogized to the change of velocity of a moving object. Each is due to collision or some other energetic interaction. Both a system in the spontaneous change toward equilibrium and the spontaneous stability of a system in equilibrium are in some sense non-perturbed frames of reference. They are in this way analogous to Galilean reference frames in constant unperturbed movement or at rest, respectively. So there is, in fact, a deep commonality which derives from the fact that thermodynamic processes are themselves reflections of underlying microscopic dynamical processes.

In summary, we can draw a number of rough analogies between Newtonian dynamics and thermodynamics. First, the equilibrium state can be crudely analogized to a mass moving at constant velocity in a straight line, in the sense that a system in equilibrium is dynamically active, changing from state to state, and yet exhibiting no change in global (distributional) properties. Second, like a moving mass, a thermodynamic system at equilibrium will tend to maintain its dynamics within the same distributional parameters until perturbed: its maximum entropy state. Third, like the inertia of a massive body, a thermodynamic system at equilibrium will resist being modified, with a degree of resistance proportional to the size of the collection of molecules that constitute it.

Like Aristotle's conception of constant motion, it is not surprising that there is a tendency to conceive of spontaneous thermodynamic change toward equilibrium as also being in some sense "pushed" to change. So, despite what we have been taught about the slowing of a projectile's velocity due to friction, Aristotle's commonsense interpretation still influences thinking about causality in other domains and in more abstract ways. It is still a tacit intuition that lurks behind many of the difficulties with notions of causality. For example, when schoolchildren are told that atoms are composed of particles (electrons) continually in motion "orbiting" a nucleus, this intuition prompts many students to question why these little dynamos don't eventually run down. What keeps the electrons orbiting forever? Or for that matter what keeps planets orbiting forever (at least in the ideal case of not being influenced by other objects), or what would keep a top spinning in space once set in motion?

The reason that these phenomena are a bit counterintuitive is that we tend to confuse spontaneous change that happens irrespective of any other influence with non-spontaneous change that requires something else from outside to perturb things. If we want to say that any change must have a cause, then spontaneous change (e.g., constant linear motion, or the settling of an unequally heated gas into an equilibrium state) and non-spontaneous change (e.g., altering the velocity or direction of movement of a massive object, or unevenly heating the air in a room) cannot have the same kind of cause.

Although Aristotle's physics is often blamed for this fallacious denial of the spontaneity of continuous movement, this is only partly accurate. On careful reflection, there is another way to interpret Aristotle that is a bit more charitable—and suggestive. Since the Enlightenment, it has become the doctrine of Western science that there is only one form of causality in the world: that driven by energy and force. But, as we have already seen in chapter 1, Aristotle's view of causality was more pluralistic, involving four distinct conceptions of causality (material, efficient, formal, and final causality). The modern sense of what might best be called "motive" cause, implicit in Newton's mechanics, captures only one sense of Aristotle's way of thinking about change and causality. Most would equate it with his notion of efficient cause. To understand the other three, it is necessary to imagine ourselves living in his time before most of the discoveries of science that we now take for granted.

Aristotle's conception of causality assumed a cosmology that is quite different from our modern view. It was based on the four kinds of basic substance in the cosmology of his day—earth, water, air, and fire. Each form of matter was thought to have the intrinsic property of sinking or rising until it reached its natural level in his cosmic hierarchy (i.e., layered in the above order from the Earth upward). As a result, each was thought to exhibit a tendency to change position relative to other forms of matter until it eventually reached this locus of stability. By virtue of being in mixtures with other forms of matter, and being impeded in numerous other ways, these basic tendencies could produce intermediate effects, or remain potential and able to express themselves later, when less constrained. Though for modern science this whole fanciful cosmology is of merely historical interest, this very different method of approaching the question of why things change suggests a way to partially reconcile the Aristotelean pluralistic view of cause with the modern view, and to give his notion of causality a more charitable (though perhaps too charitable) reading.

For Aristotle, although any change could be attributed to a cause, this cause did not have to be efficient cause, in the modern sense of a force. Consider again Aristotle's cosmology. His explanation for why fire tended to rise upward through the air, and rain tended to fall down through it, did not require some efficient push or pull compelling these motions. These tendencies were, according to him, intrinsic to each substance with respect to its location in the cosmic hierarchy. Each of the four prime elements of matter had its own natural position in this hierarchy in comparison to the others, and each substance naturally tended to settle in a position that was most consistent with its composition. Each possessed an intrinsic disposition to change position, until it reached its natural position of balance relative to other forms of matter. It was their natural tendency, their spontaneous nature. If one wanted to mix things in ways that deviated from these natural tendencies, or move things in directions opposite to these natural tendencies—for example, raising stones upward on a hill or forcing air downward into water under a bowl—an efficient cause would be required. Otherwise, the cause was intrinsic to the material and its relation to the hierarchic geometry of the world.

In Newton's mechanics, however, all causes were defined in external terms, with the possible exception of gravitation. In this system, the explanation for persistent linear motion became effectively "no cause." It was a spontaneous tendency, no different than being at rest—a state that would persist until it was perturbed by an externally originating force. The reduction of causality to one type only eliminated the need for a causal theory of rectilinear movement. Thus, whereas Aristotle's notion of causality distinguished causes of spontaneous change from causes of forced change, modern thinkers since Galileo and Newton have assumed that spontaneous change, such as maintaining constant movement in a straight line, is not caused. Aristotle was, of course, conceiving of "cause" in a very different and more general sense. He probably would not have been satisfied with the response that this persistence needs no explanation, even if the effects of friction were to be explained. He would not merely have been satisfied to learn that in the absence of friction, there need be no push or pull. He would have wanted to know why.

To ask what kind of "cause" is responsible for the persistence of linear movement might better be phrased: What explains why linear motion persists when there is no interference? To reply, as Galileo and Newton reasoned, that all internal relationships within that frame of reference can't be distinguished from what would be the case at rest doesn't quite answer Aristotle's question. What would answer it?

Notice that what Galileo recognized was that the geometry of the trajectories and dynamical interactions (like collisions) of events that take place within such an inertial system are the same, irrespective of this movement. The ball tossed up from your seat in a moving jet plane drops back down into your lap as if you were sitting still. In fact, sitting still on the surface of the Earth is also to be rotating at faster than the speed of sound. So to speculatively extrapolate from Aristotle's analysis, considering this fact, we might describe the cause of this indistinguishability in something like geometric terms, as being *formally* caused.

Indeed, it wasn't until Einstein's theory of relativity that modern science was able to more precisely explain why a constantly moving inertial frame is effectively no different than one that is at rest. Specifically, Einstein's account of gravitation in his General Theory of Relativity can be seen as resolving both issues in geometric terms. The geometry of space-time is effectively "curved" in an accelerated frame of reference, whether by gravitation or the application of a force, but linear otherwise. Thus in Einstein's famous analogy of standing in an elevator in space that is being constantly accelerated at 32 feet per second every second, throwing a ball would cause it to behave as though in the Earth's gravitational field. Watching the ball's movement from outside one would see it moving in a straight line, while observed from inside it would follow a parabolic path. In these frames of reference, the motion is described as following a *geodesic trajectory*. This means that it follows the most minimal path between two points, whether the space is linear or warped in some way. So, observing events in an accelerated frame of reference transforms the geometry of space-time from a Euclidean (linear) to a non-Euclidean (curved) form. The patterns of change within that frame are systematically curved, as though the geometry is warped. This makes them clearly distinguishable from what would occur at rest or in constant linear motion. Toss up a ball in an accelerating car and it won't simply fall back into your lap, because at the point it was tossed you were going slower than when it falls back down.

The problem of explaining gravity was key to Einstein's insight because, unlike the acceleration of an object by the constant application of energy, a spacecraft orbiting the Earth is following a curved path, and yet inside things behave as though it is at rest in empty space. Even though the orbit follows a curved trajectory around the Earth, there is no sensation of being accelerated off a linear path, as would be experienced if the same curved trajectory were to be generated by rocket propulsion. As Einstein reasoned, if one considered the geometry of the Earth-gravitational-frame to be warped, this curvilinear motion would be equivalent to linearity in a Euclidean frame.

Whereas the straight trajectory case could be ignored, this could not. It begged for a theory of what causes curved trajectories of change to persist spontaneously in this context. Newton's explanation, that it was a force that acted at a distance, was troublesome because mechanical forces are generated at the cost of work, but gravity just is, so to speak.[1] The explanation that Einstein provided was essentially geometric. Space-time is itself warped near a gravitational mass.[2] According to Einstein's account, falling in a gravitational field is motion that is linear with respect to the curved space of unforced trajectories. So, following a curved path (such as an orbit) due to gravity is not intrinsically distinguishable from moving in a straight line unperturbed by any force. In reflection, then, the same can be said of linear motion in a non-warped (Euclidean) space of possibilities for change. It is in this sense (taking advantage of a bit of revisionist license) that we might say that Einstein has provided something very much like a formal causal account of spontaneous dynamical tendencies.

For Aristotle, fire doesn't need to be efficiently pushed upward, because that is where it belongs in the geometry of his cosmos, and it possesses an intrinsic tendency to change location until it settles into this natural position. For Einstein, no application of force is needed to bend the space vehicle's path around the Earth because that too is consistent with the local geometry of the cosmos. In ancient Chinese philosophy, this might be described as following the *Tao*, the natural path, an unforced trajectory.

But when things change in non-spontaneous ways, they must be caused to do so extrinsically. When objects are accelerated, or bumped off course by collision, or heated to a different temperature, or broken up from previously stable configurations, they have been acted upon from the outside. All forms of non-spontaneous change require a non-geometric form of causality—one that is more familiar to modern minds. This is the realm of force, energy, and work. Surprisingly, these concepts are often misunderstood, and to some extent are even mysterious (as you can discover by trying to find a precise account of what energy actually *is*, as opposed to what it does or how it can be transformed; see below). This is not problematic for physics or chemistry, but it becomes an impediment to explaining intentional phenomena, precisely because this requires reintegrating concepts of forced and spontaneous change into a more general logic of change. Before we can lay the groundwork for a more subtle approach to the emergence of the distinctive twists of causal logic that appear in the world with living and thinking beings, we need to unravel some of these potential sources of confusion.

A BRIEF HISTORY OF ENERGY

In hindsight, it might seem surprising that it took two centuries after Isaac Newton had described the laws of mechanical interactions in precise mathematical terms to finally make sense of the concept of energy. After all, it was implicit in the concepts of force and work, and these could be precisely measured and computed from simple equations involving masses being accelerated or stopped or moved against a resistance. And it was the taming of energy to do work through water, wind, and the heat of combustion that characterized the beginnings of the Industrial Revolution during those centuries. But as ubiquitous as its use has become today, the use of the term *energy* wasn't even coined to describe this familiar scientific concept until 1807. So why was it so counterintuitive?

Actually, despite a preoccupation with energy in our present age—as we consume it, pay for it, drill into the Earth and split atoms in search of it—it is still a difficult concept to grasp in a precise sense. Mostly, we think of energy in the context of the heat and pressure produced by fuel being burned to power machines to move or manufacture things, or else as a kind of invisible substance that flows through electrical wires from machines powered by burning fuels, the weight of falling water, or the pressure of wind, and used to illuminate dark rooms or animate computers. We find that it can be stored in elevated water or in the potential chemical reactions within electric batteries. But the problem with these conceptions is that they are not quite consistent with one of the most basic laws of physics: energy cannot be created or destroyed in the processes we are talking about. When we burn fossil fuels to propel our automobiles or cook our meals, we are "using up" the fuel, breaking its chemical bonds, and often then just dumping the exhaust into the surrounding air; but the energy is *not* used up, it is merely transformed from one form to another. What *is* depleted or lost is the ability to do more work with that energy subsequently. As the chemical bonds are broken and the heat of combustion dissipated into the surroundings after it is used, it becomes less available for another use. The energy is still there. It's just more widely distributed than before, with the result that our ability to use it to do further work with has been diminished or lost altogether.

Our contemporary folk understanding of energy is in many ways still held hostage to late eighteenth- and early nineteenth-century misconceptions that it is something that can be stored and used up, even though we no longer view it as an invisible substance that makes flammable things volatile and hot things hot. In the mid-eighteenth century, it was thought that things that could burst into flames contained an invisible substance called "phlogiston." Then, in 1783, the French chemist Antoine Lavoisier discovered that a substance in the air, oxygen, was critical for combustion, as opposed to releasing phlogiston from the burning material. Instead, he found that oxygen was being used up and combined with the burning substance when it was heated to combustion. He argued that what was needed was an explanation for the heat of this process.

But heat was even more troublesome. Since it could be transferred from solid object to solid object simply by contact, it could not merely be an invisible substance in air in the way that oxygen was. Analogous to the passing of static electric charge from object to object by contact, many scientists argued that the transfer of heat occurred due to the transfer of an intangible *ether*, and they gave this intangible substance the name "caloric." However, there was a basic problem with this way of thinking. This supposedly invisible, intangible substance could be transformed into phenomena that didn't quite fit the characterization—such as movement. For example, it was apparent that while moving objects could exchange their momentum, as Newton had precisely calculated, their momentum could also be turned into heat, and vice versa. The heat of combustion could be harnessed to produce motion, and the friction produced by the motion of one object against another could produce heat. This troublesome fact was driven home by Benjamin Thompson in 1798. Thompson was curious about the reason that so much heat was generated during the boring of cannons. In measuring this heat, he noticed that there was a correspondence between the amount of boring and the amount of heat. This led him to conclude that heat must therefore be a form of mechanical motion as well.

By the middle of the nineteenth century it was clear that moving masses, heat, chemical reactions, light, and even electricity could all be transformed one into another. There needed to be a way to describe what they all shared in common. It was the invention of the steam engine that provided scholars interested in this problem with a machine that systematically transformed the heat of combustion into mechanical forces that could be precisely measured. As this new general-purpose engine began to fuel the Industrial Revolution, it now became a practical matter to understand the relationship that linked these physical quantities. A number of theorists concerned with this issue at the beginning of the century had speculated that such transformations must be at least partially reversible. This suggested that it might be possible that nothing is actually lost in such transformations. Heat might be transformed into motion and then the motion back into heat, and so on. But it was Sadi Carnot's analysis of the cyclic transformation of heat to mechanical energy in heat engine operation in 1824 that set the stage for pulling apart the problems of the conservation and loss of the capacity to do work in these transformations. He showed that despite this theoretical intertransformability of heat and mechanical motion, in the real world there would always be an inevitable failure to achieve lossless transformation. This was the beginning of the realization that a perpetual motion machine would be impossible.

The critical step toward precisely assessing this transformation process was taken by showing how it could be precisely quantified. The man who did this was the English physicist James Prescott Joule. Joule began his studies by exploring the relative economics of steam engines and the newly invented electric motor as potential alternatives for industrial applications, specifically with respect to his family's brewing business. This analysis led him to quickly recognize that the mechanical capacity derived from burning coal was far more efficient than what could be derived from the chemistry of batteries. His interest in precisely measuring electrical, chemical, mechanical, and heating processes led him to build a simple device to study how the stirring of a liquid could raise its temperature. In 1843, he showed that this increase was precisely proportional to the amount of time and effort put into the stirring, which he measured by having the stirrer unwind a string attached to a suspended weight.

He thus identified the precise correspondence for converting units of heat to units of distance that a weight was moved with respect to gravity, and since gravity exerts a constant acceleration, this translated into a constant force times distance: Newton's measure of mechanical work. Thus Joule was able to measure what was transformed from movement into heat. This could be extrapolated into a conversion that yielded a universal unit of "economical duty," as he called it. The unit now bears his name. A "joule" is this common standard of energy: the work it takes to raise one pound (e.g., against the force of gravity) to a height of one foot: a "foot-pound."

Identifying a common unit of measure opened the door to definitively answering the question: What gets preserved and what gets used up in these transformations? The surprising answer turned out to be that nothing gets eliminated, but not everything is retrievable. What made this so counterintuitive was that while in every case there was a physical substance involved—falling water, spontaneous chemical reactions, heated gases, masses in motion—these were not being depleted in the process of being transformed one into the other. So the ability of these processes to change the state of things—to do work—must be due to something other than just the materials involved. For example, unlike the movement of pressurized gas from one container to another, heat transferred from one solid structure to another does not require the movement of some substance (e.g., caloric) from one to the other, and yet both the movement of a gas and the passive transmission of heat from place to place can turn factory wheels and move masses. What is common to both the spontaneous movement of materials from place to place and the spontaneous transfer of heat between substances is that both processes involve the reduction of *a difference in the distribution* of some quantity of some physical property. This property seemed always to be something dynamical, like movement, or potentially dynamical, like the tension of a compressed spring. We know now that in the case of heat being transferred without any material transfer, it is the redistribution of the translational and vibrational movements of the molecules of the one body to the molecules of the other, like the vibrations of a tuning fork being transferred to the body of a violin by holding them in tight contact.

So, although something does indeed "move" in these cases, it needn't be a substance, or an intangible ether. What does move? Two important hints to this answer are: (1) that it moves of its own accord; and (2) it involves a difference in some feature of the materials (more generally, of some medium, such as an electromagnetic field) at one locus compared to the corresponding feature at a neighboring locus. If this difference involves something dynamic and changeable in the one, it will tend to spontaneously redistribute, becoming more evenly distributed into the neighboring vicinity to the extent that it is not impeded from doing so. This difference can take many forms: differences in the average speed of molecules, differences in the average density of molecules, differences in the concentrations of different kinds of molecules, differences in electric potential, and so on. When things aren't equal, and aren't prevented from distributing or propagating into neighboring spaces, they will tend to spontaneously even out, through whatever pathways present themselves. You might say that *nature is ultimately unbiased.*

We often tend to describe the process of energy transformation in overly concrete terms. It is common to describe such transformations as involving "a flow of energy" from one place to another, as when describing electricity flowing from the power plant to one's home. And we talk about resources that we use as "energy sources," such as the chemical bond energy liberated by burning fossil fuels, as though it is something intrinsic to that material. This language tends to perpetuate the substance misconception, which was implicit in the concept of caloric. What "flows downhill," so to speak, in these transformations, is basically just a "difference." If it is heat in one subregion of a liquid or gas, then a change in the average velocity and vibration of molecules propagates along the gradient of difference until that gradient is eliminated. If it is a difference in pressure, it is the average distance between molecules that propagates. If it is electricity, it is a difference in relative charge that propagates along a conductor. And although there can be material that embodies these differences which may move as well, such as in the flow of water or wind, it is a difference in some attribute they embody that matters (e.g., elevation or pressure).

The term *energy* was introduced into these discussions in 1807 by Thomas Young and began to be favored over such terms as *caloric* by the mid-nineteenth century because of its more generic application beyond heat. The term was already in use in English, for example, in reference to "agitated" speech, and was originally derived from the Greek ενεργια, combining the root εργεια (for "activity" or "work") and the prefix εν– (for "in" or "to"). Although the modern technical sense of the term reentered physics in the nineteenth century, Aristotle is often credited with the first use of this Greek word in a scientific sense to mean something like "vigor." So the etymology of "energy" effectively defines it with respect to work—as its intrinsic source. Although in a slightly metaphoric sense we can say that the capacity to do work *flows* across a gradient, it might be more accurate to say that the capacity to do work *is* a gradient across which there is a tendency to even out and dissipate. Energy is more accurately, then, a relationship of difference or asymmetry, embodied in some substrate, and which is spontaneously unstable and self-eliminating—a tendency described by the second law.

In line with the figure-background shift of emphasis that I have been advocating, I suggest that the key to understanding *what* energy is is to stop focusing on the stuff that embodies it, and instead consider the *form* that is embodied. In the most abstract sense, energy is a relationship of difference that tends to eliminate itself. It can be more accurately described as a relationship of difference distributed in some substrate that will spontaneously tend to even out if unimpeded. The substrate can be anything from a heated solid object to a perturbed electromagnetic field. The difference can be as trivial as the concentration difference between water molecules in a glass and those of a drop of ink placed in it. In a typical thermodynamic system this tendency to even out, described by the second law of thermodynamics, derives from the statistical asymmetry implicit in the details of spontaneous interactions of component particles (e.g., ink molecules). Iterated over time, this generates a colossally asymmetric tendency that is eliminated only when the average motions and positions of the molecules have become sufficiently uncorrelated with respect to each other so that neither increase nor decrease in the correlations is more likely.

More important, any differences that tend to eliminate themselves can be a source of energy. But only if this spontaneous tendency is in some way constrained.

FALLING AND FORCING

The difference between spontaneous and non-spontaneous change is one of the most common features encountered in everyday experience. It is also the basis of what is often described as "broken symmetry" in the world. Some changes are symmetric in the sense that they are just as natural run forwards or backwards. This is approximately the case with the collision of two billiard balls in the absence of friction. Play the movie of a collision both forwards and backwards and it will be difficult to discern which is the "real" version. This is because Newton's laws of motion are symmetric. But if the movie involves fifteen balls colliding with one another, it will often be quite obvious which sequence is shown forwards and which is shown backwards. And this doesn't depend on energy being added or friction slowing the velocities. Even on an imaginary frictionless billiard table, this can be discovered so long as one begins with an asymmetrically organized arrangement. Thus the breaking up of a symmetrically organized triangular array of balls (as in a pool "break") so that they become scattered and careen around the table will be the obvious forward direction, whereas a reversal of this movie will appear quite unnatural. However, if we begin the movie at progressively later and later points, after each ball has had an opportunity to haphazardly interact dozens of times with different balls at different positions around the table, it will become progressively more difficult to discern forward- from backward-running movies (Figure 7.1).

The early phases of such a process are familiar to our everyday experience. Mixing things up is far easier, and far more likely, than unmixing them. And once mixed up, things tend to stay that way. This spontaneous asymmetry in the pattern of dynamical change is the essence of the second law of thermodynamics. And as we have also seen, it is with respect to this spontaneous asymmetry of change that emergent processes are typically defined. So before we can make complete sense of the dynamics that produces emergent phenomena, it is important to be more precise about what makes this form of spontaneous asymmetric change similar to and different from the kind of spontaneous change that is exhibited by a body in constant linear motion.

The conundrum that heat posed for nineteenth-century physicists was explaining how the dynamics of individually colliding billiard balls could be symmetric with respect to time, and thus theoretically reversible, and yet their collective behavior is not. The answer was first glimpsed by James Clerk Maxwell and later applied to thermodynamic processes by Ludwig Boltzmann in the last half of the nineteenth century. Basically, each collision results in changing the velocity and direction of each object (billiard ball or molecule), and as more and more collisions ensue, the velocities and directions of movement will become more and more divergent from

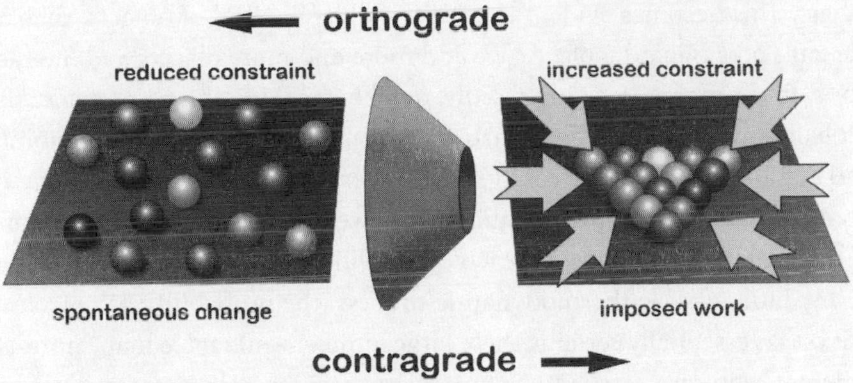

FIGURE 7.1: A cartoon characterization of the asymmetry implicit in thermodynamic change from a constrained ("ordered") state to a less constrained ("disordered") state, which tends to occur spontaneously (an orthograde process), contrasted with the reversed direction of change, which does not tend to occur spontaneously (a contragrade process), and so only tends to occur in response to the imposition of highly constrained external work (arrows in the image on the right).

original values, analogous to the way that shuffling a deck of cards causes any two cards that were once close together to progressively get more and more separated, until they vary around the average degree of separation for any two cards. Similarly, as each billiard ball or molecule interacts again and again with others, local correlations of position and movement get progressively redistributed. As the collection continues to interact (assuming no loss due to friction), it will vacillate around the average distribution; but it will be astronomically unlikely to pass through a highly regularized state (like a stationary triangular arrangement) ever again.

This too is a geometric effect, but in this case it involves the geometry of the probable paths of change, not the geometry of space-time. It exemplifies the fact that the universe has a deeply asymmetric predisposition when it comes to any process involving many components. The probability that any interaction will reflect this asymmetric bias is proportional to the number of interacting components or features.

The Newtonian collision between two objects is thus the limiting case, not the rule. It is generally possible to discern which way the movie of a dynamical interaction between many objects is being played because some large-scale distributional features become very much more likely than others. This becomes evident even with only a handful of objects, such as the balls on a billiard table. As we add more and more objects and interactions, this asymmetry grows rapidly, quickly reaching the point where the probabilities are effectively indistinguishable from 0 and 1 (impossibility and certainty). This is the case for most thermodynamic systems, since in even a minuscule volume of liquid or gas we may be dealing with billions of molecules, interacting with each other billions of times each second. So, in any human-scale thermodynamic process, the probability of reflecting this bias is essentially certain. These large numbers guarantee that "more-is-different" with near certainty in any microscopic to macroscopic comparison. This more-is-different effect will also turn out to be a key factor in the explanation of emergent phenomena, which in all cases involve significant increases in scale and a corresponding compounding of lower-level interaction effects.

Consider one of the most familiar of thermodynamic processes, dissolving some solid—say, a cube of sugar—in a container of water. At normal temperatures, a small cube of sugar dissolves naturally and effortlessly with a little patience. It is a spontaneous process. Separating it out again, however, can be incredibly complicated, laborious, and time-consuming. Even employing the most sophisticated of physical-chemical purification processes, you will never fully retrieve the amount and structure that you began with. We basically don't have to intervene in the dissolving of sugar in water, unless we want to make the process go faster (by stirring or heating) than it would occur spontaneously, but any process that exactly reverses the original mixing will be decidedly non-spontaneous, often requiring considerable outside intervention using highly contrived means. The partial exception to this takes advantage of another spontaneous process—evaporation—though it only separates the water from the sugar and doesn't produce tiny crystals organized into a cube.

We take this sort of causal asymmetry for granted, recognizing that some forms of change are spontaneous and resistant to intervention, while others require intervention to force them to occur because they are resistant to change. Both kinds of change occur because of certain forms of interactions and how they are constrained and biased by the conditions within which they occur. So it shouldn't abuse the meaning of "cause-to-happen" to say that both are caused, even though their consequences are quite dichotomous. This is made explicit in the classic thermodynamic model system: a gas in a container that can be isolated from all outside influences. Asymmetrically heat the container, using an external heat source, and the majority of molecules in one region are forced to move more rapidly than the majority at some other region. But as soon as the external influence is removed, the gas will begin an inevitable transition back to equilibrium, redistributing these local differences. In the one case the cause must be imposed from without for change to occur, in the other the cause is intrinsic; change will happen unless it is impeded by outside intervention. So, in commonsense terms, we say that some things happen "naturally," while other things don't.

For general purposes, then, it would be useful to distinguish between changes that must be forced to occur through extrinsic intervention and those that require intervention to prevent them from occurring. Surprisingly, there are no terms that characterize this difference. In order to facilitate the discussion of how they contribute to the range of dynamical processes we will encounter in the effort to explain emergent phenomena, I offer two neologisms:

I will call changes in the state of a system that are consistent with the spontaneous, "natural" tendency to change, irrespective of external interference, *orthograde* changes. The term literally refers to going with the grade or tilt or tendency of things, as in falling, or "going along with the flow." In contrast, I will call changes in the state of a system that must be extrinsically forced, because they run counter to orthograde tendencies, *contragrade* changes.

The usefulness of this distinction may appear to be minimal at this stage of the analysis, since I have merely renamed things that are already familiar. Nevertheless, it will become clear later, once we begin dealing with processes that no longer exhibit simple thermodynamic tendencies.

Because we tend to consider orthograde changes "natural," we might be tempted to describe contragrade changes as somehow unnatural. This turns out to be a misleading dichotomy because, as we'll see, even though contragrade changes are not spontaneous and intrinsic, they are in no way unnatural or artificial. They are merely the result of the interaction between contrasting orthograde processes. Because the world is structured and not uniform, and because there are many distinct dimensions of orthograde change possible (involving different properties of things, such as temperature, mass, movement, electric charge, structural form, etc.), certain of these tendencies can interact in relative isolation from others.[3] Contragrade change is the natural consequence of one orthograde process influencing a different orthograde process—for example, via some intervening medium. This implies that in one sense *all* change ultimately originates from spontaneous processes. It is simply because the world is highly heterogeneous that there can be contragrade processes. Thus, although orthograde processes are the basis for all change, the ortho/contra distinction is not artificial.

More precisely, then, we can also distinguish orthograde and contragrade processes in terms of relative causal isolation, which also distinguishes intrinsic from extrinsic causal influences. A composite system that is isolated from outside interactions will intrinsically exhibit orthograde change, but not contragrade change. Contragrade change is only possible with respect to extrinsic relationships between systems with different orthograde tendencies. This is just a restatement of the isolation conditions for the second law of thermodynamics. But as we will see subsequently, reframing thermodynamic processes in orthograde/contragrade terms will provide a much more general distinction that will be useful beyond both Newtonian and thermodynamic analyses.

How might we apply this distinction to more classical approaches to dynamical processes? Consider Newton's laws of motion. A mass moving in a straight line with a constant velocity can be described as undergoing orthograde change (of position), whereas a mass acted on by a force and altered in velocity and direction can be described as exhibiting contragrade change (for the duration of the period in which that force is acting). In a simple thermodynamic system, change toward equilibrium is orthograde while change away from equilibrium is contragrade. So the case of sugar spontaneously dissolving in a container of water exemplifies an orthograde change, whereas processes that a chemist might employ to extract this sugar again would involve contragrade processes.

The more general value of designating terms for these contrary "orientations" of change is that it can help us to distinguish the different ways we tend to use the concept of causality. While both the spontaneous dissolving of sugar in water and its extraction by chemical means are changes that are caused in a generic sense, what we mean by "cause" is quite different in these two situations. Both are consistent with the laws of physics and chemistry; they nevertheless differ radically in how these causal influences are organized.

In the case of processes like the dissolving and diffusion of sugar molecules in water, this distinction also is relative to scale. The dislodging of a sugar molecule from its crystal lattice is the result of microscopic contragrade processes. The molecular collisions and electrochemical interactions between water and sugar molecules are contragrade, because the interaction among these molecules changes their spontaneous motions and ranges of movement. The continually diverging diffusion of sugar molecules into the surrounding water is also the result of contragrade dynamics. Each of the innumerable collisions that results in changes in a molecule's velocity and direction of movement is a contragrade event.

The trajectories of sugar molecules as they interact with neighboring water molecules are more likely to expand into new territories. This tendency effectively reflects the geometry of the situation. There are more ways for molecules to diverge than to converge in their relative locations. As every molecule is bumped nanosecond-to-nanosecond onto a new path, each molecule's new position gets more and more superimposed on the others' former positions, while their velocities and directions of motion also sample values once exhibited by others. This tendency of interacting molecules to wander into each other's spatial positions and dynamical values is responsible for the orthograde dynamic that characterizes the global change toward equilibrium. In this way, contragrade dynamics at one level produce orthograde dynamics at the higher level.

Another merit of describing change in these complementary terms is that it gives new meaning to the defining property of matter—a resistance to change—and the defining property of energy—that which is required to overcome resistance to change. Since orthograde processes ensue spontaneously, they are ubiquitously present, even during processes of contragrade (forced) changes. A contragrade change must therefore derive from two or more orthograde processes, each in some way undoing the other's effects. To put this in the terms introduced in the previous chapter, each must *constrain* the other. The tendency of one orthograde process to realize the full range of its degrees of freedom (e.g., the diffusion into all potential locations) must diminish the tendency of another orthograde process to realize all its potential degrees of freedom. This is easily demonstrated for interacting thermodynamic systems otherwise isolated from other outside influences. Thus, for example, the specific heat of one object placed in contact with another with a different specific heat will result in their combined development toward a different maximum entropy value than either would reach had they remained isolated from one another. This is because there will be a net asymmetric redistribution of the molecular motions in the two materials, in which the rate of orthograde change of one will accelerate and the other will decelerate with respect to their prior rates of change, since the one is now further and the other closer to the maximum entropy state than before. One is thereby relatively deconstrained and the other relatively constrained in its domain of possible states. Resistance to change is thus a signature of the additive and canceling effects of interacting orthograde dynamics. In this respect, it is again roughly analogous to the composition of momenta of colliding objects in Newtonian mechanics.[4]

Using this insight, we can now redefine the concept of constraint in orthograde and contragrade terms. Constraints are defined with respect to orthograde maxima, that is, the point at which an orthograde dynamic change is no longer asymmetrical. A constrained orthograde process is thus one in which certain dimensions of change are not available. This can be the result of extrinsic bounds on these values such as might be imposed by the walls of a container, or the result of contragrade processes countering an orthograde change. Interestingly, as we will see in the next chapter, this means that contragrade processes at one level can generate the conditions for a higher-level orthograde process of constraint generation.

One benefit of articulating this orthograde/contragrade distinction is that it provides a language for describing the dynamical relationships that link different levels of a process. In classical emergentist terms, we could even say that in the case of close-to-equilibrium thermodynamics, the orthograde increase in entropy is supervenient on the cumulative effects of the contragrade dynamics of incessant molecular interactions. This suggests that the orthograde/contragrade distinction may offer a useful way to reframe the emergence problem. Indeed, ultimately we will discover that a fundamental reversal of orthograde processes is a defining attribute of an emergent transition.

Under some circumstances (which will be the focus of much of our subsequent analysis), an extrinsically perturbing dynamical influence can be highly stable, thus providing an incessant contragrade influence. This is, for example, the condition of the Earth as a whole, where the perturbations provided by constant low-level solar radiation over billions of years have made life a possibility. The constancy of this source of contragrade influence is a critical factor, since its stability has allowed the formation of a vast web of dissipative pathways through which energy is released. While in transit through terrestrial substrates, this persistent perturbation has been available to drive chemical reactions in a myriad of contragrade directions. The result is that otherwise unlikely molecular structures are being constantly synthesized even while orthograde thermodynamic tendencies tend to break them down.

Following the Nobel laureate theoretical chemist Ilya Prigogine, who revolutionized how we think about such systems, we call the Earth a *dissipative system*, because it is constantly both taking in and dissipating energy. During the time that this dissipative relationship has been stable—roughly for a little over 5 billion years—it has made numerous other contragrade processes possible, including nearly all of the contragrade chemistry that constitutes life. Indeed, every living organism is a constellation of contragrade processes, which continually hold local molecular orthograde processes at bay.

Of course, the history of human technology involves the discovery of ways to utilize certain orthograde processes in order to drive desired contragrade processes. This is the general logic characterizing all the many types of machines that we employ to bend the world to our wishes. Reliably being able to produce unlikely outcomes, such as the purification of minerals, can provide the conditions for producing yet other, even more unlikely outcomes, such as the assembly of refined materials to form the circuits of my computer. But as we will see shortly, systems formed by stable contragrade processes can sometimes exhibit what amount to higher-order orthograde processes that are quite different than those from which they arise.

REFRAMING THERMODYNAMICS

Although the use of the terms *orthograde* and *contragrade* to describe both Newtonian and thermodynamic processes might at first appear to be merely a convenient way to group together diverse processes that have one or the other of two superficial features in common, the contrast it identifies is fundamental. Understanding how this underlying dynamical principle is exemplified in these most basic classes of processes is the first step toward a theory of what I will call *emergent dynamics*.

In classical thermodynamic theory there is, in effect, only one orthograde tendency: the increase in entropy of a compositional system. *Entropy* is a technical term introduced in the mid-nineteenth century by Rudolf Clausius. It can be defined in a number of ways. Generally, following the work of a later theoretician, Ludwig Boltzmann, it is a measure of the disorder among parameters defining the state of a collection of components, such as the momenta of molecules of a gas. However, this way of describing entropy is easily misunderstood. As we saw in the last chapter, order understood as merely some state of things that we prefer—such as the alphabetical order of names in a list, or a particular symmetry in shape—is subjective and epiphenomenal. It has no independent role to play in physical interactions. Thermodynamic processes occur indifferently to our observation or analysis of them, and indeed are essential to our being alive and able to make such assessments. To divorce the concept of order from an epiphenomenal, model-based, subjective notion of form, and instead only consider the intrinsic model-independent and objective aspects of form, it is necessary to consider it in inverse, with respect to constraint.

Rather than order or disorder, then, I suggest that we begin to think of entropy as a measure of constraint. An increase in entropy is a decrease in constraint, and vice versa. When gas in a container is far from equilibrium, local correlations among molecular velocities and molecular types are higher than at equilibrium. They are not thoroughly mixed with respect to each other and with respect to the properties of other molecules. Their distinctive properties, relative spatial positions, and momenta occupy only a subset of what is possible in this context. So, for example, there are vastly more fast-moving (warm) CO_2 molecules near my mouth and more slow-moving (cooler) oxygen molecules distributed elsewhere in the room. But after I leave, this asymmetry of correlated movements and molecular concentrations will eventually even out. The equilibrium state of maximum entropy is one in which correlations are minimized, and the likelihood of measuring any given property, such as molecular concentration or temperature, is the same irrespective of where it is measured. The distribution of attributes is no longer constrained. We can thus describe the increase in entropy as a decrease in constraints, and the second law can be restated as follows: In any given interaction, the global level of constraint can only decrease.

Achieving a state of maximum entropy and minimum dynamical constraint—the state of equilibrium where a system is resistant to further global change—is not to bring that system to rest microscopically. Whereas a gas at equilibrium is exemplified by an absence of change in the global distribution of properties—no change in constraint—the component molecules are incessantly moving, constantly bumping into one another, and thereby contributing to the stable pressure and temperature of the whole volume. Stability at this higher level is only a reflection of the statistical smoothing and central tendency of the dynamics at the lower molecular interaction level. What *is* stable at equilibrium is the level of dynamical constraint.

In principle, from a Newtonian perspective, each individual molecular interaction in a liquid or gas is exactly reversible. If a movie of each molecular collision were run in reverse, like the reversed movie of a billiard ball collision, it would appear as lawful and physically realistic as when run forwards. And yet, as the work of Maxwell and Boltzmann demonstrated, the lack of temporal distinction quickly breaks down with larger and larger groups of interacting components, because each interaction will tend to produce progressively divergent values from the previous state, ultimately sampling more and more of the possible values distributed around the mean value for the entire collection. The dissociation of macro from micro processes that characterizes thermodynamic systems, like gases, means that we cannot attribute the asymmetry of thermodynamic change to the properties of individual molecular collisions alone. Molecular collisions are necessary for any change of state of a typical thermodynamic system, but they are not sufficient to determine its asymmetric direction of change. This is ultimately due to the highly asymmetric "geometry" of the possible distributions of molecular properties and trajectories of their movements.

It is common to represent each possible distribution of molecular properties within such a system as a point in an abstract phase space,[5] or state space. Changes in state of the system can then be represented as a continuous line within this abstract space of possibilities. One way to conceive of the asymmetry of the geometry of thermodynamic change is to imagine it in a "warped" phase space, as though at one point the plane of possible trajectories is pulled like a rubber sheet to create a deep dimple. In such a space, all trajectories will tend to bend into the region in which there are the most adjacent options, that is, toward the region of maximum curvature. It is as though that region of alternatives is somehow more dense and contains more positions for the same "volume."

As we saw earlier, in the terms of complexity theory and dynamical systems theory, such a region of values toward which trajectories of dynamical change are biased is called an *attractor*.[6] Attraction in this sense is not an active process or the result of a force. It is merely the result of the statistical asymmetry of optional states. The apparent tendency to transition to states which are closer to an attractor—e.g., toward an equilibrium state in thermodynamics—is a function of this biased probability. This might be thought of as a "warping" of the space of probable configurations. Each individual micro interaction is unbiased with respect to any other, and so it only accounts for change. It is only when the repeated interactions of many components are considered that this curious *geometry of chance* becomes evident. It is a relational property, determined with respect to possible configurations, that determines the asymmetric directionality of the change of the whole collection. Systems of interacting components tend to change toward equilibrium states because there are vastly more trajectories of change that lead from less symmetric to more symmetric distributions than the other way around.

A FORMAL CAUSE OF EFFICIENT CAUSES?

This way of understanding the microdynamics underlying the second law of thermodynamics complicates how we need to think about physical causality, but it also provides a more precise way to formulate the concepts of orthograde and contragrade change that can help unravel the tangle of confusions surrounding the nature of emergent phenomena. An intuitive sense of how this reframes the notion of causality can be gained by a comparison to two of Aristotle's notions of cause: formal and efficient causes. Orthograde thermodynamic change occurs because it is an unperturbed reflection of the space of possible trajectories of change for that system. It is in this sense a consequence of the geometric properties of this probability space. An orthograde change just happens, irrespective of anything else, so long as there is any change occurring at all. I take this to be a reasonable way to reinterpret Aristotle's notion of a formal cause in a modern scientific framework, because the source of the asymmetry is ultimately a formal or geometric principle.

Contragrade change, however, does not tend to occur in the absence of intervention. It is extrinsically imposed on a system that doesn't tend to change in that direction spontaneously. The source of a contragrade change is what we typically understand as the mechanical cause of something. In this sense, it is the result of efficient means for forcing change away from what is stable and resistant to modification. So this provides a somewhat more precise reframing of Aristotle's efficient cause, appropriate to a modern scientific framework.[7]

The value of fractionating the concept of physical cause in this way is that it helps us to avoid two of the most serious pitfalls of both crude reductionism and naïve emergentism.

The first pitfall is an effort to define emergent relationships in terms of "top-down" or "downward" causality. This is basically the assumption that emergent novel causal properties at a higher level of organization can impose an influence on the lower-level processes that gave rise to them. Although top-down causality sounds as though it could be guilty of vicious regress if understood synchronically, there is at least a general sense in which the concept can be rendered unproblematic from a process perspective. Consider again the dynamically maintained biomolecules reciprocally synthesized and constituting a living cell. They are not components independent of this metabolic network, but are created by being enmeshed in that network of chemical reactions. In this sense, they are both constituents and products of the larger dynamical network. One could say that component molecules generate the whole via a bottom-up causal logic, and the whole generates these molecules via an independent top-down causal logic. Of course, each molecule is generated via interactions among other molecules, and it is only the constraints on these patterns of interaction that are relevant. These constraints don't arise from any contragrade forcing from the whole. Rather they arise reciprocally, from the constellation of relationships that makes the synthesis of each type of biomolecule in an organism indirectly dependent on each other. It is the systemic "geometric" position within this whole dynamical network that is the source of these constraints. In this sense, I have to agree with Claus Emmeche and his colleagues, who argue that "downward causation cannot be interpreted as any kind of efficient causation. Downward causation must be interpreted as a case of formal causation, an organizing principle."[8]

The second pitfall is the assumption that what is absent cannot be a causal influence on real events. The geometry of possible trajectories of change from state to state is not anything material. While it is accurate to say that the dynamics of colliding molecules in a volume of gas is the cause of the changes from state to state as that gas transitions toward equilibrium, it would not be accurate to attribute the asymmetry of this trajectory to those collisions. In our rephrasing of formal and efficient cause, the incessant change of state is efficiently caused, but the asymmetry of those changes is formally caused. But this analysis suggests something even more radical. If all contragrade change (such as that produced by molecular collision) is the result of the interaction of orthograde processes, then in Aristotlean terms we are forced to conclude that all efficient causes ultimately depend on the juxtaposition of formal causes!

As we will discover in the following chapters, however, the relationship between contragrade and orthograde processes can be far more complicated. Orthograde and contragrade processes can also have a hierarchic dependence on one another. This is implicit in the fact that the macroscopic orthograde increase of thermodynamic entropy (change toward equilibrium) depends on microscopic contragrade processes (molecular interactions, such as collisions). But while the efficient causality of these micro interactions is not sufficient to account for the asymmetry that is intrinsic to the higher-order orthograde dynamic of increasing entropy, without this incessant flux of micro-contragrade dynamics, the higher-order orthograde process would not exist. This asymmetric geometry of possibilities would not be "explored" by the various states of the system at different moments were it not for the incessant contragrade changes occurring at the lower level.

The title of this chapter—Homeodynamics—coins a term that I think can more generally describe this most basic orthograde dynamic wherever we encounter it. It is a dynamic that spontaneously reduces constraints to their minimum and thus more evenly distributes whatever property is being changed from moment to moment and locus to locus. Distinguishing the general case from the specific case of thermodynamic entropy increase becomes indispensable as we begin to analyze the dynamic interdependencies that span more than two levels of scale. For example, it allows us to distinguish a powerful general principle: orthograde and contragrade dynamics reverse in dominance as we consider adjacent levels of process.

Consider the model thermodynamic system: a simple ideal gas.

Starting at the atomic level, we can ascribe the elasticity of collisions between atoms to the stability of their internal dynamics. This internal stability is a function of the precise symmetries of the electron orbitals that form what amounts to the outer "shell" of an atom. These orbitals tend to assume stable configurations that are essentially the expression of orthograde quantum electrodynamic tendencies. In simple terms, electron orbitals are determined by symmetrical relationships between oscillatory properties associated with specific energy levels and what might be described as resonant symmetries of specific orbital configurations. Being out of resonance (so to speak) is unstable; indeed, it takes work to perturb the electrons of an atom away from this ground state, and they will tend to spontaneously return to this state subsequently. This is vaguely analogous to tending toward a state of thermodynamic equilibrium, though for very different physical reasons. This is why it makes sense to think of this tendency more generally than merely in thermodynamic terms, though statistical analysis is likewise relevant. The stable ground state represents the most symmetric distribution of energy, given the wavelike nature of electron "movement." Indeed, this most stable state is essentially a kind of oscillatory attractor of probable locations.[9]

The possibility for contragrade interactions to take place as atoms in a gas collide with one another depends on this lower-level subatomic homeodynamic tendency. As atoms collide, they exert a contragrade influence on each other that tends to perturb the subatomic ground state of each. The orthograde-based resistance to this perturbation causes the atoms to rebound—a contragrade interaction dependent on a lower-level orthograde tendency. And, as we've seen, this contragrade interatomic dynamic enables the higher-level thermodynamic tendency to be actualized—an orthograde tendency dependent on lower-level contragrade interactions. For our purposes, what is important is that this alternation of orthograde and contragrade processes at adjacent hierarchic levels of compositional scale has no necessary upper bound, and with each hierarchic iteration the potential complexity of these interactions increases. As we will see in the next two chapters, the special contragrade relationships that can be generated by the juxtaposition of different orthograde processes can produce complex forms of constraint. By these means, simple homeodynamic tendencies can be organized into highly indirect and convoluted relationships with respect to one another, ultimately producing unprecedented higher-level orthograde tendencies. This is the engine of emergence.

8

MORPHODYNAMICS

. . . all the evolution we know of proceeds
from the vague to the definite.

—CHARLES SANDERS PEIRCE[1]

ORDER FROM DISORDER

Although the epitome of a local reversal of the second law is observed in living and thinking beings, related local deviations from orthograde thermodynamic change are also found in many non-biological phenomena. Inorganic order-producing processes are fewer and more fleeting than any found in life, nor do they exhibit anything resembling entential logic—neither end nor function—yet many physical processes share at least one aspect of this time-reversed order-from-disorder character with their biological and mental counterparts. Understanding the dynamics of these inorganic order-production processes offers hints that can be carried forward into our explorations of the causality behind life and mind.

In these processes, we glimpse a backdoor to the second law of thermodynamics that allows—even promotes—the spontaneous increase of order, correlated regularities, and complex partitioning of dynamical features under certain conditions. Ironically, these conditions also inevitably include a reliable and relentless *increase* of entropy. In many non-living processes, especially when subject to a steady influx of energy or materials, what are often called *self-organizing* features may become manifest. This constant perturbation of the thermodynamic arrow of change is in fact critical, because when the constant throughput of material and/or energy ceases, as it eventually must, the maintenance of this orderliness breaks down as well. In terms of constraint, this means that so long as extrinsic constraints are continually imposed, creating a contragrade dynamic to the spontaneous orthograde dissipation of intrinsic constraints, new forms of intrinsic constraint can emerge and even amplify.

There are many quite diverse examples of constantly perturbed self-organizing inorganic processes (several of which will be described below). Among them are simple dynamical regularities like whirlpools and convection cells, coherence-amplifying dynamics such as occurs in resonance (e.g., Figure 8.1) or within a laser, and the symmetrical pattern generation that occurs in snow crystal growth. Even computational toy versions of this logic produced by computer algorithms, such as cellular automata and a variety of recursive

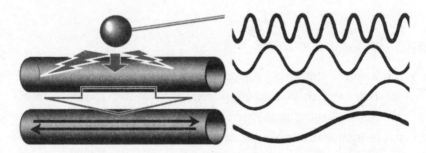

FIGURE 8.1: Resonance: a simple mechanical morphodynamic process. A regular structure that is capable of vibrating (a tubular bell: *left*) will tend to transform irregular vibrations imposed from without (depicted as a mallet striking it: *top left*) into a spectrum of vibrations (*right*) that are simple multiples of a frequency determined by the rate at which vibrational energy is transformed back and forth from one end to the other (*bottom left*). This occurs because as vibrational energy from varying frequencies "rebounds" from one end to the other, it continually interacts with other vibrations of differing frequencies. These reinforce each other if they are in phase and cancel each other if they are out of phase. Over many thousands of iterations of these vibrational interactions, it is far more likely for random interactions to be out of phase. So, as the energy is slowly dissipated, these recurring interactions will tend to favor a global vibrational pattern, where most of the energy is expressed in vibrations that coincide with even multiples of the time it takes the energy to propagate from one end to the other. This is well exemplified in a flute, where air is blown across the mouthpiece, disturbing the local internal air pressure, and this imbalance is transformed into a regularly vibrating column of air that in turn affects the flow of air across the mouthpiece. Image produced by António Miguel de Campos.

non-linear computational processes, exemplify the way that constant regular perturbation can actually be a factor that increases orderliness.

In recent decades, a focus on these spontaneous order-producing processes has galvanized researchers interested in explaining the curious thermodynamics of life. However, the sort of order-generating effect observed in these non-living phenomena falls short of that found in living organisms. These processes are rare and transient in the inorganic world, and their presence does not increase the probability that other similar exemplars will be produced, as is the case with life. An individual organism may also be a transient phenomenon; but the living process has a robust capacity to persist despite changing conditions, to expand in complexity and diversity, to make working copies of itself, to adapt to ever more novel conditions, and to progressively bend the inorganic world to its needs.

The second law of thermodynamics is an astronomically likely tendency, but not an inviolate "law." You might say that it is a universal rule of thumb, even if its probability of occurring is close to certainty. But precisely because it is not necessary, there can be special circumstances where it does not obtain, at least locally. This loophole is what allows the possibility of life and mind. One might be tempted to seize on this loophole in order to admit the possibility of an astronomically unlikely spontaneous violation of this tendency. And many have been tempted to think of the origins of life in terms of such an incredibly unlikely lucky accident. Actually, as we'll see in the next chapter, life follows instead from the near ubiquity of this tendency, not from its violation. This loophole does, however, allow for the global increase of entropy to create limited special conditions that can favor the persistent generation of local asymmetries (i.e., constraints). And it is the creation of symmetries of asymmetries—patterns of similar differences—that we recognize as being an ordered configuration, or as an organized process, distinct from the simple symmetry of an equilibrium state. What needs to be specified, then, are the conditions that create such a context.

In what follows I will use the term *morphodynamics* to characterize the dynamical organization of a somewhat diverse class of phenomena which share in common the tendency to become spontaneously more organized and orderly over time due to constant perturbation, but without the extrinsic imposition of influences that specifically impose that regularity. Although these processes have often been called *self-organizing*, that term is a bit misleading. As we will see in the next section, this process might better be described as *self-simplifying*, since the internal dynamical diversity often diminishes by vastly many orders of magnitude in comparison to being a relatively isolated system at or near thermodynamic equilibrium. However, since the term *self-organizing* is widely recognized, I will continue to use it, and when referring to the class of more general dissipative processes that build constraints, I will describe them as *morphodynamic*.

Morphodynamic processes are typically exhibited by systems or collections of interacting elements like molecules, and typically involve astronomical numbers of interacting components, though large numbers of interacting elements and interactions are not a necessary defining feature. If precise conditions are met, as they can in simulated contexts or engineered systems, it is possible for simple recursive operations to exhibit a morphodynamic character as well. Indeed, abstract model systems generated in computers have provided much of the insight that has been gleaned concerning the more complex spontaneous order-producing processes of nature (some of which were discussed in chapters 5 and 6). Morphodynamic processes are distinguished from other regular processes by virtue of a spontaneous regularizing tendency that can be attributed to intrinsic factors influencing their composite dynamical interactions, in contrast to regularities that result from externally imposed limitations and biases.

Coincidentally, the term *morphodynamic* has been independently coined to describe related phenomena in at least two quite distinct scientific domains: geology and embryology. Since coining it in my own writings to refer to spontaneous self-simplifying dynamics, I discovered that it had been in use for nearly a century. And although I independently conceived of the term and this usage, I am not even the first to use it to characterize dynamical processes that produce spontaneous regularity. Coincidentally (and thankfully), these prior uses share much in common with what I describe below, and the phenomena to which it has been applied generally fit within the somewhat broader category that I have in mind.

Most authors trace its first use to a 1926 paper bearing that title ("Morphodynamik"), written by the developmental biologist Paul Weiss. He was one of the founders of systems thinking in biology, along with Ludwig von Bertalanffy. Weiss' research focused on the processes that result in the development of animal forms. His conception of developmental processes was based on what he described as morphogenetic fields, which were the emergent outcomes of interacting cell populations and not the result of a superimposed plan. More emphatically, he believed that many details of animal morphology were not predetermined, even genetically, but rather emerged spontaneously from the regularities of cellular interactions. In a little known but prescient paper published in 1967 and enigmatically titled "1+1≠2,"[2] he described numerous examples of molecular and cellular patterns emerging spontaneously *in vitro*[3] when biological molecules or cell suspensions were subject to certain global conditions.

Though recent years have seen a shift in emphasis back toward the molecular mechanisms of cell differentiation and structural development, the term *morphodynamic* is still used in approaches that focus instead on geometric properties involved in the formation of regular cellular structures, tissue formation, and body plan. A classic example is the formation of the regular spiral whorls of plant structures, called spiral phylotaxis, where shoots, petals, and seeds often grow in patterns that closely adhere to the famous Fibonacci number series (1, 1, 2, 3, 5, 8, 13, 21, 34, 55, . . .), which is generated by adding the two previous numbers of the series to produce the next. In these patterns, the distribution of plant structures form interlocking opposite curved spirals with adjacent Fibonacci numbers of arms. Thus, for example, a pinecone can have its seed-bearing facets arranged into spirals of eight arms clockwise and thirteen arms counterclockwise (as shown in Figure 8.2). This turns out to be highly advantageous. Plant structures like leaves and branches that follow Fibonacci spirals are arranged so that they are maximally out of each other's way, for nutrient delivery, for exposure to the sun, and so forth.

Mathematical models of this process have long demonstrated that this pattern reflects growth processes in which unit structures are added from the center out in a way that depends on how previous units have been added.[4] In growing plant tips, this is regulated by the diffusion of molecular signals from previously produced buds that inhibits the growth of other new buds. Since there is a reduction of concentration with distance from each source and with the maturation of each older growing bud, new buds appear in positions where these inhibiting influences, converging from the previously erupted buds, are weakest. This indicates that the Fibonacci growth pattern is not dependent on any intrinsic template or archetypal form (e.g.,

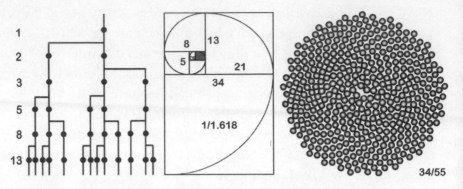

FIGURE 8.2: Three expressions of the Fibonacci series and ratio. *Left:* regular branching of a lineage in which there are regular splits (reproductive events for organisms) that occur at the same interval (distance) along each line. This produces the sequence 1, 2, 3, 5, 8, 13, 21, 34, 55 . . . that is generated by adding the two previous numbers in the series. *Middle:* dividing adjacent numbers in this series yields closer and closer approximations to the non-repeating decimal ratio 0.618 . . . which can define the adjacent sides of an indefinitely nested series of smaller and smaller rectangles. Such rectangles are self-similar to one another, and a spiral can be traced from corner to corner that is also self-similar in shape at whatever magnification it is shown. *Right:* spherical objects distributed around a central point in a closest-packed pattern also form a self-similar pattern at whatever size they are shown. As each new object is added, the next is found 137.5° around the center from the last. Depending on the size of the components, a self-similar array of this sort will demonstrate interlocking, oppositely curved spirals, such that the number of spirals in each direction corresponds to adjacent Fibonacci numbers. This is reflected in many forms of plant growth in which the addition of new components (e.g., seeds in a sunflower) occurs where there is the most space closest to the center.

encoded directly in the genome). It is induced to emerge by the interaction of diffusion effects, the geometry of growth, and the threshold level of this signal at which point new plant tissue will begin to be generated. Recently, this patterning of growth has also been demonstrated to occur spontaneously in inorganic processes. For example, Chinese scientists have demonstrated the spontaneous growth of mineral nodules on a metal surface with conical protrusions that conforms to either 5 x 8, 8 x 13, or 13 x 21 patterns of interlocking spirals, due to electrochemical effects.[5] This further confirms that the spontaneous emergent character of this patterning is not unique to biology and not merely an expression of its functional value to the plant.

More recently, self-organizing logic (though not called morphodynamics in these contexts) has been used to describe the formation of regular stripe and spot patterns as adaptations for cryptic coloration or species signaling in animals. Examples include the regular stripe patterns on tigers and zebras, the complex spiral lines and spots on certain snail shells, the spots on leopards and giraffes, and the beautiful iridescent patterns of color on butterfly wings. The logic of these processes has been well studied both by simulation and by developmental analysis of the molecular and cellular mechanisms involved. All of these pattern generation processes appear to involve a diffusion logic that is loosely analogous to that just described for Fibonacci spiral formation in plants. Each takes advantage of local molecular diffusion dynamics to generate regularity and broken symmetries, though in each case utilizing quite distinct molecular-cellular interactions.

Within cells, there are also processes that can be described as morphodynamic. These are molecular interactions that produce spontaneously forming structures like membranes or microtubules. Even the regular protein shells that surround many viruses are the result of spontaneous form generation. These molecular-level form-generating processes are often described as *self-assembly* processes, and they are responsible for much of the microstructure of eukaryotic cells. (They will be treated in greater detail in the next chapter, when we explicitly explore the morphodynamics of living processes.)

In geology, the term *morphodynamic* also has an extended history of use. It is used primarily to describe processes involved in the spontaneous formation of the semi-regular features of landscapes and seascapes, such as river meanders, frost polygons, sand dunes, and other geologic features that result from the dynamics of soil movement. It can be seen as the solid dynamical counterpart to the physics of fluid movement: hydrodynamics. The physics of particulate movement and assortment in continually perturbed collections of objects, such as gravel movement in geology and object sorting in industrial processes, is surprisingly counterintuitive and remains an area where theoretical analysis lags behind descriptive knowledge. Some of the most surprising and interesting geomorphodynamic processes are those that produce frost polygons. The repeated freezing and thawing of water within the soil in arctic regions can result in the formation of gravel that is regularly distributed around the perimeters of remarkably regular polygons (see Figure 8.3).

Examples of other physical phenomena that I would include as morphodynamic processes range from simple inorganic dynamical phenomena like the formation of vortices and convection cells in fluids to more complex phenomena like the growth of snow crystals. The dynamical processes involved in the formation of these regularities will be described in more detail below, but at this point it is worth remarking that what makes all these processes notable, and motivates the prefix *morpho-* ("form"), is that they are processes that generate regularity not in response to the extrinsic imposition of regularity, or by being shaped by any template structure, but rather by virtue of regularities that are amplified internally via interaction dynamics alone under the influence of persistent external perturbations.

In addition, I would also include a wide variety of algorithmic systems with a similar character due to analogous virtual dynamics, such as in cellular automata (like Conway's computer *Game of Life*), and computational

FIGURE 8.3: The formation of three different kinds of natural geological polygons. *Left*: soil and gravel polygons are the result of the way that the daily and seasonal expansion and contraction of ice in arctic soil causes larger stones to be pushed upward toward the surface, and outward from a center of more silty soil which more effectively holds the water, and from which ice expansion and contraction slowly expels the larger stones. The regular segregation of soil and stones is thought to result from the relatively even distribution of this freezing/melting effect, and the common rate at which the dynamics takes place in each polygon. However, multiple competing hypotheses seem equally able to explain this self-organizing effect, as is true for many self-organized, particulate segregation effects (see Kessler and Werner, 2003, for a more technical account). *Center*: regular polygonal cracks can also form for similar reasons as a result of ice crystal expansion and drying contraction of the soil. *Right*: basalt columns form in cooling sheets of lava, probably as a result of a combination of convection effects (see Figure 8.4) and shrinkage as the lava cools, with the cooler peripheries of convection columns shrinking first and forming cracks. Photos by M. A. Kessler, A. B. Murray, and B. Hallet (left); Ansgar Walk (center); L. Goehring, L. Mahadevan, and S. W. Morris (right).

network processes, such as so-called neural nets. Although computational models are not truly dynamical in the physical sense, they do involve the regular highly iterative perturbation of a given state, and the consequences of allowing these perturbations to recursively amplify in effect. In this sense, the recursive organization of these computational processes can be seen as the abstract analogue of the physically recurring perturbation of a material substrate, such as constant heating or constant growth. Ultimately, much of what we know about the logic of morphodynamic processes has come from the investigation of such abstract computational model systems.

Understanding how to take advantage of these special dynamical processes has played an important role in the development of many technologies, including the production of laser light and superconductivity. These too will be described in more detail below.

SELF-SIMPLIFICATION

The concept of self-organization was introduced into cybernetic theory by W. Ross Ashby in a pioneering 1947 paper.[6] Ashby defined a self-organizing system as one that spontaneously reduces its entropy, but not necessarily its thermodynamic entropy, by reducing the number of its potential states. In other words, Ashby equated self-organization with self-simplification. In parallel, working in physical systems, researchers like the physical chemist Ilya Prigogine explored how these phenomena can be generated by constantly changing physical and chemical conditions, thereby continually perturbing them away from equilibrium. A well-known example is the famous Belousov-Zhabotinsky reaction, which produces distinctive alternating and changing bands of differently colored chemical reaction products, which become regularly spaced as the reaction continually cycles from state to state. This work augmented the notion of self-organization by demonstrating that it is a property common to many far-from-equilibrium processes; systems that Prigogine described as *dissipative structures*.

With the rise of complex adaptive systems theories in the 1980s, the concept of self-organization became more widely explored, and was eventually applied to phenomena in all the many domains from which the above examples have been drawn. However, the precision of Ashby's conception is often lost when it is employed in complex systems theories, where it is often seen as a source of increasing complexity rather than simplification. This demonstrates that the relationships between complexity, systematicity, dynamical simplification, regularity, and self-organization are not simple, and not fully systematized, even though the field is now many decades old. More important, I fear that a recent focus on understanding and managing complexity may have shifted attention away from more fundamental issues associated with the spontaneous generation of order from disordered antecedents. Indeed, as I will argue below and in the next chapter, the functional complexity and synergy of organisms ultimately depends on this logic of self-simplification.

In general, most processes that researchers have described as self-organizing qualify as morphodynamic. So, it will typically be the case that the two terms can be used interchangeably without contradiction. However, I will mostly avoid the term *self-organization*, because there are cases where calling processes self-organizing can be misleading, especially when applied to living processes where both terms, *self* and *organization*, are highly suggestive without providing any relevant explanatory information about these properties.

The term is problematic both for what it suggests and what it doesn't explain.

First, self-organization implicitly appears to posit a sort of unity or identity to the system of interacting elements in question—a "self," which is the source of the organizing effect. In fact, the coherent features by which the global wholeness of the system is identified are emergent consequences, not its prior cause. This is of course presumed to be an innocent metaphoric use of the concept of a self, which is intended to distinguish the intrinsic and thus spontaneous source of these regularities, in contrast to any that might be imposed extrinsically. But although the term has been used metaphorically in this way, and is explicitly understood not to imply anything like agency, it can nevertheless lead to a subtle conceptual difficulty. This arises when incautious descriptions of the globally regularized dynamics of such a system are described as causing or constraining the micro dynamics. As we discussed in chapter 5, the phrase "top-down causality" is sometimes used to describe some property of the whole systemic unity that determines the behavior of parts that constitute it. This has rightly been criticized as circular reasoning, treating a consequence as a cause of itself. But even when understood in process terms, where a past global dynamical regularity constrains future microdynamic interactions which in turn contribute to further global regularity, the term fails to explain in what sense the global dynamics is in any sense unified, as the word "self" suggests.

Second, describing these processes as self-organizing tends to suggest that the system in question is being *guided* away from a more spontaneous unorganized state. Used in this sense, it is metaphorically related to a concept like self-control. The problem with this comparison is that the organization is not imposed in opposition to any countervailing tendency. Self-organizing processes are spontaneously generated. The process could even be metaphorically described as "falling toward" regularity, rather than being forced into it, as is also the case with change toward equilibrium in simpler thermodynamic conditions. The specific forms of such processes are explicitly *not* imposed; they arise spontaneously, due to intrinsic features of the components, their interaction dynamics, and the constant perturbation of the system in question. In contrast, it often takes work to disrupt the regularity of a self-organized dynamical system, while constant perturbation is actually critical to its persistence.

Consider an eddy in a stream. It can be disrupted by stirring the water in opposition to the rotation of the vortex, or in any sufficiently different pattern, but stirring in the same direction is minimally disruptive. With sufficiently vigorous disruption, the rotational symmetry can be broken and a chaotic flow can be created—at least briefly. But so long as the stream keeps flowing, when the irregular stirring ceases, the rotational regularity will re-form. This is because the vortex flow is itself a consequence of constant perturbation as water flows past a partial barrier. The circular flow of the water is disrupted only by a contrary form of perturbation. In general, an intervention that can disrupt a stable morphodynamic process must diverge from it in *form*. This will differ for each distinct morphodynamic process, because there are many ways that a process can exhibit regularity.

Finally, the regularity that is produced is a consequence, not a formative influence or mechanism. Though all of the processes I will describe as morphodynamic are identified by virtue of converging toward a particular semi-regular pattern, what counts is that this consequence is approached, but need not ever be achieved. The asymmetric orthograde directionality of change is what matters. It is the tendency toward regularity and increasing global constraint that defines a morphodynamic process, not the final form it may or may not achieve. In this sense it is analogous to the way that the increase in entropy, but not the achieving of equilibrium, defines the orthograde tendency of a thermodynamic process. This is because using the production of a stable orderly dynamic as the sole criterion for identifying a morphodynamic process would cause us to overlook many relevant types of processes that fail to fully converge toward a regular state. In fact, as morphodynamic processes become more complex and intertwined, as they do in living organisms, none may actually converge to a regular pattern. Each may be generating a gradient of morphodynamic change with respect to others, even though none of the component processes ever reaches a point of morphodynamic stability.

Nevertheless, in either a thermodynamic or morphodynamic process, the same dynamical conditions that would ultimately converge to a stable end state, if left to run unaltered, are already at work long before there is any hint of stability. The point being that both thermodynamic and morphodynamic processes are defined by a specific form of orthograde change, not the end stage that such a change might produce. Thus, a morphodynamic process can be discerned in systems that will never ultimately converge to a stable end state. Even if a dynamical system only converges to a slightly less than chaotic regularity, it may still be morphodynamic.

How are we to recognize these processes in cases where there is limited time or contravening influences preventing convergence to regularity? The answer is, of course, that we must identify a morphodynamic process by virtue of a specific form of spontaneous orthograde change. So, while we have named morphodynamic processes with respect to their tendency to converge toward regular form, we must define them in terms of the dynamic process and not the form it produces. A brief and superficial description of the common dynamical principles characterizing morphodynamic processes is that they all involve the amplification and propagation of specific constraints. Of course, this brief statement requires considerable unpacking and qualification.

FAR-FROM-EQUILIBRIUM THERMODYNAMICS

Thinking in these terms can be confusing because of a double-negative logic that is hard to avoid. For example, in typical discussions of thermodynamic processes, we tend to think of energy as a positive determinant of change. Introduce energy into a system and it will eventually be dissipated throughout the system (and the surroundings if it is not isolated). But thinking in terms of constraint and entropy, the description becomes a bit more convoluted and counterintuitive. When a thermodynamic system—such as a gas in a closed container—is disturbed, say by the asymmetric introduction of heat, a constraint on the distribution of molecular movements has been imposed. Although the system is now more energetic, it is not merely the added energy that is responsible for the directional change that will eventually take the system to a new equilibrium. This would in fact occur whether one part was heated or one part was cooled. Removing heat in an asymmetric fashion is just as effective at initiating a re-equilibration process as is adding it. So, what is the cause of the asymmetric change, if not the addition of energy?

As we saw in the last chapter, this orthograde tendency in an equilibrating gas is the result of the biased distribution of molecular motions that this perturbation created, and it occurs irrespective of whether heat is added or removed. What matters is the creation of an asymmetric distribution of molecular motions. This perturbation lowers the entropy of the system, making some molecular movements more predictable with respect to their location within it. Or, to put it in constraint terms: the distribution of molecular velocities is more asymmetric (thus constrained) just after perturbation than when in equilibrium. The second law of thermodynamics thus describes a tendency to spontaneously reduce constraint, while thermodynamic work involves the creation of constraint. This complementarity will turn out to be a critical clue to the explanation of morphodynamic processes.

To see the relationship between this most basic thermodynamic process and morphodynamic processes, consider the following simple thought experiment: Imagine attaching a device to an otherwise isolated container of gas at equilibrium that will simultaneously cool one side of the container and heat the other side equivalently, but in a way that will keep the total energy (specific heat) in the container constant. Heat will be added exactly as rapidly as it is removed. If such a perturbation occurred only at one point, and then the system was isolated, the asymmetric state would have immediately started changing toward a state of equilibrium, attaining progressively more symmetric distributions of molecular motions with time. But what happens if the perturbation is continuous? In one sense, there is no intrinsic difference in the way the second law tendency is expressed. There is still a tendency toward redistribution of faster molecular velocities to one side and slower velocities to the opposite side in a way that runs counter to the perturbation, and yet every incremental change in distribution due to this spontaneous tendency is balanced by the perturbing influence of the heated and cooled sides of the container. The result is, of course, a stable gradient of temperature from hot to cold from one side to the other that never gets closer to an equilibrium state. And if somehow one were to stir things up, in a way that disturbed this gradient, as soon as the stirring stopped the same gradient would begin to re-form and re-stabilize.

What about the entropy of the gas in the container? Here is where things can get a bit counterintuitive. Clearly, energy in the form of heat is channeled through the gas, flowing from one side to the other, generating a constant increase in the entropy in the larger system that includes the heating and cooling mechanisms. Because the system is continually far from equilibrium due to constant perturbation, however, the gas within the container doesn't increase in entropy. So, although the dynamical interactions within the gas medium are in the process of increasing entropy, the entropy of the gas remains constantly well below its maximum, because of the continual disturbance from outside. Of course, the external heating-cooling device must do continual thermodynamic work to maintain this local constancy. This extrinsically generated contragrade dynamic continually counters the spontaneous orthograde dynamic within the gas medium. It is generally assumed that in such dissipative processes, the dynamical organization of gas molecule interactions eventually stabilizes to produce an *entropy production rate* (EPR) for the whole system (including heating and cooling mechanisms) that balances the rate of perturbation.

Locally within the gas we thus have a curious dynamical situation in which contragrade and orthograde processes are in balance. If the constant heating-cooling mechanism is adjusted to produce a higher or lower differential, the EPR will also shift to a new value. In a simple thermodynamic system, entropy increase occurs more rapidly if the temperature difference is greater and more slowly if the temperature difference is less (in this sense, a temperature difference is analogous to a pressure difference).

To understand what this means in terms of constraints, and thus the organization of the system, imagine that this constant extrinsic perturbation is suddenly removed and the system is again isolated. At this initial point, the fluid medium would be in a highly constrained state (thus highly organized) and would also spontaneously destroy that constraint (order) more rapidly than if it were in a less constrained state. Reflecting on what this means for thermodynamics in general, we can now see that the rate of constraint elimination is higher the more constrained the system (e.g., exhibiting a high heat gradient), and this rate will eventually decrease to zero when the system reaches the equilibrium state. What the external perturbation accomplishes then is to increase constraint (and drive the entropy down); or, in other words, it drives the system in a contragrade direction of change, thus becoming more highly ordered.

But there is an upper limit to this diffusion process that has a critical effect on the overall global dynamic. This is a threshold at which dynamical discontinuities occur. These are often called *bifurcation points*, on either side of which distinctively different dynamical tendencies tend to develop. At this threshold, local variations can produce highly irregular chaotic behaviors, as quite distinct dynamic tendencies tend to form near one another and interact antagonistically. But exceeding this threshold, radical changes in orthograde dynamics can take place. As the heat gradient increases to a high value, the potential influence of any previously minimal interaction biases and extrinsic geometric asymmetries will also become relevant, such as the viscosity of the medium, the shape of the container, the positions of the heating and cooling sites, and any possible impediments to dissipation or conduction through the container walls. This is because these additional biases—while irrelevant in their influence on the microdynamics of molecule-to-molecule interactions—are able to bias larger-scale collective molecular movements and allow differences to accumulate regionally. This can lead to the asymmetric development of correlated rather than uncorrelated molecular movements.

RAYLEIGH-BÉNARD CONVECTION: A CASE STUDY

This is most easily demonstrated by a liquid variant of the simple non-equilibrium thermodynamic condition just described. Perhaps the paradigm example of self-organizing dynamics is found in the formation of what are termed *convection cells* in a thin layer of heated liquid. Highly regular shaped convection cells (hereafter termed *Bénard cells*) can form in a process known as Rayleigh-Bénard convection in a uniformly heated thin layer of liquid (e.g., oil).[7] In 1900, Claude Bénard observed that a cellular deformation would form on the free surface of a liquid with a depth of about a millimeter when it was uniformly heated from the bottom and dissipated this heat from its top surface. This often converged to a regular pattern of tiny, roughly hexagonally shaped columns of moving fluid, producing a corresponding pattern of hexagonal surface dimples (see Figure 8.4). These Bénard cells form when the liquid is heated to the point where unorganized (i.e., unconstrained and normally distributed) molecular interactions are

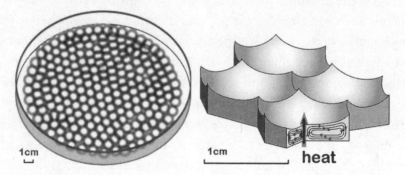

FIGURE 8.4: One of the most commonly cited forms of morphodynamic processes involves the formation of hexagonally regular convection columns called Bénard cells in shallow, evenly heated liquid. They form in liquid that is heated to a point where simple conduction of heat is insufficient to keep the liquid from accumulating more heat than it can dissipate. This creates instabilities due to density differences, and induces vertical currents due to weight differences. The heat dissipation rate increases via convection, which transfers the heat faster than mere passive conduction. The geometric regularity of these currents is not imposed extrinsically, but by the intrinsic constraints of conflicting rising and falling currents slowing the rate. These rate differentials cause contrary currents to regularly segregate and minimize this interference. Hexagonal symmetry reflects the maximum close packing of similar-size columns of moving liquid.

less efficient at conducting the heat from the container bottom to the liquid surface than if the liquid moves in a coordinated flow. The point at which this transition occurs depends on a number of factors, including the depth, specific gravity, the viscosity of the liquid, and the temperature gradient. The depth and dynamical properties of the liquid become increasingly important as the temperature gradient increases. The large-scale coordinated pattern of fluid movement reliably begins to take over the work of heat dissipation from random molecular movement when a specific combination of these factors is reached.

At moderate temperature gradients, local molecular collisions transfer heat as faster-moving molecules near the bottom bump into and transfer their momentum to slightly slower-moving molecules just slightly above them, so that collision-by-collision, molecular momentum is conveyed from the hot to the cool surface, where it ultimately gets transferred to air molecules. As the temperature gradient is increased, this process begins to be overshadowed by the direct movement of whole regions of faster-moving hot molecules toward the cooler surface because of their relative buoyancy. Hotter regions contain more energetic molecules that, as a result, maintain slightly larger distances between one another, and thus are collectively less dense and lighter than cooler regions. Cooler and thus heavier regions of liquid nearer the surface tend to descend and drive the lighter regions upward. The transfer of molecular momentum via collision inevitably follows highly indirect trajectories, whereas the collective movement of larger masses of liquid is less impeded by collision and their trajectory toward the surface is therefore more direct. So the rate that heat can be dissipated is inevitably higher if moving molecules, aligned with the direction of the heat gradient, are minimally impaired by collision. When the temperature gradient is slight, viscosity effects limit the rate at which nearby streams of liquid can move past one another, and so transmission of molecular momentum differences (heat) can occur more readily via molecular collisions. But at higher-temperature gradients, there is a reversal of this bias

Why do regularly spaced hexagonal columns of organized flow result? To see this, we need to focus on mid-scale dynamics between molecular and fluid dynamics. Because of the relatively more direct transmission of heat by aligned fluid flow at high-temperature gradients, regions of vertically aligned correlated movements will more effectively transfer heat out of the liquid than regions with less correlated molecular movements. But in the process of fluid convection, as more buoyant heated liquid rises from the bottom and heavier cooled liquid sinks back toward the bottom, there is an inevitable interaction between oppositely flowing streams. Consequently, the friction of viscosity affects the rate of convection. Regions with less correlated molecular movements will build up undissipated heat (higher local intermolecular momentum) more rapidly than neighboring regions, and create localized temperature gradients that are slightly out of alignment with the large-scale gradient. This creates a bias of movement out of regions with disoriented and less correlated flow toward regions with more aligned and correlated flow. Thus, centers of aligned and correlated convection in either direction will expand, and regions of chaotic flow will have their volumes progressively diminished. The result is that regions of countervailing flow will become minimized over time.

Hexagonal columns of fluid flow develop because the heated and cooled surfaces are planar, and the most densely packed way to fill a surface with similar-size subdivisions is with hexagons.[8] Subdividing the plane of liquid into hexagonal subregions thus most evenly distributes the inversely moving columns of flowing liquid, and represents the most evenly distributed pattern of heat dissipation that can occur via fluid movement. This action both spontaneously evens out the rate of heat dissipation from region to region, and optimizes the rate of dissipation overall. So, as convection becomes critical to maintaining a constant high gradient of heat transfer through the liquid, the whole volume becomes increasingly regularized, until hexagonal columnar convection cells form. At this point, no more efficient movement patterning is available. From this point on, the local value of entropy within the liquid will tend to remain fairly stable. But because of the highly organized global convection patterning, stable global-level constraints are also generated that do not dissipate. The total rate of entropy production, and thus constraint dissipation, is maximized as far as convection can generate it.

Bénard cells thus exhibit an effective reversal of the typical thermodynamic orthograde tendency: that is, for macroscopic constraints to be progressively eliminated through microscopic dynamical interactions. Instead, as these convection patterns form, the intrinsic constraints of fluid movement become amplified and propagated throughout the system, taking on some of the potential differential that otherwise would have been borne by microscopic intermolecular momentum differences, trading movement gradient effects for heat gradient effects.

To understand what this more complex condition means in terms of constraint (and thus the organization of the system), imagine once more that this constant extrinsic perturbation is suddenly removed and the system becomes isolated. At this point, the fluid medium is even more highly constrained, and partitioned into subregions with highly divergent parameters, than if it were just exhibiting a high heat gradient. Consequently, the rate at which these constraints would tend to spontaneously decay will also be more rapid than in a simple heat gradient of the same value. In the far-from-equilibrium constantly perturbed condition, then, where convection has taken over much of the heat dissipation from molecular collision, the extrinsic perturbation is also driving the system into an even more constrained (and thus organized) state than it would be with a simple heat gradient. In the absence of convection, this contragrade process would have resulted in a significantly higher gradient of heat difference within the fluid; but in effect, the formation of convection cells has redistributed this level of global constraint to slightly more local constraints of a different sort— constraints on correlated movement with the result that the temperature gradient is less steep than in their absence. Because correlated movement of large collections of molecules involves fewer dimensions of difference than uncorrelated local molecular movement, the system is by definition in a simpler and thus more orderly state. What the external perturbation coupled with the effects of buoyancy and the geometric constraint of hexagonal close packing together accomplish is to increase and redistribute constraint in a more symmetric way. This redistribution of constraint into other dimensions of difference provides more ways for constraints to be eliminated as well. This is what creates the higher rate of constraint elimination (and thus entropy production).

In summary, morphodynamic organization emerges due to the interaction of juxtaposed constant work opposing a homeodynamic process (e.g., processes involving both constant external perturbation and constant internal equilibration). High, uniform rates of perturbation lead to the production of second-order intrinsic constraints that eventually balance the rate of constraint dissipation to match the introduction of constraints from outside

the system. Balanced rates are achieved when the rate of entropy production increases to the point that it keeps pace with the rate of disturbance by
generating more dimensions of internal constraint that can now be simultaneously eliminated. Again, if the system is suddenly isolated, not only will
temperature begin to equilibrate, but correlated movement will also begin to
break down. This demonstrates that some of the heat energy went to create
convection. Constraint dissipation was able to occur at a higher rate because
fluid movement was doing work to increase the rate of heat transfer.

This may sound counterintuitive, but notice that the correlated movement of the convection cells is itself capable of being tapped to do work, and
to generate this level of correlation in the first place, work was required, and
that must have come from somewhere. It cannot have come for free. It cannot have come from gravitation, because although the falling, heavier liquid
does work to push hotter, more buoyant liquid upward, this is the result of
the density differential that was produced by the asymmetry of the heating
and cooling. Gravitation merely introduces a bias, as does hexagonal close
packing, which organizes the flow but doesn't initiate it. The only available
source of work to organize convection is the constantly imposed temperature differential. Inevitably, then, some of the microscopically distributed
molecular momentum differential must be reduced as it is transferred into
the correlated momentum of the convection flows. Heat in the form of
microscopic motion differences thus is transformed into global correlated
movement differences.

The morphodynamic process that results is an orthograde process,
because it is an asymmetric orientation of change that will spontaneously
re-form if its asymmetry is somehow disturbed and so long as lower-order
dynamics remain constant. But this higher-order asymmetry is dependent
on the persistence of lower-order thermodynamically contragrade relationships: thermodynamic work. It is because of this lower-order work that the
higher-order orthograde dynamic exists.

What does this tell us in terms of morphodynamic processes in general?
Using the Bénard cell case as an exemplar, it demonstrates that if there are
intrinsic interaction biases available (buoyancy differences, viscosity effects,
and geometric distribution constraints in this case), the persistent imposition of constraint (constant heating) will tend to redistribute this additional
constraint into these added dimensions of potential difference. Moreover,

these additional dimensions are boundary conditions, to the extent that they are uniformly present across the system. This includes the geometric constraint, which is not derived from any material feature of the system or its components. Because these additional dimensions are systemwide and ubiquitous, they are also of a higher level of scale than the constraints of molecular interaction. So this transfer of constraints from molecular-level differences to global-level differences also involves the propagation of constraint from lower- to higher-order dynamics.

The distinct higher-order orthograde tendency that characterizes the morphodynamics of Bénard cell formation thus emerges from the lower-order orthograde tendency that characterizes fluid thermodynamics. This tendency to redistribute constraint to higher-order dimensions is an orthograde tendency of a different and independent kind than the spontaneous constraint dissipation that characterizes simpler thermodynamic systems. Yet without the incessant process of lower-order constraint dissipation, it would not occur. The higher-order orthograde morphodynamic tendency is in this sense dynamically supervenient on the lower-order orthograde thermodynamic (homeodynamic) tendency, though it is not supervenient in a mereological sense.

Bénard cells are entirely predictable under the right conditions, and will thus re-form spontaneously, like a whirlpool in a stream, if disturbed by stirring the fluid and temporarily disrupting the regular flow patterns. This spontaneous tendency to return to a dynamical regularity that will persist unless again disturbed is what defines an *attractor* in the space of dynamic possibilities (in the terms of dynamical systems theory). The combination of an up-level shift in the form of constraint and the generation of a distinct higher-order orthograde attractor is the defining characteristic of an emergent dynamical transition. This is because, ultimately, the capacity to do work to change things in a non-spontaneous way depends on the presence of constraints, and so a new domain of orthograde constraint provides a new domain of causal powers to introduce change.

THE DIVERSITY OF MORPHODYNAMIC PROCESSES

This detailed exploration of Bénard cell formation demonstrates that in order to realize the potential of emergent causal power at the morphodynamic level, the interaction of contragrade dynamics is required at the next lower level of scale. Just as the spontaneous increase in the rate of constraint dissipation (and entropy production) can be realized in diverse ways, and not merely via the jostling of molecules in a heated gas or liquid, the amplification-propagation of higher-order constraints can also be realized in diverse ways, producing quite different sorts of attractors. This second-order orthograde logic characterizes all morphodynamic processes, including those that are not strictly speaking thermodynamically embodied (e.g., computational processes); and while they only occur when special conditions are met, they typically stand out as surprising apparent violations of the tendency toward disorder that is the most ubiquitous tendency in the universe.

First, consider a couple of close cousins to Bénard cells: geological analogues to convection, that include the formation of basalt columns and soil polygons.

Basalt columns are hexagonal columns that form in molten rock due to convection, as heat dissipates upward toward the surface.[9] As the molten lava cools and hardens, it also tends to contract. The columns break apart at the boundaries between the hexagonal cells where the rock is coolest and least plastic. The large size of the hexagons in comparison to Bénard cells is in part a consequence of the very different fluid properties of lava.

Soil and frost polygons (shown in Figure 8.3) form in multiple ways, due to different mechanisms driven by the expansion and contraction of ice particles in the soil, or alternatively as a consequence of the drying and shrinking of mud. Cracks formed into roughly rectangular polygons often result from contraction of soil due to drying. This does not involve regular movement of soil components, and so cracks form more or less along fracture lines, creating crisscross patterning with fissures that are roughly evenly spaced because of the relatively equal rates of drying. In contrast, more or less hexagonally arranged roughly circular rings of stones form due to the differential upwelling of larger particulates (such as stones), driven by cycles of freezing and thawing of the surrounding soil. In this respect, there is a loose analogy to Bénard convection, as the larger stones are driven upward in part because they do not expand and contract with the surrounding soil. A related sorting of particulates occurs on rocky seashores, where the action of waves helps to arrange different-sized, -shaped, and -weighted objects in different strata close to or further from the shore. The sorting of particulates according to shape, size, and/or weight is not only a common geological feature, but is also used in the development of mechanical agitation mechanisms employed to sort different objects for various industrial purposes.

A quite different example of morphodynamic change is exhibited by the amplification and propagation of constraints that takes place in the growth of snow crystals. The structure of an individual snow crystal reflects the interaction of three factors: (1) the micro-structural biases of ice crystal lattice growth, which result in a few distinct hexagonally symmetric growth patterns; (2) the radially symmetric geometry of heat dissipation; and (3) the unique history of changing temperature, pressure, and humidity regimes that surrounds the developing crystal as it falls through the air (see Figure 8.5).

Snow crystal growth occurs across time in the process of traversing these variable atmospheric conditions. But an ice crystal lattice tends to grow according to only a few quite distinct patterns (spires, hexagonal sheets, and hexagonal prisms, to name the major forms), depending on the specific combination of temperature, pressure, and humidity of the surrounding air. As a result, the history of the differences in atmospheric conditions that a

FIGURE 8.5: Selected snow crystals from the classic collection of photographs made by Wilson Bentley (1865–1931), obtained from the NOAA National Weather Service Collection Catalog of Images. All exhibit an elaborate hexagonally symmetric form, made up of plate, spire, and surface-etching components. This remarkable regularity is a result of the way temperature, humidity, and pressure determine ice crystal lattice formation, radial heat dissipation, and the compounding of geometric constraints on growth surfaces as the crystal grows. Snowflakes typically consist of many crystals stuck together.

growing crystal encounters as it falls to earth is expressed in the variants of crystal lattice structure at successive diameters and branch distances in its form. In this way, the crystal is effectively a record of the conditions of its development. But snow crystal structure is more than merely a palimpsest of these conditions. Prior stages of crystal growth progressively constrain and bias the probability of further growth or melting at any given location at subsequent stages. Thus, even identical conditions of pressure, temperature, and humidity, which otherwise produce identical lattice growth, can produce different patterns depending on the prior growth history of the crystal. In this way, the global configuration of this tiny developing system at each stage of growth plays a critical causal role in its microscopic dynamics. The structure present at any moment in its growth history will strongly bias where molecular accretions are most and least likely, irrespective of the surrounding conditions, which will instead influence the mode of growth that will occur at these growth points. Since partial melting and refreezing are also possible, and this too will depend on the constrained distribution of heat at any given position, the resultant crystal shape will not necessarily exhibit precise angular crystalline structure, but may also yield highly regular symmetrically distributed amorphous patterns as well, including ridges and trapped bubbles (shown in Figure 8.5). This is what contributes to the proverbial individuality of each crystal.

Snow crystals are self-simplifying in a quite different way from convection cells, and yet both share a deep commonality. A snow crystal's growth is analogous to Bénard cell formation in the way it too is a consequence of persistent non-equilibrium dynamics. It is continually made thermodynamically unstable by the accretion of new molecules which release heat into the crystal lattice as they fuse, and from which heat is continually being dissipated through the highly constrained geometry of the crystal. In the process of exporting heat as fast as it is accumulated during growth conditions, the crystal tends to spontaneously regularize at a macroscopic level (i.e., macroscopic with respect to the scale of molecular interactions) because asymmetric growth leads to asymmetric heat dissipation, which tends to slow overgrown and accelerate undergrown regions. And as the process continues, the constraints of prior growth and heat dissipation further constrain the possibilities for growth. Along with the geometrically limited growth possibilities of ice crystal lattice, these factors amplify contingent events in the growth history of the crystal. Because of the compounding of constraints from the prior growth history, a snow crystal incorporates and amplifies the unpredictable influence of these random accretions into its complex symmetry, including even the effects of melting and refreezing. The quite exquisite symmetries that can spontaneously form are thus the result of a complex interplay of many different kinds of constraints, amplifying the extrinsically acquired biases introduced by a historically unique sequence of changing boundary conditions.

Laser physics provides an example of morphodynamic logic in a very different domain (see Figure 8.6). It is based on a constraint amplification effect that is due to the temporal regularity of quantum resonancelike effects involving the atomic absorbance and emission of radiation. Lasers produce

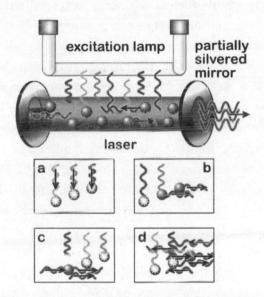

FIGURE 8.6: Laser light is also generated by a morphodynamic process that is roughly analogous to resonance. It is generated when broad-spectrum (white) light energy is absorbed by atoms (a), making them slightly unstable. This energy is re-emitted at a specific wavelength, corresponding to the energy of this discrete quantum level (b). Re-emission is preferentially stimulated if an unstable atom interacts with light of the same wavelength (c and d). The critical effect is that the emitted light is emitted with the same phase and wavelength as the incident light. If this emitted light is caused to recycle again and again through the laser material by partially silvered mirrors, this recursive process progressively amplifies the alignment of phase and wavelength of the light being emitted.

intense beams of monochromatic light such that all the waves are in precise phase alignment—that is, with the peaks and troughs of the waves emitted from different atoms, all aligned. Light with these precisely correlated features is called *coherent light*. It is generated from white (polychromatic) light, which contains mixed wavelengths aligned in every possible phase. The conversion of white light to coherent light is accomplished by virtue of the recurrent emission and resorption of light by atoms whose emission features correlate with their excitation features. When the energy of out-of-phase polychromatic light is absorbed into the electron shells of the atoms of the laser material, it is incorporated into a system with very specific (quantized) energetic regularities. The energy is incorporated in the form of a shift of electron energy level to a higher and less stable quantized state.

An unstable excited atom reverts to its relaxed ground state (able to be re-excited) by dissipating the excess energy as re-emitted light. The light it emits carries a discrete amount of energy corresponding to the discrete quantum difference in energy levels of stable electron shell configurations, and is thus emitted at a specific wavelength. If the laser material is uniform, the excitation results in uniform color output. Amplification of the regularity of the emission of light by different atoms in this medium is achieved by causing the emitted light to reenter the laser by virtue of partially silvered mirrors. Thanks to the common quantized character of the light-absorbing-and-emitting spectra of atoms capable of lasing, light at the emission frequency is also most likely to induce an energized atom to revert to its more stable ground state, and emit its excess energy as light. When it does this in response to light emitted by another atom at the same precise wavelength, it does so in a phase that is precisely correlated with the exciting light. Repeated charging with white light and recycling of emitted light thus amplifies this correlation relationship by many orders of magnitude.

The continual pumping of energy into the atoms of the laser in the form of white light makes them unstable and predisposed to dissipate this energy. It is the analogue of asymmetrically heating the fluid in Bénard cell formation or accreting water molecules to the snow crystal lattice. A laser is similarly a dissipative system. The orthograde tendency of the atoms is to offload this excess energy and return to a more stable configuration; but the excited state is also somewhat stable, though far more easily disrupted even by the random fluctuations of the electrons themselves. Although spontaneous emission tends to be entirely uncorrelated, and is in this way analogous to the random dissipation of heat in a thermodynamic system via molecular interactions, emission that is stimulated by interaction with photons with precisely the same energy (and thus wavelength) is in effect biased (or constrained) to be emitted in resonance with the stimulating light. So, although light-emitting atoms are distributed in space throughout the laser, as previously emitted light travels recurrently back through the population of continuously excited atoms, it recruits more and more phase-matched photons. Since the system is being continually perturbed away from its ground state, and constantly dissipating this disturbance in the form of emitted light, the facilitation of emission results in an amplification of the matched constraints of stimulation and emission.

To summarize, let's reiterate the common dynamic features that characterize each of these morphodynamic phenomena, and which make them an emergent level removed from subvenient homeodynamic processes, whether at a thermodynamic or subatomic level. In each case we find a tangled hierarchy of causality, where micro-configurational particularities can be amplified to determine macro-configurational regularities, and where these in turn further constrain and/or amplify subsequent cycles of this process, producing a sort of compound interest effect. Although mechanistic interactions, aggregate thermodynamic tendencies, or quantum processes (as in the case of lasers) constitute very different domains with different dynamical properties, the specific reflexive regularities and the recurrent causal architecture of the cycles of interaction have in these cases come to overshadow the system's lower-order orthograde properties. These systems must be open to the flow of energy and/or components, which is what enables their growth and/or development, but they additionally include a higher-order form of closure as well. Such flows propagate constraints inherited from past states of the system, which recurrently compound to further constrain the future behaviors of its component interactions. The new higher-order orthograde dynamic that is created by this compounding of constraints is what defines and bounds the higher-order unity that we identify as the system. This centrality of form-begetting-form is what justifies calling these processes morphodynamic. And the generation of new orthograde dynamical regimes is what justifies describing morphodynamic processes as emergent from thermodynamic processes.

THE EXCEPTION THAT PROVES THE RULE

As we have seen, the emergence of a morphodynamic process is necessarily supervenient (in the dynamical sense described above) on persistent far-from-equilibrium conditions. In thermodynamic terms, this means that in order to maintain constraints that do not tend to form as a result of relaxing to thermodynamic equilibrium, it is necessary to "freeze" them in some sense, as literally occurs in snow crystals and figuratively in frost polygons. But otherwise it seems that the spontaneous higher-order organization of dissipative systems should spontaneously undermine the very gradient that sustains this condition. And yet, the maintenance, reconstruction, and reproduction of dynamical constraints is a core characteristic of life. How could a process that is dependent on constraint dissipation and maximizing the rate of entropy production ever become a means for reducing this rate? A clue is provided by recognizing that, besides being a product of work, constraint is also the precondition for work.

To see how it is possible to use constraint dissipation against itself, consider the simple problem of trying to cool a building. If the building is hotter than the air outside, it will tend to cool spontaneously by heat being conducted through the walls and windowpanes. When a door or window is opened, a constraint on the dissipation of heat is removed, and assuming that there is a temperature gradient between inside and outside, the rate of heat exchange will increase. The inside of the building will cool, and the whole system (indoors plus outdoors) will approach equilibrium more rapidly. This can even be improved by opening a door on an upper floor since heat rises. This takes advantage of a bias introduced by gravity: hot air is less dense and therefore lighter than cool air. This is, of course, the analogue of the way that Rayleigh-Bénard convection more effectively reduces a heat gradient and increases the rate of heat dissipation, by redistributing this microscopically embodied temperature constraint to macroscopic movement constraints.

Such a convection "breeze" will spontaneously converge on a pattern that depends strongly on the placement of these openings with respect to local biases, including the effects of gravity, the position of the heat source, and the positions of the paths open to the outside. Indeed, these otherwise passive and spatially distributed constraints will become increasingly influential as the temperature differential is increased.

Once we add convection to the process, however, an interesting additional twist becomes possible. Air flow is a new source of potential work, because unlike the globally distributed heat dissipation pattern embodied in microscopic differences, moving streams of air are composed of molecules in coordinated local movement, such that their momenta are all oriented in a similar direction (i.e., constrained). Consider the effect on an open door or window. If the convection breeze is sufficiently powerful, it could blow the door closed. One possible consequence of the tendency for a system to dissipate a heat gradient by shifting some of this work to higher-order constrained dynamics is that it can produce conditions which can act to decrease that rate by organizing local work to introduce a new constraint that incidentally impedes dissipation. Thus higher-order work can undermine the lower-order work that it depends on.

There is a loose analogy between the dissipative processes of life and the blowing closed of the door in the hot building scenario sketched above. Life is characterized by the use of energy flowing in and out of an organism to generate the constraints that maintain its structural-functional integrity. Since organisms are subject to the incessant dissipative effects of the second law of thermodynamics, they additionally need to constantly impede certain forms of dissipation. Organisms take advantage of the flow of energy through them to do work to generate constraints that block some dissipative pathways as compared to others. This is important, because it shows that living organisms don't necessarily increase the rate of entropy production over some background inorganic rate. Indeed, here is where the simple thermodynamic analogy breaks down even further, because the work generated by an organism doesn't just block constraint breakdown, it also builds new organisms constituted by new constraints. So we additionally will need to understand how the production of new forms (new constraints) and the production of newly partitioned systems can affect both the homeodynamic and the morphodynamic processes that take place in organisms. This is one of the main challenges of the next chapter.

9

TELEODYNAMICS

It does seem odd . . . that just when physics
is . . . moving away from mechanism, biology and
psychology are moving closer to it. If the trend
continues . . . scientists will be regarding living and
intelligent beings as mechanical, while they suppose
that inanimate matter is too complex and subtle to
fit into the limited categories of mechanism.

—DAVID BOHM, 1968[1]

A COMMON DYNAMICAL THREAD

The fact that there are spontaneous inorganic processes that generate macroscopic order is seen by many as a missing link between living and non-living processes. Even the simplest life forms are remarkably complex in their microphysical structure and chemical dynamics, and all forms of life exhibit the uncanny ability to resist the relentless eroding influence of thermodynamics. So, demonstrating that structural and dynamical regularity can arise spontaneously in the absence of life provides some hope that the life/nonlife threshold could be understood in terms of a special chemical self-organizing (morphodynamic) process.

Indeed, there are many who believe that life is essentially a highly complex form of self-organizing (i.e., morphodynamic) process. The question that will occupy this chapter is whether life is *nothing but* complex morphodynamics, or whether instead there is something more to life than this, an additional emergent transition. As the title of this chapter suggests, I will argue the latter, but like the transition from thermodynamics to morphodynamics, this additional emergent transition turns out to be dynamically supervenient on morphodynamics and therefore also on thermodynamics. Explaining this doubly emergent dynamical logic is not just relevant to life, however. As we will see below, this additional emergent dynamical transition is necessary to account for ententional phenomena in general. Morphodynamical processes generate order, but not representation or functional organization, and they lack any normative (or evaluative) character because there is nothing like a self to benefit or suffer.

I will call this particular way of organizing causal processes *teleodynamics*, because of the characteristic end-directedness and consequence-organized features of such processes. By describing the emergence of this mode of dynamical organization from the simpler modes of thermodynamic and morphodynamic processes, and its dependency on them, we will build the base foundation upon which the emergence of more highly differentiated forms of ententional relationship can be understood, including even human consciousness.

Evidence that life involves morphodynamic processes comes from two obvious attributes that are characteristic of all organisms. First, organisms are incessantly engaged in processes of creating and maintaining order. Their chemical processes and physical structures are organized so that they generate and maintain themselves by continually producing new appropriately structured and appropriately fitted molecular structures. Second, to accomplish this incessant order generation, they require a nearly constant throughput of energy and materials. They are in this respect dissipative systems. Together, these two characteristics give life its distinctive capacity to persistently and successfully work against the ubiquitous, relentless, incessantly degrading tendency of the second law of thermodynamics. Individual organisms do this via metabolism, development, repair, and immune response. Lineages of organisms do it by reproduction and evolution. In this chapter, I will argue that these are all products of a common dynamical logic.

Though life is characterized by its success at circumventing the near inevitability of thermodynamic degradation, this does not mean that the global thermodynamic trend is reversed—only that living processes have created protected local domains in which the orthograde increase in entropy is effectively reversed by virtue of contragrade processes that generate order and new structural components at the expense of a net entropy increase in their surroundings. But living processes are not just local pockets of resistance against entropy increase. They also persistently decrease it, within themselves and their progeny over the course of evolution, by developing and evolving complex supportive correlations between structures and processes for maintaining bodies and ecosystems. They do this in the context of specific and often variable environmental conditions. This is the feature of organisms that we describe as *adaptation*. In terms of thermodynamics, an adaptation can be defined as any feature of an organism or lineage of organisms that directly or indirectly plays a role in compensating for spontaneous entropy increase within organisms. This can be as specific and local as molecular repair and replacement processes within cells, or as general as organism behaviors that compensate for changes in essential resources or critical physical conditions present in the environment.

Without question, the most dramatic expression of the emergent nature of life's distinctive dynamic is the generation of increasingly diverse and complex forms of organisms that have evolved during the past 3.5 billion years of Earth history and have adapted to an ever-increasing range of environmental conditions. In the general dynamical terminology that we have been using, this evolutionary process must be understood as an orthograde process because of the spontaneity of its dynamical asymmetry. In other words, although evolution depends on the work done by organisms resisting degradation long enough to reproduce, the global asymmetric dynamic of evolution (constantly adapting organisms to changing environments, creating and expanding into novel niches, and producing increasingly complex forms over time) is only an indirect higher-order product of this work. There is no work that forces this particular trend in contrast to the opposite trend, and indeed evolutionary degradation is common. The asymmetry of the evolutionary process is the result of formal asymmetries in the space of options for adaptation that spontaneously tend to arise over time for living ecosystems. These options (niches) are like the formal features of the space of opportunities for change that characterize orthograde processes in both homeodynamic and morphodynamic processes. Like the transition from thermodynamics to morphodynamics, the higher-order break in orthograde symmetry exemplified by the evolutionary process is one of the defining characteristics of an emergent dynamical transition.

The thermodynamically exceptional and mechanically counterintuitive reversal of the causal logic of natural selection stands out as one of the most significant dividing lines in natural science. Described in terms of thermodynamics, living processes superficially appear to exhibit mirror-image tendencies compared to what is common in the non-living world. On the non-living side, we find processes that (a) have wide generality in their dynamical tendencies; (b) are describable with good predictive power by fairly simple dynamical equations or statistical methods; (c) exhibit higher-order aggregate properties that can typically be extrapolated from properties of their components in interaction; and (d) exhibit a tendency to dissipate constraints, simplify complex interdependencies, and redistribute free energy in ways that decrease any capacity to do work. On the living side, we find processes that (a) consistently partition thermodynamic processes so that many component processes follow trajectories that run radically counter to global thermodynamic probabilities; (b) are highly heterogeneous in their structures and dynamics; (c) produce processes/behaviors that are so convoluted, divergent, and idiosyncratic as to defy compact algorithmic description; (d) generate and maintain aggregate systemic properties that are quite distinct from properties of any component; and (e) reflect the effects of deep historical contingencies that may no longer be existent in their present context.

The transition from physical systems exhibiting mechanical/thermodynamical regularity to living systems exhibiting adaptive and self-sustaining features therefore seems almost like an inversion of the most basic underlying physical tendencies of nature.[2]

Many of these features exhibit the signature of morphodynamic processes, and indeed, many theorists have equated living processes with such dissipative processes. A focus on the far-from-equilibrium dynamics of living systems was highlighted by Erwin Schrödinger's effort to bring attention to their unusual thermodynamic tendencies, and it was a significant motivating factor in the pioneering studies of cyclic chemical processes by Manfred Eigen and of dissipative structures by Ilya Prigogine.[3] Prigogine's demonstration that dissipative systems are characterized by a tendency toward a maximum entropy production rate set the stage for the development of a quite sophisticated understanding of the augmentation of classical thermodynamic theory, which was needed to explain chaotic and morphodynamic processes, as well as non-linear dynamical processes in general.

Living systems are characterized both by far-from-equilibrium thermodynamics and order production—the hallmarks of morphodynamics. So it is not surprising that living processes have often been equated with these processes. This has led to a widespread tendency to equate organisms with non-linear dissipative processes exhibiting merely morphodynamic features, and to assume that a dynamical systems theory account framed in these terms provides a sufficient explanation for living dynamics. This identification of life with self-organization has proved to be both a major step forward and yet also an impediment to developing a conception of living processes sufficient to account for their ententional characteristics.

This view has been eloquently articulated by a number of theorists. Thus Roderick Dewar argues that "Maximum entropy production is an organizational principle that potentially unifies biological and physical processes,"[4] while A. Kleidon argues that "biological activity increases the entropy production of the entire planetary system, both living and non-living,"[5] and Rod Swenson claims that "evolution on planet Earth can be seen as an epistemic process by which the global system as a whole learns to degrade the cosmic gradient at the fastest possible rate given the constraints."[6] As we will see shortly, this last caveat ("given the constraints") is a critical qualification that ultimately becomes the tail wagging the dog, so to speak, in that it entirely undermines the universality of this claim when applied to life.

As we saw in the last chapter, it is well established that morphodynamic processes develop in persistent far-from-equilibrium conditions because the increase in internally generated dynamical constraints more efficiently (requiring less work) depletes the energetic and/or material gradient that is driving the system away from equilibrium. In this way, morphodynamic processes accelerate the destruction of whatever gradient is responsible for generating them. They are, in this respect, self-undermining, and are only maintained when this gradient is constantly replenished by some extrinsic source. This principle is eloquently stated by two major proponents of the view, E. D. Schneider and J. J. Kay, in an influential paper arguing for the relevance of maximum entropy production in characterizing organic and evolutionary processes. They state that

As systems are removed from equilibrium, they will utilize all avenues available to counter the applied gradients. As the applied gradients increase, so does the system's ability to oppose further movement from equilibrium. . . . No longer is the emergence of coherent self-organizing structures a surprise, but rather it is an expected response of a system as it attempts to resist and dissipate externally applied gradients which would move the system away from equilibrium.[7]

Living systems do indeed incorporate morphodynamic processes at nearly every level of their organization, from complex cycles of catalytic molecular reactions to the embryonic cellular interactions which determine the elaborate theme-and-variation organization of plant and animal body architecture. And yet in many respects organisms as whole dynamical systems, evolving lineages, or embedded within ecosystems also exhibit properties that differ radically from those characterizing morphodynamic processes. These include at least four ways that organisms invert the logic of morphodynamics:

1. Organisms depend on and utilize energetic and material gradients in their environment in order to perform work to sustain the constraints of their persistent, far-from-equilibrium dynamics, and to maintain constraints that are critical for countering the tendency toward thermodynamic decay.
2. Organisms actively reorganize their internal dynamics and relationships to the environment in ways that specifically counter or compensate for any depletion of the gradients that is necessary to maintain their dynamical integrity and their capacity to so respond.
3. Many organisms have evolved means of gradient assessment and spatial mobility that enable them to anticipate and avoid conditions of depleted gradients and to seek out more optimal gradients.
4. Organisms and ecosystems evolve toward forms of organization that increase the indirectness of the "dissipation-path length" of energy and material throughput in order to extract more work from the available gradients.

In general, the ways that these processes all serve to maintain the capacity for self-repair and self-replication exemplify a clear inversion of the most typical features of morphodynamic systems. This indicates that dynamical systems approaches limited to morphodynamic and chaotic non-linear dynamics are insufficient to account for living dynamics. Life's paradoxical dependence on morphodynamics, but its inversion of its most characteristic consequences, suggests that an additional dynamical inflection separates living processes from morphodynamic processes. This phase change thus exemplifies a further emergent transition (summarized in Figure 9.1).

Described in these general terms, we can begin to define the signature of this common higher-order dynamical logic. It is *a dynamical form of organization that promotes its own persistence and maintenance by modifying this dynamics to more effectively utilize supportive extrinsic conditions*. Importantly, such a dynamical organization is not identifiable with any particular constituents, or even any constituent morphodynamic process. It is this higher-order dynamical organization itself that is organized with respect to its own persistence, and precisely because it is *not* bound to specific material substrates or component dynamical process.

As we will see below, this fundamentally open and generic nature of living processes also means that they can additionally entrain and assimilate any number of intermediate supportive components and processes. This generic

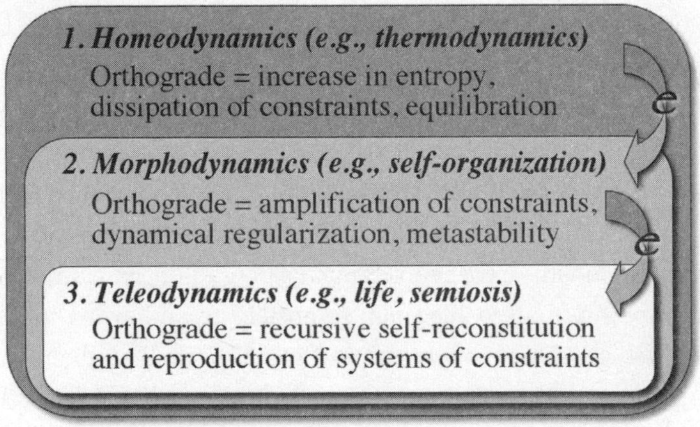

1. **Homeodynamics (e.g., thermodynamics)**
Orthograde = increase in entropy, dissipation of constraints, equilibration

2. **Morphodynamics (e.g., self-organization)**
Orthograde = amplification of constraints, dynamical regularization, metastability

3. **Teleodynamics (e.g., life, semiosis)**
Orthograde = recursive self-reconstitution and reproduction of systems of constraints

FIGURE 9.1: The nested hierarchy of the three emergent levels of dynamics; their typical exemplars; and their emergence (*e*) from subvenient dynamical processes.

openness is what allows new functions and (in more complex organisms) new end-directed tendencies to evolve. By giving this general dynamical logic the name *teleodynamics*, we are highlighting this consequence-relative organization.

LINKED ORIGINS

It is a central hypothesis of this argument that the threshold zone between life and non-life corresponds to the fundamental boundary between teleo-dynamic processes and the simpler regimes of morphodynamic and ther-modynamic processes. This does not necessarily mean that the origin of life is the only threshold leading to teleodynamics, or that only life can be teleo-dynamic. However, although the nature of this dynamical transition might prove to be quite generic—i.e., achievable in many ways in diverse kinds of substrates—there are good reasons to believe that the simplest (and per-haps the only) way that this threshold could be *spontaneously* crossed likely involves a simple molecular system; a first precursor to life as we know it.

So, although explaining the emergence of teleodynamics and investigat-ing the origin of life are not the same enterprise, they are likely to inform one another. The non-life/life transition was both a critical threshold mark-ing a fundamental phase change in dynamical organization, and yet must also have involved only a small number of constituent structures and pro-cesses, given that their synergistic coalescence occurred "accidentally." Being both fundamental and yet necessarily simple at the same time makes this a promising point of entry for explaining the origins of teleodynamic processes. Explaining the origins of life may actually be a more difficult task. Rather than focusing on the component structural or functional features of organisms per se, we can instead turn our entire attention to the dynamical problem itself, irrespective of particular planetary conditions and molecular specifics.

This is helpful because with respect to the origin of life, the specific molecular and geochemical details are daunting. Many of the dozens of essential molecular components of simple living cells have been considered as promising prerequisites to the formation of life, including nucleic acids, amino acids, phospholipids, sugars, purines and pyrimidines, and many mineral complexes and ions, such as involve phosphorus, iron, sulfur, calcium, chlorine, sodium, and so on. For this reason, it is often quite difficult to decide which few of these could be most primitive, and which should rather be considered acquired components that became essential only later in the evolution of life on Earth. More problematic is the fact that the most critical molecular constituents of even the simplest organisms are chiefly large complex molecules—polymers—that do not easily form spontaneously, and are easily broken up. Their production in cellular metabolism requires some of the most complex combinatorial molecular processes in existence.

Discerning the essential primitive dynamics of life is similarly clouded by life's current complexity, and because the evolutionary process has likely erased most traces of life's earliest, minimally sufficient precursors. Indeed, because of this necessary simplicity, the first molecular systems able to spontaneously achieve the dynamical basis for a quasi-evolutionary process inevitably lacked many of the characteristics found in contemporary life forms, and yet they would need to be teleodynamic.

Definitions of life based on generalizations derived from functional features found in all living creatures typically cite molecular replication, cell division, energy metabolism, semi-permeable membranes, and the maintenance of far-from-equilibrium chemical processes as minimal characteristics, with perhaps a few others. Because the probability of the serendipitous chance coalescence of even a few of the molecular types and dynamic interdependencies necessary to produce these effects is vanishingly small, scenarios for the origins of life have had to envision ways that a much smaller set could be spontaneously assembled, often utilizing aspects of life that show some degree of self-association and ancillary environmental support (lipid "bubble" formation, the catalytic-template possibilities of clays, etc.).

The explosive plurality of dynamic features and molecular species to choose from has led to quite diverse scenarios and contrived arrangements between presumed early terrestrial environments and molecular processes. It is not even clear what criteria could be used to rank the plausibility of these alternatives.

Finally, organisms are both components and products of the evolutionary dynamic in much the same way as biomolecules are both components and products of the synthetic network that constitutes a cell. For the same reason, organisms and the evolutionary process are not separable. Explaining life requires explaining its evolutionary predisposition, because it must have emerged coextensive with whatever form of molecular teleodynamic process characterized the dawn of life. These conditions necessarily created a persistent natural selection dynamic that enabled the accumulation of additional supportive molecular processes. So, identifying the minimal, spontaneously probable, molecular conditions for life should also provide the minimal conditions for natural selection.

This more abstract focus on its distinct dynamical properties will enable us to ignore many incidental features of biology and molecular chemistry. Instead of attempting to discover what attributes of living cellular chemistry are dispensable, without losing essential living functions, or what environmental conditions were the most likely to have given rise to the first living systems, we can focus attention on the more general problem of accounting for the phase change in dynamical organization that necessarily occurred with the origins of life, irrespective of any specific molecular details.

Consider again the basic teleological glosses given to processes in even the simplest life forms. Living organisms are integrated and bounded *wholes*, constituted by processes that maintain persistent *self-similarity*. These processes are *functions*, not merely chemical reactions, because they exist to produce specific *self-promoting* physical consequences. These functions are *adaptive* and have evolved with respect to certain requirements in their environment that may or may not obtain. And these adaptations exist *for the sake of* preserving the integrity and persistence of these integrated systems and the unbroken chain of ancestral forms for which they are the defining links.

In other words, as Stuart Kauffman has emphasized, organisms are spontaneously emergent systems that can be said to "act on their own behalf" (though "acting" and "selfhood" must be understood in a minimal and generic sense that will be developed further below).[8] Since they may incidentally encounter favorable or unfavorable environments, organisms must embody a tendency to generate structures and processes that maximize access to the former and minimize exposure to the latter, in such a way that these capacities are preserved into the future. By virtue of these properties that we easily recognize in organisms, we see the most basic precursors of what in our mental experience we describe as self, intention, significance, purpose, and even evaluation. These attributes, even in attenuated form, are significantly unlike anything found spontaneously in the non-living world, and yet they inevitably emerged in their most basic form first in systems far simpler than the simplest known organisms—systems characterized by an unprecedented form of dynamical organization.

COMPOUNDED ASYMMETRIES

As we have seen, both orthograde and contragrade processes are asymmetric processes of change, irrespective of whether cast in thermodynamic or morphodynamic terms. Living and mental processes are problematic not only because of their causal asymmetry, but also because of the unusual nature of their particular asymmetries. Their orthograde tendencies are triply complicated compared to the orthograde asymmetry of thermodynamic processes. This is what provides their final causal character.

Consider, first, the end-directed processes associated with life. Even the simplest organisms exhibit organizational features for which there are no non-living counterparts. They tap external energy sources to do work to transform raw materials from their surroundings into the components to construct their own bodies, and they involve far-from-equilibrium chemical reactions that collectively oppose thermodynamic degradation by continually maintaining critical boundary conditions and by synthesizing and replacing components that have degraded.

H. Maturana and F. Varela, focusing on the generative nature of the process, have called this process *autopoiesis*, which literally means "self-forming."[9] Mark Bickhard, focusing on the stabilization of dynamical self-similarity, called this form of dynamical organization *recursive self-maintenance*, to distinguish it from the simpler self-maintenance exhibited by non-equilibrium processes that persist in a similar dynamic state because of deviation-minimizing boundary conditions. Bickhard's paradigm example of simple self-maintenance is a candle flame because the heat of the flame melts and vaporizes a limited volume of wax so that it is capable of ignition to create heat, and so forth.[10] Stuart Kauffman has additionally highlighted the fact that maintenance of systemic self-similarity constitutes an autonomous individual, or as he terms it, an *autonomous agent*.

Teleodynamics is not identical with any one of these. Each in a slightly different way attempts to characterize in abstract terms the dynamical characteristics distinctive of life, with an eye toward their possible wider application. Teleodynamics can also be understood as characterizing the distinguishing dynamics of life. However, rather than being an abstract description of the properties that living processes exhibit, it is a specific dynamical form that can be described in quasi-mechanistic terms. Although it is the distinguishing characteristic of living processes, it is not necessarily limited to the biological domain. Teleodynamic processes can be identified with respect to the specific end-directed attractor dynamics they develop toward. And they are characterized by their dependence on and emergence from the interactions of morphodynamic processes. Teleodynamics is the dynamical realization of final causality, in which a given dynamical organization exists because of the consequences of its continuance, and therefore can be described as being self-generating. Specifically, it is the emergence of a distinctive realm of orthograde dynamics that is organized around a self-realizing potential, or—to be somewhat enigmatic—it is *a consequence-organized dynamic that is its own consequence.*

As we discovered in the last chapter, the signature of an emergent transition is the appearance of an unprecedented form of higher-order orthograde change—a pattern of asymmetric change that occurs spontaneously, due to the amplification of interaction constraints at a lower level producing a discontinuous break from the lower-order orthograde asymmetries that contribute to it. This is doubly true of teleodynamics. Not only do we observe the emergence of living dynamics and the functions that constitute it, but also this gives rise to the emergence of natural selectionlike processes. So, while individual organisms may be doing local work with respect to their own survival and reproduction, these spontaneous processes do not lend their asymmetric character to the larger evolutionary dynamic. The process of evolution, rather than merely maintaining and reproducing dynamical form, exhibits a spontaneous tendency for its dynamics to diversify and complexify these forms, both intrinsically and in their relationships to their contexts. Thus evolution is neither pushed by intrinsic forces of development nor pulled toward a goal of greater complexity and efficiency. It effectively "falls" in these directions. It is, as noted above, an orthograde process.

Both Varela's notion of autopoiesis and Kauffman's characterization of autonomous agency are exemplified by self-repair and self-reproduction. As we will see, these are not independent processes, but two sides of the same dynamic. The synthesis of replacement components and the stabilization of system organization are critical to both organism self-maintenance and reproduction. Together, they constitute the crucial counterdynamics to the relentless tendency toward thermodynamic degradation. To resist this spontaneous degrading influence requires work. The maintenance of non-equilibrium conditions is thus essential, both for stabilization and for the generation of replacement components, analogous to the constant effort that is required to keep an office in order or an engine running despite constant use and wear and tear. For this reason, the need for constant repair is the inevitable predicament of life in general. And the work required is not merely energetic. The incessant need to replace and reconstruct organism components depends upon synthetic form-generating processes, not merely resistance to breakdown.

For this reason, teleodynamic processes are inevitably dependent on morphodynamic processes for their form-generating capacity. As we've seen, morphodynamic processes are dependent on the maintenance of far-from-equilibrium thermodynamic conditions. So the dependence of teleodynamics on morphodynamics and morphodynamics on thermodynamics constitutes a three-stage nested hierarchy of modes of dynamics, which ultimately links the most basic orthograde process—the second law of thermodynamics—with the teleodynamic logic of living and mental processes. That being said, it remains to be shown exactly how this additional phase transition in causal dynamics is accomplished. Just as we have traced the logic of the transition from thermodynamic to morphodynamic processes, it is now necessary to do the same from morphodynamic to teleodynamic processes.

There are reasons to suspect that this additional emergent phase transition exhibits features that are analogous to those involved in the transition from thermodynamic to morphodynamic processes. The orthograde dynamics that characterize a morphodynamic process emerge from thermodynamic processes that are in effect pitted against one another, in complementary ways. In the case of Bénard cells, for example, continuous heating constantly undermines the increase in entropy of the fluid medium and its tendency to destroy this heat gradient. So, although the spontaneous thermodynamic orthograde process never ceases, it can never converge toward its attractor: equilibrium. Like Sisyphus in Greek mythology, who finds himself doomed to continually roll a boulder uphill, or Alice and the Red Queen in Lewis

Carroll's *Through the Looking-Glass*, who must run fast just to stay in the same place, the conditions underlying this new orthograde attractor are in a delicate balance. This analogy suggests that the transition to teleodynamic organization is likely the result of a higher-order balance between persistent morphodynamic tendencies arranged in some kind of opposing but complementary relationship with respect to one another—one that also keeps both from ever reaching their ultimate thermodynamic end state.

The simplest systems we know of that exhibit teleological properties are simple organisms, like bacteria. It might seem prudent to look to the simplest life forms for clues to how such complex convoluted dynamics might form. But living organisms are immensely complex. The organisms currently living are the end products of 3.5 billion years of evolution, in which there have been vast changes to almost every aspect of their molecular components and functional organization, as well as massive complexification. The differences between physical systems lacking these key features of life and even the simplest living organisms are immense.

As we have also noted, a spontaneous transition from chemical interactions lacking these features to an integrated chemical system constituted by them cannot have involved even dozens of the hundreds of specialized interdependent molecules and chemical reactions that currently play the core roles in simple organisms. This illustrates an important point.

Although the transition from a non-teleological chemical dynamic to the first teleodynamically organized molecular system constituted a fundamental transition in the cosmic history of causality (on Earth or wherever else it also occurred), it necessarily involved a simple combination of molecular processes. It is not the complexity of the transition that was critical, but the appearance of a fundamentally different orthograde dynamical organization. The initial crossing of the dynamical threshold to a primitive form of teleodynamics should therefore not be beyond our current scientific tools, though it may involve overcoming certain tacit assumptions.

SELF-REPRODUCING MECHANISMS

A core feature of life that in many respects exemplifies the most basic entential properties is its self-reproductive capacity. Indeed, it will turn out that in a more completely analyzed form, this capacity is one of the defining features of a teleodynamic process. But reproduction in this sense is not merely pattern replication or copying. It is rather the construction of a dynamical physical system which is a replica of the system that constructed it, in both its structural and functional respects, though not necessarily a faithful replica in every detail. Self-reproduction is thus an end-directed dynamic in which the end is only a potential general form represented within the dynamical system that produced it, but which is a physical system with the same general properties of its progenitor.

One of the first formal analyses of the physical requirements for a mechanism to be considered capable of self-reproduction came from the work of the mathematician John von Neumann, a pioneer of modern computer design. In a short 1966 monograph, he explored the logical and physical problem of machine self-reproduction. His explorations of the logical requirements of self-reproduction followed the insights of the then-new DNA genetics by conceiving of reproduction as an instruction-based process that could be modeled computationally. He argued that a device capable of constructing a complete replica of itself (that also inherits this capability) must include both assembly instructions and an assembly mechanism capable of using these instructions to assemble a replica of itself, including the tokens encoding the instructions. In simple terms, his criterion for self-reproduction is that the system in question must be able to construct a copy of itself, that also possesses this constructive capability. But specifying how to construct a mechanism with this capability turns out to be much more difficult than this simple recipe might suggest.

Von Neumann and subsequent researchers explored the formal requirements of self-reproduction almost exclusively via simulation—for example, in cellular automata—because it was quickly recognized that specifying what he called a "kinetic" model of self-reproduction with autonomous means for substrate acquisition and structural assembly adds highly problematic material constraints and energetic demands that rapidly expand the problem into the realms of physics, chemistry, and mechanical design. He envisioned the process as involving two distinct components: a universal constructor and a universal description. As he outlined the process, the universal constructor would somehow read off instructions for building another constructor device and then transfer a copy of the instructions to the new mechanism. The description could therefore be embodied in a passive artifact, like a physical pattern. But while he was able to specify the self-replication process in formal terms (laying the groundwork for how one would construct an algorithm capable of replicating itself within a computer) the design of the kinetic constructor was another story. Indeed, von Neumann considered the problem of the material basis of this process to be far more critical than its formal implementation.[11]

In the decades since von Neumann outlined the problem, many software simulations of natural selection processes have been developed. The field that has grown up around this work is often described as artificial life (or A-Life). But the goal of physically implementing machine self-reproduction—von Neumann's kinetic self-reproducing machine—as opposed to merely simulating it, has not been achieved. So, while the field of A-Life has had an explosive and productive growth in the decades since his death, little progress has been made toward implementing this idea in kinetic terms. The big difference is, of course, that the production of computer code is entirely parasitic on the physical work done by the computer mechanism. Issues of component synthesis, access to free energy and substrates, and translating the instructions into actions can thus be entirely bracketed from consideration. Of course, these material-energetic entailments are what critically matter. Without them, instructions are merely inert physical patterns.

Physically embodied (rather than simulated) self-reproduction requires one physical system to construct another like itself. This requires physical work to modify physical materials, which requires a mechanism able to access free energy, a mechanism to use this energy to manipulate substrates to transform them into self-components, a mechanism that organizes and assembles these components into an appropriately integrated system, and a means to maintain, implement, and replicate the constraints that determine this integrated organization. And if the system is to be capable of evolving, it must also be sufficiently tolerant of structural-functional degeneracy to allow some degree of component replacement and dynamical variation without losing these most basic capacities.

This is, of course, only a list of what must be accomplished by a kinetic self-reproduction process. It is not an account of how it can be physically accomplished. And this, as we have seen, is the real problem: explaining how living processes do their end-run around the ubiquitous second law of thermodynamics. Knowing what has to be accomplished is a first step; but it is far from knowing how to do it, or even whether it is physically possible. In the biological world (in contrast to the virtual world of computer simulations), life requires the constant acquisition of energy and raw materials from its environment, and an incessantly active, tightly orchestrated use of these to stay ahead of the ravages of thermodynamic decay.

Only in the abstract can we ignore how reproduction is accomplished and focus on the forms being copied. Artificial life simulations, for example, can simply assume the availability of an assembly mechanism and resources to supply the material and energy to reproduce their representations, as all these physical details are built into the computer's most basic operations of copying and modifying memory register patterns from one location to another. Although it should be possible for a simulation to take the computer's energetic usage, storage space requirements, and the physical features of the medium into account in determining differential replication—features that might be useful for software designers—these physical correlates of computing are not generally of interest to modelers of evolutionary processes. Moreover, the specific relevant thermodynamic features of interest (food, protection from predation, etc.) can themselves be represented in data form, as can energy use and development. Even so, all the physical attributes of the computing process itself are ignored. Even though thermodynamic requirements and ecological resources can be incorporated into a model, this crucial feature makes a simulation less than reproduction and evolution. What is most fundamental to these processes is that they are at base thermodynamically grounded. In the course of evolution, any incidental physical attributes of the mechanisms that embody these operations and substrates can become subject to selection, and thus part of what must be represented and reproduced.

For example, although the redness of hemoglobin is not critical to its oxygen-ferrying capacity, and is merely a side effect of this being done using an iron atom, if the redness of blood became a feature that a deadly parasite used to locate its victims, there would almost certainly be selection to alter or mask this color. In this way, evolution constantly transforms incidental physical attributes of organism dynamics into functionally relevant information that must be passed from generation to generation. This is a fundamental feature that distinguishes biological evolution from evolution modeled *in silico*—a distinction that is lost when only focusing on the infor-

mation side of evolution. Ignoring this energetic-material-formative half of the explanation effectively traps us in a dualistic conception of life.

WHAT IS LIFE?

Erwin Schrödinger is the physicist responsible for the equation that describes the statistical logic of quantum mechanics: a wave function of probability that effectively marks the doorway between quantum and classical physics. But he is also responsible for a very different and equally revolutionary insight in another science—biology—though he would never have claimed to be a biologist. In the early 1940s, he gave a series of lectures that resulted in a small book with the title *What Is Life?* In it he asked two quite general and interesting questions about the physical nature of life, and offered some general speculations about what sort of answers might be expected.

The first question concerned the energetics of life. Living organisms, individually and in the grand sweep of evolution, appear to make an end run around one of the most fundamental tendencies of the universe, the second law of thermodynamics. From a physicist's point of view, this ubiquitous feature of life is more than just curious. It demands an explanation of how such a counterintuitive trick is performed by the chemistry of organisms, when chemistry by itself seems to naturally follow the thermodynamic law to the letter.

The second question was indirectly linked to the first. How could the chemistry of life embody the information necessary to instruct the development of organisms and maintain them on a path that defends against the incessant increase of entropy? In other words, what controls and guides the formation and repair of organism structures to compensate for the ravages of thermodynamics and just plain accident, and what makes it possible for living processes to be organized with respect to specific ends, such as survival and reproduction?

Schrödinger's general answer to the first question was that living chemistry must somehow be continuously performing work to stave off the pressures of thermodynamic decay, and to do this it needs to feed off of sources of free energy available in the outside world that he enigmatically called "negentropy," or sources of order. His general answer to the second question was that there must be a molecular storage medium available to

organisms, which embodies a chemically arbitrary and interchangeable but functionally distinguishable molecular pattern that can be preserved and copied and passed to succeeding generations. He envisioned a medium that he metaphorically described as an "aperiodic crystal." He speculated that at the heart of every organism, and indeed every living cell, there must be something akin to a crystal, because it would have to have repeatable units, and yet it would have to be non-crystalline in another way, since these units need to be slightly varying, unlike mineral crystals, which are highly regular. He reasoned that in order to be able to store information, each unit cell of this crystallinelike molecule would need to exhibit one of a number of structurally different but energetically equivalent states, not unlike the way that the typographical characters of this sentence are interchangeable.

This turned out to be a remarkably prescient generic description of the properties of DNA molecules. And not surprisingly, Watson and Crick's account of DNA structure, described a few years later, was influenced by this vision. Though many biologists proclaimed that the discovery of DNA and its chemistry "solved" the riddle of life, or constituted the "secret of life," this was only half the mystery that Schrödinger had envisioned. It did not address the thermodynamic problem.

In comparison to his contribution to the search for the informational basis for genetic inheritance, Schrödinger's attention to the thermodynamic riddle of life, and his intuition that *these two mysteries must be intrinsically linked*, fell into the background of biological discussion, to be followed up by a comparatively small group of theoretical biologists. In the generations since Schrödinger, the exploration of the riddle of life has mostly been characterized by parallel and often non-interacting threads of research in these two domains; and the progress has been mostly one-sided. For the most part, the vast majority of evolutionary and molecular biologists have followed the trail of molecular information, while a much smaller cadre of physicists and physical chemists have followed the trail of far-from-equilibrium thermodynamics. This is ultimately an untenable segregation, because information turns out to be dependent on thermodynamic relationships. An explicit effort to reunite these two theoretical domains will have to wait until chapters 12 and 13; but even without an explicit theory of information, the relationship of these processes to the origins of teleodynamics can be explored in some detail.

What Schrödinger's questions do not approach is how this physically atypical state of affairs came about, and why they must be linked. Exploring this riddle will provide the first steps toward reconstructing the bridge leading from the non-living, non-feeling, non-knowing world to the world of life and mind.

FRANKENCELLS

One way to approach the question of what is minimally necessary to define life is to start with what we have, and then subtract features until the removal results in failure to function. Attempting to produce artificial exemplars of minimal living cells has currently relied on two general approaches. Laboratory-based approaches to identify the minimal conditions for life are generally distinguished as top-down approaches. Top-down approaches attempt to extrapolate backwards, theoretically and experimentally, from current organisms to simpler precursor organisms. Prominent in this paradigm is the attempt to describe and produce "minimal cells" by stripping a simple bacterium of all but its most critical components.[12] Though considerably simpler than naturally occurring organisms, these minimal cells still contain hundreds of genes and gene products,[13] and so turn out to be vastly more complex than even the most complex spontaneously occurring non-living chemical systems, implying that they cannot be expected to have formed spontaneously.

The alternative bottom-up approach attempts to generate key system attributes of life from a minimized set of precursor molecular components. An increasing number of laboratory efforts are underway to combine molecular components salvaged from various organisms and placed into engineered cellular compartments called *protocells*.[14] The intent is to produce an artificial "cell" with the capacity to maintain the molecular processes necessary to enable autonomous replication of the entire system. To accomplish this, protocells must obtain sufficient energy and substrates and contain sufficient molecular machinery to replicate contained nucleic acid molecules and to produce the protein molecules comprising this machinery from these nucleotide sequences—and probably much more. Though vastly simpler than what is predicted for a minimal bacterial cell, even the most complex protocell systems fall considerably short of the goal of achieving

persistent autonomous reproduction, even though they incorporate dozens of complex biomolecules (nucleic acids, complex proteins, nucleotides,
phospholipids, etc.). In most cases these molecules are already organized
into working complexes (such as ribosomes) by the living systems from
which they have been removed. Given that they are composed of cellular
components that accomplish the relevant functions in their sources, it is
expected that protocell research will accomplish its goal as more complex
protocells are constructed. However, even the simplest protocells currently
imagined are sufficiently complex to raise doubts that such systems could
coalesce spontaneously from pre-organic substrates.

An implicit assumption of the great majority of these approaches is
that replication and transcription of molecular information, as embodied
in nucleic acids, is a fundamental requirement for any system capable of
autonomous reproduction and evolution. This is a natural assumption since
nucleic acid–based template chemistry is the most ubiquitous attribute of
all known forms of life. But despite its intuitive attraction and laboratory
accessibility, assuming the spontaneous appearance and both the replicative
and protein template functions of nucleic acids brings with it impossible
demands.

First, including nucleic acid replication in an artificial cell is extremely
demanding in terms of critical support mechanisms. In order to replicate
nucleic acid sequences and transcribe them into protein structure, some of
life's most complex multimolecular "machines" are required. Each of the
macromolecular complexes supporting the various aspects of this process
(replication, translation, transcription, etc.) typically consists of over a half
dozen different interlocking and synergistically interacting molecules. The
precise stereochemical demands placed on the major protein units of these
complexes are reflected in the highly conserved gene sequences for each,
which have undergone little change since before the epoch of eukaryotic
cells. They are almost certainly the product of long evolutionary fine-tuning
for this function. Though some of these complications may be avoided by
molecules serving multiple functions, such as the capacity for RNA to
also exhibit catalytic functions,[15] an impressive number of these complex
molecular structures and interactions must still be provided to support such
processes. For this reason, protocells based on these known core molecular processes of life provide valuable test beds for exploring how subsets of

these molecular mechanisms contribute to replication. But we should be wary of using this approach to model the earliest forms of life.

If the goal is to engineer life as we currently understand it, then building a protocell capable of reproducing itself constitutes success, no matter how its parts are obtained. But engineered cells built from molecules found in living organisms are likely to provide a rather misleading model for studying how life began. Before life got a foothold in our solar system, few if any of these components were present, and of course there were no engineers to select the critical molecules and appropriately assemble them into working complexes within a cellular container. Stripped-down, modified viruses are already in widespread use in biomedical research, and artificial bacterial protocells are almost certainly just around the corner. Creating a stripped-down artificial bacterium or protocell that can reproduce itself will certainly provide important insights about life in general. It is quite likely that such cells could provide useful vehicles for replicable nanomachines functioning at a cellular scale. But protocells are unlikely to be useful models for the missing link between physics and biology.

Protocells are Frankencells, like the unfortunate "monster" of Mary Shelley's now prophetic tale of a revivified cadaver. They are reconstructed by recombining components extracted from once-living cells, in the hopes of revivifying the experimental combination. But like the cadaver pieces stitched together by the fictional Dr. Frankenstein, these cellular components were the product of billions of years of evolutionary fine-tuning for their respective interdependent roles. Starting with these end products is like using computer components to explain the construction of the first abacus.

So, in three respects, making sense of the spontaneous emergence of a teleodynamic molecular system from non-ententional antecedent conditions is more challenging than simply reverse engineering simple organisms. First, unlike approaches using complex biomolecules as building blocks, we cannot assume to take advantage of the products of prior evolutionary (and thus teleodynamic) processes to explain how the various components arose and came to work together. Second, the components we can consider must be synthesized by spontaneously occurring non-biological processes. And third, it is not enough to have components merely brought into proximity with each other; they must reciprocally produce one another and maintain their synergistic relationships. Merely collecting the critical molecular components of cells into a cell-like container will not suffice, even if they perform the chemical functions they would provide to a naturally evolved organism. This even applies to the self-replication of nucleic acids. In the absence of the synergistic co-production of all components, we have nothing more than an organic chemistry experiment carried out in a tiny lipid reaction vessel.

One difficulty in thinking about this issue more broadly derives from the lack of exemplars of extraterrestrial life for comparison. This limits our ability to discern essential from incidental features of life. Since many of the widespread or even universal features of living chemistry might well have been incidentally acquired due to unique features of the Earth environment, focusing too narrowly on the chemistry of specific biomolecules—even DNA—could overly limit the scope of the search. A related difficulty arises due to the absence of life forms that are more primitive than the simplest contemporary species (e.g., of bacteria). It is certain that even the simplest contemporary organism is far more complex than could form spontaneously. Over the course of 3.5 billion years of evolution on Earth, it is almost inevitable that innumerable early stages of life's precursors have been replaced so that nearly all traces of these stages are gone. Finally, efforts to produce artificial systems with these properties, but composed of constituents not found in organisms, are still in their infancy. So the range of possible alternative mechanisms for the key properties of living systems remains largely unexplored.[16] These limiting circumstances make the deconstruction of living organisms for clues about the earliest transitional stages unlikely to succeed.

Ultimately, the cosmos must have achieved this transition with a system of molecular processes that is far simpler than anything resembling life as we know it today. As will be discussed in chapter 12, the property that is usually considered most central to life—information—is unlikely to be a primary player in this early transition, and as we will further develop the concept of information in chapters 13 and 14, it will become apparent that information in the full sense of the term (which is both about something and has normative characteristics) is not a primary property of matter, but is dependent on underlying teleodynamic processes. Moreover, DNA-based processes are far too complex to have appeared spontaneously and fully developed. The possibility that molecular structure could be information about something (e.g., the chemical dynamics of a cell with respect to its likely environments) is necessarily a higher-order emergent function abstracted from and dependent on more basic teleodynamic processes. To put this in terms of Schrödinger's dual characterization of life as a marriage between transmission information and non-equilibrium thermodynamics, the informational functions of life are emergent from and dependent on more basic non-equilibrium dynamical processes.

Hints about how this might be possible have come from work by research-ers exploring the physical chemistry of the synthetic and metabolic pro-cesses that ultimately support the replicative and informational functions of cells.[17] Rather than thinking in terms of life's current well-honed tool kit, these researchers ask what general dynamical operations are accomplished by these mechanisms? Generalizing the notions of metabolism and con-tainment, respectively, we can describe these processes in terms of their roles in countering two thermodynamic challenges. First, there must be a mechanism that counters the incessant tendency for component elements of the system to degrade—either by repairing damaged components or by syn-thesizing new ones. And second, there must be a mechanism that resists the degradation of constraints on potential interactional relationships among components, such that the critical synthetic processes are reliably achieved. The compatibility of this approach with the goal of explaining the origins of teleodynamics is that both approaches assume that the informational func-tions of life emerge from simpler dynamical foundations.

10

AUTOGENESIS

... an organized natural product is one in which
every part is reciprocally both end and means.

—IMMANUEL KANT'S PRINCIPLE OF "INTRINSIC FINALITY," 1790
(ITALICS IN THE ORIGINAL)[1]

THE THRESHOLD OF FUNCTION

Even though there is a close interrelationship between them, investigating the mystery of the origins of life and the origins of teleodynamics requires quite different approaches. All current approaches to the origins of life on Earth begin from tacit and unanalyzed assumptions about the nature of the basic unit of life: an organism, with its physical boundedness, functional organization, and heritable information. These assumptions are typically glossed as though they can be defined by intuitively obvious physical-chemical properties such as containment in a lipid membrane, metabolic processes powered by ATP, and information intrinsically embodied in nucleic acids. These glosses allow us to assume what we should be explaining: why these same structures and relationships do not have intrinsic functional or informational character except when embodied in living dynamics. It is their interdependent contribution to constituting the dynamical system that is an organism, and not any of their intrinsic properties that matters.

So, even before the question of how these components collectively constitute these properties, we need to understand the nature and origins of this form of collective organization itself. But whereas life demands an account of its molecular peculiarities, teleodynamics merely demands an account of its dynamical requirements. Instead of considering which compositional features matter, we can be somewhat agnostic about composition and begin to pay attention to this dynamical organization, and inquire into how it constitutes this transition. In other words, to simply assume that entential issues can be operationally ignored or assumed as given, so long as we have the right molecules, is to fall prey to the homunculus fallacy. These properties are not intrinsic to the material constituents of life. Entential properties like function and information will only be explained when we can demonstrate how they emerge from non-entential precursors, and so the teleodynamic processes that characterize life must be explained (though not reduced) by reference only to thermodynamic and morphodynamic precursors.

To explain the emergence of teleodynamics, we must also resist the temptation to attribute it to similarly mysterious and counterintuitive phenomena, such as the strange and spooky properties exhibited by quantum-level processes, which similarly appear to violate our familiar notions of cause and effect. Although these might in some cases superficially resemble entential relationships, such as those involving measurement, action at a distance, or entanglement, these must be excluded for three reasons. First, if we were to make use of any such extraordinary physical phenomena, we would at best only have substituted one mystery for another. And even so, this would not explain the difference between non-living and living systems in this respect. As we earlier noted with respect to panpsychic or pansemiotic assumptions, it would still require us to come up with an explanation for why living and mental processes are associated with such radically different causal dynamics than their non-living counterparts. Of course, were someone to show that it was *in principle* impossible to explain these higher-order properties and processes using the tools of known macroscopic physics and chemistry, that might force us to consider the possible role of these very different quantum properties. But this brings us to the second reason to avoid such an appeal. If we *can* explain this shift in causal organization without invoking strange causality, it will render the justification for appealing to strange quantum properties irrelevant.

Lastly, although quantum phenomena often seem strange, they are not strange in the sense of exhibiting the kind of causality that organisms and minds do. No one describes any of the strange quantum properties, such as the Casimir effect or quantum entanglement, as consequence-organized behavior. More important, the scale at which we do unambiguously recognize entential properties is vastly larger than the scale of quantum events, and in between there are only thermodynamic and chemical processes. The basic physical and chemical forces that dominate at our level of scale, from cells to ecosystems, appear to be the only place we find the emergence of the entential properties—exhibited in living and mental and communicative processes. In short, entential phenomena are not strange in the way that quantum processes can be strange, and they are manifest only at a much higher level of scale.

With these caveats in mind, we can make a first pass at identifying *the minimal requirements for the emergence of teleodynamic processes*; then we will describe a molecular thought experiment that embodies these requirements. Providing a realistic, empirically testable, constructive account of the emergence of teleodynamics in a simple molecular system is only the barest starting point of a theory of the origins of life. It is a long way from explaining even the features exhibited by the simplest known bacteria. It is only a minimal proof of principle. Many more details (some to be explored in subsequent chapters) will need to be filled in to even approach a methodology sufficient to begin to account for the evolution of the higher and more complex forms of teleodynamic processes, such as are found in living organisms and mental processes.

So, in this first step out of mere physics, we will only offer evidence that a spontaneous transition from non-entential to entential forms of causal organization is possible. The goal is solely to develop a minimally complicated, thoroughly described, physically plausible model system exemplifying this transition. With such a model system at our disposal, we can begin to ground our accounts of at least simple entential processes and relationships in more precise physical, chemical, and thermodynamic details. The basic principles developed in this simple system can then later be applied at higher levels of analysis, where we encounter entential phenomena of considerably more subtle and complex form. To the extent that we develop such a plausible and non-mysterious bridge analysis, it will open the door to a methodology for examining how these higher-level entential phenomena might have emerged and evolved from the lower-level ententional phenomena exhibited by the first precursors to life.

To conceive of the first lifelike process, we must remind ourselves that it was not reproduced, it had no parent, and therefore it did not evolve. It emerged. It was fitted to its environment by chance alone, not by virtue of having been tested by natural selection, and yet the very fact of its emergence indicates a consistency with its environment. Indeed, it would be circular to argue that the process of natural selection itself evolved by natural selection, or that the reproductive process arose by being reproduced. On the other hand, it is quite likely that the first ancestral counterpart to today's genetically mediated process of reproduction was quite simple and inefficient, and that the first ancestral counterpart to natural selection was considerably less effective at achieving and maintaining favorable variations within a lineage.

With ultimate simplicity, we can expect only a very crude first exemplar of that which modern organisms accomplish with comparatively flawless elegance. In comparison to this first form, even the genetically undisciplined and promiscuous nature of bacterial and viral evolution would appear like the essence of clockwork precision. This is because the process we now see enshrined in these mechanisms of reproduction is the outcome of billions of years of fine-tuning by natural selection, and the entire web of contemporary Earth life descends from the one winner in a billion-year competition among possible alternatives. This contemporary near uniformity is almost certainly the outcome of an older period of disorderly inefficient experiments in adaptation and inheritance. The tricks that life currently employs to generate its many lineages and diverse adaptations likely evolved to be *ever more evolvable* over these many billions of years. How these critical and powerful semiotic and functional capacities arose will be the subject of a later chapter. But the processes that honed the evolutionary process itself cannot have been any form of evolution as we now know it, and as we will shortly see, the very logic of evolution must have also emerged in this transition.

The challenge of this book is to explore the possibility that teleodynamic processes can emerge spontaneously by natural laws from physical and molecular processes devoid of these properties. But we should be wary of looking to contemporary biology for clues. The debates that raged in past centuries over the plausibility of spontaneous generation suggest a useful caution. Surrounded by teeming life, so imperceptibly minuscule as to require extreme care to avoid contamination, spontaneous generation theorists were fooled again and again into thinking they had demonstrated the emergence of life from non-life. Like the early Enlightenment biologists immersed in a world teeming with invisible life, we are immersed in a world of ubiquitous entential assumptions that we barely recognize as potential contaminants. Trying to explain this emergent transition, from the perspective of hindsight, and within our currently well-honed biological context, risks analogous contamination. Setting our thought experiments in a teleologically barren context is thus the equivalent of testing spontaneous generation in a sterile environment. This is not as easy as it sounds.

AUTOCATALYSIS

One of the most significant insights concerning the nature of life has come from studies of thermodynamic processes that are far from equilibrium, which I have collectively referred to as *morphodynamic processes*. In organisms, far-from-equilibrium conditions are maintained by the complex cycles of molecular reactions constituting cell metabolism. Normally, in a closed system, each instance of a given chemical reaction will slightly decrease the future probability of similar reactions, due to the depletion of precursors, the increase of products, the dissipation of the energy of the reaction, and how all these changes affect the relative probability of the reaction running forwards versus backwards. So, in general, a closed molecular system will tend toward some steady state with fewer and fewer asymmetric reactions occurring over time, as the overall distribution of reactions runs in directions that offset each other—the second law of thermodynamics at work in chemistry. In contrast, in chemical systems maintained far from equilibrium, where the conditions for asymmetric reaction probabilities are not reduced, non-equilibrium dynamics can produce some striking features. The most relevant are morphodynamic chemical processes.

One of the most relevant classes of non-equilibrium chemical process is autocatalysis. It is relevant both because there are many analogues to this in living cell metabolism and because it can arise spontaneously as a transient, locally deviant, non-equilibrium process. A catalyst is a molecule that, because of its allosteric geometry and energetic characteristics, increases the probability of some other chemical reaction taking place without itself being altered in the process. It thus introduces a thermodynamic bias into a chemical reaction as a consequence of its shape with respect to other molecules. Autocatalysis is a special case of catalytic chemical reactions in which a small set of catalysts each augment the production of another member of the set, so that ultimately all members of the set are produced. This has the effect of producing a runaway increase in the molecules of the auto-catalytic set at the expense of other molecular forms, until all substrates are exhausted. Autocatalysis is thus briefly a self-amplifying chemical process that proceeds at ever-higher rates, producing more of the same catalysts with every iteration of the reaction.

Because of the widespread presence of chemical cycles in living systems, autocatalysis offers a suggestive starting point for discussing the origins of life, and it has been promoted by many as the chemical analogue of self-replication. Unfortunately, it requires the remarkably coincidental coalescence of reciprocally interlocking molecules, making it a fairly rare and spontaneously improbable chemical condition. But is its spontaneous occurrence sufficiently probable to serve as a one-in-a-billion-years spring-board for life?

The theoretical biologist Stuart Kauffman has, however, argued that this intuition about the improbability of spontaneous autocatalysis is mistaken.[2] Using principles from graph theory to model the networks of reaction of catalytic interactions, he argues that autocatalysis will be almost inevitable under not too extreme chemical conditions. Because catalysis is a more-or-less effect, based on only the relative precision of "fitting" between different molecules, a polymeric "soup" with sufficient diversity of molecular forms will tend to include many molecules able to at least weakly catalyze reactions among others in the soup. This plurality of potential catalytic possibilities can be envisioned as comprising a sort of network of relationships between molecular types (forming the nodes in the network), and their synthesis and breakdown relationships (constituting branching relationships in the network). As molecular diversity increases (or with high average catalytic potential between any two molecules), the effective connectivity of the corresponding network will increase as well. With increasing network connectivity, re-entrant circles of reactions become more common. So, as molecular shape diversity increases, the probability of spontaneous circles of catalysis will increase as well. We may then expect that multiple inter-linked autocatalytic loops will be present in sufficiently "rich" polymeric soups that are not so concentrated as to form glue, and that this provides for a sort of initial exploration for the most robust and sustainable variants. But although autocatalysis is a function of molecular diversity, it turns around and generates molecular homogeneity, since molecules that tend to produce other molecules from a linked autocatalytic set rapidly replace other forms that do not.

At the point where the molecular diversity in the solution reaches the stage where circles of reactions of this sort start to become common there is a discontinuity in the dynamics of chemical reactions. With the introduc-tion of autocatalysis, there is a break in the symmetry of the distribution of probabilities of the different types of chemical reactions. Members of this autocatalytic constellation will rapidly increase in proportion to all other molecular types, and as they do so, this very process will speed up too. It is analogous to the rapid multiplication and growth of new organisms intro-duced into a pristine environment. But also like over-reproduction, it will be a transient dynamic that quickly undermines its base.

A higher-order special case of autocatalysis was investigated by the phys-ical chemist Manfred Eigen in which two or more autocatalytic cycles are interdependently linked to create cycles of cycles.[3] These were dubbed *hyper-cycles*. In general, a hypercycle is a second-order autocatalytic cycle formed when a product of one autocatalytic cycle is an essential element in another autocatalytic cycle, which produces a product that is essential to the first; thus forming an autocatalytic cycle of autocatalytic cycles. There are prob-ably many ways this kind of higher-order autocatalysis can be constituted, but the obvious cases involve autocatalytic cycles contributing substrates to one another, or autocatalytic cycles partially overlapping by sharing cata-lysts in common. Both occur in living cells, where higher-order networks of linked catalytic reactions produce highly complex and interwoven meta-bolic and synthetic cycles of cell biochemistry that are mutually reinforcing.

Of course, all catalysis—and especially autocatalysis, because of its run-away dynamic—is constrained by the availability of substrate molecules and energy. This is even more critical for hypercycles, because they include mul-tiple autocatalytic cycles, each of which is subject to such extrinsic dependencies, as well as on the continual productivity of all the other linked cycles. When the raw materials are used up, or energy for molecular reactions is no longer available, reactions stop and the set of interdependent catalysts dissipates. Autocatalysis is thus self-promoting, but not self-regulating or self-preserving. And hypercycles will be even more sensitive to substrate conditions and dissipation, because they are subject to weakest-link effects. Substrate limitations on any one subcycle of a hypercycle will limit all, and the slowest catalytic process will exert a rate-limiting influence on all other linked subcycles.

In addition to these problems, there is an intrinsic limitation on the initiation and persistence of any autocatalytic process. Such factors as the relative concentrations of the interdependent catalysts with respect to any other molecules not involved in autocatalysis, their tendency to react with other molecules in ways that are non-conducive to autocatalysis, and the ubiquitous effects of spontaneous diffusion, all will limit autocatalysis. These are not unlikely limiting conditions, even in laboratory settings, and they pose significant impediments to spontaneous autocatalysis, especially if spontaneously achieving autocatalysis among a random assortment of molecules depends on significant molecular diversity. Moreover, these difficulties increase geometrically with the complexity of the autocatalytic network of reactions, making even modest catalytic complexity highly unlikely. Nevertheless, complex reciprocally catalytic networks of molecular interactions characterize the metabolic processes of living cells, so it is far from impossible.

CONTAINMENT

The second ubiquitous feature of life is containment. All units of life on Earth, whether single cells, multicelled organisms, or viruses, are self-contained. All cells have lipid cell membranes that prevent indiscriminant diffusion of molecules between inside and outside. Even viruses, which lack other sorts of molecular systems, are encased by protein shells or more complex coverings that both protect their nucleic acid payload and aid in getting the virus insinuated into a host cell. Self-containment is a ubiquitous feature of life because life depends on the structural contiguity of its molecular systems remaining intact and unchanging over long periods of time. Containment is the most obvious expression of the importance of constraint for the successful persistence of life.

Containment creates physical individuality. A boundary that distinguishes inside and outside is almost synonymous with the self/other distinction, both functionally and metaphorically. The characteristic unit of contained individuality in life is a cell, and although there is individuation at higher levels in multicellular organisms (e.g., organism bodies), and at lower levels with respect to cellular subdivisions (e.g., organelles), it is the cellular level of life that exemplifies the most robust and omnipresent unit of functional individuality throughout the living world. The cell membrane is a boundary distinguishing a continuously maintained self-similar milieu inside from a varying and unconstrained outside world. Though neither impermeable nor inert, its role is vested in what cannot happen as a result. The constraint on molecular movements and interactions that containment provides is a necessary constitutive factor in all living systems.

Finiteness of the contained materials also makes the state space of their interrelationships manageable. Insulation from extrinsic factors that are too many, too distributed, and too unpredictable likewise is an essential contributor to the maintenance of self-similarity across time. And simply maintaining material proximity of potentially diffusible components that must interact regularly and predictably to maintain living processes is an important constraint on the increase in entropy. For all these reasons, and because of its ubiquity in the living world, many scenarios concerned with the origins of life treat containment as primary.

But containment is a double-edged sword. Maintaining components in continual proximity and excluding new interactions increases similarity and predictability from moment to moment, but this happens for the same reason that a closed thermodynamic system quickly and inevitably runs down to a maximum entropy state. Complete closure rapidly leads to stasis. Self-similarity rather than stasis is the basis for life's individual units, however. Partial or periodic containment, on the other hand, is able to contribute to both. Without spatial constraint on the correlated loci of interdependent molecules and processes, neither life nor evolution seems possible.

Interestingly, most abstract conceptions of Darwinian processes ignore this factor. In simulations, for example, self-replication is almost universally represented without any consideration of containment, but this may again reflect more on the abstraction process than on the constraint requirement. A-Life, evolutionary programming, genetic algorithms, and social evolution theories may not need to deal with the problem of containment because the symbolic medium provides it implicitly in another form. Conventionalization of sign vehicles allows the selective identification of each representative unit by distinctive characteristics. Von Neumann's replicating mechanism can also ignore containment because the physical mechanism is identical to its container. It is presumed to be a permanently interconnected device that by design coheres as a physical machine without diffusion tendencies. Still, the measurement of what must be replicated at least requires the assessment of essential boundaries to determine what must be "contained" in the representation, as "self" (to be copied) versus other. But where we cannot take for granted the self-identification or intrinsic structural coherence of all essential functional elements, some coherence-maintaining, proximity-maintaining mechanism must be present. For any molecular system whose elements are necessarily available for chemical interaction, there must be some additional barrier to diffusion and interaction consisting of relatively non-reactive molecules that limits this potential. Whether in the form of extrinsically maintained proximity and coherence or intrinsically integrated binding mechanisms, these functional roles must somehow be fulfilled to sustain a functional unity.

Containment of organisms and viruses is generally achieved by a molecular process called *self-assembly*. This is a ubiquitous feature of all organisms and viruses, and is responsible for more than mere containment. Virtually all multimolecular complexes that constitute the molecular machines doing the molecular work of the cell spontaneously assemble. So their structure is not explicitly constructed in the way engineered devices or fabricated artifacts must be. As noted in previous chapters, organisms and machines differ critically in this respect. A machine must have its parts chosen, gathered together, aligned, interconnected, and set in motion by extrinsic processes. The process of assembly of molecular building blocks into large multicomponent structures in living cells, in contrast, seldom requires the existence of an external means of accomplishing this feat. These structures tend to spontaneously self-assemble, and in effect "fall" together. This makes life radically unlike the self-reproducing device that von Neumann envisioned.

Recall that his self-reproducing device had to include both instruction-copying mechanisms and a machine-fabrication mechanism, with the latter being capable of fabricating a copy of itself from the instructions. The engineering paradigm takes as given the need to impose structure from without; but this is not an intrinsic requirement, and it is almost entirely irrelevant for life. In many ways, this is more than just an advantage that life has over engineering. It is a necessity. This critical constitutive feature helps to explain a good deal about what makes life unusual, and makes evolution possible.

Spontaneous self-assembly is neither as esoteric nor as special as the term implies. One does not need to imagine parts that are animated and seeking each other out or the existence of special intrinsic assemblers. Self-assembly of macromolecular structures is essentially a special case of crystallization. Like crystal lattice formation, the growth of multi-unit macromolecular structures is an expression of the intrinsic geometry of component molecules, the collective symmetries these offer in aggregate, and the lower energy state of the crystallized forms. It is in this sense an expression of a thermodynamic orthograde tendency to reduce total entropy. It's just that molecular shape and charge features, and their relationships to other molecules—including especially to the surrounding water—are responsible for the spontaneity of their growth. The accretion of molecules into crystallinelike arrangements is thus a function of thermodynamics, but the influence of their structural and charge characteristics may contribute to the amplification of constraints, thereby generating regularities in the ways they form into aggregates; a morphodynamic consequence.

The combination of these features ultimately determines which among the possible macroscopic arrangements is most likely. Macromolecular structures within cells result from the same symmetry logic that produces the geometrically regular growth of crystal lattices. Individual molecules can align with each other with respect to complementary structural symmetries in their shapes and charges so long as there are nearly complementary symmetries in these molecular features. Crystal-like growth of cellular structures, composed of hundreds or thousands of identical or iterated complementary elements, is common and plays a central role in cellular-scale architecture.

Paradigm examples of self-assembling macromolecular structures in cells include a wide array of laminar and tubular structures within cells. These include various forms of cell membrane that are found both inside and outside cells, and tubular structures which provide the internal three-dimensional support for cells, the thoroughfares along which molecules and vesicles are moved, the capacity for differential mobility and reshaping of cell geometry as in pseudopodium extension, and the structural elements that form the core of flagella and cilia.

Consider one of the most ubiquitous of these forms: microtubules. Most microtubule formation is based on the spontaneous, spirally symmetric packing of component molecules (tubulin) to produce crystal prism shapes that can be indefinitely extended. Spontaneous precipitation onto the end of a tubule occurs when there is a sufficient concentration of component tubulin molecules and an appropriate ionic concentration for their (orthograde) precipitation. Extension and collapse of these tubular structures at specific loci within a cell can thus be regulated by a variation in substrate availability, solute factors, and other relevant conditions. This coupling of spontaneous formation and conditionality is the key to their functional usefulness. Note that the structure of these tubes is not a function of genetically specified protein structure alone, because the genetic code only incompletely constrains its range of possible three-dimensional structure (which also depends on contextual conditions), and even this structure may be consistent with many different semi-regular packing configurations. Ultimately, both the specification of the structure and the symmetries that drive the self-assembly process are context-derived—emergent from laws of symmetry and thermodynamics. Much of the structural information and construction work thus comes "for free," so to speak, in the sense that it is neither maintained by natural selection nor strongly determined by genetic information. The genetic information might rather be thought of as maintaining all these variables within constraints that make such spontaneous tendencies highly likely.

Another paradigmatic class of self-assembling structures are the sheets and surface structures that constitute the walls and partitions of cellular architecture, including lipid bilayers, protein and carbohydrate matrices, and the protein capsid containers of viruses. These also tend to form spontaneously via the symmetric aggregation of large numbers of identical components. Probably the simplest constitute the lipid bilayer sheets that form the external and internal membranes of all cells. These form as a result of the polar hydrophilic-hydrophobic structure of lipid molecules. In an aqueous solution, the hydrophobic "tails" of lipid molecules tend to aggregate together, exposing only the hydrophilic ends to the surrounding aqueous solution. Aggregate lipid balls and bubbles are energetically more "relaxed" than individual lipid molecules in water because aggregation allows neighboring molecules' hydrophobic tails to hide each other from water molecules. Again, this is an expression of the interaction between thermodynamic and structural features of the molecules, which bias how they interact as a consequence of congregating due to settling into a lower energy state. But it is also possible for lipids to form sheets, either at the interface of water and a non-aqueous medium, or with a complementary sheet of lipid molecules, resulting in the formation of a bilayer sheet in which the hydrophobic tails of each lipid molecule faces inward from both sides. This is again a more relaxed geometry than forming irregular aggregates in water; and so free lipid molecules will spontaneously "prefer" to aggregate into sheets when conditions are favorable to this, and these sheets will tend to form into closed spheres if they grow large enough. Because of the highly specific symmetry of the molecular orientation in a sheet, and the structural instability of growing sheets only a couple of molecules thick, lipid bilayers form more effectively and predictably when there is some surface to provide an appropriate planar template to bias this orientation and growth.

Other molecular types are also capable of forming into surfaces with slightly different properties. For example, the formation of viral shells is more analogous to crystal growth than lipid sheet formation, and yet is also quite distinctive from both in that it often involves the formation of regular polyhedral structures. The capsid units that form these shells are typically composed of proteins. Individual protein molecules or protein multimers fold into shapes that tend to bind edge-to-edge with one another due to structural symmetries. Proteins that assume regular polygonal or polyhedral shapes will tend to aggregate spontaneously into tessellated sheets, also due to the lower energy of the close packing enabled by this geometric symmetry. Polyhedral prisms created by tessellated surfaces of protein molecules will be the most stable of these forms. This is because all molecules in a three-dimensional polyhedron are bound to one another and surrounded on many sides in a low-energy symmetrical configuration that is thereby

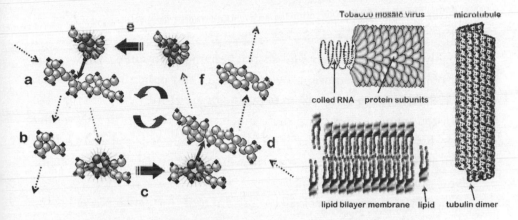

FIGURE 10.1: Two self-organizing molecular processes common to all life: autocatalysis and self-assembly. *Left*: in autocatalysis, one molecule catalyzes a reaction that produces a second catalyst as a byproduct, which in turn catalyzes the first. In this depiction, energy is released by the breaking of bonds of split substrate molecules and causes the process to be self-sustaining so long as substrate molecules are present. *Right*: three different self-assembling molecular process are depicted: self-assembly of the protein and RNA components of a virus; self-assembly of a lipid bilayer due to hydrophobic and hydrophilic affinities of these molecules' polar structure (hydrophilic "tails" are forced together); and self-assembly of tubulin molecules into a microtubule, one of the major components forming the flagella that propel bacteria and other mobile cells. Details of each of these processes are described in the text.

structurally resistant to external forces. The regular or semi-regular polyhedrons thus formed typically enclose a DNA- or RNA-filled core.

After incorporation into a host cell, viral genes are released from this core and repeatedly transcribed by host cell mechanisms to generate high concentrations of capsule proteins. At the same time, viral genes are being replicated in high numbers as well. The spontaneous formation of hundreds of viral shells in the context of hundreds of gene replicas thus has a high probability of encapsulating the genes that produce them, even without additional packaging mechanisms (though packaging is typically aided by other molecules synthesized from viral genes). The spontaneous self-assembly of viral capsule molecules into regular containers allows viruses to be elegantly simple and minimalistic, and consequently highly efficient replicating systems.

SYNERGY

Probably the most prescient and abstract characterization of the dynamic logic of organism design was provided by the philosopher Immanuel Kant. In a 1790 critique of the problem of teleology in nature, he argued that "An organized being is then not a mere machine, for that has merely *motive power*, but it possesses in itself *formative power* of a self-propagating kind which it communicates to its materials though they have it not of themselves" (italics in the original).[4]

Implicit in Kant's abstract characterization of "formative power" is the fact that organisms are organized so as to resist dissolution by replacing and repairing their degraded components and structural characteristics, and eventually replacing themselves altogether by reproduction. More important, as described in the epigraph to this chapter, he emphasizes that this is a reciprocal process. No component process is prior to any other. Kant's characterization is prescient in another way that is relevant to our enterprise. In this essay, he is puzzling over the question of whether there is something like intrinsic teleology in organisms. Kant concludes that this formative reciprocity constitutes what he calls "intrinsic finality." Although modern accounts can be far more concrete and explicit than Kant's, by virtue of their incorporation of over two hundred years of biological science, this knowledge can also be a source of distraction. In these intervening centuries we have of course discovered a vast hidden world of molecules, chemical processes, and cellular interactions, as well as forms of life and half-life (e.g., viruses) that no one in Kant's time could have dreamed of. But it is not clear that contemporary definitions of life are actually any more fundamental and general. Only able to reason about life in the abstract, Kant focused on life's distinctive dynamical organization, and so it is the synergy of living processes that stands out for him. Today, it is possible to add flesh to Kant's skeletal definition and in so doing demonstrate its prescience.

As we have seen, both autocatalysis and self-assembly are morphodynamic molecular processes that are capable of occurring spontaneously in a wide variety of conditions. What they share in common is a dependency on molecular shape-effects and a propensity for promoting rapid self-amplifying regularities. They are both non-linear processes, like compound interest, and are as a result intrinsically constraint-amplifying. What is amplified in each case is a bias based on symmetry and shape complementarities between molecules, which together produce increasing self-similarity (of molecular types or geometric regularity, respectively) over time. These processes are achieved and sustained by a constant availability of raw materials and energy in the case of catalysis, or local concentrations of identical molecules in the case of self-assembly.

These requirements, however, make autocatalytic cycles (and especially hypercycles) intrinsically susceptible to diffusion and side reactions, and make self-assembly a self-limiting process due to its intrinsically concentration-depleting dynamic.

Thus, despite the spontaneous potential for formation of autocatalytic sets, autocatalysis is a fleeting and transient occurrence in the non-living world because it is self-undermining and self-limiting. Precisely because of its deviation-amplifying dynamic, autocatalysis will rapidly deplete the local environment of required substrates. As the process continues and catalysis slows with declining concentration, the diffusion tendencies of the second law of thermodynamics will no longer be compensated by replacement of newly synthesized molecules, and the component catalysts will tend to diffuse away from available substrates and away from other members of the set with which they could interact, if substrate molecules were available.

Similarly, self-assembly is dependent on a local concentration of free-floating component molecules that will tend to give up their kinetic energy to settle into a lower-energy state within a growing sheet, matrix, or tube. Changes in concentrations of components, changes in the rest of the chemical milieu, or the presence or absence of different external structural biases can halt growth, or even trigger the shift to alternative binding symmetry, disrupting large-scale structure growth. And growth itself depletes this local substrate concentration. Without the continual availability of unbound substrates, growth ceases. Growth must also outpace forces that tend to disrupt the coherence of the enlarging—and therefore more fragile—structure. As sheets expand, or molecules link up into larger surfaces or longer tubes, they may decrease in their overall energetic stability to the point where further growth becomes limited by structural fragility. For this reason, sheets tend to form into stable closed configurations (e.g., polyhedrons or tubes) that are both more robust to external perturbation and nearly as energetically relaxed. As closed structures, however, they may cease to be able to grow further.

But these intrinsic limitations of autocatalysis and self-assembly processes are also a source of potential synergy. The conditions produced by each of these processes and their limitations together comprise a complementary and reciprocally supportive effect.

Self-assembly provides the conditions that are most critical for sustaining autocatalysis: the proximity of reciprocally interdependent catalysts. The major consequence of self-assembling containment is local blockage of molecular diffusion. Spatial proximity of all constituents of an autocatalytic set is an essential necessary precondition for sustained autocatalysis, but in solution, spontaneous diffusion will undermine this requirement. Containment can, however, maintain proximity irrespective of whether catalysis is or is not taking place, and even in the absence of substrates.

And reciprocally, autocatalysis complements self-assembly. The major consequence of autocatalysis is the continual production of identical molecules in the same region, whereas self-assembly is most robust if the concentration of component molecules is maintained despite depletion due to this process. So self-assembly is most reliable in conditions where there is continual replenishment of component structural molecules up to the point of structural closure, at which point additional components are irrelevant. Thus rapid production of identical molecules at an accelerating pace by autocatalysis is specifically consistent with the conditions promoting self-assembly.

The reciprocal complementarity of these self-organizing processes means that spontaneous linkage of autocatalysis with self-assembly containment is a possibility. This would occur in the case that an autocatalytic cycle produced a byproduct molecule that itself was conducive to spontaneous self-assembly into a shape that could act as a container. In this configuration, each self-organizing process would reciprocally contribute to conditions promoting the stability, persistence, or recurrence of the other. Raw materials would be produced that are conducive to container formation. An enclosing structure would tend to form in the vicinity of the most active autocatalysis and thus would spontaneously tend to enclose autocatalytic set molecules within it. And closure of the container would prevent dissipation of the components of the autocatalytic set. Although this containment would also eventually halt autocatalysis by limiting availability of substrates, it would do so only after closure.

AUTOGENS

This complementarity of morphodynamic processes can produce far more than merely the creation of a kind of molecular bottle. Their reciprocity produces a special kind of emergent stability, unavailable to either process in isolation. Although continuous catalysis is prevented in the enclosed state, it remains *potentiated* by the proximity of all the essential molecules of the autocatalytic set. In the event that such a container were to be broken up or breached by agitation or chemical disruption in the vicinity of new substrate molecules, autocatalysis would recommence and new container-forming molecules would be synthesized. This could lead to either re-formation of the original or formation of two or more new containers of catalysts in its place. In other words, *the reciprocal complementarity of these two self-*

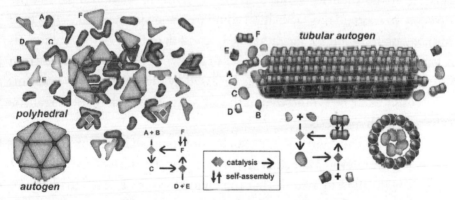

FIGURE 10.2: Two forms of simple autogenic molecular processes. *Left:* the formation of polyhedral capsules which contain catalysts that reciprocally catalyze the synthesis of each other and also produce molecules that tend to spontaneously self-assemble into these polyhedral capsules thereby likely to enclose the catalysts that generate them. *Right:* the formation of a tubular form of encapsulation, which although not fully closed will tend to restrict movement of contained catalysts along its length, but will tend to be increasingly susceptible to partial breakage and release of reciprocal catalysts as it grows longer. Both will tend to re-form or replicate additional copies if disrupted in the presence of appropriate catalytic substrates.

organizing processes creates the potential for self-repair, self-reconstitution, and even self-replication in a minimal form.

Importantly, this is not a property that is likely restricted to just a tiny set of molecular forms. It is in effect a generic class of chemical dynamics that may be achievable in numerous quite diverse ways. In general terms, the key requirements are only a reciprocal coupling of a spontaneous component production process and a spontaneous proximity maintenance process that encompasses all essential components. Though these reciprocal relationships are modestly restrictive, they are not extreme, and their spontaneous occurrence is made more probable because they are likely realizable in quite diverse chemical environments.

Candidate molecular conditions will be discussed in a later chapter, but the co-facilitation between two such morphodynamic processes could likely be realized by a wide variety of molecular substrates, and in the laboratory it could probably even be achieved using non-organic materials. Although the stereochemical and energetic requirements for such a system to form spontaneously are nevertheless quite limiting, similar limitations apply to any set of molecular interactions proposed as precursors for living processes. It is the relative simplicity of this system of dynamical relationships, and the diversity of ways this co-facilitation could be achieved, that makes this type of molecular complex a far more likely candidate for a first spontaneous self-reproducing system. With respect to known processes extracted from life, protein-based autocatalytic sets enclosed by the analogues of viral capsules could likely be synthesized in the laboratory. So this is also a realistically testable hypothesis.

Because of the familiarity of such component classes of molecular processes, and despite the currently hypothetical nature of the process, I think it is justified to assume that the characteristics I have attributed to these simple molecular systems are realistic extrapolations, which justify considering the implications for emergent dynamic effects. Indeed, I offer this as the potentially simplest class of teleodynamic systems. Because of its simplicity and the non-mysterious nature of the properties that such a molecular process would exhibit, if we can demonstrate that its dynamical properties have resulted in a clear, higher-order orthograde dynamic than its component processes with unambiguous ententional properties, even if these are minimal, then we have at our disposal a model system capable of

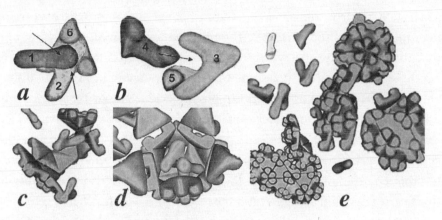

FIGURE 10.3: A cartoon depiction of various stages of polyhedral autogenic structures (as in Figure 10.2) self-assembling. The final image (*e*) shows the breakup and reassembly due to collision.

fully demystifying this realm without reducing it to mere thermodynamics or sneaking homuncular properties into the account.

Elsewhere, I have called this sort of hypothetical molecular system an *autocell*.[5] Unfortunately, I have found this term to be somewhat limiting and misleading, since the components described are not necessarily cellular, and the term is not mnemonically descriptive of its most distinctive properties. For the remainder of the book, I will adopt the more descriptive term *autogen* for the whole class of related minimal teleodynamical systems. This term captures what is perhaps its most distinctive defining feature: being a self-generating system. In this respect, it is closely related to Maturana and Varela's *autopoiesis*, though referring to a distinct dynamical unit process rather than a process more generally, for which I have reserved the more general term *telodynamic*. The term *autogen* is also easily modified to apply to a broader class of related forms by describing any form of self-encapsulating, self-repairing, self-replicating system that is constituted by reciprocal morphodynamic processes as *autogenic*, and describing the process, appropriately, as *autogenesis*.

Autogenesis is not, however, meant to be descriptive of just any process of self-generation. It is conceivable that a looser interpretation of this term could refer for example to replication of identical molecules in autocatalytic processes or to the hypothetical self-replication of RNA molecules.[6] The term is reserved for simple dynamical systems that accomplish self-generation by virtue of harnessing the co-dependent reciprocity of component morphodynamic processes. Though I am here provisionally assuming that this is only a property exhibited by simple molecular systems, there is no theoretical reason to assume that it is impossible in radically different forms. Experimentation with simple molecular autogenic systems will ultimately be required to determine the parameters for formation, persistence, and replication of molecular autogens, as well as to determine the supportive properties required of substrate molecules and the surrounding molecular environment. But the logical and theoretical plausibility of this form of dynamical organization provides sufficient justification to consider its implications for the emergence of ententional phenomena. So, in advance of definitive laboratory creation of autogenic molecular processes, and irrespective of whether this property can be realized in other kinds of substrates, many reliable extrapolations concerning the properties of such systems can be explored; and in particular, those that exemplify teleodynamic organization.

Autogenic theory is superficially similar to Manfred Eigen's hypercycle theory to the extent that each of these two self-promoting processes also promotes the other in some way, forming the analogue of a causal circle of causal circles, so to speak. But the resemblance to hypercycle architecture stops there, and in other respects autogenic theory is fundamentally different. As we saw above, an autocatalytic cycle is susceptible to self-undermining and self-limiting dynamics, and a hypercycle is doubly (or multiply) susceptible to this (depending on the number of substrate-dependent subcycles that constitute it). Each is a potential weak link that if broken will be catastrophic for the larger synergy. In contrast, the reciprocal linkage of the two complementary morphodynamic processes constituting an autogen has the opposite effect. Though each component process is self-undermining in isolation and co-dependent, together they are reciprocally self-limiting, so that their self-undermining features are reciprocally counteracted. Thus, whereas substrate exhaustion leads to both autocatalytic and hypercycle cessation and component dispersion, an autogenic system will establish its capacity to re-form before exhausting substrates, so long as closure is completed before this point is reached. Each process contributes essential boundary conditions for the other, providing the equivalent of a continual supportive environment for both processes. One might then describe an autogen as a hierarchic hypercyclic system, with each self-organizing component acting as supporting environment or context for the other.

To the extent that autogenesis provides the possibility for self-reconstitution after partial disruption, it also provides a potential mechanism for self-reproduction. By the same process that enables an autogen to form in the first place, fractional components of a disrupted autogen—including shell and catalytic molecules in close proximity and new substrate molecules from the surrounding medium—will be able to reconstitute a new complete whole. In other words, a disrupted autogen will be as likely to produce two identical autogens as one. So an autogen can accomplish in molecular terms—and with considerably more compactness than previously envisioned—what von Neumann demanded of self-reproduction: it can reproduce itself *as well as its physical capacity to reproduce itself*. Remarkably, it can accomplish this without many of the attributes normally assumed to be essential for life: for example, molecular template-based replication of components and of the template molecules, incessant far-from-equilibrium thermodynamics, semi-permeable membrane containment, and so on.

This self-reconstituting dynamics provides an active self-similarity-maintaining quality which constitutes a form of individuality, or "self," that does not otherwise exist outside of living processes. There is both an individual autogenic identity, as a closed, inert, but potentially self-reconstituting unit, and a self-maintaining lineage identity, due to the transmission of relatively invariant intrinsic dynamical constraints and molecular types from generation to generation as a result of replication. Self-reconstitution does not completely maintain material identity across time because it allows for molecular replacement, and it does not maintain energetic or dynamical continuity across time either, since it may persist in a static phase for extended periods. But this self-reconstitution capacity does maintain a persistent and distinctive locus of dynamical organization that maintains self-similarity across time and changing conditions. And yet ultimately there is no material continuity, as autogens are disrupted only to be reconstituted and replicated with newly synthesized components. Only the continuity of the constraints that determine the autogenic causal architecture is maintained across repeated iterations of dissolution and reconstitution.

So an autogen has an identity only with respect to this persistent general pattern of constraint maintenance and replication, and irrespective of any particular molecular constituents. Indeed, it is the continuity of the inheritance of constraints on its molecular dynamics that constitutes this individuality. But identity may vary as an autogen lineage evolves variant forms of this defining dynamics (see the section on autogenic evolution below). For all these reasons, an autogen self and autogen lineage identity are examples of efficacious general types, in a philosophical sense (see chapter 6). To be more specific, an autogen is an empirical *type* determined only by the continuity of these dynamical constraints, which are themselves expressions of dynamical limitation—potential modes of change not expressed.

Autogenic organization only exists with respect to a relevant supportive environment. So autogenic individuation is also only defined with respect to a particular *type* of environment. Identity and environment are thus reciprocally defined and determined with respect to each other, because the same molecular configuration in a non-supportive environment lacks any of the defining properties of autogenesis. Indeed, the very possibility for autogen existence can be described as one of the possible micro configurations of a certain class of environments with the molecular constitution conducive to autogen formation.

This is the ultimate basis for what Jacob von Uexküll called an *Umwelt*: the organism-relevant external world.[7] This critical autogen-environment relationship is, however, curiously at once a sufficient but not a constant necessary condition for autogenesis to persist. The potential for autogenesis can be maintained even in non-permissive conditions. Although such a molecular configuration is only an autogen with respect to a specifically supportive environment, it can nevertheless persist structurally across diverse and unsupportive environments, to later be reproduced in a supportive one. In this way, autogenic identity maintenance transcends any specific context dependence. In its inert closed state, an autogen can maintain this potential across vast epochs of time and through diversely non-supportive contexts. This then provides a degree of autonomy from context that is again a distinctive quality of living but not non-living dynamical systems.

An autogen's individuality is strangely diaphanous in one interesting sense. When closed and complete, an autogen is inert and yet when broken open and actively forming new constituents, it is merely a collection of molecules dispersed into a larger molecular milieu. In other words, it is a bounded individual only when inert, and actively self-generating only when it is no longer a discretely bounded material unit. This further demonstrates that what constitutes an autogenic "self" cannot then be identified with any particular substrate, bounded structure, or energetic process. Indeed, in an important sense, the self that is created by the teleodynamics of autogens is only a virtual self, and yet is at the same time the locus of real physical and chemical influences. This was also a feature recognized by Francisco Varela in his conception of an autopoietic system, even though he did not recognize how synergistic reciprocity of self-organizing processes could produce this. He describes this virtual self as follows:

> What is particularly important is that we can admit that (i) a system can have separate local components [for] which (ii) there is no center or localized self, and yet the whole behaves as a unit and for the observer it is as if there was a coordinating agent "virtually" present at the center.[8]

The generic, autonomous, and diaphanous character of autogenic systems makes this a functional property, not a material, chemical, or energetic property. An autogen is a precisely identifiable source of causal influence because it generates and preserved dynamical constraints—the basis for thermodynamic work. But, as we've seen, constraint is the exemplification of something that does not occur. So, in this sense, an autogenic system confronts us finally with an unambiguous absential quality. This is the essence of teleodynamical causality. By examining this model system more closely, we will be able to demonstrate how ententional processes acquire this seemingly paradoxical character of efficacious absence.

AUTOGENIC EVOLUTION

Because autogens are capable of self-replication, they are also potential progenitors of autogen lineages. An autogen lineage will increase in numbers so long as there are sufficient substrate molecules in the surrounding environment, along with sufficient molecular agitation to periodically disrupt autogen integrity, but not so much agitation as to disperse their contents more rapidly than they can reassemble. This is the first condition for natural selection.

The second condition is competition among these lineages. This will occur spontaneously as well because the multiplication of each lineage is dependent on the same (or catalytically similar) molecules in the surrounding environment. Different lineages are therefore in competition for these substrates, as well as for persistence against the relative disruptive influences potentially present in the environment—and to an extent necessary for lineage growth.

The third condition is variation among these lineages. Let's begin with the simplest case: a single type of autogen. Although autogens tend to self-reconstitute and thus reestablish the molecular structures predisposed by the catalysts and shell molecules that produced their progenitor, because this must occur via partial breakup and reclosure of the shell, it is likely that each new shell will additionally incorporate other molecules present in the local environment, some of which may continue to be passed on via future divisions. For the most part, these will tend to be innocuous inclusions that might only decrease the relative concentrations of active catalysts. Higher or lower concentrations of these relatively neutral molecules will tend to produce slight differences in rates of reconstitution or sensitivity to dissociation between lineages. If, however, incidentally included molecules are uncorrelated with any allosteric specificity of autogen functions, their addition to or loss from a lineage will be a matter of chance. Some fraction of incidentally incorporated molecules will have allosteric similarities with functional catalysts, their products, or their substrates. These will have a tendency to interact directly with them and will affect *functional* attributes of autogen chemistry. Their possible effects include directly interfering with or augmenting the rate of catalysis, affecting container formation, altering container stability, influencing molecular diffusion during disruption, providing linked parallel catalytic steps in the cycle that enables utilization of alternate or slightly variant molecular substrates, and so on.

Incorporation of functionally interactive molecules will make a given lineage relatively more or less successful in propagation. Autogen lineages containing different types of molecules that augment function in any of these ways will tend to out-reproduce others. This will only have a permanent selective effect on a lineage when another condition is met. Persistent inclusion of these divergent molecular types within a lineage will depend on being synthesized as part of an extended or parallel autocatalytic cycle. By these means, increased complexity and reliability of autocatalysis, improved containment, and utilization of special structural and energetic features of other molecular types (e.g., metals) could evolve by differential lineage propagation.

Together, these considerations show that although autogens are incredibly simple molecular systems, their self-reconstitution properties in favorable environments spontaneously bring into being the systemic conditions that are sufficient to initiate a persistent, if weak, form of natural selection. So, in an environment where autogen reproduction is likely, autogen evolution is also likely. In molecularly complex environments, autogen lineage competition for resources will tend to lead to the evolution of variant lineages differentially "fitted" to their local environments. This satisfies all necessary and sufficient material and logical conditions for natural selection, despite occurring in a system lacking many core attributes of life, including genetic inheritance in the biological sense.

Having established the logic of autogen evolvability, we can briefly consider certain special cases and features of this process that diverge from the norm of natural selection among organisms.

One potential source of lineage difference is independent emergence. It is possible that entirely independent and distinct types of autogens could arise in slightly differing environments, each giving rise to lineages that eventually overlap in their environments. Such independently arising lineages need not have overlapping substrate needs, and so would not necessarily be in competition. Multiple parallel threads of autogen evolution are possible, and perhaps likely, since a molecular context that is sufficiently rich to give rise to one type probably has the potential to give rise to other variants as well. This would not necessarily produce a natural selection dynamic, except to the extent that there was chance overlap of some substrate. Even so, their independent effects on other aspects of the environment might eventually lead to indirect effects that influence relative propagation rates in some more generic way (e.g., by effects on molecular diffusion).

This is crudely analogous to the way different species in an ecosystem affect each other's evolution, even if not directly interacting. But the passive and partially destructive nature of autogen propagation, which is quite unlike most living cellular reproduction, also introduces other potential avenues of evolutionary interaction. The most direct form of interaction could arise by mixing, as a result of inclusion of the components of one type of autogen within another. Since containment and co-localization of essential catalysts are somewhat generic features of autogen dynamics, there are many opportunities for linkages to develop between lineages with very different origins, and even for inclusion of the entire catalyst set of one autogen type within another. This is like endosymbiosis, now recognized as an important, if perhaps rare, source of novel synergistic functions in evolution, or lateral gene exchange, which is common in bacterial and viral evolution. The incorporation of complementary self-reconstituting systems within other self-reconstituting systems with somewhat different substrate requirements, catalytic dynamics, and structural self-assembly conditions could be a significant source of complexification and adaptation. The somewhat unregulated cycles of breakup and re-enclosure that autogen propagation depends on are in this way more analogous to virus reproduction than to cellular reproduction, and so may share other characteristics with virus evolution as well.

Natural selection is the aggregate dynamic that arises out of the interactions of large numbers of unit systems with these properties. Different lineages of autogens each maintain a thread of mnemonic continuity and identity, even when mixture occurs. Irrespective of such lineage convergence effects, when they are separated in different lineages, variant autogenic mechanisms will be in competition with respect to resources available in surrounding conditions. This makes each individual autogenic mechanism also a representative of a specific correlation between its intrinsic dynamic topology and features of the larger environment.

So, with autogens, there are multiple interlocking units of evolutionary individuation: the lineage, the autogen as autonomous system, and the reciprocally self-maintaining dynamical synergy, though the latter will likely grade into an autogenic unity once independent reciprocal dynamics intertwine into a single integrated system. Living organisms derive their peculiar causal dynamics—their seeming end-directedness, functional logic, superficial reversal of the second law of thermodynamics, and adaptability—from this upward shift in what constitutes a unit of reproductive and evolutionary individuation.

The foregoing analysis has shown that natural selection emerges from the dynamics of reciprocally reinforcing self-organizing processes. This self-reconstitutive dynamical synergy is the essential ingredient that precedes and underlies life. This in turn suggests that natural selection is ultimately an operation that differentially preserves certain alternative forms of morphodynamic processes compared to others, with respect to their synergy with one another, and with respect to the boundary conditions that enable them. Selection is not then fully defined only with respect to replication of genetic information. As Kant recognized, a self-maintaining "formative power" is critical. And this requires processes that generate, preserve, and propagate constraints. Morphodynamic processes are the only spontaneous processes that generate and propagate constraints, and autogens demonstrate that reciprocity between morphodynamic processes can preserve and replicate constraints.

THE RATCHET OF LIFE

Autogens may be subject to natural selection and evolution but they are not alive in most senses. They lack the majority of attributes associated with living organisms today. Most significantly, they are effectively passive—though not inert—structures, because they do not actively accumulate and mobilize energy within themselves that can be used for self-repair and reproduction. The energy driving their self-perpetuating dynamics may derive from an environmental source (such as the heat of volcanic vents), or it may be obtained by breaking the molecular bonds of large substrate molecules. In any case, autogens are parasitic on their environment, even for the initiation of replication, in much the same ways that viruses are. Unlike viruses, however, autogens reproduce without the help of other organisms.

In this respect, autogens are not quite "autonomous agents," in Stuart Kauffman's sense. For Kauffman, the most basic characteristic of living systems is their ability to "act on their own behalf." Although an autogen does not "act" in a self-animated sense, it nevertheless fulfills the two criteria that Kauffman uses to define what he means by autonomous agency—criteria that he argues are the basis for this self-directed activity. It has the *capacity to reproduce* (and/or reconstitute) itself, and it *completes a work cycle* in the process. In other words, with each disruption and self-reconstitution cycle, an autogen performs sufficient work to also reconstitute the capacity to repeat this cycle all over again. Because of this, an autogen can be the progenitor of lineage that can continue indefinitely.

Perhaps the most striking difference between autogens and organisms is that autogens do not maintain persistent non-equilibrium dynamics. They nevertheless effectively out-maneuver the second law with respect to their own structure and function. They do so via what amounts to a higher-order ratchet effect. As we have seen, morphodynamic processes maximize entropy flow—that is, the dissipation of constraints—but as fast as entropy increase diffuses the introduced distributional asymmetry, external perturbation reestablishes it. The result is a constant throughput of energy and a constant rate of entropy generation. In dissipative systems (discussed in chapter 8), those processes that dissipate constraints more slowly and less efficiently will tend to be spontaneously supplanted by those that do so more rapidly and more efficiently. The buildup of local constraints creates conditions where a fraction of their potential to do work is diverted into the generation of global constraints that progressively increase global dissipation rates. In this way, continual dynamical perturbation causes local impediments to dissipation to become self-eliminating. For this reason, self-organization is often described as a process that works in the service of the second law—i.e., maximum entropy production (see chapter 9).

A number of theorists have argued that life too should be considered as an entropy flow-maximizing process. Could the evolution of life on Earth be seen as a dynamical process in service of efficiently generating entropy to balance the constant flux of solar energy bombarding the planet? If life were merely a morphodynamic process, this would indeed be likely. But this is not the case for teleodynamics. Even though teleodynamic processes are constituted by the interactions between morphodynamic processes, the entropy flow maximization of these component processes is not additive in this interaction. This is because the attractor basins toward which morphodynamic processes tend are specifically structured, and thus constitute constraints in their own right. The teleodynamic features that emerge in an autogenic process are the result of reciprocal constraint generation. Autogen formation must indeed involve an increase in entropy, and each component morphodynamic process naturally develops toward a state of maximum entropy production rate. But the complementary constraints that each generates with respect to the other are self-undermining. Their progression toward optimal entropy production rate is also toward a state where dynamical change ceases. Regularity is built up, only to be frozen by closure. Full dissipation is prevented at a point where optimal conditions for rapidly reinitiating the process are achieved.

An autogen is thus effectively a negentropy ratchet. This ratchet effect, which conserves constraint at the cost of stopping entropy generation prematurely, is the secret to life's tendency to preserve information about past adaptive organization. This enables both the adaptive fine-tuning and the complexification of life by making it possible to build on previous successes. This retained foundation of reproduced constraints is effectively the precursor to genetic information (or rather, the general property that genetic information also exhibits). As will become clearer in subsequent chapters, whether it is embodied in specific information-bearing molecules (as in DNA) or merely in the molecular interaction constraints of a sim-

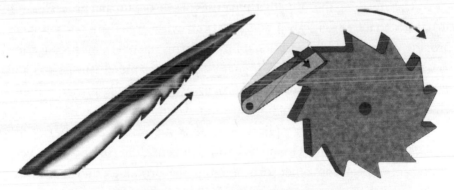

FIGURE 10.4: Typical examples of the ratchet effect. *Left*: the barb of a honeybee stinger redrawn from a scanning electron micrograph magnified 400 times. The barb structures make penetration easy in one direction and nearly impossible in the other. *Right*: the structure of a typical ratchet gear and movable catch. Clockwise rotation of the gear is easy, but counterclockwise motion is prevented; thus random forces tending to rotate the gear in both directions will only result in one direction of movement. This same logic determines that an autogenic system will tend to support the generations of constraints by morphodynamic and thermodynamic work, but will prevent their dissipation.

ple autogenic process, information is ultimately constituted by preserved constraints.

It should come as no surprise that an organism does not maximize the rate at which it generates entropy or the throughput of energy. Instead, an organism uses the flow of entropy to build constraints that ultimately divert and slow this process, increasing the amount of local work it can extract. Our ability to tap the fossil fuel reserves of the Earth for the energy to run our industries and ease our lives is the result of the buildup of complex molecules, using the energy of sunlight. Ancient organisms accumulated this surplus in ways that the inorganic Earth could not have. They slowed this dissipation process and redirected the flows of energy down byzantine molecular pathways, capturing a fraction in newly synthesized molecular bonds (constraints on atomic mobility) over the course of millions of years. This temporarily sequestered, unreleased energy was not allowed to dissipate at the death of many of these organisms because they were buried under layers of sediment, which further impeded the dissipation process. The comparatively simplistic chemistry of combustion that we use to extract this ratcheted energy is a way to rapidly generate the entropy increase that the process slowed for so long.

By transforming the formative dynamics of self-organizing processes into structures that transiently resist degradation, living organisms and autogens provide the foothold for additional linked forms of constraint propagation to develop. So, whereas morphodynamic processes merely propagate and amplify constraints, teleodynamic processes additionally preserve them. This is the common theme of both life and evolution.

There is one further complication to this physical twist of thermodynamics provided by life. Organisms and autogens differ in one other very important respect from other sorts of dissipative processes: they reproduce. In this way, they not only can use local flows of energy to reconstitute the constraints that individuate them, and thus resist full dissipation; they can also generate additional individuated replicas. This multiplies paths of energy dissipation and loci of constraint generation. Reproduction therefore multiplies regional rates of constraint dissipation, even as each individual autogenic unit ratchets it. In the course of evolution, more rapidly reproducing species tend to replace slower-reproducing species by degrading the available energy gradient more rapidly. An interesting balance tends to develop between organism and ecosystem rates of entropy production. Since a resource niche will tend to be rapidly invaded, further reproductive competition will tend to favor species with a higher efficiency of constraint generation. This, of course, is a crucial driver of natural selection.

THE EMERGENCE OF TELEODYNAMICS

Autogens mark the transition from maximum entropy production to constraint production and preservation, and from orthograde processes characterized by self-simplification (morphodynamics) to the orthograde processes exemplified by self-preservation and correlative complexification (teleodynamics). This transition from morphodynamic to teleodynamic organization can be described as an emergent transition, in the same sense as the transition from thermodynamic to morphodynamic organization is emergent. Each of these transitions is characterized by the development of an orthograde disposition that is contrary to what preceded and gave rise to it.

Thus morphodynamic organization emerges due to the interaction of opposed thermodynamic processes (e.g., perturbation and equilibration), and it results in constraint amplification rather than constraint dissipation (i.e., increase in entropy). Analogously teleodynamic organization emerges due to reciprocally organized morphodynamic processes, and results in constraint stabilization rather than constraint amplification, and entropy ratcheting rather than entropy production. In this respect, autogen formation exemplifies the defining feature of an emergent phase transition—the appearance of a new form of orthograde organization.

Autogen dynamics also demonstrates that the chasm between thermodynamics and living dynamics cannot be crossed in a single step, but requires an intermediate morphodynamic bridge: a synergistic arrangement of non-equilibrium self-organizing processes. The evolutionary processes that result are in this way two emergent levels removed from simple thermodynamic processes. Teleodynamics emerges from morphodynamics emerges from thermodynamics.

Being able to provide a complete description of an extremely simple molecular system, capable of self-reproduction and susceptible to natural selection, puts us in a unique position to reflect on the emergent status of living and evolutionary process and their relationships to other natural processes. The plausibility of autogens provides a simple model system for exploring the requirements of ententional properties in general. This is an important first step on the way to an augmented physical theory adequate to explain living and mental processes, as well as non-living processes. In simple terms, it is the scaffolding from which to build an emergent rather than an eliminative methodology for studying all consequence-organized phenomena, from metabolism to mental representation (though this in no way suggests that they are equivalent).

Theoretically, this analysis also helps to distinguish between self-organized and selection-based features of complex adaptive systems, and shows how they are intrinsically linked. The generality of this type of dynamical relationship suggests that the specific molecular features of life on Earth may be far less universal than might otherwise have been guessed, and that autogenlike processes may be present in forms and planetary conditions with very different features than ever were present on Earth. Beyond explaining the linked contribution of self-organization and Darwinian selection to phylogenetic evolution, this analysis may also shed light on their interaction in other biological and even non-biological processes, such as epigenesis, neural signal processing, and language evolution. Self-organizing processes can arise in many dynamic systems and can be constituted by many substrates. The dependence of evolutionary processes on self-organizing processes is not necessarily confined to molecular processes. This analysis should be general. Evolutionary dynamics should emerge spontaneously in any domain where analogous reciprocal self-organizing conditions are met.

The emergence of teleodynamic processes includes more than merely consequence-organized phenomena. Even within a teleodynamically organized system as simple as an autogen, we can identify component relationships that are the minimal precursors of many end-organized features we associate with life.

For example, although the higher-order orthograde dynamic that characterizes autogens provides their robustness to perturbation and their maintenance of integrity, this is not in itself sufficient to constitute something we could justifiably call autonomous individuality. Robustness to perturbation is one of the defining features of any orthograde dynamic. Thus systems in thermodynamic equilibrium resist being driven further from that state, and morphodynamic (self-organizing, dissipative) systems near their most "relaxed" attractor dynamics will also tend to resist disruption of their characteristic regularity. Because autogens are constituted both by thermodynamic and morphodynamics processes, they inherit both of these forms of resistance. But they don't merely resist perturbation. In the face of catastrophic perturbation, the physically dissociated components nevertheless retain their systemic identity sufficiently well to be able to reassociate into identically organized unit structures. This indicates that these components are to some extent present *because* of their contributions to a higher-order whole. It is in this respect that self and function are interdependent concepts that are intrinsically defined with respect to consequences.

Although the concept of a function is also sometimes applied to inanimate systems—such as when describing the gravitation of the Sun as functioning to counter the expansive effect of fusion—this is largely a metaphoric use, treating some physical relationship as though it were organized to accomplish that end. Of course, with the exception of man-made devices, inanimate non-living systems are not organized for the sake of achieving a given consequence. But autogens are. For example, it is appropriate to describe the self-assembling container of an autogen as functioning for the maintenance and perpetuation of the autocatalytic set's capacity to cycle, and the autocatalytic set can likewise be described as functioning for the sake of supporting the self-assembly process. So, in Kant's terms, each of these component processes is present *for the sake of* the other. Each is reciprocally both end and means. It is their correlated co-production that ensures the perpetuation of this holistic co-dependency.

Functions are normative to the extent that they fail or succeed to achieve some end. The autocatalysis, the container, and the relationship between them *are generated in each replication precisely because they are of benefit* to an individual autogen's integrity and its capacity to aid the continuation of this form of autonomous individual. Although one could say that, like the gravitation and fusion forces in the Sun, the component processes in a first spontaneously generated autogen just happened to accidentally co-occur to produce this metastable form, one couldn't say this about the replication of this organization in succeeding autogens. Unlike accidental correlations, the organization that is created anew via autogen replication occurs precisely *because it had this consequence.* But to repeat: it was the *teleodynamic organization* that had this consequence, and not merely some collection of inter-reacting molecules, because these are replaceable while the organization is not. The identity and the beneficiary is not a thing but a dynamical form.

So, even these simple molecular systems have crossed a threshold in which we can say that a very basic form of value has emerged, because we can describe each of the component autogenic processes as there for the sake of the autogen integrity, or for the maintenance of that particular form of autogenicity. Likewise, we can describe different features of the surrounding molecular environment as "beneficial" or "harmful" in the same sense that we would apply these assessments to microorganisms. More important, these are not merely glosses provided by a human observer, but intrinsic and functionally relevant features of the consequence-organized nature of the autogen itself.

Adaptive functions are, however, more than just elements of an entity that respond to the entity's environment. They embody in their form and dynamic potential—as in a photo negative—certain features of this environment that, if present, will be conducive or deleterious to the persistence of this complex dynamic. The presence of these conditions may or may not obtain. Thus adaptations may be appropriate or inappropriate to a given context, to the extent that the consequence with respect to which they are organized may or may not be achieved. This is another indication of functional normativity: the possibility of dysfunction.

In the case of dysfunction, the correspondence relationship between internal organization and extrinsic conditions no longer exists. In a crude sense, then, we can describe this as an erroneous prediction based on a kind of physical induction from past instances. It is in this respect—of possible but fallible correspondence—that we can think of an adaptation as embodying information about a possible state of the world. And like more obvious forms of representation, this projection can be in error; there may be nothing in the immediate environment to which it corresponds. So, in this very basic sense, autogens could be said to represent their environment, in roughly the same sense as a shoeprint could be said to represent a shoe.

We are far from a full explanation for the sorts of teleological processes experienced at the level of human consciousness, and a considerable distance from what is found in the simplest living forms on Earth. The "proof

of principle" is in this regard quite minimalistic. Nevertheless, it is a definitive exemplar of crossing the fundamental gap that separates the mechanical
world from the functional and normative world.

In summary, then, in this simple dynamical molecular system we can
discern the minimal precursors of function, adaptation, teleology, valuation, and even the dim anticipation of information about the environment
and a self with respect to which all this matters. Each of these attributes is
implicitly ententional, and yet their emergence can be precisely understood
without the need either to attribute them to mysterious or weird forms of
causality or to argue that there isn't any fundamental threshold crossed at
this point. This demonstrates that, at least in principle—and ultimately that's
all that matters—*real teleological and intentional phenomena can emerge
from physical and chemical processes previously devoid of these properties.*

Now that we have thoroughly explored this simple teleodynamic system,
and have demonstrated how it can emerge from less convoluted morphodynamic processes, and ultimately from thermodynamics, we are in a
position to begin to reframe the way we understand the physics of consequence-organized processes (i.e., Aristotle's final causality). At a minimum,
this analysis demonstrates that there can be no simple one-to-one mapping
of teleodynamic relationships to mechanistic relationships. The link between
basic mechanics and ententional processes and relationships must necessarily be bridged by an intervening level of morphodynamic processes. The
persistent failures to map living dynamics and cognitive processes directly
to simpler physical processes, and thus to reduce mental to mechanistic processes, was always doomed to fail, though not because of some fundamental
discontinuity or dualism. The supervenient relationship between them is
indeed necessary, but it is both doubly indecomposable and doubly negative.
It is doubly indecomposable because it is based on two emergent transitions,
each defined by the intrinsic generation of higher-order holistic constraints,
that cannot be decomposed to any lower-level components or relationships.
It is doubly negative because each level of dynamics emerges from the interaction of lower level constraints, which are themselves absential properties.
So, in effect, each is the expression of a form of higher-order absence generated by relationships between lower-order absences. This doubly negative,
doubly absential character of teleodynamic processes is almost certainly one
reason that we find them so mechanistically counterintuitive.

The analysis of this additional level of emergent transition also enables us to identify a common logic for unambiguously defining emergence more generally. *An emergent dynamical transition is signaled by a change in the topology of the phase space of probable dynamical trajectories.* Using the term that we have applied to asymmetries of probable changes of state, each emergent dynamical transition involves *the appearance of a new mode of orthograde attractor logic.* Thus the transition from a simple thermodynamic regime to a morphodynamic regime is marked by the emergence of orthograde tendencies toward highly regularized global constraints that run counter to the constraint-dissipation orthograde tendencies of the underlying thermodynamic processes.

Emergence is, in effect, defined by a polarity reversal in orthograde dynamics with ascent in scale. Thus the orthograde signature of thermodynamic change is constraint dissipation, the orthograde signature of morphodynamic change is constraint amplification, and the orthograde signature of teleodynamic change is constraint preservation and correlation. The polarity reversal that defines the emergence of teleodynamics from morphodynamics is what characterizes life and evolution. A fit or interdependent correspondence between constraints in different domains is the essence of both biological adaptation and the relationship characterizing representational relationships. So this provides the first bridge across the "epistemic cut" that has been the dividing line between the two sides of the Cartesian dilemma—the no-man's-land that has divided both science and metaphysics into seemingly incommensurate universes.

So far this analysis has been mostly confined to processes in the range from molecular thermodynamics to the very simplest lifelike processes. Emergent dynamical transitions are, however, ubiquitous in nature, and although they are necessarily simpler at lower levels of scale, they can be correspondingly far more complex at levels of scale where multicelled organisms, brains, and human social phenomena emerge. Nevertheless, there is a general principle exemplified by the simplest autogenic system, described here, that applies at all higher levels of emergent dynamics: all teleodynamic processes must be constituted by reciprocally synergistic morphodynamic relationships, and all morphodynamic processes must be constituted by competing homeodynamic processes. Although higher-order teleodynamic processes may exhibit properties that are more elaborate than those exhibited by basic autogenic systems, they must arise by a recapitulation of this same hierarchic emergent dynamic logic, even if the components are themselves teleodynamic systems. Teleodynamic systems can interact homeodynamically; homeodynamic relationships between teleodynamic systems can produce morphodynamic relationships; and synergistically reciprocal morphodynamic relationships constituted by interacting teleodynamic systems can produce higher-order teleodynamic relationships.

With each such emergent transition, there will be a characteristic new level of orthograde geometry of causality, but the generation of each emergent transition must necessarily depend on this homeo-morpho-teleo emergent logic. Like basic autogentic systems, higher-order teleodynamical systems will also be self-creating, self-maintaining, self reproducing individuated systems; but they may also exhibit emergent teleodynamic properties not exhibited by their lower-order teleodynamic components. Terming such second- and third-order teleodynamic systems *autogens* and describing their properties as *autogenic* would therefore be too restrictive and falsely reductive. So I will call such higher-order teleodynamic systems *teleogens*, in order to designate both their individuality and their capacity to generate additional forms of teleodynamic processes. We will survey some of the properties of such higher-order emergent teleodynamic relationships (such as sentience) in the final chapters of this book, and discuss the implications of homeodynamic, morphodynamic, and even teleodynamic processes constituted by the interaction of lower-order teleodynamic systems, as is found in ecosystems, complex organisms, brains, and even social systems. But before we can apply this logic to ever more complicated systems, we need to reflect on the generality of the analysis beyond this basic model system, which we have used to exemplify this third realm of dynamics.

11

WORK

What is energy? One might expect at this point a
nice clear, concise definition. Pick up a chemistry
text, a physics text, or a thermodynamics text, and
look in the index for "Energy, definition of,"
and you find no such entry. You think this may be an
oversight; so you turn to the appropriate sections
of these books, study them, and find no help at all.
Every time they have an opportunity to define
energy, they fail to do so. Why the big secret? Or is
it presumed you already know? Or is it just obvious?

—H. C. VAN NESS, 1969[1]

FORCED TO CHANGE

The theory of emergent dynamics that we have outlined in the previous three chapters does not in any way conflict with the basic principles of physical dynamics. Indeed, it is based almost entirely on well-established pre-twentieth-century physical theory. However, there is an interesting way that it can amplify and broaden these principles to extend into organizational and entential realms that have until now remained mostly disjointed from the physical sciences. In this chapter, then, we will reexamine the common notions of energy, power, force, and particularly work (from which these other concepts are abstracted), in an effort to understand how they can be generalized and reformulated in emergent dynamic terms.

It took the geniuses of Galileo and Isaac Newton to show definitively that, neglecting the effects of friction, objects moving in a straight line will maintain their velocity and direction of movement indefinitely, so long as they are not affected by gravity, impeded by collision, or otherwise forced off course. And as a result, constant rectilinear motion is now understood to be equivalent to being at rest. It tends to persist as a spontaneously stable pattern of change. A thermodynamic system in equilibrium is also in continuous change, even though at a macroscopic level it appears to be unchanging. In this respect, it is analogous to an object in constant undisturbed movement. Extending the analogy, we can likewise compare the interactions of differently moving objects to the interactions between thermodynamic systems with different specific heat or energy levels. The differences in relative momentum when objects interact result in accelerated changes in their motions. The differences in the total heat or energy content between different thermodynamic systems at equilibrium can translate into changes to both if brought into interaction. These correspondences form part of the bedrock on which classical physics was built, and later found even more subtle expression in relativistic and quantum theories.

How much things change from what would have occurred spontaneously is a reflection of the amount of *work* exerted to produce this change. So long as contragrade change persists, work is involved, and thus it can accumulate over time and distance. The amount of change also reflects the amount of energy exchanged during the transition from the prior condition to the changed condition in the interaction. Energy is related to the capacity to do work, irrespective of whether it is exhibited in the collision of two masses or held in the electrical potential compressed into the covalent bonds of a hydrocarbon molecule. As we have discovered, however, the concept of energy is remarkably difficult to pin down. This is in part because it is a notion abstracted from the concept of work. In one sense, it could be considered merely a way of balancing the books. It is what has not changed in any physical change, spontaneous or not.

Work, on the other hand, is intuitively more tractable, as it is directly reflected in the extent of non-spontaneous change that results. And, unlike energy, the capacity to do work does not remain constant. This is what we intuit when water has drained downhill and we can no longer use its motion to turn a waterwheel, or when heat is fully transferred between two systems so that they are at equilibrium. During the falling of water, we say that energy is transferred from the movement of the water to the turning of the wheel; and during the exchange of heat, we say that energy is transferred from the hotter to the cooler container. After reaching ground level, the water can do no more work; and after reaching equilibrium, no more work can be extracted from the thermodynamic system. We say in this case that there is no more "free energy," which is also to say that there is energy that is no longer free to be transferred in such an interaction. Of course, some of the energy implicit in the elevated water could become "freed" were we to find a way to get it to flow to an even lower elevation, and some of the energy of a thermodynamic system at equilibrium would again become free if that system were subsequently placed in contact with an even cooler system. Clearly, at some point, there is no place lower for water to flow to, and no cooler system to take on some of the heat of a thermodynamic system. The latter condition is called absolute zero (and not surprisingly matter begins to act strangely near this temperature). This ability to "free" and "trap" energy tells us that, whatever it is, energy is not in itself the source of change. Rather (as we discussed in chapter 7), it is its availability for "movement" from place to place or transition from form to form that provides the potential for contragrade change. This is why constraint on this "movement" is so central to notions of causality.

Moreover, the generation of higher-order emergent dynamics also depends on work at a lower order. What we will discover in this chapter is that with each emergent transition, a novel capacity to do work emerges as well. And a new mode of work introduces new causal possibilities into the world. So, to understand the emergence of novel forms of causality, we need to explain the emergence and nature of these higher-order forms of work.

EFFORT

Newton's precise analysis of the concept of mechanical work was only the beginning of the physical analysis of this explanation for change in state. To make all forms of physical work consistent and to explain their interconvertability, it was necessary for nineteenth- and twentieth-century scientists to discover how to equate this Newtonian conception with the change-producing capacities of heat engines, chemical reactions, electromagnetic

interactions, and nuclear transmutations. By the third decade of the twentieth century, this project was largely complete. But behind this towering
achievement of theoretical physics is a broader conception of work, one
that is generally ignored in the natural sciences because it is assumed to be
merely a colloquial analogue to this more technical understanding. Indeed,
the technical analysis of work was long preceded by this more generic
conception that we use to describe a vast array of effortful enterprises.

The question that we will address here is whether these other not-
exactly-energetic conceptions of work can also be brought into a more precise formal relationship with the physical science notion. In other words, to
highlight just one exemplar, can the work of conceiving of these thoughts,
transforming them into sentences, and selecting the most appropriate words
to express them be understood to be as real and measurable as the work of
an internal combustion engine? I am not merely inquiring about the metabolic support for the neural and muscular biochemical reactions involved,
though these are relevant, but the higher-order work of forming and interpreting the concepts involved—that which makes daydreaming effortless
but metabolically equivalent problem solving difficult. Though often they
are not energetically equivalent, I will argue that it is not the energy use that
makes a difference, but the manipulation of the content of these thoughts—
something that we have argued is not physically present.

This is not a merely academic problem to be solved. Measuring the
amount of work needed for a given task is often an important factor in
determining the minimum requirements for success, or for determining the
relative efficiency of different ways of achieving the same goal. In the assessment of physical tasks, for example, engineers need to routinely calculate
how much work must to be done to move objects around on a factory floor,
to lift a heavy object to a certain height, or to accelerate a vehicle to a given
velocity. But the sort of work that we are often interested in analyzing is not
always simply physical in this sense. Perhaps the most common uses of the
term *work* refer to activities that do not neatly fit into this physical schema at
all, but involve making difficult decisions, analyzing unknown causal effects,
and exploring mysteries.

We even tend to describe our occupations as work. We ask strangers,
"What kind of work do you do?" and talk about "driving to work" in the
morning, where work is either considered a class of human activity or even

the location of that activity. We consider it work to keep one's home clean, to organize and manipulate food and cooking utensils to prepare a meal, or to convince colleagues of a counterintuitive theoretical idea. What these more colloquial notions share in common with the Newtonian conception of work is the superficial implication that the activity being described makes things happen that wouldn't come about without it, or else that would happen unless work is done to prevent it. In general terms, then, we can describe all forms of work as activity that is necessary to overcome resistance to change. Resistance can be either passive or active, and so work can be directed toward enacting change that wouldn't otherwise occur or preventing change that would happen in its absence. This also means that work may be activity pitted against some other form of competing work, as when law enforcement officers work to counter the work of criminals.

This more generic use of the concept of work is also employed to describe activities merely requiring mental effort, in which the linkage with physical activity and energetic processes can be quite obscure. We know that producing a novel software routine that is able to accomplish some computational task, or solving a difficult crossword puzzle, or conceiving and organizing the steps necessary to efficiently construct a garment, all take work. But it is not at all obvious how to measure or compare these forms of work to each other or to the general physical concept of work.

Nevertheless, we are often adept at coming up with relative measures of mental work. Difficult jigsaw puzzles take more work to solve than easy ones. This might be simply because they have more parts. And they can be more difficult if the pieces are all turned upside down so that there is no picture to suggest which pieces are likely adjacent to one another (a constraint that reduces the amount of work required?). To make it even more difficult, some puzzle makers cut the pieces into very nearly the same shape, so that constraints of shape incompatibility are made unavailable. These complications do not necessarily require more physical manipulation of the pieces (although this might also follow), but almost certainly they will require more "mental manipulation." This simple example indicates that numbers of optional arrangements of things that aren't desirable or functional, and the difficulty of distinguishing among them, contributes to this notion of the amount of mental work required.

How might this relate to the more familiar notion of mechanical work?

Though some mechanical work is involved in moving puzzle parts from place to place, and more movements are necessary to discover the correct fittings by trial and error in many-component problems than those with fewer components, for the most part this is not how we intuitively estimate the amount of work that will be required. Probably the best estimator is some measure of the number of operations that will likely be required to reach a solution. Of course, each operation—even each mental operation—requires energy, since these too are not likely to happen spontaneously. But this is often the most trivial source of resistance to be overcome.

In writing this book, for example, the finger work of typing that is necessary to input text into the computer is trivial in comparison to the work required to conceive of and express these ideas. This agrees with the intuition that the re-analysis and reformulation of otherwise widely accepted ways of thinking—particularly the effort to craft a convincing critique and alternative explanation—and that which is involved in discovering how best to communicate ideas that are counterintuitive or alien or otherwise go against received wisdom, is particularly difficult work, though it may be no more energetically demanding that walking a mile. This suggests that the sources of resistance that are the focus of the work to be done also include many tendencies not generally considered by physicists and engineers; for example, tendencies of thought that contribute to the difficulty of changing opinions or beliefs.

One might be tempted to object that the family resemblances between these various ententional notions of work and the physicist's concept of work are merely superficial, and that the use of the same term to refer to a form of employment or a creative mental effort is only metaphorically related to the Newtonian notion. But if there is a deeper isomorphism linking them all, there could be a great benefit in making sense of this connection. It is becoming increasingly important to discover how best to measure and compare all these diverse forms of work, especially in an era in which vast numbers of people spend their days sorting and analyzing data, organizing information in useful ways, and communicating with one another about how they are doing it. If it were possible to identify some unifying principles that precisely express the interdependencies between the physicist's conception of work and the computer programmer's experience of work, for example, it might have both profound scientific and practical value. This is not just because such knowledge could help to assess the relative efficiency of management strategies, aid Wall Street agencies in discerning optimal advertising campaigns, or even contribute to political efforts to manipulate public opinion, but because work is the common denominator in all attributions of *causal power*, from billiard ball collisions to military coups to the creative outputs of genius.

More generally, what we mean by *causality* and what we mean by *work* are deeply interrelated. One of the main reasons scientists and philosophers still argue about the kind of causality that constitutes our ability to initiate goal-directed activities is that we can't figure out how to link this mental form of work to the physical forms of work that are also necessarily involved. To finally cut through the tangle of confusions that surround the mysteries of mental agency and the efficacy of representations, then, we first need to develop a *general theory of work*: one that explicitly demonstrates the link between the ways that both energy and ideas can introduce non-spontaneous change into the world.

AGAINST SPONTANEITY

The anthropologist and systems thinker Gregory Bateson is well known for his relentless effort to expose the widespread fallacy of describing informational relationships in biology and the human sciences using energetic metaphors. In an effort to define information in its most general sense, and to distinguish it from energy, he described it as "a difference that makes a difference."[2] This makes explicit a conception of information that is central to Claude Shannon's *Mathematical Theory of Communication*, and to which Bateson added an implied cybernetic aspect by virtue of the double meaning of "making a difference." We will return to the problem of defining information in the next chapters, but this way of talking about difference is somewhat ironically also relevant to the concept of work.

Bateson was trying to distill the essence of the logic of information and control theory by highlighting the fact that according to this theory, information was merely a measure of variety (e.g., of letters or signal patterns) or difference, and not some "thing." He was emphasizing the fact that a difference is an abstract relationship, and as such behaves quite differently from material substances and their interactions. As an example, he points out that a switch is neither within nor outside an electric circuit. It mediates a relationship between events outside and those inside the circuit. When the switch is thrown by an external difference in some feature (e.g., a rise in temperature) it creates an internal difference in the circuit (e.g., breaking the circuit and cutting power to the furnace), which in turn causes an external difference (e.g., a drop in temperature), and so on.

Bateson was reacting against the misleading metaphorical use of energetic concepts to talk about informational processes, such as show up in concepts like the "force" of ideas, the "power" of ideology, or the "pressure" of repressed emotions (due to impeding the flow of libido in Freudian theories of neurosis). He was struggling against an entrenched substance terminology (analogous to the eighteenth-century conceptions of phlogiston and caloric), which obscured the critical differences between physical principles and those beginning to be articulated by the infant fields of information theory and cybernetics. This misleading conflation of energy with information often leads us to treat information as though it is a physical commodity; a kind of stuff that one can acquire, store, sell, move, lose, share, and so on. Indeed, as this list makes obvious, this is precisely the colloquial understanding of the concept. Bateson's point suggests that, as in the case of energy, progress could only be made when this was replaced by a dynamic relational conception of information. This was finally achieved (though incompletely) in the 1940s. In the next chapter, we will reanalyze the concept of information in some detail, and both explain this insight and explore how it still falls short of a full conception of information. But for now, it is sufficient to recognize that this substantializing tendency is similar to the substance conceptions of energy that dominated the eighteenth and early nineteenth centuries. So, although Bateson's phrase captures a number of important features that characterize information, and which show it to be different from mere stuff, much is left ambiguous as well. In fact, it doesn't quite disambiguate energy and information, as Bateson had intended.

Recently, a colleague (Tyrone Cashman) recounted a discussion he had some years ago with the influential systems ecologist Howard Odum. Cashman was attempting to explain the distinction that Bateson was making between energy and information by the use of his epigrammatic phrase "a difference that makes a difference." But Odum objected that his phrase did not in fact uniquely demonstrate this distinction, because it could equally well be applied to the concept of energetic work: a difference in the distribution of energy in one system that can be used to produce a difference in the distribution of energy in another. This objection is well taken. As we saw above, this is a quite accurate abstract definition of the concept of mechanical or thermodynamic work. Was Bateson mistaken? Is this a poor definition of information?

On the one hand, I have to agree with Odum that this epigram does not do a very good job of picking out the distinguishing feature of information processes that make them different from energetic work. On the other hand, if it is nevertheless a useful characterization of information (which I think it is), then this parallelism suggests something quite interesting. It suggests that the generation of information might also be understood as a form of work, perhaps related to, though not merely, energetic work. So, comparison to the development of the energy concept might offer clues about the kinds of misconceptions that tend to arise when analyzing information. This will be the topic of chapters 12 and 13; but before embarking on this issue in greater depth, we can for now explore the implications for a technical conception of work that is as precise, but more generalizable than just what we describe with Newtonian physics and thermodynamics; generalizable even to processes as diverse as order creation, information production, and decision making.

How might this Batesonian conception, treated as a description of physical work, point the way to a precise general conception of work? Consider how it might apply to a physical process. When energy is transformed from one form to another, it is a difference that is being transferred from substrate to substrate (a gradient of non-equivalence, an asymmetry of distribution, say, of molecular momenta), but there is inevitably some resistance involved in this transfer. Systems "resist" being shifted away from a state of equilibrium (though as will become clear in a moment this resistance is not a simple concept). This is also implicitly captured in Bateson's phrase, since it implies that the second difference is compelled—or made—to come into existence by the first difference. The difference that is "made" depends on a difference that is provided as a given. The implication is that the new difference that is created in this process would not have occurred had the first difference not existed. So another way to describe work, using this Batesonian characterization, is that it involves something that doesn't tend to happen spontaneously being induced to happen by something else that is happening spontaneously. In the dynamical terms introduced in chapter 7, we can describe work as the organization of differences between orthograde processes such that a locus of contragrade process is created. Or, more simply: *Work is a spontaneous change inducing a non-spontaneous change to occur.* With this first approximated generic definition, we can now begin to unpack the logic that links energy, form, and information.

Isaac Newton had already provided a precise definition of mechanical work prior to the nineteenth century. The importance of Joule's experiment a few centuries later was to show that there was a precise relationship between this accepted notion of mechanical work and the generation of heat. In both cases, work was being defined with respect to a change of something that would otherwise tend to stay the same. But the relationship between mechanical and thermodynamic work is deeper and more thoroughly interrelated than merely parallel.

Recall the fact that thermodynamic properties are macroscopic reflections of Newtonian dynamics at the molecular level. From a Newtonian perspective, each collision of molecules in an ideal gas involves a minute amount of mechanical work, as the colliding molecules are each altered from their prior paths. Even at equilibrium there is constant molecular collision, and thus constant work occurring at the molecular level. Indeed, the overall energy of the average collision does not differ, whether the system is at equilibrium or far from it. This incessant "Brownian motion" is the means by which—even at thermal equilibrium—the molecules in a drop of ink dripped into water are eventually diffused throughout the solution. Without this molecular work, there can be no change in state of any kind. But notice that while it is possible to get work from a system that is in a state far from equilibrium, as it spontaneously develops toward equilibrium, this capacity rapidly diminishes, *even if the total amount of molecular level work remains constant* throughout. And at equilibrium, the vast numbers of collisions and the vast amount of microscopic work that is still occurring produce no "net" capacity for macroscopic work. So, although the potential for macroscopic work depends upon an incessant process of microscopic work, macroscopic work doesn't derive from it. Rather, macroscopic work depends on microscopic work being distributed in a very asymmetric way throughout the system. This shows us that the two levels of work—microscopic-molecular and macroscopic-thermodynamic—are not directly correlated. Microscopic work is a necessary but not sufficient condition for macroscopic work.

Any interaction with this system that shifts it from equilibrium will also be the result of changes in these micro collisions. Speeding up a subset of molecules that are in contact with a heat source, or slowing down some molecules in contact with a cold surface, both produce the capacity for thermodynamic work. Irrespective of adding or removing energy from the system, it is the degree of spatially asymmetric difference in average molecular velocities that matters. Ultimately, the capacity of the perturbed system as a whole to be tapped to perform work at the level above that of molecular collision is a consequence of the *distributional features* of the incessant micro work, not the energy of the component collisions, which as a whole can increase, decrease, or remain unchanged. In other words, in thermodynamics the macro doesn't simply reduce to the micro, even though it is dependent upon it. The macroscopic form of the distribution is the critical factor.

So also, like a mass in rectilinear motion, a state of incessant change can also be a stable state in thermodynamics. This suggests another interesting analogy between Newton's notion of work and the thermodynamic conception of work. In Newtonian terms, a mass can only be perturbed from rest or from a linear trajectory by the impositions of an extrinsic force, such as by interaction with another mass with different values of velocity and direction, or under the influence of some field of force, like gravity. In other words, its *resistance* to change is reflected in the amount of work required to produce a given change. Resistance is also characteristic of thermodynamic systems. A thermodynamic system at equilibrium can only be driven away from its dynamically symmetric basin by being coupled to a second system with a different value of some system variable, such as temperature or pressure. When two systems with different equilibrium values are coupled, stability gives way to change, as the coupled system changes toward a new equilibrium point. The transient phase, during which the now-combined larger thermodynamic system changes to a new global equilibrium state, is thus analogous to the brief period during which colliding objects in a Newtonian world are being accelerated or decelerated due to their interaction.

In the real world, even Newtonian interactions are not instantaneous, especially if the colliding objects are elastic. Elastic effects underscore the thermodynamiclike basis of even Newtonian interactions, since the elastic rebound of real colliding solid objects involves an internal asymmetric destabilization, in the form of compression of some molecular distances, followed by re-equilibration as both objects' internal energies redistribute. Of course, the analogy becomes increasingly stretched (not to make a pun) at this point, because unlike the Newtonian analogue—in which the objects' internal states return to where they were before collision and the objects permanently decouple—interacting thermodynamic systems do not have such a neat distinction between internal state and external relational features, such as momentum. Whatever the source of resistance and stability, however, the change toward the new equilibrium values and the new dynamical stability is spontaneous. This means that the intrinsic pattern of spontaneous change is itself also the source of a system's resistance to change. Thus, two thermodynamic systems which are either at different equilibria or are both undergoing spontaneous change at different rates can be considered contragrade to one another. Because of this, they will do work on one another if they become coupled.

This allows us to propose an even more general definition of work: it is simply *the production of contragrade change*. This way of describing work with respect to a spontaneous tendency of change shows us that the possibility of doing work to change the state of things is itself dependent on the relationships between processes that do *not* require work. In other words, differences in spontaneous processes of change, and the resistance of these to deviation from the specific parameters of that change, are the source of work. Or to put it succinctly, *contragrade processes arise from the interaction of non-identical orthograde processes.*

So, somewhat paradoxically, interactions between systems' different spontaneous tendencies are responsible for all non-spontaneous changes. Given that composite systems, with inherently iterative dynamics, display statistical asymmetries due to variations in their component interactions, a given composite system will generally have quite different asymmetric spontaneous tendencies than a second system. This is particularly likely in cases where the substrates of this interaction are of radically different form (e.g., interactions between light radiation and thermal motion).

In previous chapters we borrowed the distinction between efficient and formal causes from Aristotle, to argue that the capacity to produce non-spontaneous change could be loosely analogized to Aristotle's efficient cause and the conditions that produce spontaneous change could be loosely analogized to his formal cause. We now are in a position to be more explicit about this comparison.

First, let's recap what we have concluded about orthograde processes. If the global distribution of lower-order (micro) work is not symmetric in a thermodynamic system then it will tend to change in an asymmetric direction, to symmetrize this distribution, and will resist any tendency to reestablish a new asymmetry. It is this distributional feature, and the statistical asymmetry of interaction possibilities at the micro level, irrespective of the total work occurring at that lower level, that is responsible for the directionality of change. This property of the whole is effectively a geometric property—both of the spatial distribution and of the statistical distribution of work at a lower level. This distributional basis is why it makes sense to think of this property as a kind of formal cause. But formal constraints and biases do no work. They do not bring about change away from what would occur spontaneously. Yet as the foundation of any asymmetries that may arise between systems, they become the basis for work at a higher level.

Now consider the capacity for work at these two levels. The non-correlation of relative molecular movements is the basis for the work done by molecular collisions within a gas, but it is the global and statistical distributional character of this lower-level work that is responsible for higher-level orthograde properties. Analogously, the non-correlation (i.e., non-equivalent values) of the orthodynamic properties of linked thermodynamic systems (e.g., creating a temperature gradient) is responsible for higher-level work. So, without micro work, there can be no orthograde tendency to change or resist change at a higher level and no capacity for work at a yet higher level. In this regard, orthograde and contragrade processes provide the necessary conditions each for the other, but at adjacent levels in a compositional hierarchy.

If we think of work as the analogue of Aristotle's efficient causality, then, we can treat efficient and formal cause as interdependent and inseparable counterparts to one another. But there is another interesting distinction that must be added, relating to how they inversely react to the symmetry relationships that are their basis: their symmetry with respect to time. Both Newtonian interactions and the transformations produced by thermodynamic work can be run in reverse with approximate symmetry. For example, the movie of a simple billiard ball collision run in reverse does not appear odd, and a heat engine can be run in reverse to produce refrigeration. Yet a billiard ball break that scatters the balls previously racked into a triangle and the diffusion of a drop of ink into a glass of water would appear quite unnatural if shown in reverse. We can describe this in abstract terms as follows: when one asymmetry is transformed into another asymmetry, they are symmetric to one another in their respective deviations from symmetry, which is to say that they are interconvertable; but an asymmetric state transformed to a more symmetric state involves a fundamental change in symmetry, and so the two states are not interconvertible. But because the production of reversible contragrade change—work—is always based on lower-order irreversible orthograde processes, the reversibility of work is never completely efficient. There must always be some loss in the capacity to do work with each transformation. To put this in the form of a mnemonic pun: efficient cause is never 100 percent efficient.

In summary, then, we have identified what appears to be a quite general principle of causality that shows how work and the constraints that make it possible are to be understood in terms of levels of scale and supervenient organization. More generally, this dissects the logic of physical causality into two component aspects, roughly compared to Aristotle's notions of efficient and formal causality. It also distinguishes their complementary roles in the production and organization of change, and shows how they complement one another at adjacent levels in a supervenient hierarchy.

TRANSFORMATION

Normally, for human purposes (and for living organisms in general), we focus on only one direction of this interactive effect. That is, we are typically interested in changing the spontaneous status of some one thing in particular, to make it more useful, and so endeavor to bring some other influence to bear to accomplish this. The change this reciprocally imposes on the other system's spontaneous tendency is often not of any consequence to us and tends to be ignored. So, for example, in the familiar case of a simple internal combustion engine used to raise a weight off the ground (see Figure 11.1), we take advantage of the spontaneous expansion of the ignited fuel to over-

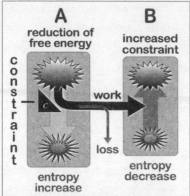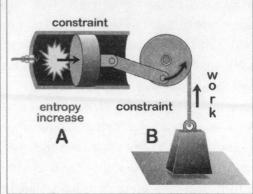

FIGURE 11.1: The left diagram schematically depicts the logic of thermo-dynamic work in which one physical system (*A*), which is changing in an orthograde direction (in which the reduction of free energy and the increase of entropy are depicted as an arrow from higher potential to lower potential), is coupled to another system (*B*) via constraints that cause the second system to change in a contragrade direction (depicted as a reversal of *A*). A familiar example of this relationship is depicted on the right where the exploding air-fuel mixture in a cylinder is constrained to expand in only one direction, and this is coupled to a simple mechanical device that raises a weight.

come the spontaneous inertia of the vehicle or the weight, though in the process, the rate of expansion of the gas is considerably impeded compared with what would have occurred in the absence of this coupling. The raising of the weight, the slowing of the expansion of the gas, and the constraint that limits where it is able to expand are all different than they would have been in the absence of this coupling. But the expansion of the gas, though slowed by contragrade action, still proceeds in an orthograde direction, while the change in the position of the weight proceeds in an entirely contragrade direction due to the imbalance in the forces involved. For this reason, we typically describe work being performed on the weight by the exploding gas and not the other way round, even though the slowing of the rate of this expansion is also contragrade and constitutes an equal and opposite amount of work.

Within the mechanical and thermodynamic realms, of course, there are as many diverse forms of work as there are heat engines. And this variety is only a fraction of the possibilities, which are as diverse as the possible kinds of substrates and couplings that can be realized. Ultimately, every transformation of energy from one form to another involves work as we have defined it so far. In the transformation, organization matters. How the interactions are constrained is a critical determinant of the nature of the work that results, because ultimately all such transformations involve a change in the dimensions and degrees of freedom (i.e., mode of dynamics and constraints) while the total energy remains unchanged. This inevitably requires work, because it is a process of restructuring constraints.

Over the past few centuries we have learned to build all variety of devices that can convert energy from one form to another. Thus, a temperature gradient can be transformed to generate mechanical work in a heat engine, an electric current can be transformed to generate a temperature gradient in a refrigerator, light energy can be transformed into electric current in a photovoltaic cell, and all these processes can be organized to run in the other direction as well. All involve the production of contragrade changes, by coupling these otherwise largely uncoupled domains.

In the case of a heat engine, these domains are separated by the radical differences in scale between microscopic molecular motions on the one hand and large-scale mechanical motions on the other. But the distinctions can be even more qualitatively extreme, and as a result, the limitations on possible interactions can be quite restrictive. In order to overcome such natural partitioning, it often takes highly specific materials organized in precise forms. This substrate-based limitation on the possibility of interaction is the basis for the diversity of forms of work that can be generated, given the appropriate mediating dynamical linkage. So, for example, a photovoltaic cell requires metals in which electrons (and their complementary absences, called "holes") are easily displaced by light to mediate the transformation of light energy into the energy of an electric charge gradient, and the immense compression force of the Sun's gravitation is required to transform nuclear mass into light and heat, via nuclear fusion.

The real value of this abstract conception of work, however, is not that it provides another (perhaps intuitively more transparent) way to describe forms of work that are already well understood in contemporary physics, but that it provides an abstract general characterization that can even be applied outside these mechanical realms. What is provided by this approach—and is not at all obvious from the physics alone—is the possibility of extending an analytically precise concept of work into domains that we would not identify with the physics of energetic processes: namely, the domains of form generation and semiotic processes. In other words, this abstract characterization can be applied to all three of the levels of dynamics that we have been exploring in this book. On this basis, we can begin to frame a theory of morphodynamic work and of teleodynamic work.

Such formulations are possible because at each level of dynamics there are classes of orthograde tendencies—conditions of broken symmetry that tend toward greater symmetry—which correspondingly define classes of "attractor" states analogous to thermodynamic equilibrium. Thus, the coupling of morphodynamic systems (or teleodynamic systems) can, analogously, bring their distinct and often complex orthograde tendencies into interaction, causing contragrade tendencies to emerge from their net differences. This opens the door to an emergent capacity to generate ever more complex, unprecedented forms of work, at progressively higher-order levels of dynamics, thereby introducing an essentially open-ended possibility of producing causal consequences that wouldn't tend to arise spontaneously. That is, we can begin to discern *a basis for a form of causal openness* in the universe. To frame these insights in somewhat more enigmatic and cosmic terms, we might speculate that whereas the conservation laws of science tell us that the universe is closed to the creation or destruction of the amount of possible "difference" (the ultimate determinate of what constitutes mass-energy) available in the world, they do not restrict the distributional possibilities that these differences can assume, and it is distributional relationships which determine the forms that change can take.

Having said this, we must keep in mind that the relationship between these different dynamical paradigms of work is complex. In one sense, they are but analogues of one another, and it is important that they not be confused. They involve very different substrates and conceptions of what is spontaneous or not, very different conceptions of what constitutes the differences that generate spontaneous change, and quite distinct notions of what constitutes a locus or system and how they may be linked together. And yet they are also more than mere analogues of one another. They play the same functional role at each of dynamical level, and more important, they are hierarchically interdependent and nested in the same way as are these three levels of dynamics. Teleodynamic work is dependent on morphodynamic work is dependent on thermodynamic work. At each level there is a class of orthograde and contragrade tendencies in which contragrade change can only be effected by pitting orthograde processes against each other.

The change of a thermodynamic system toward a state of increased entropy, if totally unperturbed by extrinsic influences, is the classic and most basic example of orthograde change. In the real world, of course, nothing is ever completely isolated, and no physical system remains forever unperturbed by external influences (though it is often argued that this is true of the whole universe, even if it is finite). This means that thermodynamic orthograde change is often resisted, modulated, or reversed, depending on the relative degree and duration of interaction with other contrary constraints or processes. In any context, orthograde processes will continue until they reach a state in which there is symmetry in the probable directions of change, or until the supportive conditions change. This dynamical terminus of an orthograde process is its *attractor*, which may or may not be a quiescent state.

In chapter 8, we extended the concept of orthograde dynamics to include intrinsically asymmetric forms of dynamics that arise in persistently far-from-equilibrium contexts. Such morphodynamic change is comparable in form to the orthograde asymmetry described by the second law in three important respects: (1) it exhibits a highly probable, characteristic, intrinsically generated asymmetric bias in the way that global properties change; (2) the direction of this change converges toward a common attractor irrespective of initial conditions; and (3) this asymmetry is dynamically supervenient on a balanced (i.e., symmetric) lower-order contragrade dynamic (work).

Recall that thermodynamic orthograde processes depend on incessant lower-order molecular interactions whose contragrade effects (work) are time symmetric, and in an isolated system the total energy of the system (e.g. measured as its specific heat) remains constant[3] throughout the orthograde change (and in this sense is balanced), though entropy increases. This incessant lower-order contragrade activity guarantees constant change from state to state, but the asymmetric directionality of the trajectory of global change is not a direct consequence of this work. The constant lower-order

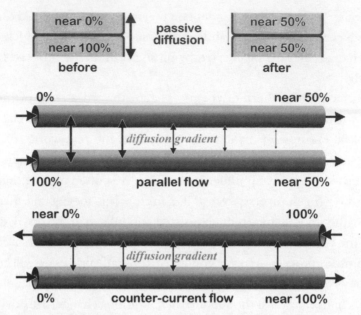

FIGURE 11.2: Countercurrent exchange demonstrates how formal constraints can be harnessed to do thermodynamic work. By causing media like coolant liquids (in a heat exchanger), or blood and environmental water (in fish gills), to flow in opposed directions, the asymmetries created can locally drive the system far beyond passive thermodynamic equilibrium (e.g., passive diffusion or parallel flow, above), so long as the movement continues. Though most naturally occurring countercurrent processes involve fluids and heat transfer, or chemical diffusion (as in fish gills and kidneys), this is a generalizable relationship. It can apply to any process involving entropy increase/decrease, including information processes.

contragrade activity determines constant change, but not its directionality. This global asymmetry has a "formal" origin, because it develops within the biased geometry of the space of possible trajectories of change.

An interesting example of the relationship between formal constraints and thermodynamic work is provided by processes that involve counter-current diffusion (also often described as counter-current flow; see Figure 11.2). This is a common mechanism found in the living world for driving systems beyond the point of thermodynamic equilibrium using oppositely directed fluid flow. It is characteristic of the ways fish gills extract oxygen from water, kidneys extract metabolites from blood, sea turtles cool themselves, and many birds regulate their core body temperature via blood flow in their legs. It is also an important trick used in engineering applications, such as in cooling systems for nuclear reactors and desalination processes.

Similarly, morphodynamic orthograde processes are also dynamically supervenient on incessant lower-order stable contragrade dynamics: the balance (symmetry) between incessant extrinsically introduced forms of destabilization, which introduce constraints, and the incessant spontaneous (orthograde) dissipation of these constraints. This constitutes a dissipative system, in Prigogine's terms. An asymmetric trend toward amplification and propagation of constraints to higher levels of scale can result if these constraints are not dissipated as quickly as they are introduced. This asymmetry is not, however, simply an elaboration of this lower-order symmetric dynamic, but rather again reflects global formal biases of the available trajectories of global property change. The asymmetry arises under these conditions as constraints compound non-linearly. This occurs because any dynamical option that is impeded from occurring due to the introduction of extrinsic constraint cannot lead to increased dynamical variation via any further contragrade interactions that the constrained region of the system has with others. Thus, as long as new extrinsic constraint is introduced faster than it is dissipated (e.g., due to incessant disruption, as in heating), subsequent stages of change will exhibit progressively reduced ranges of variation. In other words, they will self simplify and become more "orderly."

Teleodynamic orthograde processes are more complex because they dynamically supervene on morphodynamic processes. Nevertheless, these general principles still apply with respect to their lower-order dynamical support. Teleodynamic patterns of change emerge from the contragrade interactions between morphodynamic processes. Thus in the model autogenic system described in the last chapter, two morphodynamic orthograde processes with entirely different attractor dynamics—autocatalysis and self-assembly—interact, and as a result do work with respect to one another. Importantly, this interaction occurs in both the thermodynamic and morphodynamic domains. Assuming that both processes are thermodynamically orthograde in supportive conditions, each can only be sustained if there is constant work to maintain the thermodynamic imbalance that supports this asymmetry. In the case of each morphodynamic process, both autocatalysis and self-assembly require continuously maintaining locally high levels of substrate molecules. This could, for example, be extrinsically provided in laboratory conditions. But when these processes are linked by virtue of a product of the one (autocatalysis) serving as a substrate for the other (self-assembly), such supportive boundary conditions can in part be intrinsically generated. Thus while autocatalytic substrates need to be available, the process of autocatalysis provides substrates for self-assembly and generates more in this local region even as they are taken up by accretion to the growing shell. Reciprocally, as we saw in the last chapter, self-assembly minimizes diffusion of the interdependent catalysts, which is critical for its continuation. However, it also eventually halts autocatalysis via enclosure.

So autocatalysis is doing thermodynamic work by asymmetrically increasing the local concentration of substrates for shell self-assembly, and self-assembly is doing thermodynamic work in impeding catalyst diffusion. In addition, this interaction involves countervailing constraint generation processes, and thus morphodynamic work. (This higher-order form of work is described in detail in the next section.) Each morphodynamic process generates increasing constraints, though in different domains. But the development of constraints on diffusion that result from self-assembly of a shell increasingly limits the substrate availability that is required for autocatalysis, to the point that ultimately autocatalysis halts inside a container entirely depleted of substrates. Self-assembly of a closed shell is additionally self-limiting. But the likely enclosure of a complete set of interdependent catalysts also involves morphodynamic work in creating the constraints to prevent dissolution of these interdependencies, and possibly to replicate them. The resulting self-reconstituting capacity is a teleodynamic orthograde dynamic that is persistently potentiated by this underlying morphodynamic and thermodynamic work.

Although attempting to outline a "general theory of work" may sound like a tedious and dull enterprise, it turns out to be as critical for clarifying the concepts of form and information as it was to clarifying the concept of energy in nineteenth-century physics.

MORPHODYNAMIC WORK

Thermodynamic orthograde processes are vastly more likely to appear spontaneously in the universe than morphodynamic orthograde processes.

Correspondingly, examples of spontaneously occurring morphodynamic work are rare in comparison to thermodynamic work, and are also easily missed because their form is unfamiliar. To help identify them, we can begin by defining our search criteria by considering some thermodynamic analogies and disanalogies.

Any change of state is ultimately a thermodynamic change, but some thermodynamic changes are more complex than others. In describing forms of work that are more complex than thermodynamic work, we are not implying the existence of some new source of energy or a form of physical change that is independent of thermodynamic change, and certainly not a ineffable influence. Higher-order forms of work inevitably also involve—and indeed require— thermodynamic work as well. It's just that the account of certain processes in thermodynamic terms is too simple, completely ignoring whole categories of phenomena that play critical roles in organizing what is occurring. This is because thermodynamic descriptions only provide an accounting of the energies involved, and assume as given the boundary conditions and structural constraints within which the thermodynamic transformations take place. But thermodynamic work can also change these boundary conditions; and since these are the essential determinants of the organization of the causal possibilities in a system (its orthograde and contragrade organization), this higher level of change is often the most critical factor for understanding a system's dynamics. As we have just seen in our considerations of various machines, the formal regularities and symmetry conditions of these constraints, and the correlations between them, provide the framework within which thermodynamic work is made possible. Thermodynamics can be considered independently of these enabling constraints because analysis is focused on constant features of energy conservation, energy transformation, and the quantity of work involved. But, as we will see, the way that constraints in one system can influence constraints in other linked systems follows a logic that exhibits invariant features irrespective of diverse thermodynamic conditions. Because the rate of thermodynamic change is a function of the properties of the phase space of possible microstates, and these are a function of the constraints and biases of dissipation and micro interaction, any alteration of these constraints implicitly involves thermodynamic work. But to the extent that constraints on thermodynamic tendencies are required to perform thermodynamic work, it is also the case that prior constraints are being used to create new constraints whenever thermodynamic work is performed.[4] *This capacity of constraints on dynamical change to propagate new constraints to other linked dynamical systems is the capacity for morphodynamic work.*

In the case of thermodynamic work, it was first necessary to describe those processes that tend to happen spontaneously, and then identify the ways that this can be resisted and/or used to overcome this resistance in coupled systems. Similarly, in the case of morphodynamic work, we need to begin with an analysis of spontaneous tendencies of constraint formation and then explore the ways that these tendencies generate resistance to perturbation, and how that can be used to perturb spontaneous constraint formation in coupled systems. The second law of thermodynamics describes what tends to happen spontaneously—orthograde change—in the thermodynamic domain. Morphodynamic processes also exhibit powerful tendencies to develop toward stable dynamical patterns. Therefore, juxtaposing or resisting these tendencies can be analogously understood as involving a form of work: contragrade effects with respect to orthograde tendencies.

That said, the dependency of morphodynamic processes on thermodynamic processes makes the distinction between them subtle and complicated. Morphodynamic processes are not only dependent on underlying thermodynamic dissipative processes, they are specifically the result of constant thermodynamic work continually countering the entropy-increasing tendency within some restricted or partially isolated domain. Whether the whistle of the wind blowing across wires and tree branches or the hexagonally symmetric rolling of convection cells transferring heat to the surface air, all morphodynamic regularities arise in the context of a *balance* between the rate at which the system would spontaneously develop toward higher entropy and the externally imposed thermodynamic work impeding this process—a balance, that is, of thermodynamic orthograde and contragrade influences. In this regard, morphodynamic regularity establishes a new level of dynamical equilibrium between the rate of entropy increase and the rate of work countering this tendency. This is why self-organized regularity of a low order of complexity is a special case, not the rule in nature. Most often, constantly perturbed systems remain in semi-chaotic turbulent states. The balance of independent parameters that is required for convergence toward dynamical regularity is often quite specific, and for this reason is an unlikely coincidence.

The basic distinction between thermodynamic work and morphodynamic work can be easily recognized because of the specificity of form involved in the latter. For example, a vortex formed in a stream behind a boulder (or a sharp bend in the stream bank) is a morphodynamic system produced by incessant thermodynamic work. Water molecules that have acquired momentum as they (spontaneously) flow downhill encounter the resistance of the boulder, and the work that results from their interaction systematically shifts the momentum of many molecules angularly with respect to their original trajectories. But this redirection is contragrade to the orthograde flow and tends to impede downstream movement, potentially causing local instabilities as water builds up here and there faster than it moves on, creating increased turbulent flow. Vortex formation regularizes the redistribution of this instability, with the result that such chaotic imbalances are minimized. The rotational symmetry of the vortex that accomplishes this redistribution of angular momentum in the flow is thus a consequence of (1) the re-symmetrizing orthograde thermodynamics that causes chaotic flow to be self-undermining and (2) the geometric constraints on how the global asymmetry introduced by the boulder can be globally re-symmetrized, to again produce more nearly linear flow.

What kind of work can be done with respect to this morphodynamic orthograde tendency? Notice that stirring with a paddle in parallel with this rotation will not disrupt it. But stirring in any other pattern, or merely impeding this pattern of flow, will tend to disrupt it. These disturbing *patterns* of interaction are in this way contragrade to the orthograde tendency of the system to regularize. The non-parallel patterns of interaction are doing morphodynamic work against this orthograde tendency, whereas the parallel pattern is not. Notice also that any pattern that results in morphodynamic work also involves thermodynamic work, whereas the parallel pattern does not. But this is relative to the orthograde tendency that is intrinsic to the system in question. Thus, if the flow tends to be chaotic because the geometric organization of the stream pattern with respect to the flow rates are not conducive to vortex formation, stirring in the appropriate direction can aid vortex formation, and decrease turbulence. This would also be morphodynamic work, since vortex formation was not the intrinsic orthograde tendency.

Consequently, the introduction of morphodynamic work also requires thermodynamic work. And notice that the amount of mechanical/thermodynamic work involved is strongly dependent on the *form* of the paddle-induced disturbance with respect to the form of the flow. This suggests a general rule: in order to perturb a dissipative self-organized dynamical form away from its spontaneous attractor tendency, a conflicting *form* must be introduced, and the combined amount of thermodynamic and morphodynamic work involved will be a function of the number of dimensions of asymmetry that are reversed in the process, adjusted according to their relative magnitudes in the two alternative dynamical processes. In this way, the parameters defining the higher-level morphodynamic work play a significant role in determining the correlated amount of thermodynamic work that is required.

Recognizing that there are both parallels and asymmetric dependency relationships involved, we need to be clear about the analogies and disanalogies between morphodynamic and thermodynamic work. For example, we might be tempted to describe the regular dynamics of simple self-organized processes as morphodynamic equilibria, on the analogy of thermodynamic equilibria. Such an analogy is complicated by the fact that many morphodynamic processes remain partially chaotic (like the stream example above)—describable only in terms of constraints, not geometric regularities—and even simple morphodynamic systems may have more than one quasi-stable dynamical attractor. This complicated attractor logic, which has become the hallmark of complexity studies, also complicates the analysis of morphodynamic work.

Because of the potential for explosive symmetry breaking in morphodynamic systems, describing the way that the interaction between morphodynamic processes can transform their orthograde dynamics into contragrade change in the morphodynamic domain (i.e., the description of a morphodynamic engine, if you will) is far more difficult than for thermodynamic systems. This is in part due to the hierarchic complexity of morphodynamic processes. It takes thermodynamic work to drive morphodynamic attractors, so morphodynamic interactions cannot undermine this thermodynamic base and still do morphodynamic work. Precisely organiz-

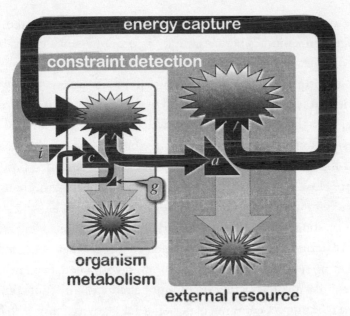

FIGURE 11.3: A diagrammatic depiction of the thermodynamic work performed by an organism to maintain its integrity with respect to thermodynamic degradation, and to support its higher-order orthograde (teleodynamic) capacity to replicate the constraints that support this process. Organisms must extract resources from their environment, e.g., by doing work (a) to constrain some energy gradient in order to access free energy to maintain their metabolisms (which maintain a persistently far-from-equilibrium state). Because the environment is often variable, they must also obtain information (i) about this variability in order to use it as a source of constraints (c) to regulate the work they perform. Constraints are depicted as right triangles deviating energy flows (arrows), and the constraints inherited genetically (g) are depicted as both within and outside the organism (since they are inherited from a parent organism).

ing a mediating mechanism that is able to take advantage of the interactions between different morphodynamic orthograde attractors is thus limited by the need to align both thermodynamic and morphodynamic processes. Moreover, since morphodynamic attractors are not merely defined by quantitative parameters (e.g., energy gradients) but also by formal symmetry properties, the possibilities for contragrade alignment of different morphodynamic processes are very much more restricted.

Despite these conceptual difficulties, we have already described one special case of morphodynamic work: a simple molecular autogenic system. It is a special case, because of its precise recursive synergistic organization. But before reconsidering this special case in terms of the work involved, we need to examine some more generic examples in order to gain a general conception of what gets transformed during morphodynamic work.

Consider, for example, a resonating chamber such as a flute or pennywhistle that is continually supplied with energy by air passed across an aperture. With this steady turbulent flow at one end, the air along the length of the resonating tube settles into a stable vibratory pattern, heard as a continuous tone. Changes in the effective length of the chamber, produced by opening or closing holes at various positions along the length, change which patterns of vibration (tones) are stable, even if the flow of blown air remains the same. Resonant vibrations are remarkably robust in linear chambers like the tube of a flute, but in irregular-shaped chambers they become less reliable and more sensitive to changes in the energy supplied. Even in a musical instrument like a flute, fluctuation of input energy can disrupt convergence to a stable vibration. A common experience for a novice flutist is to blow too hard, too soft, or at the wrong angle across the mouthpiece, with the result that the sound warbles between alternate tones, interspersed with the hissing sound of chaotic air flow. This demonstrates that the morphodynamic regularity exhibited by the tone being produced is sensitive to the rate and the form of the perturbing energy being introduced.

Once stable vibration is established, however, the sound of the flute can induce other objects (e.g., a wineglass) to vibrate sympathetically, to the extent that they too are capable of regular vibration in the frequency range of the flute tone. Interestingly, if this sympathetic resonator has its own distinctive resonant frequency, an unstable interaction may result, with the consequence that it vibrates in a pattern that is different from but typically attracted to some regular multiple of the frequency of the flute tone: a harmonic. Not only constant energy but constantly dissonant vibrations must be provided to induce the glass to assume a vibration pattern that is different from its most robust spontaneous resonant frequency. Thus there are two levels of work that are necessary to maintain this non-spontaneous regularity: (1) thermodynamic work, which is responsible for the energy necessary to induce the sympathetic resonator to vibrate; and (2) morphodynamic work, which is necessary to cause the sympathetic resonator to vibrate at something other that its spontaneous frequency.

The role of morphodynamic work is demonstrated by the fact that even with the same energy, different driving frequencies will have very different capacities to push the resonant response away from its spontaneous frequency. In other words, the differential in regularity between the vibratory patterns of the two resonating objects results in a pattern in the one that would not occur if this specific pattern of excitation were different or random. Moreover, if these two differently resonant objects are rigidly connected, the effect can be bidirectional: the total system will likely assume a vibrational state that is a complex superposition of the two resonant patterns (in proportion to other relative properties, such as shape, relative mass, and vibrational rigidity) as each structure becomes a source of morphodynamic work affecting the other.

Let's be more specific about this subtle distinction between the two levels of work in this example. Separately, each of the resonating structures tends to converge to a different, relatively stable, global vibrational state when mechanical energy is introduced and allowed to dissipate. Resonance is a morphodynamic attractor: the resultant stable form of an orthograde tendency. It is produced because of the geometry of the resonating chamber, the vibration-propagating characteristics of the material, and the level and stability of the input energy. These are the boundary conditions responsible for the morphodynamic attractor tendency. For differently resonant bodies, the boundary conditions are different, and will determine different orthograde tendencies. The thermodynamic work—blowing—that induces vibration is potentially able to produce an unlimited number of vibratory patterns. What actually gets produced is dependent on the specific boundary conditions imposed by the flute. Although no vibration will occur without the introduction of a stable airflow to contribute the energy of vibration (constant thermodynamic work), the properties of the flute will constrain the domain of the possible spontaneous (orthograde), stable vibrational patterns. And this will be robust to modest changes in the flow of air, so that a range of input energies will converge to a single resonant frequency.

This many-to-one mapping of thermodynamic work to morphodynamic work is a characteristic feature of this dependency relationship. But it is not simply a many-to-one relationship; it is the mapping of a continuum to discrete states. We will return to this feature later, because it turns out to be a critical contributor to the discontinuity of emergent effects as we move up the hierarchy from thermodynamic to morphodynamic to teleodynamic processes.

The morphodynamic work produced by linking oscillators results from one set of boundary conditions affecting another. Specifically, their differences in geometry, mass, and the way they conduct vibratory energy all contribute to the total work of this transfer of form. The thermodynamic work component is roughly the same irrespective of whether the coupled oscillators reinforce each other's vibrations or rapidly damp all regular vibrations, transferring most of the energy into the irregular micro vibrations of heat. To the extent that their resonant features interact to produce a shift in global regularities compared to the uncoupled condition, morphodynamic work is also involved, and can be judged more or less efficient on the basis of this transfer of regular global dynamics. But whereas thermodynamic coupling yields a combined system that dedifferentiates toward a state of global equilibrium—determined by the mean boundary condition of the total—morphodynamic coupling does not. There is nothing quite analogous to a "mean" value, because of the relative discreteness of the morphodynamic attractors involved. The coupling of boundary conditions must be such that each reinforces the other in some respect in producing a third discrete orthograde tendency: one that is both amplified and amplifies each of the other two. For two oscillators, this can be a simple common multiple of the two resonant frequencies; but with additional couplings, the probability of simple and discrete dynamics quickly diminishes, and dynamical chaos results. And beyond the domain of simple oscillators this is far more likely to be the case.

This means that the ability to perform morphodynamic work can be quite easily disrupted. In coupled physical resonators, for example, if one structure is more regular in shape and form, and therefore more effective at form amplification, it will tend to drive the vibratory activity of the coupled system, though this will be resisted and constrained by the vibratory regularities or irregularities of the second structure to which it is linked. In this case, we can say that morphodynamic work is continually bringing the less resonant system to a non-spontaneous semi-regular vibratory state. In the case where both have different but nearly equally efficient resonant tendencies, the resulting vibratory state of the coupled system may converge to a pattern that combines the two, amplifying common harmonics and producing complex waveforms, or may never resolve a chaotic state, because of the incompatibility of their orthograde tendencies.

In such cases of competing resonant tendencies, we can discern another parallel with thermodynamic work: some systems are more difficult to perturb than others. In other words, the orthodynamic tendencies of different systems may be of different "strengths." Just as objects with greater momentum or inertia and thermodynamic systems with greater total specific energy require more mechanical work to produce equal changes of motion compared to less massive or extensive systems, morphodynamic systems can differ greatly in the relative strength of their attractor dynamics. There are two potential contributors to this morphodynamic "inertia." First, one system may simply be more susceptible to thermodynamic work because it is physically smaller, less massive, or better at conducting energy. This follows from the simple fact that morphodynamic work is entirely dependent on thermodynamic work. But second, one system may be more regular, such as the shaped body of a resonant musical instrument, or it may be more easily regularized, as is the minimally constrained flow of fluid in a Bénard convection cell or vortex. This combination of factors will determine both the potential to do morphodynamic work and the tendency to resist morphodynamic change.

The potential to perform either thermodynamic or morphodynamic work is proportional to the divergence from an attractor maximum. But this can be a problem for the capacity to do morphodynamic work because systems with complex attractors tend not to exhibit consistent extended spontaneous change in any single direction. This means that only systems with highly reliable and relatively simple attractor dynamics are able to contribute any significant amount of morphodynamic work. This makes it difficult to find spontaneous examples of morphodynamic work, and makes it very rare for highly complex morphodynamic transformations to occur without highly sophisticated forms of human intervention. So, although simple examples such as the coupled resonators described above represent exceptions in nature, not the rule, simplicity is an advantage when it comes to making use of morphodynamic work. This is not, however, an absolute impediment, since elaborate webs of morphodynamic work are found in the metabolic networks of living organisms.

What about more complex morphodynamic processes that produce regularities with more dimensions of regularity? Consider these somewhat fanciful Rube Goldberg uses for Bénard cell formation. The regular hexagonal tessellation of the surface of the water could, for example, be utilized to sort small floating objects into discrete collections of similar numbers, each collection sitting within the tiny hexagonal bowl of a Bénard cell. Or the concave shape of these regular surface depressions could be used to focus incident light to dozens of individual points just above the surface. In these cases, there is very little thermodynamic work linking the two interacting substrates (especially in the case of reflected light), but the morphodynamic work occurs as the spontaneous regularization of fluid convection similarly regularizes something else that otherwise would never assume this configuration. Thus, via morphodynamic work two otherwise independent thermodynamic systems accomplishing thermodynamic work can be coupled.

More practical examples of morphodynamic work include the use of specially shaped vessels or vibrating containers for sorting different shapes, weights, or sizes of particulate materials, such as pills or grains. Depending on the shape of the vessel, the way it is rotated or shaken to induce the contained objects to move with respect to one another, and the differences in object features (such as shape or weight), it is possible to automatically separate objects, transforming a well-mixed, uniformly distributed collection into a highly asymmetric distribution in which different types of objects occupy distinct positions relative to one another. Natural examples of this particulate sorting process occur with pebbles on ocean beaches and stones rising to the surface in soil as a result of periodic freezes and thaws; but other uses include ways of separating pills and minerals, as well as the classic method of separating gold nuggets from sand and other pebbles, by "panning."

Morphodynamic work shares one very significant attribute with thermodynamic work: the law of diminishing returns. As a consequence of the first and second laws of thermodynamics, and the constraints of doing work, perpetual motion machines are impossible. There are always some degrees of freedom of increasing entropy that cannot be fully constrained, and so the capacity to do work in one direction, and then reverse this organization and use that gradient to do work in the opposite direction, decreases with each step, making full reversal unobtainable. The potential to do iterated morphodynamic work also diminishes rapidly with increasing degrees of freedom and thus also with each interaction. There is something analogous to nature's prohibition of perpetual motion machines when it comes to morphodynamic work as well. In fact, the efficiency problem is much worse, because of the discreteness issue. In most instances of coupled morphodynamic processes, the interactions between their distinct regularities result in complex dynamics that appear highly chaotic. Classic examples of so-called deterministic chaos reflect the complexity that can result even as a result of coupling three otherwise quite simple morphodynamic processes into a larger system, as for example happens when different length pendulums are coupled with one another. Whereas the recursive dynamics in a simple self-organizing system amplify regular dynamic features, strongly coupled self-organizing processes can recursively amplify both concordant and non-concordant boundary conditions, producing complex and often extreme divergence and damping effects. This is especially true if thermodynamic energy is continually introduced, as in dissipative systems. This kind of coupling of organized dynamical processes is probably one of the factors contributing to the unpredictable and almost turbulent character of human social and economic systems; though, as we will see, this tendency to complexity becomes amplified to a far greater extend when we consider the superimposition of teleodynamic processes.

Given these limitations, and since morphodynamic regularities even with robust simple attractors only form under very limited boundary conditions, interactions between morphodynamic systems with different boundary conditions end up producing larger systems with complicated and irregular boundary conditions. So, for many reasons, morphodynamic work of any significant complexity and magnitude will tend to occur quite rarely under natural circumstances.

TELEODYNAMIC WORK

There is, however, one class of phenomena that presents glaring exceptions to this rarity: living processes. Indeed, self-organizing processes in living organisms and ecosystems defy the apparent problem of the chaos that should tend to result from coupling self-organizing processes to one another—and to an astounding degree—since even the simplest bacteria are composed of hundreds of strongly coupled cycles of chemical processes. Life appears to have cornered the market on morphodynamic work, and to have done so by taming the almost inevitable chaos that comes with morphodynamic interactions. Not only are living organisms themselves enormously complex webs of self-organizing processes, but they also tend to evolve to complement higher-order complex dynamical regularities made up of the large numbers of other organisms comprising their ecosystem, all embedded in semiregular patterns of climatic and resource change. So, it is within living processes that we must turn to find the greatest number and diversity of exemplars of morphodynamic work.

Besides energy and raw materials to maintain their far-from-equilibrium thermodynamics, living organisms also require incessant form production processes: production of specific molecular forms, specific patterns of chemical reactions, and specific structural elements. Morphodynamic work must be reliable and constant for life to be possible. This requires both thermodynamic and morphodynamic work cycles—engines of form production that are analogous to human engines designed to perform thermodynamic work cycles. The process of biological evolution has not merely "discovered" and "remembered" how to set up a vast array of morphodynamic work processes; it has discovered complex synergies and reciprocities between them that enable repeatable cycling. We have encountered a simple example of this in the case of autogens, but to understand how the evolutionary process is able to mine the morphodynamic domain for these sorts of reciprocities and complementarities, we will first need to understand a yet-higher-order form of work: teleodynamic work.

Teleodynamic work can be defined analogously to the prior levels of work we have described. It is the production of contragrade teleodynamic processes. Since this must be understood in terms of orthograde teleodynamic processes, the first step in describing this level of work is to define and identify examples of orthograde teleodynamics. In general terms, an orthograde teleodynamic process is an end-directed process, and more specifically, one that will tend to occur spontaneously. Although teleodynamic processes are incredibly complex, and an explanation of the structure of teleodynamic work is by far the most elaborate—since it is constituted by special relationships between forms of morphodynamic work—it is also the most familiar. So it may be helpful to first consider the human side of teleodynamic work before delving into the underlying dynamical structure of this process.

Teleodynamic work is what we must engage in when trying to make sense of an unclear explanation, or trying to produce an explanation that is unambiguous. It is what must be produced to solve a puzzle, to persuade resistant listeners, or to conduct scientific investigations. It is also the sort of work that goes on in board meetings and in domestic arguments, and which leads to the design of machines and governments. And it characterizes what is difficult about creative thought processes. Although these examples could mostly be considered forms of mental work, they have a natural kinship with the simple process of communicating, and with biological adaptive processes as well. All share in common the work of generating new information and new functional relationships, or of changing thought patterns or habits of communication and human intentional actions.

If you have read to this point, you have probably found some parts of the text quite difficult to follow. Perhaps you have even struggled without success to make sense of some claim or unclear description. But unless you are a very easily agitated reader, you will probably not have found yourself running out of breath or breaking a sweat because of the energy you have exerted to do this. While writing this chapter, I took a break to cut and split some wood for a fire. Doing so worked up a sweat. Though only a small fraction of writing time was devoted to this process, I no doubt expended far more energy chopping wood than in all my writing for the day. But which was the more total work? Obviously, the energy expended isn't the most useful means of assessment. Nevertheless, engaging in the effort of writing or reading does require metabolically generated energy, and the more difficult the task of creation or interpretation or the more stimulating or frustrating, the more thermodynamic work tends to be involved.

We have no trouble recognizing the capacity to do this kind of work. In an individual, we may describe it as intelligence. In a simpler organism, we may describe it in terms of its adaptability. This is the "power" that we recognize in great insights, influential ideologies, or highly developed analytical tools. It is the power to change minds and organize human groups. It can ultimately translate into "the power to move mountains," as the old adage implies, though its capacity to do this is necessarily quite indirectly implemented. It is commonly described as "the power of ideas."

FIGURE 11.4: Reading exemplifies the logic of teleodynamic work. A passive source of cognitive constraints is potentially provided by the letterforms on a page. A literate person has structured his or her sensory and cognitive habits to use such letterforms to reorganize the neural activities constituting thinking. This enables us to do teleodynamic work to shift mental tendencies away from those that are spontaneous (such as daydreaming) to those that are constrained by the text. Artist: Giovanni Battista Piazzetta (1682–1754).

So, engaging in teleodynamic work is one of the most familiar mental experiences of being a human agent. It is what characterizes what we describe as "intentionally willed action." It is naturally understood as work because of the resistances that it can overcome, and because it is required in order to modify otherwise spontaneous mental or communicative processes, such as an otherwise unquestioned belief or habitual process of reasoning. But as familiar as this experience is in everyday life, its relationship to physical work is often quite ambiguous, even though it is intuitively taken for granted. No one would confuse what must be done to raise a weight with what must be done to raise a question, but both involve effort, that experience of promoting contragrade dynamics. And while physicists and engineers can be incredibly precise at comparing the amount of work that is done to accelerate a car versus a baseball to 60 miles per hour, it seems almost a matter of idiosyncratic opinion how one should compare the work of solving a mathematical equation to that of solving a crossword puzzle clue. So we might be forgiven for thinking that the latter cases of mental work are merely metaphorical comparisons to physical work.

The production of teleodynamic work is, of course, totally dependent on the other forms of work we have been considering. Teleodynamic work depends on morphodynamic work depends on thermodynamic work. Mental work, for example, requires physiological processes doing thermodynamic work to maintain the constant activity of neurons producing and transducing signals, and also the production of spontaneous and non-spontaneously self-organized patterns of neural activity that recursively amplify and damp as they traverse complex changing neural networks. What is additionally involved, however, is the generation of new semiotic relationships and new end-directed tendencies in the face of spontaneous habitual interpretive tendencies and countervailing end-directed processes. In the performance of difficult interpretive processes, for example, we directly experience both the energetic demands and the morphological demands of the supporting lower-level forms of work that are also involved. There is the fatigue that is generated by struggling to maintain attention on the interpretive problem until it is solved, and the challenge of trying to find an appropriate translation into other words or to conjure up an appropriate mental image among the many spontaneously arising but inadequate alternatives. But although this domain of work necessarily involves the others, it is intuitively not merely reducible to these simpler processes alone, as mere "mindless" physical labor demonstrates. Thus, the free association of daydreams or the nearly unconscious performance of a highly familiar task are not experienced as mentally effortful, even though they do require metabolic support and the generation of distinct appropriate patterns of neural activity. So we intuitively recognize the distinctive effortful character of teleodynamic work.

Although the exercise of mental effort is unquestionably its most familiar exemplar, teleodynamic work occurs in many forms that are neither cognitive nor even associated with brains. Before we can hope to understand the processes responsible for the teleodynamic work that occurs in and between human brains, we will need to analyze its organizational details and its dependency on other forms of work in some far simpler systems. To do this, we can back down in complexity and familiarity, and at the same time incrementally unpack the complex contributions of other forms of work.

One of the more fundamental forms of teleodynamic work is that which occurs in the process of biological evolution. This is epitomized by novel functional adaptations and their heritable representations being constantly created where none previously existed. In other words, new teleodynamic systems and relationships are generated from the interactions of prior teleodynamic systems with respect to their shared environmental contexts. Using natural selection as an exemplar, let's analyze what makes it teleodynamic work, using the generic logic that we have outlined above for thermodynamic and morphodynamic work.

To begin the analysis, we need to identify what constitutes orthograde and contragrade processes in this teleodynamic domain. In the case of biological evolution, there are two very general classes of orthograde teleodynamic processes. First, there are the actions of organisms that function to maintain them against degrading influences, such as thermodynamic breakdown of macromolecules and degradation of metabolic networks. Second, there are processes of growth, differentiation, and reproduction, which are involved in producing what amount to backup copies of the organism, in the form of daughter cells or offspring. These are teleodynamic processes because they are end-directed toward specific target states. They are orthograde because they are what organism dynamics produce naturally and spontaneously (given supportive underlying forms of morphodynamic and thermodynamic work). Organisms are of course highly complicated synergistic constellations of teleodynamic processes that each collectively contribute some fraction of the global teleodynamics of the organism. But for the purpose of this first level of analysis, we can lump all together as though the life history strategy of the organism is a single orthograde teleodynamic process. In this context, we can identify contragrade teleodynamic processes as those that are organized in such a way that they impede or contravene these orthograde processes. In other words, contragrade biological teleodynamic processes are those that are in some way bad for the organism, in that they are detrimental to survival and reproduction.

With respect to evolution, where the critical end is reproductive success that is sufficient to guarantee continuation of one's lineage, reproductive competition from other members of the species is the most directly relevant source of precisely contragrade influences. In this sense, each organism is doing teleodynamic work against its reproductive competitors. So far, this is intuitively familiar. The work required to compete with other organisms at many levels, over many kinds of resources, in order to achieve diverse ends, is in many ways the hallmark experience of being a living organism. But like other forms of work, teleodynamic work can also be used to transform one form of teleodynamic process into another, and to generate emergent phenomena at a higher order. This is what happens in the process of natural selection.

So now let's describe natural selection in terms of teleodynamic work. In the standard model of natural selection, variants of the same teleodynamic process (adaptive traits) that are represented in the different members of a species in one generation are brought into competition over resources critical to reproduction. In this competition of each against all (in the simplest case), work done to acquire resources, mates, and so on is also work that degrades the teleodynamic efficacy of competitors with respect to these same requirements. This work is both directly and indirectly a source of distributed contragrade effects on other organisms. Because the teleodynamics of organisms is supported by extensive morphodynamic and thermodynamic work, it is also the case that teleodynamic competition ramifies to all these lower-level processes as well. And since all ultimately depend on thermodynamic work, this is the final arbiter of teleodynamic success. Analogous to the way the coupling and juxtaposition of non-concordant orthograde thermodynamic and morphodynamic processes can be utilized to generate specific contragrade patterns of entropy decrease and form production, respectively, the complex juxtaposition of non-concordant teleodynamic processes can generate teleodynamic systems that would not otherwise occur. Because the widespread integration of diversely contragrade teleodynamic interactions in one generation is mediated through an environment that is also the source for the resources supporting their underlying morphodynamic and thermodynamic work processes, the constraints and regularities intrinsic to that environment become the analogue to the constraints of an engine for channeling the teleodynamic work into forms consistent with that environment.

In biological evolution, this ultimately results in an increasing asymmetry of the presence of these teleodynamic traits in succeeding generations. The differential reproduction and elimination of the less fitted variants from the population in each generation thus has a recursive influence on all levels of work involved, maintaining them in concordance with those constraints. More significantly, due to the inevitable spontaneous (thermodynamic) degradation of the capacity to do work at all these levels, new variant teleodynamic complexes continually arise and are entered into this evolutionary work cycle. The familiar result is the production of increasingly integrated, increasingly diverse, increasingly complex, increasingly well fitted teleodynamic systems.

What we can conclude from this is that evolution is a kind of teleodynamic engine, powered by taking advantage of spontaneous thermodynamic, morphodynamic, and teleodynamic processes, and which continually generates new teleodynamic processes and relationships. Additionally, because teleodynamic processes are supported by the synergistic organization of morphodynamic and thermodynamic work cycles, these too evolve, as do the synergies that bind them together into the individual causal loci we know as organisms.

EMERGENT CAUSAL POWERS

This brings us back to issues discussed in the introductory section of this chapter. We can now make a somewhat more detailed assessment of why there is a difference in the amount of work required to solve a difficult puzzle versus a simple one. Returning to the comparison between two jigsaw puzzles—one with fewer parts, distinctively different-shaped pieces, and forming a recognizable and heterogeneously organized picture when assembled, compared to one with more pieces, much more subtle shape differences, and a very homogeneous surface image—we can recognize that much more work at all three levels will be required to assemble the second, harder puzzle. More pieces require more physical movements, less shape distinction between pieces will mean more errors of fit and will also translate into more movements, as will less pictorial distinctions. Hence, more thermodynamic work is required.

But now let's consider the cognitive challenge. The puzzler must generate mental images of each comparison, and will almost certainly generate
many times more mental comparisons than actual trial fits. Each mental
comparison requires the generation of at least two distinctive forms of
neural activity to represent the compared pieces. Since there are inevitably other cognitive and mnemonic tendencies that would otherwise express
themselves and could be potential distractions, this is a non-spontaneous
process of form generation. This is the source of the feeling of resistance
to focusing attention on the problem at hand and generating these highly
constrained mental images rather than others. This then is a form of morphodynamic work: contragrade form generation against a background of
competing orthograde form generation tendencies (we will talk about how
this might be generated in brains in chapter 17). But of course this requires
metabolic energy (thermodynamic work), and presumably more energy
than is required to let one's thoughts wander in orthograde stream of consciousness patterns. In fact, this is made unambiguously evident in *in vivo*
imaging studies of regional changes of brain metabolism when comparing
active versus passive interaction with stimuli (this too will be discussed in
more detail in chapter 17).

But these forms are not *merely* forms; they are mental representations
of objects being encountered in the context of analyzing and physically
assembling the puzzle. The forms as representations are not then merely the
result of morphodynamic work, but must be generated with respect to this
context and tested against it. As these mental images are being generated,
biases from past memory and from expectations will also compete with the
generation of representations that are faithful to the task at hand, so there
will be innumerable ways that the generation of adequately correspondent
representations can go awry. Almost certainly, in the generation of these
mental images, there will be multiple partial "drafts" that are produced and
compared against the information provided by the senses. In this process,
all but one will be rejected as not sufficiently constrained to correlate with
incoming patterns. And this mental generation and testing of imagined
comparisons will itself be iterative in advance of actually picking up a piece
and trying its fit, looking for physical feedback to be the final arbiter of
accurate representation. Indeed, this process also must be proceeding at

many other levels in parallel, such as the generation of the teleodynamic predisposition to continue working on the problem, despite distractions and competing demands on one's time.

This generation and selection process is an expression of teleodynamic work. And given that it requires the generation, comparison, and context-sensitive elimination of vastly many partially developed early "drafts" of the mental representations involved, each of these was a product of morphodynamic work, and so on. So the level of work that must take place is significantly amplified with the puzzle difficulty. In short, it takes more work at all levels. But notice that this is far more efficient than randomly picking up pieces and testing their physical fit. This shift of work up-level, so to speak, significantly decreases the total amount of thermodynamic work to achieve this end. And thermodynamic work is what supports the base of the whole hierarchy.

As this one simple example illustrates, the domain opened by teleodynamic work is enormous. The diversity of forms of teleodynamic work is as extensive as the fields of biology, psychology, human social behavior, and all the arts combined. Not only is it far beyond the scope of this book, much less this chapter, to do more than cherry-pick a few illustrative examples of the process; a thorough analysis of teleodynamic work in such cases would likely be redundant and superficial compared to analyses pursued in these many specialized fields. Indeed, one might argue that throughout this chapter we have merely redescribed selected phenomena drawn from the domains of mechanics, acoustics, evolutionary theory, and now cognitive and behavioral science in terms of the concept of work. But if this were merely redescription and the renaming of otherwise well-understood phenomena, there would be little gain over knowledge gained in these already well explored territories. What I hope this analysis accomplishes is not the introduction of conceptual tools to *replace* those already employed in the various sciences we have touched upon, but rather a road map for following the common thread that links these hitherto disconnected and independent realms. What has been missing is a full understanding of how the power to effect change at all levels is interconnected and interdependent. I believe that something like this expansion and generalization of the concept of work is the critical first step.

One important contribution of this perspective is that it untangles the notion of causal power from undifferentiated notions of causality. Whereas our commonsense conceptions of cause tend to be applied willy-nilly to all manner of physical changes, attributions of causal power are typically invoked in a more restricted sense. I believe that this distinction parallels the distinction explored in this chapter between changes due to work and changes in general. The addition of the term *power* is the clue that the issue has to do with work and not just causality in general—in other words, it invokes a sense of overcoming resistance, of forcing things to change in ways that they wouldn't otherwise change. And this is what is important to us. It is the emergent capacity to reorganize natural processes in ways that would never spontaneously occur, which is what we have been struggling to understand with respect to life, and with respect to mind.

The term *causal power* has become particularly associated with debates about emergence and mental agency. For example, the contentious debates about the status of human agency and the efficacy of mental representations are not merely concerned with whether physical changes of state occur in conditions associated with these phenomena; they concern the question of their status as sources of work, that is, the production of non-spontaneous change in some conditions. So, while there is no serious debate over whether mental events are associated with such physical events as the release of neurotransmitter molecules, or the propagation of action potentials down neuronal axons, there is concern about whether the effects of mental experiences are anything more than this, and whether the experience of mental effort really plays any role in initiating physical changes that would not otherwise be attributable to the spontaneous consequences of physiological chemistry and locomotor physics. Putting these issues in the context of an expanded theory of work, we can appreciate that although all the physiological changes do in fact involve (for example) molecular interactions, the real challenge is determining in what way a person's spontaneous molecular, formative, and end-directed processes have combined to produce non-spontaneous mechanical, organizational, and semiotic changes. The issue is not whether the changes introduced into the world due to a person's considered actions are caused or not—of course they are—but whether there is something physically non-spontaneous about the effect. Debates concerning the status of human conscious agency are about the proper locus of this causal power.

Causal power is also a code word for what is presumed to be added to the causal architecture of the universe as a result of an emergent transition. But as we've seen, when this idea is conflated with more generic notions of causality, it yields a troubling implication: that such phenomena as life and cognition might be changing or adding to the fundamental physical laws and constants, or at least be capable of modifying them. The presumed restriction against this is the postulate of causal closure, discussed in chapter 5. We can now re-address this issue more precisely. Although the fundamental constants and laws of physics do not change, and there is no gain or loss of mass-energy during any physical transformation process, there can be quite significant alterations in the organizational nature of causal processes. Specifically, work can restructure the constraints acting as boundary conditions that determine what patterns of change will be orthograde in some other linked system. This is the generation of new formal causal conditions, and because the resulting orthograde dynamics will determine the possible forms of work that can result, it sets the stage for the emergence of unprecedented organizations of efficient causality, and so forth, with the generation of yet further new constraints, and new forms of work. As we have seen, this can also occur in ascending levels of dynamics, with a correlated increase in the possibilities of organizational complexity. So to restate the closure or conservation laws a bit more carefully: the universe is closed to gain or loss of mass-energy and the most basic level of formal causality is unchanging, but it is open to organizational constraints on formal cause and the introduction of novel forms of efficient cause. Thus we have causal openness even in a universe that is the equivalent of a completely isolated system. New forms of work can and are constantly emerging.

The concept of causal power is also of particular interest to social scientists, since it has become a critical issue for discussing the capacity for semiotic processes to control behavior and shape the worldview of whole cultures. Indeed, it is often argued that the very nature of the interpretive process is subject to the whims of the power of hegemonic semiotic influence. The problem has been that although this sort of power is an intuitively recognized and commonplace feature of human experience, it is at least as difficult to define as the concepts, like interpretation, that are supposed to be grounded in it. This difficulty may be mitigated by recasting this notion of power in terms of teleodynamic work.

Specifically, as we have seen, teleodynamic work is defined contra to orthograde teleodynamic processes. In cognitive terms, orthograde teleo-dynamic processes may be expressed as goal-directed innate adaptive behaviors, spontaneous emotional tendencies, learned unconscious pat-terns of behavior, stream-of-consciousness word associations, and so forth. In social terms, orthograde teleodynamic processes may be expressed as common cultural narratives for explaining events, habits of communication developed between different groups or classes of individuals, conventional-ized patterns of exchange, and so on. As a result, we can easily recognize how it is that these orthograde predispositions to change might come into conflict, why they might encounter resistance, and how they might become specifically linked to afford transformation of work from one domain to another or become engaged in a form of semiotic and social evolutionary dynamic.

As we will see in subsequent chapters, these predispositions and the work that can be generated thereby are the basis for the generation of new forms of teleodynamic relationships: that is, new information and new representa-tions. In fact, the very concept of interpretation can be cashed out in terms of teleodynamic work. This will be the subject of the next two chapters.

In conclusion, being able to trace the thread of causality that links these domains avails us of the ability to discern whether methods and concepts developed in different scientific contexts are transferable in more than merely analogous forms. It also makes it possible to begin the task of for-malizing the relationships that link energetic processes, form generation processes, and social-cognitive processes. Most important, it shows us that what emerges in new levels of dynamics is not any new fundamental law of physics or any singularity in the causal connectedness of physical phenom-ena, but rather the possibility of new forms of work, and thus new ways to achieve what would not otherwise occur spontaneously. In other words, with the emergence of new forms of work, the causal organization of the world changes fundamentally, even though the basic laws of nature remain the same. Causal linkages that were previously cosmically improbable—such as the special juxtapositions of highly purified metals and semiconduc-tors constituting the computer that is recording this text—become highly predictable.

This causal generativity is a consequence of the fact that higher-order forms of work can organize the generation of non-spontaneous patterns of physical change into vast constellations of linked forms of work, connecting large numbers of otherwise unrelated physical systems, spanning many levels of interdependent dynamics. Although I have only described three major classes of work, corresponding to thermodynamics, morphodynamics, and teleodynamics, it should be obvious from previous discussions of levels of emergent dynamics that there is no limit to higher-order forms of teleodynamic processes. Thus, the possibilities of generating increasingly diverse forms of non-spontaneous dynamics can produce causal relationships that radically diverge from simple physical and chemical expectations, and yet still have these processes as their ground. This is the essence of emergence, and the creative explosion it unleashes.

12

INFORMATION

The great tragedy of formal information theory is
that its very expressive power is gained
through abstraction away from the very thing that it
has been designed to describe.

—JOHN COLLIER, 2003[1]

MISSING THE DIFFERENCE

The current era is often described as "the information age," but although
we use the concept of information almost daily without confusion, and
we build machinery (computers) and networks to exchange, analyze, and
store it, I believe that we still don't really know what it is. In our everyday
lives, information is a necessity and a commodity. It has become ubiquitous
largely because of the invention, perfection, and widespread use of computers and related devices that record, analyze, replicate, transmit, and correlate
data entered by humans or collected by sensor mechanisms. This stored
information is used to produce correspondences, invoices, sounds, images,
and even precise patterns of robotic behavior on factory floors. We routinely
measure the exact information capacity of data storage devices made of silicon, magnetic disks, or laser-sensitive plastics. Scientists have even recently
mapped the molecular information contained in a human genome that cor
responds to proteins, and household users of electronic communication are
sensitive to the information bandwidth of the cable and wireless networks
that they depend on for connection to the outside world.

Despite this seeming mastery, it is my contention that we currently are working with a set of assumptions about information that are only sufficient to handle the tracking of its most minimal physical and logical attributes, but are insufficient to understand either its defining representational character or its functional value. These are serious shortcomings that impede progress in many endeavors, from automated translation to the design of intelligent Internet search engines.

The concept of information is a central unifying concept in the sciences. It plays critical roles in physics, computation and control theory, biology, cognitive neuroscience, and of course the psychological and social sciences. It is, however, defined somewhat differently in each, to the extent that the aspects of the concept that are most relevant to each may be almost entirely non-overlapping. The most precise technical definition of information has come from the work of Claude Shannon, who in the 1940s made precise quantitative analysis of information capacity and transmission possible. As we will see, however, this progress came at the cost of entirely ignoring the representational aspect of the concept that is its ultimate base. As the epigraph to this chapter hints, this reduced definition of information is almost devoid of any trace of its original and colloquial meaning. By stripping the concept of its links to reference, meaning, and significance, it became applicable to the analysis of a vast range of physical phenomena, engineering problems, and even quantum effects. But this reduction of the concept specifically obscures distinctions which are critical to identifying the fundamental features that characterize living and mental processes from other physical processes. To begin to make sense of these world-changing transitions, then, we need a concept of information that is both precise and yet also encompasses the referential and functional features that are its most distinctive attributes.

In many ways, we are in a position analogous to the early nineteenth-century physicists in the heyday of the industrial age (with its explosive development of self-powered machines for transportation, industry, time-keeping, etc.), whose understanding of energy was still laboring under the inadequate and ultimately fallacious conception of ethereal substances, such as caloric, that were presumably transferred from place to place to animate machines and organisms. Even though energy was a defining concept of the early nineteenth century, the development of a relational rather than substantive concept of energy took many decades of scientific inquiry to clarify. The contemporary notion of information is likewise colloquially conceived of in substancelike terms, as for example when we describe the "purchase" and "storage" of information, or talk about it being "lost" or "wasted" in some process.

The concept of energy was ultimately tamed by recognizing that it was a quite general sort of physical difference, which could be embodied in any of many possible substrates (heat, momentum, magnetic attraction, chemical bonds, etc.), and which could give rise to non-spontaneous change if able to interact with another physical substrate. Eventually, scientists came to recognize that the presumed ethereal substance that conveyed heat and motive force from one context to another was an abstraction from a process—the process of performing work—not anything material. Importantly, this abandonment of a substance-based explanation did not result in the concept of energy becoming epiphenomenal or mysterious. The fallacious conceptions of an ineffable special substance were simply abandoned for a dynamical account that enabled precise assessment.

A complete account of the nature of information that is adequate to distinguish it from merely material or energetic relationships also requires a shift of focus, but the figure/background shift required is even more fundamental and more counterintuitive than that for energy. This is because what matters is not an account of only its physical properties, or even its formal properties. What matters in the case of information, and produces its distinctive physical consequences, is a relationship to something not there. Information is the archetypical absential concept.

Consider a few typical examples. Information about an impending storm might cause one to close windows and shutters, information about a stock market crash might cause millions of people to simultaneously withdraw money from their bank accounts, information about some potential danger to the nation might induce idealistic men and women to face certain death in battle, and (if we are lucky) information demonstrating how the continued use of fossil fuels will impact the lives of future generations might affect our patterns of energy use worldwide. These not-quite-actualized, non-intrinsic relationships can thus play the central role in determining the initiation and form of physical work. In contrast, the material sign medium that mediates these effects (a darkening sky, a printed announcement, a stirring speech, or a scientific argument, respectively) cannot. The question is: How could these non-intrinsic relationships become so entangled with the particular physical characteristics of the sign medium that the presence of these signs could initiate such non-spontaneous physical changes?

Answering this question poses a double problem. Not only do we need to give up thinking about information as some artifact or commodity, but we must also move beyond an otherwise highly successful technical approach that explicitly ignores those aspects of information that are not present (i.e., referential content and significance).

The currently dominant technical conception of information was important for the development of communication and computing technologies in response to the need to precisely measure information for purposes of transmission and storage. Unfortunately, it has also contributed to a tendency to focus almost exclusively on only the tangible physical attributes of information processes, even where this is not the only relevant attribute. The result is that the technical use of the term *information* is now roughly synonymous with difference, order, pattern, or the opposite of physical entropy, and thus is effectively reduced to a simple physical parameter entirely divorced from its original meaning. Indeed, something like this was implicit in Boltzmann's account of the concept of entropy, because at maximum entropy, we have minimal information about any given microstate of the system. This redefinition of the concept of information as a measure of order has, in effect, cemented the Cartesian cut into the formal foundations of physics. More perniciously, the assumption that information is synonymous with order has been used to implicitly support the claims of both eliminativism and panpsychism. Thus, if it is assumed that we can simply replace the term *information* with *order*, thought becomes synonymous with computation, and any physical difference can be treated as though it has mentalisitc properties.

To step beyond this impasse and make sense of the representational function that distinguishes information from other merely physical relationships, we will need to find a precise way to characterize its defining non-intrinsic feature—its referential capacity—and show how the content thus communicated can be causally efficacious, despite its physical absence. And yet this analysis must also demonstrate its compatibility with the precise technical conception of information, which although divorced from this key defining attribute, has been the basis for unparalleled technical advances in fields as different as quantum physics and economics.

Tacit metaphysical commitments have further hindered this effort. This more encompassing concept of information has been a victim of a philosophical impasse that has a long and contentious history because of its entanglement with the problem of specifying the ontological status of content of thought. We have progressively chipped away at this barrier in previous chapters, so that at last it can be addressed directly with fresh conceptual tools. The enigmatic status of this relationship was eloquently, if enigmatically, framed by the German philosopher Franz Brentano's use in 1874 of the curious word "inexistence" when describing mental phenomena:

> Every mental phenomenon is characterized by what the Scholastics of the Middle Ages called the intentional (or mental) inexistence of an object, and what we might call, though not wholly unambiguously, reference to a content, direction toward an object (which is not to be understood here as meaning a thing), or immanent objectivity.
>
> This intentional inexistence is characteristic exclusively of mental phenomena. No physical phenomenon exhibits anything like it. We can, therefore, define mental phenomena by saying that they are those phenomena which contain an object intentionally within themselves.[2]

As I will argue below, neither the engineer's technical definition of information as physical difference nor the view that information is irreducible to physical relationships confront the problematic issues raised by its not-quite-existent but efficacious character. The first conception takes the possibility of reference for granted, but then proceeds to bracket it from consideration to deal only with physically measurable features of the information-bearing medium. The second conception disregards its physicality altogether and thus makes the efficacy of information mysterious. Neither characterization explicitly accounts for the relationship that the "inexistent" content has to existing physical phenomena.

OMISSIONS, EXPECTATIONS, AND ABSENCES

The importance of physical regularity to the analysis of information is that it is ultimately what enables absence to matter. Where something tends to regularly occur, failing to occur or in other ways diverging from this regularity can stand out. A predictable behavior (and more generally any tendency to change in a redundant way or to exhibit regularity or symmetry) is presupposed in many circumstances where something absent makes a difference. The source of this regularity need not be exemplified by human action. It could, for example, be exemplified by the operation of a machine, as when stepping on the brakes fails to slow one's vehicle and thus provides information about damage. It could be provided by some naturally occurring regularity, like the cycling of the wet and dry seasons, which in a bad year may be interpreted as information that certain deities are angry. Even systems that appear mostly chaotic and only partly constrained in their behavior can offer some degree of predictability against which deviation can be discerned. So, although financial markets are highly volatile and largely defy prediction on a day-to-day basis, there are general trends with respect to which certain deviations are seen as indicative of major changes in the economy that cannot be discerned otherwise.

In the realm of social interactions, we are quite familiar with circumstances where omissions can have major consequences. For example, after April 15, if a U.S citizen has not prepared and submitted a U.S. tax return, or an extension request, serious consequences can follow. In a legal context that requires producing a tax return by this date, its non-existence will set in motion events involving IRS employees coercing the delinquent taxpayer to comply. They may write and send threatening letters and possibly contact banks and credit agencies to interfere with the taxpayer's access to assets. Similarly, so-called sins of omission can also have significant social consequences. Consider the effect of the thank-you note not written or the RSVP request that gets ignored. Omissions in social contexts often prompt deliberations about whether the absence reflects the presence of malice or merely a lack of social graces. Failure to initiate an intervention can also provide information about the actor. A failure to prepare or to attend to important warnings can be the indirect cause of a disaster, and this failure can be grounds for punishment, if intervention is expected. Even ignorance or the absence of foresight due to lack of appropriate information or analytic effort can be blamed for *allowing* "accidents" to occur that might otherwise have been avoided.

Intuitively, then, we are comfortable attributing real-world consequences to not thinking, not noticing, not doing, and so on. In these human contexts, we often treat presence and absence as though they can have equal potential efficacy. But these omissions are only meaningful and efficacious in a context of specific expectations or processes that will ensue if certain conditions are not met, or in the context of undesirable tendencies that will ensue in the absence of opposition. Where there is a habit of expectation, a tendency that is intrinsic, a process that needs to be actively opposed or avoided, or a convention governing or requiring certain actions, the failure of something to occur can have definite physical consequences.

These familiar examples are, of course, special cases that invert the general case of representation—something present that is taken to be about something absent—and yet they exemplify a critical point: being about something need not be based on any intrinsic properties of the signal medium. Indeed, this fact suggests that specific details of a sign's embodiment can be largely irrelevant.

One might object that although this is our colloquial way of describing these relationships, we don't actually believe that absence is causative in these cases. We tend to consider only the habit itself as causally relevant. Nevertheless, it is because the habit in question is organized around a likely future change of state—that is, something not currently present which in some way is likely to follow—that we correspondingly develop a habit of expectation. This indicates that deviation from regularity alone isn't enough. Ultimately, this something missing with respect to a tendency must also stand out with respect to another tendency that interprets it; and (as we will see in the next chapter) for this relationship to be informative, at some point there must be a teleodynamic regularity involved, which is the ground of this expectation; *a projection into the future.*

As we have repeatedly argued, regularity is the expression of constraint, and it is with respect to some constraint on variety that something absent or deviant can stand out. This absence or deviance is ultimately what is conveyed. So, not surprisingly, besides being the formal requirement for a transformation of energy to result in work, constraint is one of the defining attributes of information. Information is in this respect a feature of certain forms of work: teleodynamic work. As we will see, even the most reduced conception of information, as it is defined in mathematical information theory, is a function of constraint and the possibility for work that this embodies.

TWO ENTROPIES

To the extent that regularity and constraint provide a necessary background for deviation and absence to stand out, nature's most basic convergent regularity must provide the ultimate ground for information. This regularity is of course the spontaneous tendency for entropy to increase in physical systems. Although Rudolf Clausius coined the term *entropy* in 1865, it was Ludwig Boltzmann who in 1866 recognized that this could be described in terms increasing disorder. We will therefore refer to this conception of thermodynamic entropy as *Boltzmann entropy*.

This reliably asymmetric habit of nature provides the ultimate background with respect to which an attribute of one thing can exemplify an attribute of something else. The reason is simple: since non-correlation and disorder are so highly likely, any degree of orderliness of things typically means that some external intervention has perturbed them away from this most probable state. In other words, this spontaneous relentless tendency toward messiness provides the ultimate slate for recording outside interference. If things are not in their most probable state, then something external must have done work to divert them from that state.[3]

A second use of the term *entropy* has become widely applied to the assessment of information, and for related reasons. In the late 1940s, the Bell Labs mathematician Claude Shannon demonstrated that the most relevant measure of the amount of information that can be carried in a given medium of communication (e.g., in a page of print or a radio transmission) is analogous to statistical entropy.[4] According to Shannon's analysis, the quantity of information conveyed at any point is the improbability of receiving a given transmitted signal, determined with respect to the probabilities of all possible signals that could have been sent. Because this measure of signal options is mathematically analogous to the measure of physical options in thermodynamic entropy, Shannon also called this measure the "entropy" of the signal source. I will refer to this as *Shannon entropy* to distinguish it from thermodynamic entropy (though we will later see that they are more intimately related than just by analogy).

Consider, for example, a coded communication sent as a finite string of alphanumeric characters. If each possible character can appear with equal probability at every point in the transmission, there is maximum uncertainty about what to expect.[5] This means that each character received reduces this uncertainty, and an entire message reduces the uncertainty with respect to the probability that any possible combination of characters of that length could have been sent. The amount of the uncertainty reduced by receiving a signal is Shannon's measure of the maximum amount of information that can be conveyed by that signal.

In other words, the measure of information conveyed involves comparison of a received signal with respect to possible signals that could have been sent. If there are more possible character types to choose from, or more possible characters in the string, there will be more uncertainty about which will be present where, and thus each will potentially carry more information. Similarly, if there are fewer possible characters, fewer characters comprising a given message, or if the probabilities of characters appearing are not equal, then each will be capable of conveying proportionately less information.

Shannon's notion of entropy can be made quite precise for analysis of electronic transmission of signals and yet can also be generalized to cover quite mundane and diverse notions of possible variety. Shannon entropy is thus a measure of how much information these media can possibly carry. Because it is a logical, not a physical, measure, it is widely realizable. It applies as well to a page of text as to the distribution of objects in a room, or the positions that mercury can occupy in a thermometer. Since each object can assume any of a number of alternative positions, each possible configuration of the collection of objects is a potential sign.

Shannon's analysis of information capacity provides another example of the critical role of absence. According to this way of measuring information, it is not intrinsic to the received communication itself; rather, it is a function of its relationship to something absent—the vast ensemble of other possible communications that could have been sent, but weren't. Without reference to this absent background of possible alternatives, the amount of potential information of a message cannot be measured. In other words, the background of unchosen signals is a critical determinant of what makes the received signals capable of conveying information. No alternatives = no uncertainty = no information. Thus Shannon measured the information received in terms of the uncertainty that it removed with respect to what could have been sent.

The analogy to thermodynamic entropy breaks down, however, because Shannon's concept is a logical (or structural) property, not a dynamical property. For example, Shannon entropy does not generally increase spontaneously in most communication systems, so there is no equivalent to the second law of thermodynamics when it comes to the entropy of information. The arrangement of units in a message doesn't spontaneously "tend" to change toward equiprobability. And yet something analogous to this effect becomes relevant in the case of real physically embodied messages conveyed by real mechanisms (such as a radio transmission or a computer network). In the real world of signal transmission, no medium is free from the effects of physical irregularities and functional degradation, an unreliability resulting from the physical effects of the second law.

So both notions of entropy are relevant to the concept of information, though in different ways. The Shannon entropy of a signal is the probability of receiving a given signal from among those possible; and the Boltzmann entropy of the signal is the probability that a given signal may have been corrupted.

A transmission affected by thermodynamic perturbations that make it less than perfectly reliable will introduce an additional level of uncertainty to contend with, but one that decreases information capacity. An increase in the Boltzmann entropy of the physical medium that constitutes the signal carrier corresponds to a decrease in the correlation between sent and received signals. Although this does not decrease the signal entropy, it reduces the amount of uncertainty that can be removed by a given signal, and thus reduces the information capacity.

This identifies two contributors to the entropy of a signal—one associated with the probability of a given signal being sent and the other associated with a given signal being corrupted. This complementary relationship is a hint that the physical and informational uses of the concept of entropy are more than merely analogous. By exploring the relationship between Shannon entropy and Boltzmann entropy, we can shed light on the reason why change in Shannon entropy is critical to information. But the connection is subtle, and its relationship to the way that a signal conveys its information "content" is even subtler.

INFORMATION AND REFERENCE

Warren Weaver, who wrote a commentary article that appeared in a book presentation of Shannon's original paper, commented that using the term *information* to describe the measure of the unpredictability reduced by a given signal is an atypical use of the term.[6] This is because Shannon's notion of information is agnostic with respect to what a signal is or could be about. This agnosticism has led to considerable confusion outside of the technical literature, because it is almost antithetical to the standard colloquial use of the term. Shannon was interested in measuring information for engineering purposes. So he concentrated exclusively on the properties of transmission processes and communication media, and ignored what we normally take to be information, that is, what something tells us about something else that is not present in the signal medium itself.

This was not merely an arbitrary simplification, however. It was necessary because the same sign or signal can be given any number of different interpretations. Dirt on a boot can provide information on anything from personal hygiene to evidence about the sort of geological terrain a person recently visited. The properties of some medium that give it the potential to convey information don't determine *what* it is about, they merely make reference possible. So, in order to provide a finite measure of the information potential of a given signal or channel, Shannon had to ignore any particular interpretation process, and stop the analysis prior to including any consideration of what a sign or signal might be about. What is conveyed is not merely a function of this reduction of Shannon entropy.

This is where the relation between Shannon and Boltzmann entropy turns out to be more than merely analogical. A fuller conception of information requires that these properties be considered with respect to two different levels of analysis of the same phenomenon: the formal characteristics of the signal and the material-energetic characteristics of the signal. Consider what we know about Boltzmann entropy. If, within the boundaries of a physical system such as a chamber filled with a gas, a reduction of entropy is observed, one can be pretty certain that something *not* in that chamber is causing this reduction of entropy. Being in an improbable state or observing a non-spontaneous change toward such a state is evidence of extrinsic perturbation—work imposed from outside the system.

Despite their abstract character, information transmission and interpretation are physical processes involving material or energetic substrates that constitute the transmission channel, storage medium, sign vehicles, and so on. But physical processes are subject to the laws of thermodynamics. So, in the case of Shannon entropy, no information is provided if there is no reduction in the uncertainty of a signal. But reduction of the Shannon entropy of a given physical medium is necessarily also a reduction of its Boltzmann entropy. This can only occur due to the imposition of outside constraints on the sign/signal medium because a reduction of Boltzmann entropy does not tend to occur spontaneously. When it does occur, it is evidence of an external influence.

Openness to external modification is obvious in the case of a person selecting a signal to transmit, but it is also the case in more subtle conditions. Consider, for example, a random hiss of radio signals received by a radio antenna pointed toward the heavens. A normally distributed radio signal represents high informational entropy, the expected tendency in an unconstrained context, which for example might be the result of random circuit noise. If this tendency were to be altered away from this distribution in any way, it would indicate that some extrinsic non-random factor was affecting the signal. The change could be due to an astronomical object emitting a specific signal. But what if instead of a specific identifiable signal, there is just noise when there shouldn't be any?

Just such an event did occur in 1965 when two Bell Labs scientists, Arno Penzias and Robert Wilson, aimed a sensitive microwave antenna skyward, away from any local radio source, and discovered that they were still recording the "hiss" of microwave noise no matter which way they pointed the antenna. They were receiving a more or less random microwave signal from everywhere at once! The obvious initial assumption was that it probably indicated a problem with the signal detection circuit, not any specific signal. Indeed, they even suspected that it might be the result of pigeons roosting in the antenna. Because it exhibited high Shannon entropy and its locus of origin also had high Shannon entropy (i.e., equal probability from any location), they assumed that it couldn't be a signal originating from any object in space.

Only after they eliminated all potential local sources of noise did they consider the more exotic possibility: that it did not originate in the receiving system itself, but rather from outside, everywhere in the cosmos at once. If the signal had consisted only of a very narrow band of frequencies, had exhibited a specific oscillatory pattern, or was received only from certain directions, that would have provided Shannon information, because then compared with signals that could have been recorded, this would have stood out. They only considered it to be information about something external, and not mere noise, when they had compared it to many more conceivable options, thus effectively increasing the entropy (uncertainty) of potential sources that could be eliminated. As they eliminated each local factor as a possible source of noise, they eliminated this uncertainty. In the end, what was considered the least reasonable explanation was the correct one: it was emanating from empty space, the cosmic background. The signal was eventually interpreted as the deeply red-shifted heat of the Big Bang.

In simple terms, the brute fact of these deviations from expectation, both initially, then with respect to possible sources of local noise, and much later with respect to alternative cosmological theories, was what made this hiss information about something. Moreover, the form of this deviation from expectation—its statistical uniformity and lack of intrinsic asymmetries—provided the basis for the two scientists' interpretation. Heat is random motion. The form of this deviation from what they expected—that the signal would be irregularly distributed and correlated with specific objects—ultimately provided the clue to its interpretation. In other words, this became relevant for a second kind of entropy reduction: analytic reduction in the variety of possible sources. By comparing the form of the received signal with what could have been its form, they were able to eliminate many possible causes of both intrinsic and extrinsic physical influence until only one seemed plausible.

This demonstrates that something other than merely the reduction of the Shannon entropy of a signal is relevant to understanding how that signal conveys evidence of another absent phenomenon. Information is made available when the state of some physical system differs from what would be expected. And what this information can be about depends on this expectation. If there is no deviation from expectation, there is no information. This would be the case for complete physical isolation of the signal medium from outside influence. The difference is that an isolated signal source cannot be *about* anything. What information can be about depends on the nature of the reduction process and the constraints exhibited by the received signal. How a given medium can be modified by interaction with extrinsic factors, or how it can be manipulated, is what is most relevant for determining *what* information can be conveyed by it.

Not only is the Shannon entropy of an information-bearing process important to its capacity; its physical dynamics with respect to the physical context in which it is embedded is important to the determination of what it can be about. But in comparison to the Shannon entropy of the signal, the physical constraints on the forms that a change in signal entropy can take contribute another independent potential for entropy reduction: *a reduction of the entropy of possible referents.* In other words, once we begin considering the potential entropy of the class of things or events that a given medium can convey, the physical characteristics of the medium, not merely its range of potential states, become important. This reduces the Shannon entropy of the range of possible phenomena that a given change in that medium can be about. In other words, what we might now call *referential information* is a second order form of information, over and above Shannon information (Figures 12.1, 12.2).

A first hint of the relationship between information as form and as a sign of something else is exemplified by the role that pattern plays in the analysis. In Shannon's terms, pattern is redundancy. From the sender's point of view, any redundancy (defined as predictability) of the signal has the effect of reducing the amount of information that can be sent. In other words, redundancy introduces a constraint on channel capacity. Consider how much less information could be packed into this page if I could only use two characters (like the 0s and 1s in a computer). Having twenty-six letters, cases, punctuation marks, and different-length character strings (words), separated by spaces (also characters), decreases the repetition, increases the possibility of what character can follow what, and thereby decreases redundancy. Nevertheless, there is sufficient redundancy in the possible letter combinations to make it possible to fairly easily discover typos (using redundancy for error correction will be discussed below).

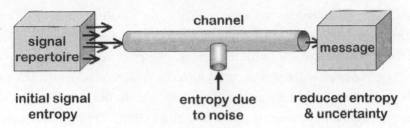

FIGURE 12.1: Depiction of the logic that Claude Shannon used to define and measure potential information-conveying capacity ("Shannon information").

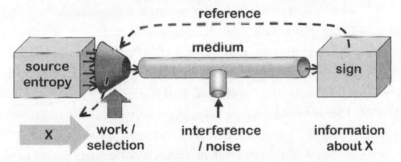

FIGURE 12.2: Depiction of the way that Shannon information depends on the susceptibility of a medium to modification by work. It derives its potential to convey information about something extrinsic to that medium by virtue of the constraints imposed on that medium by whatever is responsible for this work.

Less information can get transmitted if some transmissions are predictable from previous ones, or if there are simply fewer alternatives to choose from. But the story is slightly more complicated when we add reference. From the receiver's point of view, there must be some redundancy with what is already known for the information conveyed by a signal to even be assessed. In other words, the *context* of the communication must already be redundantly structured. Both sender and receiver must share the set of options that constitute information.

Shannon realized that the introduction of redundancy is also necessary to compensate for any unreliability of a given communication medium. If the reliability of a given sign or signal is questionable, this introduces an additional source of unpredictability that does not contribute to the intended information: noise. But just as redundancy reduces the unpredictability of signals, it can also reduce the unreliability of the medium conveying information. Introducing expected redundancy into a message being transmitted makes it possible to distinguish these two sources of Shannon entropy (the variety of possible signals that could have been generated versus the possible errors that could have arisen in the process). In the simplest case, this is accomplished by resending a signal multiple times. Because noise is, by definition, not constrained by the same factors as is the selection of the signal, each insertion of a noise-derived signal error will be uncorrelated with any other, but independent transmissions of multiple identical signals will be correlated with one another by definition. In this way, noisy components of a signal or a received message can be detected and replaced.

Error-reducing redundancy can be introduced by means other than by signal retransmission. English utilizes only a fraction of possible letter combinations, with very asymmetric probabilities and combinatorial options. Grammar and syntax further limit what is an appropriate and inappropriate word string. Last but not least, the distinction between sense and nonsense limits what words and phrases are likely to occur in the same context. This internal redundancy of written English makes typos relatively easy to identify and correct.

Redundancy as a means of information rectification is also relevant to assessing the reliability of second-order (i.e., referential) information. For example, when one hears multiple independent reports by observers of the same event, the redundancy in the content of these various accounts can serve an analogous function. Even though there will be uncorrelated details from one account to another, the redundancies between independent reports make these more redundant aspects more reliable (so long as they are truly independent, i.e., don't reflect the influence of common biases). So, whereas redundancy decreases information capacity, it is also what makes it possible to distinguish information from noise, both in terms of the signal and in terms of what it conveys.

IT TAKES WORK

What a signal medium can indicate is dependent on the possibility of physical relation with some relevant features of its physical context, and the possibility that this can result in a change of its Shannon entropy. Since reduction in the Shannon entropy of a physical medium is also often a reduction of its physical entropy, such a change is evidence that work was done to modify the signal medium. Recall that work in some form is required to change something from its spontaneous state or tendency to a non-spontaneous state. So, for any physical medium, a contragrade change in its state is an indication that work has been performed. Moreover, a contragrade change can only be produced by extrinsic perturbation; the medium must be an open system in some respect. The reference conveyed by a reduction in Shannon entropy is therefore a function of the ways that the medium is susceptible to outside interference.

Though the external factors that alter a system's entropy (variety of possible states) are not intrinsic features of that medium, the signal constraint is an intrinsic feature. Referential information is in this sense inferred from the form of the constraints embodied in the relationship between unconstrained possibility and received signal. In this way, Shannon information, which is assessed in terms of this constraint, embodies a trace of the work that produced it.

But openness to the possibility of an extrinsic influence involves specific physical susceptibilities, which also constrain what kind of work is able to modify it, thus constraining the domain of potential extrinsic phenomena that can be thereby indicated. Because of this explicit physical constraint, even the absence of any change of signal entropy can provide referential information. No change in entropy is one possible state. This means that even unconstrained fluctuations may still be able to convey referential information. This is information about the fact that of the possible influences on signal constraint, none were present. It is the mere possibility of exhibiting constraints due to extrinsic influence that is the basis of a given medium's informative power.

This possibility demonstrates that reference is more than just the consequence of physical work to change a signal medium. Although such a relation to work is fundamental, referential information can be conveyed both by the effect of work and by evidence that no work has been done.[7] This is why no news can be news that something anticipated has not yet occurred, as in the examples of messages conveyed by absence discussed above. The informative power of absence is one of the clearest indications that Shannon information and referential information are not equivalent. This is again because the constraint is both intrinsic and yet not *located in* the signal medium; it is rather a relationship between what is and what could have been its state at any given moment. A constraint is not an intrinsic property but a relational property, even if just in relation to what is possible. Only when a physical system exhibits a reduction of entropy compared to some prior state, or more probable state, is extrinsic influence indicated. This ubiquitous tendency can, conversely, be the background against which an unaltered or unconstrained feature can provide information about what hasn't occurred. If the sign medium exhibits no constraint, or hasn't diverged from some stable state, it can be inferred that there has been no extrinsic influence even though one could have been present. The relationship of present to absent forms of a sign medium embodies the openness of that medium to extrinsic intervention, whether or not any interaction has occurred. Importantly, this also means that the *possibility of change due to work*, not its actual effect, is the feature upon which reference depends. It is what allows absence itself, absence of change, or being in a highly probable state, to be informative.

Consider a typo in a manuscript. It can be thought of as a reduction of referential information because it reflects a lapse in the constraint imposed by the language that is necessary to convey the intended message. Yet it is also information about the proficiency of the typist, information that might be useful to a prospective employer. Or consider a technician diagnosing the nature of a video hardware problem by observing the way the image has become distorted. What is signal and what is noise is not intrinsic to the sign medium, because this is a determination with respect to reference. In both, the deviation from a predicted or expected state is taken to refer to an otherwise unobserved cause. Similarly, a sign that doesn't exhibit the effects of extrinsic influence—for example, setting a burglar alarm to detect motion—can equally well provide information that a possible event (a break-in) did *not* occur.

In all these cases, the referential capacity of the informational vehicle is dependent on physical work that has, or could have, altered the state of some medium open to extrinsic modification. This tells us that the link between Shannon entropy and Boltzmann entropy is not mere analogy or formal parallelism. More important, it demonstrates a precise link to the concept of work. Gregory Bateson's description of information as "a difference that makes a difference" is actually a quite elegant description of work. Now we can see why this must be so. *The capacity to reflect the effect of work is the basis of reference.*

TAMING THE DEMON

This should not surprise us. The existence of an intimate relationship between information and work has been recognized since almost the beginning of the science of thermodynamics. In 1867, James Clerk Maxwell initially explored the relationship in terms of a thought experiment. He imagined a microscopic observer (described as a demon), who could assess the velocity of individual gas molecules on either side of a divided container and could control a door between them to allow only faster- (or slower-) than-average molecules through one way or the other. Using this information about each molecule, the demon would thus be able to decrease the entropy of the system in contradiction to the second law of thermodynamics. This would normally only be possible by doing thermodynamic work to drive the entire system away from thermodynamic equilibrium; and once in this far-from-equilibrium state, this thermodynamic gradient itself could be harnessed to do further work. So it appears on the face of things that the demon's information about molecular velocity is allowing the system to progressively reverse the second law, with only the small amount of work that determines the state of the door. This seems consistent with our intuition that information and entropy have opposite signs, in the sense that a decrease in entropy of the system increases the predictability of molecular velocities on either side of this divide. For this reason, information is sometimes described as "negentropy,"[8] and has been equated with the orderliness of a system.

Maxwell's demon does not have to be a tiny homunculus. The same process could conceivably be embodied in a mechanical device able to link differences of detected molecular velocity and the correlated operation of the door. In the century that followed Maxwell's presentation of this thought experiment, many sophisticated analyses probed the question of whether the information gleaned by such an apparatus would in fact be able to cheat the second law of thermodynamics. As many analyses were subsequently to show, the mechanisms able to gather such information and use it to open the pass-through would inevitably require more work than the potential gained by creating this increase of heat gradient, and would therefore produce a net increase in total entropy of the system. The increase in entropy would inevitably exceed the reduction of entropy produced by the demon's efforts.

Although purely theoretical, this analysis validated the assumptions of thermodynamics, and also made it possible to measure the amount of Shannon information required to produce a given amount of Boltzmann entropy decrease. So, although the demon's activity in transforming differences of molecular velocity into differences of local entropy doesn't ultimately violate the second law, it provides a model system for exploring the relationship between Shannon information and work.

But if we replace the demon with an equivalent mechanical apparatus, does it make sense to say that it is using information *about* velocity to effect this change in entropy, or is it merely a mechanistic linkage between some physical change that registers velocity and whatever physical process opens the door? Although as external observers we can interpret a signal whose changes correlate with molecular velocity as representing information about that property, there is nothing about the mechanism linking this signal state to the state of the door that makes it more than just a physical consequence of interacting with the signal.

What enables the observer to interpret that the signal is about velocity is the independent availability of a means for relating differences of molecular velocity to corresponding differences of signal state. The differential activation of the door mechanism is merely a function of the physical linkage of the signal-detection and door-operation mechanisms. Molecular velocity is otherwise irrelevant, and a correlation between signal value and molecular velocity is not in any way necessary to the structure or operation of the door-opening mechanism. A correlation with molecular velocity is, however, critical to how one might design such a mechanism with this Maxwellian outcome in mind. And it is cryptically implied in the conception of an observing demon. Unlike its mechanical substitute, the demon must respond *because of this correlation* in order to interpret the signal to be about molecular velocity. The designer must ensure that this correlation exists, while the demon must assume that it exists, or at least be acting with respect to this correlation and not merely with respect to the signal. Thus the demon, like an outside observer, must already have information about this habit of correlation in order to interpret the signal as indicating this missing correlate. In other words, an independent source of information about this correlation is a precondition for the signal to be about velocity, but the signal-contingent door-opening mechanism has no independent access to this additional information.

More important, correlation is not a singular physical interaction, but rather a regularity of physical interactions. A mechanism that opens the door in response to a given signal value is not responding to this regularity but only to a singular physical influence. In this respect, there is an additional criterion besides being susceptible to extrinsic modification that constitutes the referential value of an informing medium: this modifiability must have a general character.

The analysis so far has exposed a common feature of both the logic of information theory (Shannon) and the logic of thermodynamic theory (Boltzmann). This not only helps explain the analogical use of the entropy concept in each, it also explains why it is necessary to link these approaches into a common theory to begin to define the referential function of information. Both of these formal commonalities and the basis for their unification into a theory of reference depend on physical openness. In the case of classic information theory, the improbability of receiving a given sign or signal with respect to the background expectation of its receipt compared to other options defines the measure of potential information. In the case of classic thermodynamics, the improbability of being in some far-from-equilibrium state is a measure of its potential to do work, and also a measure of work that was necessarily performed to shift it into this state. Inversely, being in a most probable state provides no information about any extrinsic influence, and indeed suggests that to the extent that this medium is sensitive to external perturbation, none was present that could have left a trace.

The linkage between these two theories hinges on the materiality of communication (e.g., the constitution of its sign and/or signal medium). So, in a paradoxical sense, the absent content that is the hallmark of information is a function of the necessary physicality of information processes.

13

SIGNIFICANCE

The first surprise is that it takes constraints on
the release of energy to perform work, but it
takes work to create constraints. The second
surprise is that constraints are information and
information is constraint.

—STUART KAUFFMAN, PERSONAL COMMUNICATION[1]

ABOUTNESS MATTERS

As we have seen, nearly every physical interaction in the universe can
be described in terms of Shannon information, and any relationship
involving physical malleability, whether exemplified or not, *can* be interpreted as information *about* something else. This has led some writers to
suggest that the universe is made of information, not matter. But this tells
us little more than that the universe is a manifold of physical differences and
that most are the result of prior work. Of course, not every physical difference is interpreted, or even can be interpreted, though all may at some point
contribute to future physical changes. Interpretation is ultimately a physical process, but one with a quite distinctive kind of causal organization.
So, although almost every physical difference in the history of the universe
can potentially be interpreted to provide information about any number of
other linked physical occurrences, the unimaginably vast majority of these
go uninterpreted, and so cannot be said to be information *about* anything.
Without interpretation, a physical difference is just a physical difference,

and calling these ubiquitous differences "information" furthermore runs the risk of collapsing the distinction between information, matter, and energy, and ultimately eliminating the entire realm of ententional phenomena from consideration.

Although any physical difference *can* become significant and provide information about something else, interpretation requires that certain very restricted forms of physical processes must be produced. The organization of these processes distinguishes interpretation from mere physical cause and effect. Consider again Gregory Bateson's aphorism: "a difference that makes a difference." Its meaning turns on the ambiguity between two senses of to "make a difference." The more literal meaning is to cause something to change from what otherwise would have occurred. This is effectively a claim about performing work (in any of the senses that we have described). In idiomatic English, however, it also means to be of value (either positive or negative) to some recipient or to serve some purpose. I interpret Bateson's point to be that *both* meanings are relevant. Taken together, then, these two meanings describe work that is initiated in order to effect a change that will serve some end. Some favored consequence must be promoted, or some unwanted consequence must be impeded, by the work that has been performed in response to the property of the sign medium that is taken as information.

This is why an interpretive process is more than a mere causal process. It organizes work in response to the state of a sign medium and with respect to some *normative* consequence—a general type of consequence that is in some way valued over others. This characterization of creating/making a difference in both senses suggests that the sort of work that Bateson has in mind is not merely thermodynamic work. To "make a difference" in the normative sense of this phrase is (in the terms we have developed) to support some teleodynamic process. It must contribute to the potential to initiate, support, or inhibit teleodynamic work, because only teleodynamic processes can have normative consequences. So to explain the basis of an interpretation process is to trace the way that teleodynamic work transforms mere physical work into semiotic relationships, and back again.

As our dynamical analysis has shown, teleodynamic work emerges from and depends on both morphodynamic and thermodynamic work. Consequently, these lower forms of work must also be involved in any process of interpretation. If this is the case, then an interpreting process must depend on extrinsic energetic and material resources and must also involve far-from-equilibrium self-organizing processes. In this roundabout way, like the dynamical processes that characterize organisms, the interpretation of something as information involves a form of recursive organization whereby the interpretation of something as information indirectly reinforces the capacity to do this again.

BEYOND CYBERNETICS

Perhaps the first hint that the 2,000-year-old mystery of interpretation might be susceptible to a physical explanation rather than remaining forever metaphysical can be attributed to the development of a formal theory of regulation and control, which can rightfully be said to have initiated the information age. This major step forward in defining the relationship between information and its physical consequences was provided by the development of cybernetic theory in the 1950s and 60s. The term *cybernetic* was coined by its most important theoretician, Norbert Wiener, and comes from the same Greek root as the word "government," referring to steering or controlling. Within cybernetic theory, for the first time it became possible to specify how information (in the Shannonian sense) could have definite physical consequences and could contribute to the attractor dynamics constituting teleonomic behaviors.

In chapter 4, we were introduced to the concept of teleonomic behavior and the simple mechanistic exemplar of negative feedback regulation: thermostatic control circuit. This model system not only demonstrates the fundamental principles of this paradigm, and the way it conceives of the linkage between information as a physical difference and a potential physical consequence; it also provides a critical clue to its own inadequacy.

A thermostat regulates the temperature in a room by virtue of the way that a switch controlling a heating device is turned on or off by the effects of that temperature. The way these changes correlate with the state of the switch and the functioning of the heating device creates a deviation-minimizing pattern of behavior. It's a process whereby one difference sets in motion a chain of difference-making processes that ultimately "make a difference" in keeping conditions within a desired range for some purpose.

Thus a difference in the surrounding temperature produces a difference in the state of the switch, which produces a difference in the operation of the heater, which produces a difference in the temperature of the room, which produces a difference in the state of the switch, and so forth. At each stage, work is done on a later component in the circuit with respect to a change in some feature of the previous component, resulting in a circularity of causal influences. Thus it is often argued that each subsequent step along the chain of events in this cycle "interprets" the information provided by the previous step, and that information is being passed around this causal circuit. But in what sense do these terms apply? Are they merely metaphoric?

We can dissect this problem by dissecting the circuit itself. A classic mechanical-electrical thermostat design involves a mercury switch attached to a coiled bimetallic strip, which expands when warmed, thus tipping the switch one way, and contracts when cooled, tipping the switch the other way (cf. Figure 4.1). The angle of the switch determines whether the circuit is completed or interrupted. But let's consider one of these steps in isolation. Is the coiling and uncoiling of a bimetallic strip information about temperature? It certainly could be used as such to an observer who understood this relationship and was bringing this knowledge to bear in considering the relationship. But what if this change of states goes unnoticed? Physically, there is no difference. The change in state of the coiled strip and of the room temperature will occur irrespective of ether being observed. Like the wax impression of a signet ring, it is merely a physical phenomenon that *could* be interpreted as information about something in particular. Of course, it could also be interpreted as information about many other things. For example, its behavior could be interpreted to be information about the differential responsiveness of the two metals. Or it could be mistakenly interpreted as magic or some intrinsic tendency to grow and shrink at random. Is being incorporated into a thermostatic circuit sufficient to justify describing the coiling behavior as "information" about temperature to the circuit? Or is this too still only one of many possible things it could provide information about? What makes it information about anything rather than just a simple physical influence? Clearly, it is the process of interpretation that matters, not merely this physical tendency, and that is an entirely separate causal process.

Consider, in contrast, a single-cell organism responding to a change in temperature by changing its chemical metabolism. Additionally, assume that some molecular process within the organism, which is the equivalent of a simple thermostatic device, accomplishes this change. In many respects, it is more like a thermostat installed by a human user to maintain room temperature than a feedback process that might occur spontaneously in inorganic nature. This is because both the molecular regulator of the cell and the engineered thermostat embody constraints that are *useful to some superordinate system* for which they are at the same time both support-ive and supported components. In a thermostat, it is the desired attractor dynamics (desired by its human users), and not any one specific material or energetic configuration, that determines its design. In organisms, such convergent behaviors were likely to have been favored by natural selection to buffer any undesirable changes of internal temperature. Indeed, in both living and engineered regulators, there can be many different ways that a given attractor dynamics is achieved. Moreover, analogously functioning living mechanisms often arise via parallel or convergent evolution from quite different precursors.

This drives home the point that it is this pattern of behavior that deter-mines the existence of both the evolved and engineered regulatory systems, not the sharing of any similar material constitution or a common accidental origin. In contrast, Old Faithful was formed by a singular geological acci-dent, and the regularity of its deviation-minimizing hydrothermal behavior had nothing to do with its initial formation. Nor does its feedback logic play any significant role in how long this behavior will persist. If the geology changes or the source of water is depleted, the process will simply cease.

We are thus warranted in using the term *information* to describe the physical changes that get propagated from component to component in a designed or evolved feedback circuit *only because the resultant attrac-tor dynamics itself played the determinate role in generating the architecture of this mechanism.* In such cases, we also recognize that its physical com-position and component dynamical operations are replaceable so long as this attractor-governed behavior is reliably achieved. In contrast, it is also why, in the case of Old Faithful or any other accidentally occurring non-living feedback process, it feels strange to use information terminology to describe their dynamics, except in a metaphoric or merely Shannonian sense. Although they too may exhibit a tendency to converge-toward or resist-deviation-away-from a specific attractor state, the causal histories and future persistence of these processes lack this crucial attribute. Indeed, a designed or evolved feedback mechanism and an accidentally occurring analogue might even be mechanistically identical, and we still would need to make this distinction.

WORKING IT OUT

As Shannon's analysis showed, information is embodied in constraints, and, as we have additionally shown, what these constraints can be about is a function of the work that ultimately was responsible for producing them (or could have produced them, even if they are never generated), either directly or indirectly. But as Stuart Kauffman points out in the epigraph at the beginning of this chapter, not only does it take work to produce constraints, it takes constraints to produce work. So one way in which the referential content of information can indirectly influence the physical world is if the constraints embodied in the informing medium can become the basis for specifying further work. And differences of constraint can determine differences in effect.

This capacity for one form of work to produce the constraints that organize another, independent form of work is the source of the amplifying power of information. It affords a means to couple otherwise unrelated contragrade processes into highly complex and indirect chains. And because of the complementary roles of constraint and energy-gradient reduction, it also provides the means for using the depletion of a small energy gradient to create constraints that are able to organize the depletion of a much larger energy gradient. In this way, information can serve as the bridge linking the properties of otherwise quite separate and unrelated material and energetic systems. As a result, chains of otherwise non-interacting contragrade processes can be linked. Work done with the aid of one energy gradient can generate constraints in a signaling medium, which can in turn be used to channel work utilizing another quite different energy gradient to create constraints in yet some other medium, and so forth. By repeating such transfers step by step from medium to medium, process to process, causal linkages between phenomena that otherwise would be astronomically unlikely to occur spontaneously can be brought into existence. This is why information, whether embodied in biological processes, engineered devices, or theoretical speculations, has so radically altered the causal fabric of the world we live in. It expands the dimensions of what Kauffman has called the "adjacent possible" in almost unlimited ways, making almost any conceivable causal linkage possible (at least on a human scale).

In this respect, we can describe interpretation as the incorporation of some extrinsically available constraint to help organize work to produce other constraints that in turn help to organize additional work which promotes the maintenance of this reciprocal linkage between forms of work and constraint. So, unlike a thermostat, where the locus of interpretive activity is extrinsic to the cycle of physical interactions, an interpretive process is characterized by an entanglement between the dynamics of its responsiveness to an extrinsic constraint and the dynamics that maintains the intrinsic constraints that enable this responsiveness. Information is in this way indirectly about the conditions of its own interpretation, as well as about something else relevant to these conditions. Interpreting some constraint as being about something else is thus a projection about possibility in two ways: it is a prediction that the source of the constraint exists; and also that it is causally relevant to the preservation of this projective capacity. But a given constraint is information to an interpretive process regardless of whether these projected relationships are realized. What determines that a given constraint is information is that the interpretive process is organized so that this constraint is correlated with the generation of work that would preserve the possibility of this process recurring under some (usually most) of the conditions that could have produced this constraint.

For this reason, interpretation is also always in some sense normative and the relationship of aboutness it projects is intrinsically fallible. The dynamical process of interpretation requires the expenditure of work, and in this sense the system embodying it is at risk of self-degradation if this process fails to generate an outcome that replenishes this capacity, both with respect to the constraints and the energy gradient that are required. But the constraint that serves as the sign of this extrinsic feature is a general formal property of the medium that embodies it, and so it cannot be a guarantee of any particular specific physical referent existing. So, although persistence of the interpretive capacity is partly conditional on this specificity, that correlation may not always hold.

The interpretive capacity is thus a capacity to generate a specific form of work in response to particular forms of system-extrinsic constraints in such a way that this generates intrinsic constraints that are likely to maintain or improve this capacity. But, as we have seen, only morphodynamic processes spontaneously generate intrinsic constraints, and this requires the maintenance of far-from-equilibrium conditions. And only teleodynamic systems (composed of reciprocal morphodynamic processes) are capable of preserving and reproducing the constraints that make this preservation possible. So a system capable of interpreting some extrinsic constraint as information relevant to this capability is necessarily a system dependent on being reliably correlated in space and time with supportive non-equilibrium environmental conditions. Maintaining reliable access to these conditions, which by their nature are likely to be variable and transient, will thus be aided by being differentially responsive to constraints that tend to be correlated with this variability.

Non living cybernetic mechanisms exhibit forms of recursive dynamical organization that generate attractor-mediated behavior, but their organization is not reflexively dependent on and generated by this dynamics. This means that there is no general property conveyed by each component dynamical transition from one state of the mechanism to the next. Only a specific dynamical consequence.

As Gregory Bateson emphatically argued, confusing information processes with energetic processes was one of the most problematic tendencies of twentieth-century science. Information and energy are distinct and in many respects should be treated as though they occupy independent causal realms. Nevertheless, they are in fact warp and weft of a single causal fabric. But unless we can both clearly distinguish between them and demonstrate their interdependence, the realms they exemplify will remain isolated.

INTERPRETATION

For engineering purposes, Shannon's analysis could not extend further than an assessment of the information-carrying capacity of a signal medium, and the uncertainty that is reduced by receipt of a given signal. Including referential considerations would have introduced an infinite term into the quantification—an undecidable factor. What is undecidable is where to stop. There are innumerable points along a prior causal history culminating in the modification of the sign/signal medium in question, and any of these could be taken to be the relevant reference. The process we call interpretation is what determines which is the relevant one. It must "pick" one factor in the trail of causes and effects leading up to the constraint reflected in the signal medium. As everyday experience makes clear, what is significant and what is not depends on the context of interpretation. In different contexts and for different interpreters, the same sign or signal may thus be taken to be about very different things. The capacity to follow the trace of influences that culminated in this particular signal modification in order to identify one that is relevant is in this way entirely dependent on the complexity of the interpreting system, its intrinsic information-carrying/producing capacity, and its involvement with this same causal chain.

Although the physical embodiment of a communication medium provides the concrete basis for reference, its physical embeddedness also opens the door to an open-ended lineage of potentially linked influences. To gain a sense of the openness of the interpretive possibilities, consider the problem faced by a detective at a crime scene. There are many physical traces left by the interactions involved in the critical event: doors may have been opened, furniture displaced, vases knocked over, muddy footprints left on a rug, fingerprints on the doorknob, filaments of clothing, hair, and skin cells left behind during a struggle, and so on. One complex event is reflected in these signs. But for each trace, there may or may not be a causal link to this particular event of interest. Each will also have a causal history that includes many other influences. The causal history reflected in the physical trace taken as a sign is not necessarily relevant to any single event, and which of the events in this history might be determined to be of pragmatic relevance can be different for different interpretive purposes and differently accessible to the interpretive tools that are available.

This yields another stricture on the information interpretation process. The causal history contributing to the constraints imposed on a given medium limits, but does not specify, what its information can be about. That point in this causal chain that is the referent must be determined by and with respect to another information process. All that is guaranteed by a potential reduction of the Shannon entropy of a signal is a possible definite linkage to something else. But this is an open-ended set of possibilities, only

limited by processes that spontaneously obliterate certain physical traces or
that block certain physical influences. Shannon information is a function of
the potential variety of signal states, but referential entropy is additionally
a function of the potential variety of factors that could have contributed to
that state. So what must an interpretive process include in order to reduce
this vast potential entropy of possible referents?

In the late nineteenth-century world of the fictional detective Sher-
lock Holmes, there were far fewer means available to interpret the physical
traces left behind at a crime scene. Even so, to the extent that Holmes had a
detailed understanding of the physical processes involved in producing each
trace, he could use this information to extrapolate backwards many steps
from effect to cause. This capacity has been greatly augmented by modern
scientific instruments that, for example, can determine the chemical constitution of traces of mud, the manufacturer of the fibers of different fabrics,
the DNA sequence information in a strand of hair, and so on. With this
expansion of analytic means, there has come an increase in the amount of
information which can be extracted from the same traces that the fictional
Holmes might have encountered. These traces contain no more physical
differences than they would have in the late nineteenth century; it is simply
that more of these have become interpretable, and to a greater causal depth.
This enhancement of interpretive capacity is due to an effective increase in
the interpretable Shannon entropy. But exactly how does this expansion of
analytic tools effectively increase the Shannon entropy of a given physical
trace?

Although from an engineer's perspective, every possible independent
physical state of a system must be figured into the assessment of its potential
Shannon entropy, this is an idealization. What matters are the distinguishable states. The distinguishable states are determined with respect to an
interpretive process that itself must also be understood as a signal production process with its own potential Shannon entropy. In other words, *one
information source can only be interpreted with respect to another information production process.* The maximum information that can be conveyed
is consequently the lesser of the Shannon entropies of the two processes.
If the receiving/interpreting system is physically simpler and less able to
assume alternative states than the sign medium being considered, or the
relative probabilities of its states are more uneven (i.e., more constrained),

or the coupling between the two is insensitive to certain causal interactions, then the interpretable entropy will be less than the potential entropy of the source. This, for example, happens with the translation of DNA sequence information into protein structure information. Since there are sixty-four possible nucleotide triplets (codons) to code for twenty amino acids, only a fraction of the possible codon entropy is interpretable as amino acid information.[2] One consequence of this is that scientists using DNA sequencing devices have more information to work with than does the cell that it comes from.

This limitation suggests two interesting analogies to the thermodynamic constraints affecting work that were implicit in Shannon's analysis. First, the combined *interpretable* Shannon entropy of a chain of systems (e.g., different media) through which information is transferred can be no greater than the channel/signal production device with the lowest entropy value. Each coupling of system-to-system will tend to introduce a reduction of the interpretable entropy of the signal, thus reducing the difference between the initial potential and final received signal entropy. And second, information capacity tends to be lost in transfer from medium to medium if there is noise or if the interpreting system is of lower entropy (at least it cannot be increased), and with it the specificity of the causal history that it can be about. Since its possible reference is negatively embodied in the form of constraints, what a sign or signal can be about tends to degrade in specificity spontaneously with transmission or interpretation. This latter tendency parallels a familiar thermodynamic tendency which guarantees that there is inevitably some loss in the capacity to do further work in any mechanical process. This is effectively the informational analogy to the impossibility of a perpetual motion machine: interpretive possibility can only decrease with each transfer of constraints from one medium to another.

This also means that, irrespective of the amount of Shannon information that can be embodied in a particular substrate, what it can and cannot be about also depends on the specific details of the medium's modifiability and its capacity to modify other systems. We create instruments (signal receivers) whose states are affected by the physical state of some process that we wish to monitor and use the resulting changes of the instrument to extract information about that phenomenon, by virtue of its special sensitivities to its physical context. The information it provides is thus limited by the instrument's material properties, which is why the creation of new kinds of scientific instruments can produce more information about the same objects. The expansion of reference that this provides is implicit in the Shannon-Boltzmann logic. So, while the material limits of our media are a constant source of loss in human information transmission processes, they are not necessarily a serious limitation in the interpretation of natural information sources, such as in scientific investigations. In nature, there is always more Boltzmann entropy embodied in an object or event treated as a sign than current interpretive means can ever capture.

NOISE VERSUS ERROR

One of the clearest indications that information is not just order is provided by the fact that information can be in error. A signal can be corrupted, its reference can be mistaken, and the news it conveys can be irrelevant. These three normative (i.e., evaluative) assessments are also hierarchically dependent upon one another

A normative consideration requires comparison. This isn't surprising since it too involves an interpretation process, and whatever information results is a function of possibilities eliminated. Shannon demonstrated that unreliability in a communication process can be overcome by introducing a specified degree of redundancy into the signal, enabling an interpreter to utilize the correlations among similar components to distinguish signal from noise. For any given degree of noise (signal error) below 100 percent, there is some level of redundant transmission and redundancy checking that can distinguish signal from noise. This is because the only means for assessing accuracy of transmission irrespective of content is self-consistency. If a communication medium includes some degree of intrinsic redundancy, such as involving only English sentences, then errors such as typos are often easy to detect and correct irrespective of the content. Because this process is content-independent, it is even possible to detect errors in encrypted messages before they are decoded. Errors in transmission or encoding that result from sources such as typing errors, transmission errors, or receiving errors will be uncorrelated with each other in each separate transmission, while the specific message-carrying features will be highly correlated from one replica to another.

This logic is not just restricted to human communication. It is even used by cells in cleaning up potentially noisy genetic information, irrespective of its function. This is possible because the genetic code is redundant, such that nucleotides on either side of the double helix molecule must exactly complement one another or the two sides can't fully re-anneal after being separated during decoding. Thus a mechanism able to detect non-complementarity of base pairing can, irrespective of any functional consequence, be evolved to make functional repairs, so long as the damage is not too extensive.

There is a related higher-order logic involved in checking the accuracy of representation. Besides the obvious utility of being able to determine the accuracy of information about something, this issue has important philosophical significance as well. The assessment of referential error has been a non-trivial problem for correspondence and mapping theories of reference since at least the writings of the philosopher David Hume in 1739–40. This is because a correspondence is a correspondence, irrespective of whether it is involved in a representational relationship or is of any significance for any interpretive process. In some degree or other, it is possible to find some correspondence relation between almost any two facts. What matters is the determination of a specific correspondence, and this requires a means for distinguishing accurate correspondence relationships and ignoring spurious ones. The solution to this problem has a logic that is analogous to correcting for signal noise in Shannon's theory.

In everyday experience, we often use the redundancy of interpretive consequences as a means for detecting representational error whenever this is available. In practical terms, this is the widely employed method of fact checking. Comparing multiple independent reports of the same event can help to reduce interpretive error. For example, multiple witnesses to a crime who may have only observed some of the relevant events, and who may have poor memories of the details, or who may be withholding or falsifying evidence, can provide accounts that can be compared and cross-checked to reconstruct the most probable course of events. Those accounts that have concordant reference are taken to provide the most likely and most accurate representations of what occurred.

This is also the essence of the method employed in the empirical sciences. When an independent researcher replicates the results of another researcher's experiments, it reinforces confidence in the prior claim. To do this, the second researcher provisionally assumes the accuracy of the prior claim and operates accordingly. There is an interesting asymmetry to this process. Failure to replicate prior results can lead to serious theoretical revisions, while discovering consistency between results is only a minimal guarantee of the correctness of a theory. Using many independent methods, analogous to obtaining reports from multiple independent witnesses, and finding consistent results between them all, is even more convincing. This is why developing new tools for investigating the same empirical phenomenon in different ways provides for a considerable increase in the representational confidence that is generated.

The logic of fact checking differs in an important respect from introducing signal redundancy, to distinguish signal from noise in the Shannonian sense. This is because it actually involves a means for *increasing* the potential entropy of the signal, not decreasing it, as happens when signal redundancy is increased. To detect representational error in this way, it is necessary to compare different and to some extent independent sources of representational information, and to instead take advantage of their otherwise *uncorrelated* diversity to overcome error. Each source of information will have its own idiosyncrasies to contribute, analogous to noise, but all will share in common being generated with respect to, and under the influence of, the same extrinsic events. So correcting representational error entails both an increase in the entropy of the signal—e.g., by multiplying interpretive processes—and taking advantage of redundancies in the constraints imposed on these processes by the common object of interest which is the extrinsic factor they all share in common.

This is a higher-order variant on the same theme, because it effectively treats each interpretive process as though it is a replica of the signal to be compared to the original, and relies on the probability of independent sources of variation to highlight redundancies of reference. Alternative interpretations that share prominent features in common despite being independently generated are likely to have been influenced by the same extrinsic cause. Unfortunately, assuming a common source of redundant constraints in independent interpretations is never an infallible inference, because it involves an assessment of similarity and difference, which is itself an interpretive process. There can be many reasons for not detecting difference, particularly when the Shannon entropy—reflected in the number, complexity, and diversity of interpretive sources—is not large.

To explore this more carefully, consider the example of the detective who compares many sources of information and uses their correlations to infer a common event, which they may or may not each indicate. Over time, as more interpretive techniques have become available for this purpose—such as DNA sequencing, materials analysis, and trace-elements detection—both the effective signal entropy and the sources of interpretive redundancy available to law enforcement have increased, with an attendant increase in interpretive confidence. The detective's problem, or that of a jury listening to a welter of potentially untrustworthy evidence, is to reduce the uncertainty of interpretation—to get at the "truth." They must generate an interpretive response to the whole ensemble of sources of evidence and counterevidence that best corresponds with what actually occurred beyond direct observation. The consistency (redundancy) and inconsistency (non-redundancy) of the evidence is not itself a guarantee that a given interpretation is accurate. Faced with the problem of comparing alternative interpretations of the same events, one is often forced to analyze other features of the source of the information to determine if there are systematic biases that might be introducing spurious or intentionally skewed levels of redundancy. Creating the false appearance of independent sources of information is, for example, a major tool employed in propaganda and confidence schemes.

Error checking is most effective when the interpretive challenge involves objects or states of affairs that remain available for subsequent exploration. In these cases, the redundancy analysis can be proactive and fairly straightforward. As we have already noted, testing a particular scientific hypothesis involves behaving as though a given trial interpretation (the initial report) is accurate, and observing the consequences of continuing to act in accordance with that interpretation (attempting to replicate it) to see if the consequences remain consistent with it. By acting in accord with a given interpretation, causal consequences of this process can be generated to act as virtual new interpretations, each of which can be compared. This approach is also relevant for criminal investigation. So, for example, on suspicion that a given business is corrupt, a law enforcement agency might set up a sting operation that will proceed as though their suspicion is true, and observe the consequences. If the hypothesis is true, the actions of the perpetrators will parallel the prediction.

The concept of producing actions to test for interpretive error was also hinted at in Bateson's aphorism about information, and again involves the performance of work: acting to change circumstances to produce predictable results. All of these approaches to the problem of representational error checking reinforce the claim that interpretation is a dynamical process, which inevitably involves the generation of new information in the form of new signals and new interactions that do work with respect to those that were generated previously. Although this shifts our analysis upwards to a higher order of the information generation process, the same core logic that we have seen at work in Shannon's classic analysis still applies: the information conveyed is determined with respect to the alternatives eliminated—whether about reliability of the signal, reliability of reference, or reliability of interpretation.

DARWINIAN INFORMATION

In many respects, this process of error detection is crudely analogous to the logic of natural selection, with an hypothesis as the analogue of a variant phenotype, and the selective exclusion of certain of these based on their non-concordance with others as the analogue of selection. Indeed, many theorists have compared scientific research and other truth-seeking enterprises to Darwinian processes,[3] and a number of contemporary philosophers have developed theories of function and mental content based roughly on the logic of natural selection.[4] But a number of problems with these approaches have been uncovered, mostly having to do with information only being defined with respect to past conditions, not current conditions. There are also questions about whether these processes can account for and detect error.[5] To sort out these problems of the interpretive generation of information, then, it may be helpful to consider analogous problems posed by evolutionary theory.

In the standard Darwinian account of evolution by natural selection, many individual organisms with variant forms constitute a pool of options from which only a subset is able to successfully reproduce, to pass on its characteristics to the next generation. This subset succeeds because of its comparatively better fittedness to prevailing environmental conditions; and as a result of genetic inheritance, the new pool of variant individuals that is produced inherits features from the parent generation which functioned best in that environment.

By analogy to Shannon's model of the transmission of information, the initial variety of genotype and phenotype forms in the prior generation provides a measure of the potential entropy of the lineage, and the extent of the reduction in "transmitted" forms due to differential reproduction and elimination provides a measure of the information generated by that process. So, in theory, one should be able to quantify this entropy reduction for a given population of organisms for a given number of generations, and estimate the amount of Shannon information produced per time in the evolution of that lineage. It is this parallelism that warrants talking about evolution in informational terms, and ultimately for describing evolution as a process that produces new information. But what is it information about? (Figure 13.1)

In the general case, reduction in signal entropy is evidence of work imposed on the signal medium from some extrinsic source, and the basis of what that signal can be about. Can we analogously conclude that "genetic" and/or phenotypic entropy is about the environment of a given lineage? In biological evolution, however, the outside source of influence may not be the environment within which this process takes place. This may be a passive context, a boundary condition such as available sunlight, ambient temperature, humidity, and so on.[6] So the work performed to "make a difference" in the generation of this Darwinian information cannot be assumed to be extrinsic to the organism. It is instead mostly intrinsic to the source of the "signal"—the work required of organisms in order to persist and reproduce. Constant work is required to maintain the far-from-equilibrium conditions that characterize life. This makes organisms constantly dependent on successfully extracting resources from their environment. But such dependency also makes them selectively sensitive to the availability of these extrinsic resources. So the information that organisms transmit to future generations will reflect the relative fit between the specific forms of work

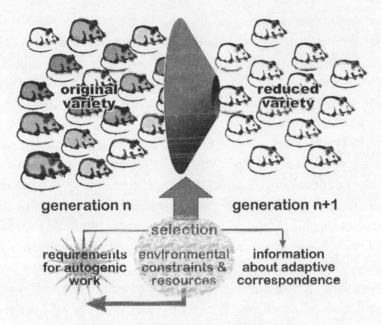

FIGURE 13.1: A cartoon depiction of the process of natural selection, showing its parallel with the logic of information generation. However, in this case the work responsible for reducing the variety is generated not by an outside influence but by the individual organisms within the population. Each organism's teleodynamic work is ultimately responsible for utilizing the constraints of the organism-environment dependency. The variety of organism traits in succeeding generations is thus reduced in ways better fitted to the environment.

they tend to engage in, the inherited constraints that make this possible, and the features in the environment that are critical to this process. In this way, the constraints implicit in this organism-environment relationship can become represented in the selective preservation of some living dynamics and not others.

This inversion of the locus of work and source of constraint in the Shannon-Boltzmann information relationship is also a characteristic shared by many scientific instruments that serve as detectors. By incessantly generating a far-from-equilibrium process, a device's intrinsic instabilities do work that can be used to exemplify their highly sensitive reactivity to certain contextual factors. A process that must continuously do work to maintain an unstable state requires specific conditions in its environment; its state can therefore be used as an indication of the presence, absence, or change of these conditions. A change in these conditions can thus generate a large difference in signal constraint because the far-from-equilibrium state of the device affords a significant potential to do work. So it can amplify even a tiny difference in some critical parameter into a large difference in the dynamics of the signal medium.

Though highly specific sensitivity reduces the diversity of what can thereby be conveyed by a change in this dynamic signal, it can instead provide high precision of reference. The canary in the mine can tell the miner that although he is not yet gasping for air, that possibility is incrementally close. Similarly, a metal detector is highly sensitive to the presence or absence of an object capable of being attracted to a magnet, but little else, because only conductive metals can disrupt the detector's constantly generated magnetic field. Despite the low entropy of the signal, this specificity is what the treasure hunter or minesweeper wants. While this is intuitively obvious, it often gets ignored in technical discussions that do not distinguish these two levels of information. It again demonstrates the difference between assessments of information and entropy at these different levels of analysis.

Because an organism is the locus of the work that is responsible for generating the constraints that constitute information about its world, what this information can be about is highly limited, specific, and self-centered. Like the treasure hunter with his metal detector, an organism can only obtain information about its environment that its internally generated dynamic processes are sensitive to—von Uexküll's *Umwelt*, the constellation of self-centered species-relevant features of the world. For the most part, these are statistical features and general tendencies, like the average density of food resources, the probability of predation, or the cyclicity of the seasons, and specific molecular or microdynamic features of the immediate physical context. They are typically either directly or indirectly relevant boundary conditions, which support the far-from-equilibrium processes constituting the various components of the organism's metabolism with a variety and indirectness of effect that correlates with organism complexity. So, although the information embodied in organism adaptations can be understood to be *about* its environment, this aboutness is always and only *with respect to* the constraints of organism dynamics, and not with respect to just any arbitrary properties of things.

FROM NOISE TO SIGNAL

Before we can turn our attention to the implications of this analysis of information and work for understanding the evolution of life and mind, we need to consider one further feature exemplified by evolutionary processes: their ability to create new forms of information. This is due to a fundamental difference between the abstract logic of communication theory and the evolutionary process: the shifting status of signal versus noise in evolution. If we liken the transmission of traits from generation to generation via reproduction to signal transmission over a communication channel, then mutation in biology is the analogue of noise introduced into a communication channel. In most communication processes, noise is a nuisance. It degrades the information by introducing new, uncorrelated entropy into the signal, and this increases uncertainty about what is signal and what is not, thereby potentially corrupting the message. But whereas the introduction of noise decreases the potential Shannon information capacity of a channel, it paradoxically increases the capacity for reference, because it increases total Shannon entropy. It is as though an additional information channel is available, because noise is also a consequence of the openness of the physical system that is being used as a sign medium, and so it too reflects some source of signal modification besides that which the sender provides. Of course, noise is just noise if you are only interested in what was originally sent, and not in the cause of the degradation of that signal. And yet, this normative decision depends on the interpretation process. Noise can be signal to a repairman.

From the perspective of Shannon information, noise is a source of equivocation or ambiguity in the signal. A noisy signal, like a text containing typos, contains signals replaced by uncorrelated alternatives. Shannon's analysis showed that it is possible to compensate for equivocation between signal and non-signal if the transmission and interpretation processes can take advantage of signal constraints or redundancies. In the evolution of adaptive phenotypes, however, there is no such shared expectation to go on. Understanding how this is accomplished when there is no context of introduced redundancy to rely upon is the critical clue to explaining how evolution ultimately transforms noise into signal.

What if there is no information available in the signal to help discern transmitted from randomly substituted bits? Consider the case of a set of instructions in which there are word substitution errors, but in which there is no violation of meaning, spelling, or grammar to indicate that it is inappropriate. This sometimes happens with foreign-made devices which come with assembly or use instructions that have been poorly translated from an unfamiliar language. In these circumstances, we often provisionally assume that the instructions are accurate and attempt to accomplish the task described. If in the process we find that something doesn't work out as described, we may suspect error in the instructions, and careful attention to the task described can often provide clues to the locus of this error.

This trial-and-error approach is also a form of hypothesis testing, as described above. The redundancy being relied upon is between the referential information in the communication and the constraints of the application context. If the information accurately represents features of some physical system (e.g., the instructions about operating some mechanical device), its interpretation in terms of the actions performed on (or interactions with) that system will correlate well with physical constraints required to achieve a given expected result. Thus the reference of a sign or signal is also susceptible to error correction via redundancy at a higher level than signal organization. The reference of a signal implicitly makes a prediction about certain extrinsic causal possibilities. Physical or logical interactions with these extrinsic conditions will be constrained either to conform or not conform to this prediction, and if not, this will disconfirm the represented state.

So the logic of natural selection is analogous in many ways to a trial-and-error process, except that in natural selection there is no extrinsic source of representation to check against. Success or failure to reproduce is all that distinguishes representational accuracy of the information embodied in the genotype and phenotype. But reproduction allows for further iterative testing of these interpretive consequences: the succeeding generations effectively stand in for the outside source of comparison necessary for error checking. If one's genetic inheritance contributes to producing a body with appropriate adaptations, it is because the constraints it embodies are in some degree of correspondence with constraints of the environment. Unfortunately, dead men tell no tales, as the cruel aphorism suggests. So it would seem that there is no recording of the many failures-to-correspond; no independent representation of which were the errors and which were not.

What counts as useful genetic information in biological evolution is determined *after the fact* with respect to its ability to pass through the functional error correction mechanism of natural selection. The significance of the inheritance signal is both "tested" and refined by the way the far-from-equilibrium dynamics of organism development, maintenance, and reproduction conform to environmental constraints and opportunities. Evolution is thus a generator of information for the organism and a process that rectifies this information with its reference. Thus, although the evolutionary process is itself non-normative (i.e., is not intrinsically directed toward a goal), it produces organisms which are capable of making normative assessments of the information they receive. In this sense, evolution generates and rectifies referential information. In this way the evolutionary process can progressively increase the functional correspondences between genetics, organism dynamics, and contextual preconditions. To the extent that the constraints of this dynamics conform to environmental constraints that are consistent with its continuation, these intrinsic constraints embody this correspondence in the ongoing dynamics. In this way, natural selection exemplifies the logic of error correction.

Students of evolution have not usually insisted that the absence of the lineages that go extinct is what determines the functionality of the traits that persist. One could see the surviving lineages and their adaptations through the lens of engineering design in terms of identified functions that were designed to achieve a previously specified purpose. But although this analogy has a superficial attractiveness, it is undermined by the fact that few if any biological structures can be said to have only one distinguishing function. Their fittedness, internally and externally, is irreducibly systemic, because adaptations are the remainders of a larger cohort of variants selected with respect to one another and their environmental context. There is no simple mapping of genetic-phenotypic information and adaptive function. That which constitutes the reference of the inherited information is ultimately defined only negatively (i.e., by constraint). Biological function is not, then, positively constructed, but is rather the evolutionary remainder that occupies the constrained space of functional correlations that have not been eliminated. This is the basis for novel functions to emerge in evolution, as well as the possibility for evolutionary *exaptation*—the shift from one adaptive function to another. In this respect, genetic information is neither merely retrospective—about successful adaptation in the past—nor does it anticipate future novel adaptations. It is not an aspect of a static relationship, but emerges in process, as its interpretive consequences perform work that may or may not turn out to support this process continuing.

INFORMATION EMERGING

Although the account so far has been framed in terms of the information involved in biological evolution, this model is generalizable to other domains. The nested dependencies of the three levels of entropy reduction—here characterized by Shannon's, Boltzmann's, and Darwin's variations on this theme of entropy reduction—define a recursive architecture that demonstrates three hierarchically nested notions of information. These three very roughly parallel the classic hierarchic distinctions between syntax (Shannon), semantics (add Boltzmann), and pragmatics (add Darwin). They also roughly parallel the relationship between data, content, and significance, though to understand how these semiotic levels are interrelated, we must carry this analysis out of the realm of biology and into the domain of communication (Figure 13.2).

The appeal to Darwinian selection as the ultimate mechanism for the generation of new information relationships might suggest that we should take a strictly etiological view of information. In other words, we might be tempted to argue that information is only discernable *post hoc*, after selection, and after entropy reduction. But this is misleading, both for biology and for information relationships in general. As with biological function, the specific selection history of a given representational capacity may be necessary to explain present usage, but past correspondences are not what it is currently about. Past correspondences have improved the chances for reliable and precise predictive correspondence, but it is a relationship to the present condition that matters. The very fact that information and noise are not intrinsically distinguished, and that mutational noise can become biological information in the course of evolution, exemplifies this property.

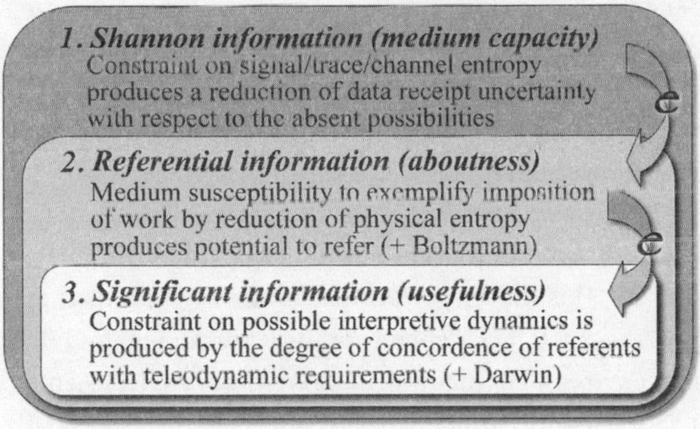

> **1. Shannon information (medium capacity)**
> Constraint on signal/trace/channel entropy
> produces a reduction of data receipt uncertainty
> with respect to the absent possibilities
>
> **2. Referential information (aboutness)**
> Medium susceptibility to exemplify imposition
> of work by reduction of physical entropy
> produces potential to refer (+ Boltzmann)
>
> **3. Significant information (usefulness)**
> Constraint on possible interpretive dynamics is
> produced by the degree of concordance of referents
> with teleodynamic requirements (+ Darwin)

FIGURE 13.2: Three nested conceptions of information. *Shannon* information is the most minimal and the most basic. *Referential* information is emergent (*e*) from Shannon information, and *significant*—or useful—information is emergent from referential information.

This is the problem with simple etiological explanations of adaptive function and representation, which treat information and function as retrospectively determined by their selection history. Because information-generating processes emerge in systems constituted by a pragmatic selection history, the ground of the correspondence between information and context is determined negatively, so to speak, by virtue of possible correspondences that have been eliminated, but it leaves open the issue of correspondences never presented. No *specific* correspondence is embodied with full precision and present correspondence is not guaranteed. With functional correspondence underdetermined, novel functions can arise *de novo* in unprecedented contexts, and incidental properties of the sign or signal may come to serendipitously serve emergent functions. In short, while the possibility of information generation and interpretation depends on a specific physical selection history, the present influence of this information on the persistence of the system that enables it may be serendipitously unrelated to this history. This is the basis for the evolution of new function, but it is also why information is always potentially fallible.

The evolutionary process is not, however, a normative process. Conditions can be good or bad for an organism, or for life in general; an organism's responses to the world can be effective or ineffective in achieving its intrinsic ends, and its adaptational dynamics can accurately or inaccurately link changes in organism activity with changes in extrinsic conditions contributing to the persistence of this dynamical organization; but evolution just occurs. So, although the evolutionary process can further the pragmatic convergence between interpreted content and extrinsic reference, information is not in any sense available *to* evolution, only to the organisms that are its products. Evolution generates the capacity to interpret something *as* information. This capacity is intrinsic to a self-perpetuating, far-from-equilibrium system, which depends on its environment and does work to modify that environment in a way that reinforces its persistence. Information is a relational property defined with respect to this persistently unstable dynamical regularity; or, as the philosopher Charles Sanders Peirce would have said, with respect to a "habit"[7]—understood in its most generic sense; specifically, a self-perpetuating self-rectifying habit.

So genetic information is about cellular chemical-reaction possibilities, their roles in constituting the organism, and how this relationship between genes and their effects also correlated with extrinsic conditions that supported the maintenance of these possibilities in the past. It is information about organism design and function because it introduces critical constraints into the non-equilibrium processes which may ultimately contribute to the perpetuation of that relationship. It is interpreted by the persistence of the self-perpetuating process that it contributes to. But although it is not information about the present world—only extrinsic signs can fill this role—the ability to make use of environmental features as information nevertheless depends upon this closed interpretation of genetic information for the ability to obtain information from extrinsic sources.

This ability to use extrinsically generated events and objects as information derives from the special dynamics of living processes. Because organisms are constituted by specially organized, persistent, far-from-equilibrium processes, they are intrinsically incomplete. In this regard, they are processes organized around absence. Not only are biological adaptations evolved and defined with respect to features of the world extrinsic to the organism, but in many respects these are only potential features which may also be absent from the current environment. Thus adaptations of an organism that have to deal with unusual conditions, like high altitude or extremes of heat or cold, may never be expressed in a lifetime. For this reason, the maintenance of intrinsically unstable, far-from equilibrium conditions entails mechanisms that effectively anticipate the possible variations of environmental conditions by simply not excluding them. They do so with respect to a living process that is at the same time incessantly asymmetrically directed contrary to high probability states in multiple ways: they do work (a) to maintain their far-from-equilibrium state (which supports persistence of the ability to do work); (b) to generate specific organic forms (i.e., they constrain dynamical processes and generate structures which have highly constrained, low-probability features); and (c) to achieve the specific outcome of maintaining themselves long enough to reproduce the global organization supporting processes a, b, and c. So, with respect to these three improbably asymmetric dynamics, there are many critical extrinsic factors that may be relevant. This combination of absence and necessary relevance to an asymmetric process, incessantly interacting with and modifying the world, is what projects the property of information into otherwise merely physical states and events.

Consider a non-mentalistic example: a deciduous tree which alters its metabolism in response to decreasing day length and cooling temperatures in the early months of autumn, resulting in the eventual withdrawal of metabolic support for its leaves, so that they dry up and eventually become severed from the branches they grew from. This adaptation to the difficulties of winter involves a mechanism which treats these environmental changes as information about likely future events that would have an impact on survival and effective reproduction. Insofar as this response has, in previous generations, resulted in persistence of the lineage compared to others lacking it, the mechanism has acquired interpretive reliability. The reliability of the seasonal changes in these factors provides constrained variation to which the constraints of the tree's metabolic mechanisms have become tuned. But it is not merely these correlations that constitute the informational property of these seasonal changes for the tree. The day length and mean temperatures are also correlated, but one is not intrinsically information about the other nor about the change of season. It is only with respect to the end-directed improbable dynamics of the tree's metabolic processes that one or the other of these is informative—and specifically informative about boundary conditions potentially affecting that dynamics.

At one point, I worked in an office near a number of trees of the same species that had been planted as part of the landscape design for the campus. A few of these trees, which were planted close to an automated streetlamp and next to the exhaust from the building's ventilation system, always were very late to change the color of their leaves and drop them, compared to the others. On the one hand, one might argue that these few trees were misinterpreting these artificial signs, because they don't accurately represent seasonal changes. On the other hand, to the extent that these artificial conditions were nevertheless reliably predictive of local factors affecting the trees' metabolism, one would be justified in arguing that the interpretation was correct, because it promoted the dynamical outcome by virtue of which the mechanism exists. This shifts the focus from the evolved function to the immediate incremental consequence of the evolved mechanism as the ground for referential information. *The evolved mechanism constrains the dynamics of possible interpretation, but doesn't determine it.* Each moment of interpretation is in some way supportive or disruptive of the self-maintenance of this dynamical trend. This means that not only is there an historical origin for the normative property of this interpretive process, there is also an ahistorical and immediately efficacious normative property as well. And this need not be consistent with its evolved function. In fact, this possibility is a necessary condition for evolution, since essentially every adaptation has evolved from prior forms and mechanisms that often served very different adaptive functions (such as feathers originally evolving as a form of insulation).

Function and representation are made possible by the way living processes are intrinsically organized around absent and extrinsic factors, and the Darwinian process inevitably generates increasingly convoluted forms of dependency on such internal absences. Information is a relational property that emerges from nested layers of constraint: constraints of signal probability (Shannon), constraints of the dynamics of signal generation (Boltzmann), and the constraints required for self-maintaining, far-from-equilibrium, end-directed dynamics (Darwin). Because information is a relationship among levels of constraint generated by intrinsically unstable physical processes, it is also normative with respect to those processes. But constraint is a negative property, and thus neither something intrinsic nor determinate. This means it is intrinsically incomplete and fallible. Yet it is these very properties that make it evolvable and indefinitely refinable.

REPRESENTATION

We can conclude that a representational relationship cannot be vested in any object or structure or sign vehicle. It is not reducible to any specific physical distinction, nor is it fully constituted by a correspondence relationship. But neither is it a primitive unanalyzable property of minds. Instead, even simple functional and representational relationships emerge from a nested interdependence of generative processes that are distinctive only insofar as they embody specific absences in their dynamics and their relationships to one another. These absences embody, *in the negative*, the constraints imposed on the physical substrates of signals, thoughts, and communications which can be transferred from one substrate to another, and which thereby play efficacious roles in the world as inherited constraints on what tends to occur, rather than acting as pushes or pulls forcing events in one direction or another. Constraints don't *do* work, but they are the scaffolding upon which the capacity to do work depends.

This is only the barest outline of an information theory that is sufficient to account for some of the most basic features of functional and representational relationships, so it cannot be expected to span the entire gap from biological function to conscious agency. But considering that even very elementary accounts of biological function and representation are currently little more than analogies to man-made machines and human communications, even a general schema that offers a constructive rather than a merely descriptive analogical approach is an important advance.

In this exploration of the relationship between information theory, thermodynamics, and natural selection, we have unpacked some of the unrecognized complexity hidden within the concept of information. By generalizing the insight captured by Claude Shannon's equation of information with entropy reduction and constraint propagation, and tracing its linkage to analogues in thermodynamic and evolutionary domains, we have been able to address some of the most vexing issues of representation, reference, and normativity (i.e., usefulness). By removing these inadequacies in current definitions of information, we may at last overcome the seemingly insurmountable obstacles to formulating a theory of representation that is sufficiently rich to serve as the basis for biology and the cognitive neurosciences, and sufficiently grounded in physics to explain representational fallibility, error checking, information creation, and the relationship between informational and energetic processes.

14

EVOLUTION

Natural selection disposes what
self-organization proposes.

—STANLEY SALTHE, 2009[1]

NATURAL ELIMINATION

The term *evolution* literally refers to an unrolling process, like the unrolling of a scroll to reveal its contents. Although it can be used in a generic sense to describe a process that develops in a particular direction, as in the "evolution of a chemical reaction" or "stellar evolution" (changes that occur during the lifetime of a star), since the time of Darwin the term has become closely associated with the biological process by which living species have come to differentiate and diversify over geological time. In biological discussions, its meaning is assumed to be synonymous with natural selection and related processes such as genetic drift; and in non-technical discussions, it is generally assumed to refer to some version of the Darwinian perspective on the origin of species. The Darwinian connotation is helpful in distinguishing what amounts to a negative or subtractive conception of change, in which certain forms are progressively eliminated or culled, from a positive conception such as we see in design processes where new modifications are added. In the standard model, both the preservation of the "more fit" and the elimination of the "less fit" are understood as the result of surviving to reproduce, or not, respectively.

The processes underlying biological evolution are not, of course, limited to a subtractive effect. Even natural selection in its simplest form requires the production of variations of form and multiplication of offspring from which the most successful are preserved. In most traditional accounts of Darwinism, the source of novel variations is considered to be the primary positive factor. The introduction of novel variation into the process, according to strict neo-Darwinism, however, is presumed to be the result of a form of damage—genetic mutational—which is essentially an expression of the second law of thermodynamics, and thus an order-destroying effect. But, as the epigraph to this chapter hints, there may be a source of increasing orderliness available as well: self-organizing processes. The recognition that there needs to be such a "positive" (order-introducing) factor, and not merely a multiplicative factor, in order to explain biological evolution is becoming more widespread. This requirement is echoed also by Peter Corning, who argues that "a fully adequate theory of evolution must encompass both self-organization and selection."[2] A failure to recognize this need has been the source of persistent theoretical problems for evolutionary theory. So precisely demonstrating how self-organization and selection processes are functionally intertwined may help resolve some of these riddles.

It might be more accurate to say that natural selection theory is explicit about its subtractive aspect but agnostic about its additive aspect. Although natural selection offers a powerful logic that can account for the way organisms have evolved to fit their surroundings, it leaves out nearly all of the mechanistic detail of the processes involved in generating organisms, their parts, and their offspring. Indeed, it is one of the virtues of this theory that it is entirely agnostic about the specific mechanisms responsible for the growth and regeneration of structures and functions, for reproducing individual organisms, for passing on genetic information, and for explaining the many possible sources of variation that affect this process. But it is precisely the process of generating physical bodies and maintaining metabolism that constitutes the coin of the natural selection economy. Variations do not exist in the abstract; they are always variations of some organism structure or process or their outcome. Organisms must compete for resources to build their parts and to maintain the far-from-equilibrium dynamics, which is the source of this self-maintenance and reproduction. It is the efficiency and context-appropriateness of these processes which determine the differential reproduction that determines what persists and does not. Or to put this more simply, what natural selection eliminates is determined with respect to the effectiveness of what gets generated in a given context.

This decoupling of function from the processes responsible for its origination means that the specific mechanisms involved don't matter. Only the consequence matters, irrespective of how it was achieved. So natural selection is a classic case of the ends justifying the means, by preserving them, retrospectively. This decoupling from specific substrates and mechanisms was the source of Darwin's most revolutionary insight: that adaptation in biology only becomes realized *ex post facto*. It helped him to recognize that variations of structure and function that arise by accident can nevertheless be functional. More important, this decoupling of consequence from cause allows for the widest possible diversity of mechanisms to be available for recruitment to serve a given adaptive function. For example, the means by which flight, photosynthesis, or thought can be accomplished in a given organism is irrelevant so long as it produces useful results. Whether flight is achieved by the fluff of dandelion seeds caught by a light breeze, the leathery sheets of Pterosaur or bat wings, or the lightness and wind resistance of feathers, what counts is getting airborne. Even though these functions depend upon underlying mechanisms that converge on certain specific mechanical constraints, it is the realization of these constraints and not their specific embodiment that matters.

In this way, natural selection is a process defined by multiple realizability, and is the paradigm exemplar for defining functionalism. This complete openness to substrate and mechanism is what opens the Pandora's box of evolvability, and makes the explosive creativity of life and mind possible. So it is with the emergence of the process of natural selection that true functional generality comes into the world. Adaptations and organisms are in this sense general types that exist irrespective of the specific details of their embodiment.

The vast power of evolvability thus is a consequence of the fact that *natural selection is a process that regularly transforms incidental physical properties into functional attributes*. An adaptation is the realization of a set of constraints on candidate mechanisms, and so long as those constraints are maintained, other features are arbitrary. But this means that with every adaptation, there are innumerable other arbitrary properties potentially brought into play. Although a given structure or process must be embodied by specific substrates with innumerable properties that are incidental to current adaptive usefulness—for example, the sound of the heartbeat or the redness of blood—any of these incidental properties may at any point themselves become substrates for selection, and thus functional or dysfunctional properties. Precisely because the logic of natural selection is mechanism-neutral, it also is minimally restricted in what kinds of properties can be recruited, except that they must be immediately usable given the constraints of organic chemistry (which is why, say, nuclear energy has never been recruited by natural selection to serve organism functions). The relevance of this for emergence arguments should be obvious. This capacity to transform the incidental and accidental into the significant and indispensable radically minimizes the causal role of any specific intrinsic physical tendency or immediate antecedent condition in determining what will or will not be the physical substrate for an evolved function.

Because an adaptation is mechanism-neutral, it is a bit like an algorithm that can be implemented on diverse machines, and in this sense it has something of the character of a description. It is a general, in philosophical terms. But an adaptation is not identical with the collection of properties constituting the specific mechanism that embodies it. It is far less than these! Only a very small subset of physical attributes of a given physical implementation of some function, such as oxygen binding, are ever relevant to the success of that adaptation. An adaptive mechanism is also something beyond any of its properties as well. It is the consequence of an extended history of constraints being passed forward from generation to generation—constraints that have perhaps been embodied in many different substrates along the way. Like the process of addition that can at various times be embodied in finger movements, the shifting of beads on an abacus, or the changing distributions of electric charge in the memory registers of a computer, the specific physical links in this chain have become incidental. It is only the transmission of conserved constraints that is critical, and the constraints are not the stuff. Of course, these constraints must be embodied in some specific physical substrate at every step, and the transfer from substrate to substrate must always be mediated by a specific physical mechanism that precisely determines the material overlap and the work of transferring these constraints. But it is the conservation of constraint, not of energy, material, or specific mechanism, that matters.

Precisely because constraint is not something positive, but is rather something *not* realized, only some of the physical details of the mechanisms recruited as adaptations are of functional relevance, even though all are part of the physical processes involved. Only those functionally relevant physical attributes—ones that guarantee the preservation of these constraints—are likely to contribute to future generations' body composition. Other physical attributes are susceptible to eventually being eliminated if the same constraints can be transferred to other substrates, or may just get degraded by variations accumulated over the generations. It is in this sense that we are justified in arguing that life's mechanisms are general types.

The power of evolutionary theory to explain much of the complexity and diversity of biological forms derives from this physical abstractness. This decoupling of material generative processes from functional consequence is responsible for the persistence of an adaptation across many generations, but also what allows for the transfer of function to progressively more suitable substrates. In a more mundane sense, it is also responsible for one of the most basic attributes of life: the successful bottom-up development of a vastly complex and well-integrated organism, regulated by a comparatively small set of constraints passed on in the genes and cytoplasm received from a parent organism.

Because the intrinsic material components and dynamical properties that constitute a given adaptation are not essential, but are determined by constraints on properties, not some finite set of specifically relevant properties, we need to think about the origin of organic mechanisms quite differently than designed mechanisms. In the evolutionary history leading up to a given form of adaptation, those constraints on materials and dynamics that were retained were merely those that were not eliminated. As scientists and engineers, we tend to focus on the properties that we discern to be most relevant to our abstract sense of a given function; but life is only dependent on excluding those that are least helpful.

This is the critical break in causal symmetry that makes the concept of biological function an *emergent general physical property* that is not determined by its lower-level constituent properties, even though we may find it useful to try to list and describe them. Those constituent properties that we biologists may find helpful in understanding and categorizing biological functions were not themselves the targets of "selection"; they represent merely part of the residue of what was not eliminated. Evolution is not imposed design, but progressive constraint.

"LIFE'S SEVERAL POWERS"

So it is essential to the power of natural selection theory that it quite explicitly declines to address how the relevant mechanisms and their variant forms are generated. The Darwinian logic quite correctly treats these issues as separable from the explanation of the determination of adaptation. Evolutionary theorists have therefore been justified in both accepting the plausibility of accidental generation of variant forms (e.g., via "chance" mutation) and rejecting the necessity of actively acquired (e.g., Lamarckian) and functionally biased variations (aka "directed mutations"). These features are neither a problem nor a necessity for the theory because the details of these mechanisms are irrelevant for the functional explanation it provides. In fact, though textbooks often suggest otherwise, Darwin himself was open to many possible sources of form generation and genetic variation, including Lamarckian mechanisms. He just recognized that the logic of natural selection was sufficient irrespective of how these generative, reproductive, and variant consequences were achieved.

Recognizing that natural selection logic is agnostic to these factors doesn't make the problem of mechanism go away, however. As we have stressed throughout much of this book, the problems of how regular dynamical processes and physical forms are generated in the face of the second law of thermodynamics, and how biological information is generated, preserved, and transmitted, are far from trivial components in biological explanation. Natural selection may occur irrespective of the specific details of form generation and reproduction, but it only occurs if these processes are reliably present. *Some* specific mechanisms are required, and the processes that are capable of producing such regularities are limited and must have very specific dynamical properties in order to thwart the ubiquity of local thermodynamic tendencies. We are not, then, freed from answering questions about how specific means of achieving these processes might influence evolution.

Organisms are highly complex systems and, as many critics of natural selection have been fond of pointing out, the number of possible variant configurations of a complicated multipart system like an organism is truly astronomical. Even in the case of the simplest organisms, an undirected, unconstrained process for sorting through the nearly infinite range of combinatorial variants of forms, and testing for an optimal configuration by the trial and error of natural selection, would fail to adequately sample this space of possible variations, even given the lifetime of the universe. In contrast to this vast space of possible combinatorial variants, the process of biological development in any given lineage of organisms is highly constrained, so that only a tiny fraction of the possible variations of mechanism can ever be generated and tested by natural selection. While this limitation may at first appear to impose an added burden on the theory, it turns out to be the most important aid to the solution of this conundrum.

Just as organism "design" by evolution must be understood negatively, development of organism form is also not a construction process, in the sense we might imagine from an engineering point of view. It is the expression of the interaction of many morphodynamic processes. Like the reciprocally end-and-means logic of the morphodynamic processes that generate the "body" of an autogen, a vastly more complex and interwoven fabric of morphodynamic processes is responsible for the generation of an organism body from its initial state following conception. The inherited constraints that we identify as genetic information are specifically constraints that make morphodynamic processes highly probable at the chemical, macromolecular, cellular, and intercellular levels. Evolution in this sense can be thought of as a process of capturing, taming, and integrating diverse morphodynamic processes for the sake of their collective preservation.

Recognizing the morphodynamic origin of organism form is a critical factor in this analysis, because it means that variations appearing at the genetic level only get expressed through a dynamical filter with highly specific tendencies to dampen and amplify constraints. The variant forms that eventually become subject to natural selection have in this sense been vetted by self-organization (as the epigraph to this chapter suggests). Even if the mechanism of the generation of variations is independent of and irrelevant to the consequences of natural selection, this does not mean that variations get expressed irrespective of the systemic integration of the organism. These quite restrictive intrinsic constraints on developmentally expressed variations significantly bias variation to be minimally discordant with respect to existing ontogenetic processes, even if they are generated irrespective of any potential adaptive outcome.

Although these morphogenetic constraints radically restrict the exploration of all possible advantageous variations of form, they nevertheless vastly increase the probability of generating variations with some degree of intra-organism fittedness—that is, concordance with the constraints and synergies that maintain the integrity of organism functions. As a result, natural selection ends up sampling phenotypic options in just a tiny fraction of the possible variation space; but these variants are much more likely to exhibit features that are not too discordant with or disruptive of already existing functions.

The significance of morphogenetic constraints on the generation of variations is only one aspect of a more global set of conditions that must be considered as antecedently relevant to natural selection theory. The importance of the constraints introduced by morphogenetic processes is that they contribute a positive counterpart to the negative logic of natural selection. Because it only provides an *ex post facto* culling influence, natural selection is only so generative as its supply of optional forms is profligate. The constrained generation of variant forms is only the most minimal expression of this presupposition of a positive process in evolution. As James Mark Baldwin argued with respect to "organic selection"—his so-called New Factor in Evolution:

> Natural selection is too often treated as a positive agency. It is not a positive agency; it is entirely negative. It is simply a statement of what occurs when an organism does not have the qualifications necessary to enable it to survive in given conditions of life. . . . So we may say that the means of survival is always an additional question to the negative statement of the operation of natural selection.
>
> The positive qualifications which the organism has arise as congenital variations of a kind which enable the organism to cope with the conditions of life. This is the positive side of Darwinism, as the principle of natural selection is the negative side.[3]

A full specification of "the positive side of Darwinism" requires significantly more than merely a theory of variations, however. It requires organisms and all of the complex properties we ascribe to them (or at least their
surrogates, e.g., in simulations), as well as the associated physical processes
that enable organisms to preserve their critical features and to reproduce.
In general terms, this means unit systems exhibiting such critical properties as self-maintenance in the face of constant thermodynamic breakdown,
continual far-from-equilibrium chemistry, mechanisms sufficient to generate complete replicas of themselves, and sufficient organization to maintain
these properties in the face of significant variations in the conditions of
existence. But if this is a requirement for evolution by natural selection, it
cannot itself have initially arisen by natural selection. Natural selection *pre-
supposes* the existence of a non-homogeneous population of individuated
systems with these thermodynamically rare and complex properties.

Darwin thus appropriately ends his "Origin of Species" with the following majestic sentence:

> There is grandeur in this view of life, with its several powers, hav-
> ing been originally breathed into a few forms or into one; and that,
> whilst this planet has gone cycling on according to the fixed law of
> gravity, from so simple a beginning endless forms most beautiful
> and most wonderful have been, and are being, evolved.

To perhaps state an obvious fact—one that is nevertheless under-
appreciated and glossed over by the majority of texts on biological
evolution—natural selection assumes the existence of processes of persistent non-equilibrium thermodynamics, self-maintenance, reproduction,
and adaptation. It cannot therefore be the complete explanation for their
origins, particularly for the origins of their teleodynamic character. Natural
selection can only improve the fit between these dynamical processes and
the various environmental conditions on which they depend or must defend
against; it cannot generate them.

The first teleodynamic systems emerged, they did not evolve. They emerged
from specific patterns of interdependency that happened to arise between
morphodynamic processes. Previously, we have used the concept of an
autogen to exemplify the requirement for this emergent transition. As we

have seen, self-organization is a distinct and vastly simpler process than natural selection. Morphodynamic processes form the ground on which teleodynamic and thus evolutionary processes have been built. In simple terms, self-organization is the expression of the intrinsic dynamics of a system that get expressed within certain non-equilibrium boundary conditions, whereas natural selection is a function of the organization of a system's internal non-equilibrium dynamics *with respect to* external conditions. In this way, natural selection involves a specific mode of intrinsic self-organization dynamics that maintains itself by virtue of the generation of effects which specifically counter certain external changes of conditions away from what is conducive to the persistence of those intrinsic processes.

ABIOGENESIS

So the first organism wasn't a product of natural evolution. The constellation of processes that we identify with biological evolution ultimately emerged from a kind of proto-evolution, supported by a kind of protolife, that ultimately must trace back to the spontaneous emergence of the first molecular systems capable of some minimal form of evolutionary dynamic. Earlier, it was shown that even a molecular system as simple as an autogen can give rise to a form of natural selection. The emergence of this constellation of properties enabling evolution, even in quite minimal form, marks a fundamental shift in the dynamical organization of the natural world. In the terms we have been developing here, this is a shift from thermodynamic and morphodynamic processes to teleodynamic processes. Wherever in the universe this occurs, it is the emergence of the first and simplest lifelike process in that region. All other teleodynamic processes and teleological relationships to develop in those environments will likely trace their origin to that crucial transition.

For this same reason, the study of the origin of life has a paradoxical status compared to the rest of biology. It violates a crucial and hard-won dogma of biology: the denial of spontaneous generation. From early Greco-Roman times, it was thought that some, if not all, life arose from inanimate matter by way of spontaneous generation. Evidence to support this theory of the genesis of life was seen in the way maggots would emerge spontaneously from rotting beef or molds would form on stale bread or overripe fruits, seemingly without these life forms being intrinsically present in these materials. In 1668, in one of the world's first controlled biological experiments, Francisco Redi challenged the theory by demonstrating that maggots did not emerge from meat when put under glass. But belief that life emerged spontaneously from non-life nevertheless died hard. Indeed, it is ultimately (but only ultimately) a necessary assumption of scientific materialism that the essence of life does not arise from a realm outside of the physical substrates of its constitution. In these first centuries of the European Enlightenment it was, in this respect, the scientifically fashionable alternative to an otherworldly spiritual origin theory, and it was one of the central fascinations of alchemical lore. Alchemical "recipes" for generating life (even homunculi, as we saw in chapter 2) were not uncommon, and considered the mark of truly perfected alchemical methods. In many respects, the central preoccupation of alchemy—transformation of matter from one form to another, including from non-life to life—was the prescientific precursor to emergentism.

Support for spontaneous generation theories of life persisted until, more than two centuries after Redi, it was famously repudiated in 1889 by Louis Pasteur.[4] Pasteur showed that sterilized rotting meat, placed in a flask with an S-shaped neck, never gave rise to maggots so long as outside contamination was prevented. Indeed, the method of sterilization, coupled with hermetic isolation of the sterilized food (e.g., in a tightly sealed canning jar), did far more than prevent new organisms from forming; it also prevented them—even those microbes still unknown to nineteenth-century science— from aiding the breakdown and putrefication of food. In this way, long-term preservation of food might be considered the most significant spin-off of origins-of-life research.

With the negative results of Pasteur's experiments, it quickly became biological dogma that the emergence of life from non-life was effectively not something to worry about. Indeed, the now ubiquitous methods of sterilizing, canning, freezing, bottling, irradiating, and yes, pasteurizing food, depend on it. The truth of the maxim "Only life begets life" is tested untold billions of times in the modern world with each can or bottle of food that is produced and consumed. But of course to accept this fact, unconditionally, leads to a conceptual paradox. Either life has been around forever in a universe without beginning, or else it originates from some other non-physical realm (e.g., of spirit) in which its analogue is somehow preformed, or it emerges from non-living materials, abiogenetically. The first two of these are *reductio ad absurdum* claims in some form or other. And so we are left with the problem of proving the spontaneous generation theorists right, at least in some limited form.

Charles Darwin clearly understood that this problem was outside of his purview when at the end of the *Origin of Species* he describes evolution commencing only after life has been "originally breathed into a few forms or into one." But this hardly set the issue aside. In fact, the debate with respect to spontaneous generation was actually quite intense among Darwinians in the latter part of the nineteenth century. Perhaps the most notable battle lines were exemplified by Alfred Russel Wallace—the co-discoverer of the principle of natural selection—and Thomas Henry Huxley—Darwin's "bulldog," as he was sometimes described. Wallace sided with theorists like the famed neurologist Henry Charlton Bastian, who supported experimentation to demonstrate the spontaneous generation of life from non-life, and who wrote three influential books on the subject in the decades following the publication of the *Origin of Species*. Bastian's interest was in part a function of his medical interest in the basis of microbial disease, but also of his critical stance against supernatural theories of life.

Although Darwin's theory appeared to specifically undermine supernatural sources for the forms and adaptive design of organisms, its silence with respect to the origins of life and the initiation of biological evolution left a gaping hole in the materialist alternative to various forms of vitalism and supernaturalism. Thus, Bastian and many others of the new generation of Darwinians felt that it was important to undertake experimentation to demonstrate that life too had a natural, purely chemical origin. Bastian distinguished this origins-of-microorganism theory from earlier spontaneous generation theories by calling it "archebiosis," referring to the original or ancient creation of life, though he thought that he could experimentally demonstrate that this was a natural process that was likely continuing in the present. He believed that general laws would eventually be discovered that governed the generation of life from inorganic matter, though he readily admitted that it would likely be a very difficult scientific challenge to discover these laws.

Although he was a staunch Darwinian, Bastian's deeper commitment was to materialism and the repudiation of spiritualistic theories. This was one of the most significant metaphysical assumptions expressed in the Darwinian perspective and clearly one of the reasons it recruited both zealous allies and vituperous critics. However, the idea of spontaneous generation also attracted vitalists, who thought that it might be a window into the secret of a hitherto undiscovered vital force, or *élan vital*, unique to life. This issue was rejoined by Huxley, who initially viewed the materialist endeavor in a favorable light, but later became convinced that the generation of life from inorganic matter was so rare and unlikely that it might only have occurred once in the ancient history of the Earth.

Huxley appears to have realized that the continuous generation of new life forms would be a problem for Darwinian theory. Both Darwin and Huxley were strongly committed to something close to a monogenesis approach to biological evolution, in which all life forms are descendant from a single or at most a very few paleo-ancestors. Were life to be continually generated over the entire epoch of life on Earth, the continuity of life forms, their interactions and interdependencies, and their deep commonalities of design would be difficult to explain. Moreover, the principle of natural selection and its corollary, "descent with modification," would be shifted from center stage; no longer the almost exclusive explanation for biological variation and adaptation. Huxley became intensely critical of Bastian's work both for its implications concerning the mechanism of evolution and for the damage that could come to Darwinian theory if it was thought that failure of experiments to demonstrate spontaneous generation of bacteria also undermined the materialist assumptions of evolutionary theory. So, although he recognized the necessity of a theory to explain how living chemistry could emerge from inorganic antecedents—a process he called "abiogenesis"—Huxley took great pains to distinguish this from spontaneous generation, and to highlight its improbability, complexity, and astronomical rarity, hinting that maybe only certain special conditions of the prebiotic Earth could explain it.

The intellectual context that nurtured this second generation of Darwinian theorists was, however, far more expansive than the study of the physical basis of life. The possibility of discovering general principles of nature that account for the abiotic generation of living processes also held the promise of completing the Darwinian assault on the special status of life and mind that seemed to exempt it from physics and chemistry as usual. If one could discover evidence in everyday processes that there was a regular path from complex chemistry directly to simple life, it would close this final explanatory gap. But if this transition was not in some way general, involving laws common to all material processes, and instead was so queer as to have only occurred once in billions of years and never again, its naturalness might be called into question.

In hindsight, although the critics of this post-Darwinian version of spontaneous generation were right in repudiating the claims of spontaneous generation, even for the simplest life forms (which turn out to be immensely complex), and in their insistence that all known life on Earth may have arisen from a common ancestral form, there may nevertheless have been a baby in the unsterilized bathwater of the theories of Bastian and others.[5]

THE REPLICATOR'S NEW CLOTHES

Although the autogenic theory proposed earlier in this book was intended primarily as a model system for elucidating the logic of the emergence of teleodynamic processes, it also offers a partial answer to Bastian's challenge and an approach to the origin of life. It is, however, something less than an exemplar of spontaneous generation, and considerably less than an account of the last universal common ancestor (often identified with the acronym LUCA) of living organisms. This is because autogens are not alive—at least not in any current sense of that concept. They lack persistent non-equilibrium dynamics, diffusible surfaces, genetic information, an autonomously implemented reproductive process, or any way to selectively react to or act upon their environment in any way that is self-supportive. Yet autogenic theory may provide something equally useful: a first building block in a theory that allows us to deduce the origins of these fundamental attributes of organisms. Below, we will see that extrapolating from the logic of autogen evolution, and utilizing the logic of emergent dynamics, it is possible to provide an account of how such protolife forms might give rise to these properties that we recognize as the hallmarks of life. Most important, a principled account of the origins of biological information is a critical step toward demystifying and de-homunculizing our understanding of the relationship between genetic information and the defining property of life.

Autogenic theory provides a glimpse of an elusive law of emergence operative at the dawn of life, exemplifying the emergence of ententional properties from non-equilibrium thermodynamics, and thus a bridge from non living to living processes. The theory also demonstrates why spontaneous generation is so exceedingly rare. This is because the conditions that make it possible are highly precise and at the same time highly atypical of the thermodynamically driven processes that are ubiquitous outside of biology.[6] But once teleodynamic systems capable of natural selection emerged, an unbounded territory opened up. The evolution of life has led to many levels of radical and unprecedented higher-order teleodynamic phenomena, including mental phenomena. But prerequisite to this capacity to evolve level upon level of more complex ententional relationships is an ability to capture the critical dynamical constraints for each lower level of teleodynamics, in a form that allows them to be preserved and transmitted irrespective of the further convolutions of dynamical organizations that may be incorporated during the course of future evolution. If the source of constraints maintaining the core critical dynamics is not somehow itself insulated from these modifications, there can be no solid foundation to build upon. Each new change will modify existing constraints, and the compounding of higher-order teleodynamic constraints on a base of preexisting constraints will be nearly impossible. There need to be separately sequestered constraints embodied in some non-dynamical attribute, which can be preserved unmodified across changes in dynamics, so that earlier dynamical achievement will not be continually undermined.

This is what genetic information provides for living organisms—and much more. It sequesters an independent source of constraints that is partially redundant to that intrinsic to the dynamics of the organism itself. This has two immediate advantages. First, it is a conservative factor. It protects against degradation of evolved adaptive constraints that might occur due to dynamical interactions with unprecedented environmentally derived factors. Second, it is an innovative factor. Its separate material properties provide a basis for indirectly modifying dynamical constraints that are physically independent of the details and limitations of these dynamics.

The molecular informational mechanism constituting genetics that is today ubiquitous to all living organisms was not just an augmentation of the autogenic process; it took it to a whole new level. The conservative effect of embodying dynamical constraint in a separate physical substrate from that doing the work to maintain the organism provided a critical foundation on which evolution could build progressively higher-order teleodynamic processes. As our analysis of the concept of information in previous chapters has demonstrated, however, genetic information cannot be simply identified with a physical substrate or pattern. Information is dependent on the propagation of constraints linking a teleodynamic system and its environmental context. That means that information is not any intrinsic property of the substrate that embodies or conveys these constraints. Although one of the crucial properties of an information-bearing medium is that it can serve as a template for copying and propagating constraints, this simple physical quality is not what defines it. The general theory of information that we explored in the two previous chapters demonstrated that information is identified with the transmission of constraints, exemplified by some physical medium linking a teleodynamic system with its environment. Information does not stand apart from this relationship, nor does it preexist the teleodynamics that it informs. Another way to say this is that teleodynamic organization is primary, and information is a special feature of some teleodynamic processes.

What does this mean for the role of genetic information in the origin of life? Basically, it suggests that genetic information is not primary, but is rather a derived feature of life. Although a molecule that has the unusual property of serving as a template for producing a precise replica of itself is without doubt potentially useful for propagating constraints, the process of structural replication by itself does not constitute information. A DNA molecule outside of an organism does not convey information about anything, and is mostly just sticky goo. And gene sequences transplanted from one sort of organism to another sort are likely to be noise in that new context.[7] So, even if DNA and RNA were abundantly synthesized and replicated in the merely geochemical environment of the early Earth, it would not under those circumstances be information about anything. It is not the template replication that is the basis for the information-conveying capacity of DNA and RNA in organisms; it is the integration of the patterns that they can exhibit into the teleodynamics of the living process that matters.

To state this claim more forcefully, DNA is just another—albeit very useful—adaptation, that itself must have evolved this function by natural

selection. This should not surprise us, because these molecules are remarkably well suited for the purpose. The almost identical bonding energies of
any given adjacent nucleotides, the nearly unlimited size of a given DNA
strand, and the precision of template specificity, among other attributes,
give the impression of a molecule that was honed by natural selection in
response to its information-carrying function. More important, replication
of a molecule like DNA is not essential for either reproduction or evolution,
since something as simple as an autogen can reproduce and evolve. Calling
it the "secret of life" is thus hyperbole. And considering genetic information
to be the defining character of life is also a bit hasty. It is without question
critical to the evolution of all higher-order forms of life, including all that
are currently available to biologists. An account of how biological information emerges from more basic teleodynamic processes is the first step to
explaining the nature of all higher order ententional properties.

This view of information poses a challenge to a widely accepted account
of the origin of life, and of evolution in general. This is the belief that life is
fundamentally just a complex kind of copying process. The most well known
version of this is replicator selection theory. The term *replicator* was coined
for this use by Richard Dawkins in 1976 in his influential book *The Selfish
Gene*. Though Dawkins gave it a name, as we will see, the core assumptions
of this theory—that the essential feature of reproduction is the copying of
template molecules—is in some form or other characteristic of nearly all
modern conceptions of the evolutionary process. Like Darwin's account
of the necessary conditions for natural selection, replicator selection also
begins by assuming some unspecified and highly non-trivial kinds of processes. But unlike Darwin, who refused to speculate about these, replicator
theories often assume that this process is so ubiquitous and uncomplicated
that it can be accepted as a defining attribute of life.

A replicator is something that gets copied. More precisely, this something
is a pattern embodied in a physical substrate. In biological systems, this is
most commonly taken to be DNA or RNA, but Dawkins suggested that
certain cultural artifacts and habits could also be replicators, and coined
the term *memes* to refer to them. Critics of replicator theories have often
argued that the concept is too narrowly reserved for genetic information,
and that other features of organisms must also be included as replicators,
such as the membrane of the cell and many organelles. These are not created

entirely anew in cell division, but physically inherited from the progenitor cell when it divides. In both views, however, whether there is only one kind of replicator or many at different levels, it is copying that is assumed to be the defining principle.

Although it is generally believed that polynucleotide chains like DNA and RNA molecules constitute life's replicators, even by Dawkins' own description, based on a characterization of DNA "replication," these molecules fail the crucial criterion. They do *not* replicate themselves. To be more explicit: polynucleotide A cannot directly produce another exact duplicate of polynucleotide A.[8] Instead, with the assistance of special contextual conditions (e.g., in company of a molecular complex made up of a number of supportive transcription molecules, or within an operating PCR machine, and including critical component molecules as raw materials), polynucleotide molecule A can produce a complementary polynucleotide molecule B, which in turn under the same conditions can produce polynucleotide molecule A.

Whereas viral, bacterial cell, and eukaryotic cell division do produce replicas (with slight variation), this involves the replication of both strands of a DNA molecule to produce two duplicated double strands. On more careful inspection, then, we can see that the idea of a replicator, which Dawkins has identified with DNA, is instead an oversimplified projection of cellular reproduction onto the process called DNA replication. Indeed, we can now easily recognize that polynucleotide "replication" is in fact a special case of what is often described as autocatalysis (also a misnomer for similar reasons). As we noted in our discussions of autogenic chemistry, the chemical process described in both cases might more accurately be called reciprocal catalysis, or even reciprocal indirect catalysis, since it always involves at least two and sometimes more complementary catalysts, each catalyzing another member of the set. So, DNA replication in a PCR machine—ignoring the role of the supportive machinery and mediating molecules—is a two-step, reciprocally catalytic loop.

Perhaps the closest analogue (though not a true example) to what can be called direct molecular replication is found in the special case of prion replication. In this process, there is no new molecular synthesis or lysis involved to form the new molecule. The precursor molecule, the so-called pre-prion protein, is made up of the same component amino acids arranged into the same polymeric sequence as the prion protein. It's just that the three-dimensional conformation of the two proteins is different. Prion formation simply involves the prion molecule binding with the pre-prion molecule in such a way as to cause the latter to deform into the prion conformation, and thus become capable of similarly deforming other pre-prion proteins. Pre-prion proteins require the molecular machinery found in mammalian brains (including the DNA code for this pre-prion sequence) in order to be synthesized. In this respect, prion "replication" does not actually generate any new material. It is merely "damage" done to proteins synthesized in mammalian brains, which subsequently propagate this damage to others.

If we return to the actual case of bacterial cell replication and try to capture its minimal logical structure, we must take into account that it is made up of a complex of reciprocally interdependent molecules. The crucial factor is the complex reciprocities that enable each part to be both end and means in forming the whole integrated organism. Specifying these details is critical to fully account for the process of self-replication. There is no subset of molecules that suffices. Ultimately, all essential components need to get replicated, if for no other reason than to replace those that have become damaged. In this sense, the nucleic acid sequences have no special claim to be *the* replicators. They are merely more central because many other molecular replication processes depend on them.

But assuming for the moment that naked nucleic acids could serve as templates for replicating identical copies, we still wouldn't be any closer to understanding the relationship between replication and genetic information. Although a replicated molecule is literally a re-presentation of its "parent" molecule's form, there is nothing but this form (or its complement) copied over and over. There is no information about something else that is copied in this process, just the molecular structure. If inserted into a living cell, this sequence might be capable of producing a protein product or some other sort of biological effect, as can be achieved by artificially generated DNA sequences, but the best we could say of this is that this sequence is "potential information" (or misinformation). It is the cellular machinery that determines that a DNA sequence has this potential, and the sequence only inherits this potential because of the existence of cells that have used similar molecules to their advantage in the past. Randomly generated DNA sequences are parasitic on this potential in the same way that a randomly mutated gene can be.

There is a significant conceptual gap to be bridged between the replication of DNA and its role as a medium of information. Although Dawkins often speaks as though a given base sequence on a DNA molecule is intrinsically a form of information (so that copying it is transmitting information), it is only information in the minimalistic Shannonian sense. As we have seen, the Shannonian conception of information is an abstraction that only considers the most minimal criteria for the possibility of carrying information. Although no molecular biologist would consider the structure of a DNA molecule to be information were it not for the fact that it contributes to the operation of other cellular-molecular processes, they may still be willing to bracket this from consideration when thinking of evolution. In life, DNA molecules do not provide information about other replicas of themselves, but rather about the molecular dynamics of the cell in relation to its likely environmental milieu. And yet many theories of the origins of life are based on the assumption that molecular replication is a sufficient defining property of living information.

Probably the most influential of the scenarios explaining the origin of life based on molecular replication is known as the RNA-World hypothesis. In this scenario, it is argued that RNA replication is the core process distinguishing the first lifelike process from other chemical processes. This view has grown in influence over recent decades because of discoveries of RNA functions in addition to its role in mediating between DNA sequences and the amino acid sequence specifying a protein. Single-stranded RNA molecules can coil back on themselves to form complex, cross-linked hairpin forms and coils, producing a complex three-dimensional structure. In the form of transfer RNA, this structure plays the critical role in binding to amino acids and aligning them with respect to a messenger RNA molecule within a ribosome. But also due to this structure, some RNA molecules are also capable of catalytic action. More recently, RNA polymer fragments have been found to play a wide variety of regulatory roles in the cell as well. These many diverse functions have suggested that RNA molecules could serve all the essential functions assumed to be requisite for an early life form. RNA can thus serve both a synthetic role and a template role. However, such scenarios treat the template-copying capacity as primary.

But does being a replicable pattern constitute information? Without the elaborate system of molecular machines that transcribes DNA or RNA sequences into amino acid sequences, without the resulting protein functions, and without the evolutionary history, there would be no information about anything in its base sequence structure. There is a special case that makes this obvious: so-called junk DNA. Over the course of the past decades, it has become clear that only a small fraction of the DNA contained in a eukaryotic cell actually codes for a protein or a regulatory function. Although there are reasons to suspect that at least some of this is nevertheless retained for other functions, it is almost certain that vast lengths get replicated in each cell division simply because they are linked to useful sections. These still qualify as replicators in Dawkins' analysis (though he would probably call them passive replicators), but it is less easy to justify calling these sequences information, precisely because they do not play any role in organizing the cellular dynamics that makes their persistence more probable.

This suggests that the information-bearing function of nucleic acids is dependent on their embeddedness in the metabolism of a cell that is adapted to its context. This means that nucleotide information is not primary. It is an adaptation, not the ultimate basis of adaptation. It is *an evolutionarily derived feature, and not a primitive one*. This also suggests that we should be able to trace an evolutionary path from a pre-DNA world to a post-DNA world, from protolife to life. In this respect, autogens don't merely provide an heuristic model of the transition to teleodynamics; they also offer a context in which to investigate how the structure of one molecule (e.g., DNA) can become information about certain patterns of chemical interaction that obtain between other molecules. DNA structure effectively represents, in concrete form, the dynamics of the chemical system that contains it and replicates it. This something that DNA is about is the source of natural selection that maintains the relatively conserved replication of certain sequences as opposed to others. The information function is thus in an evolutionary sense dependent on this prior dynamics, and so is an indirect adaptation for stabilizing the form of this molecular dynamics.

In this way, we may be able to reconstruct the steps from teleodynamics to information that constitutes the most unprecedented feature of life.

AUTOGENIC INTERPRETATION

The dependency of information on involvement in a teleodynamic process can be demonstrated by a slight complexification of the autogen model. In a discussion with two of my colleagues, Chris Southgate and Andrew Robinson, concerning the semiotic status of autogens, they proposed a modification that we all could agree involved a semiotic aspect. They argue that an autogenic system in which its containment is made more fragile by the bonding of relevant substrate molecules to its surface could be considered to respond selectively to information about its environment. Although our discussion concerned a slightly more subtle question (whether autogenic theory can help decide if iconicity or indexicality is more primary), its bearing on the nature biological of information is more illuminating.

If an autogen's containment is disrupted in a context in which the substrates that support autocatalysis are absent or of low concentration, re-enclosure and replication will be unlikely. So stability of containment is advantageous for persistence of a given variant in contexts where the presence of relevant substrates is of low probability. If, however, the surface of the containing capsule has molecular features to which the relevant substrate molecules tend to bind, and in so doing weaken its structural stability, then the probability of autogenic replication will be significantly increased. The process will tend to be more stable in environments lacking essential substrates and less stable in environments where they are plentiful. Sensitivity to substrate concentrations would likely also be a spontaneous consequence of this bonding, because if binding of substrates to container molecules weakens the hydrogen bonds between containment molecules, it would follow that weakness of containment would be a correlate of the number of bound substrates. Higher substrate concentrations would make disruption more probable, and subsequent use of local substrates would deplete their concentration and make the replicated autogens more stable and more likely to diffuse to new environments.

In evolutionary terms, this is an adaptation. Autogen lineages with this sensitivity to relevant substrates will effectively be selective about which environments are best to dissociate and reproduce in. Though such "sensitive" autogens would not exactly initiate their own reproduction—that is still a matter dependent on extrinsic disruption—their differential susceptibility to disruption with respect to relevant context is a move in this direction.

It seems to me that at this stage we have introduced an unambiguous form of information and its interpretive basis. Binding of relevant substrate molecules is information about the suitability of the environment for successful replication; and since successful replication increases the probability that a given autogenic form will persist, compared to other variants with less success at replication, we are justified in describing this as information about the environment for the maintenance of this interpretive capacity. Using terms introduced by the father of semiotic theory, Charles Sanders Peirce, we can describe these consequences as *interpretants* of this information. Peirce introduced this way of talking about the process of interpretation in terms of interpretive consequences in order to more fully unpack the somewhat opaque notion of sign interpretation. Specifically, he would have termed the decreased integrity of containment provided by bound substrates the *immediate interpretant* of the information, and he would have termed the support that this provides to the perpetuation of this interpretive habit via the persistence of the lineage the *final interpretant*. We can now unpack the notion of information in semiotic terms as well. The sign in this case is the binding of substrate, and its object is the suitability of the environment. Or again to use Peirce's more specific terminology, we might describe the presence of substrate in the environment as the *dynamical object* of the binding (that physical fact that is indicated by the sign), and the general suitability of the environment as the *immediate object* (that general property of the dynamical object that is significant for the process).[9]

What is the difference between the sensitivity in this simple molecular system and the sensitivity of a mechanical sensor, such as a thermostat used to control room temperature or a photodetector in a doorway used to detect the entrance of someone into a store? In the absence of the human designer/user, I would describe the action of these mechanical devices as providing Shannonian information only. Certain of the physical constraints embodied in these mechanisms provide the basis for potential information about particular kinds of events. But the rate of drying of a wet towel has the capacity to indicate room temperature and the tracking of dirt in from the street has

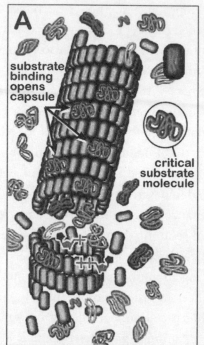

A

substrate binding opens capsule

critical substrate molecule

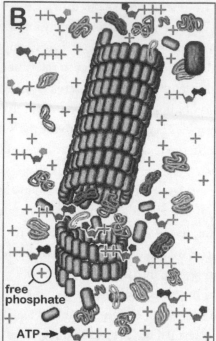

B

free phosphate

ATP →

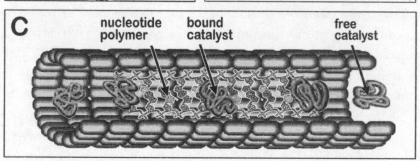

C

nucleotide polymer   bound catalyst   free catalyst

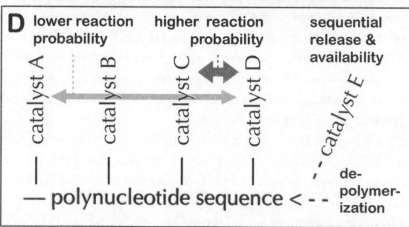

D

lower reaction probability   higher reaction probability   sequential release & availability

catalyst A

catalyst B

catalyst C

catalyst D

catalyst E

— polynucleotide sequence < - -

de-polymer-ization

FIGURE 14.1: A speculative depiction of the possible evolutionary stages that could lead from a simple autogenic system to full internal representation of the normative relationship between autogen dynamics and environmental conditions. Though somewhat fanciful, this account provides a constructive demonstration that referential normative information is supervenient on (and emergent from) teleodynamics.

A. Depiction of a tubular autogen with a simple modification that provides the capacity to assess information indicating the presence of favorable environmental conditions. This is accomplished because the exposed structure of the autogen surface includes molecular surface structures that selectively bind catalytic substrate molecules present in the environment, and where increasing numbers of bound substrates weaken containment. This increases the probability that containment will be selectively disrupted in supportive versus non-supportive environments and thus provides information to the autogenic system about the suitability of the environment for successful reproduction.

B. Depiction of a tubular autogen that produces free nucleotides as byproducts. This might evolve in environments with high concentrations of high-energy phosphate molecules as a protection against oxidative damage, and could subsequently be exapted as a means of extracting and mobilizing energy to drive exothermic catalytic reactions.

C. Within the inert state of an autogen diverse nucleotide, molecules could be induced to polymerize as water is excluded. This would both render phosphate residues inert and conserve nucleotides for future use. Although the spontaneous order of nucleotide binding will be unbiased, the resulting sequence of nucleotides can serve as a substrate onto which various free molecules within the autogen (e.g., catalysts) will differentially bind due to sequence-specific stereochemical affinities.

D. In this way, catalysts and other free molecules can become linearly ordered along a polynucleotide template, such that relative proximity determines reaction probability. Thus, for example, if this template molecule releases catalysts according to linear position (e.g., by depolymerization) they will become available to react in a fixed order. To the extent that this order correlates with the order of reactions that is most efficient at reconstituting the autogenic structure there will be favored template sequences. So long as one strand of this template is preserved, as in DNA, sequence preservation and replication are possible. Since the optimal network of catalytic reactions will be dependent on the available resources provided in the environment, this template structure is at the same time a representation of this adaptive correspondence. Although this scenario has been described using a nucleotide template in order to be suggestive of genetic information, the molecular basis of such a template could be diversely realized.

the capacity to indicate that a person has entered a room. All are potential information about particular sorts of events. But there are also numerous other things that these physical mechanisms and processes could indicate as well, probably a nearly infinite number depending on the interpretive process brought to bear. That is the difference. A person focused only on those aspects deemed relevant to some end they were pursuing would determine what it is information about. There is no intrinsic end-directedness to these mechanisms or physical processes.

This is what is provided in the most minimal sense by the autogen's tendency to reconstitute or reproduce itself after being disrupted. The autogenic process not only tends to develop toward a specific target state, it tends to develop toward a state that reconstitutes and replicates this tendency as well. So the interpretation of substrate binding is a self-constituting feature. It is a dynamical organization that is present because of its propensity to bring itself into existence. Of course, each interpretation is a unique event, so it is more accurate to say that the *general type* of this specific dynamical constraint (or organization) that we have identified as an interpretive process is self-constituting. It is only the form of this dynamical constraint that will be perpetuated by being passed on, not any specific collection of molecules, and so on. To again describe this in terms that resonate with Peirce's semiotics, the ultimate ground of interpretation is a self-sustaining habit.

There is also a necessary intrinsic normative character to this interpretive process. If by virtue of structural similarity, other molecules that are not potential catalytic substrates also tend to bind to the autogen surface and also weaken it, this would, in effect, be misinformation, or error. Sensitive autogens, which tend to respond in this non-specific way, would be less successful reproducers than those that were more selectively and appropriately sensitive. This would provide a selective influence favoring an increase in specificity. Although individual autogens and autogen lineages themselves would not detect this as error, the autogen lineage would. Over the course of evolution, such error-prone autogens will tend to be eliminated from the population. This exemplifies the fact that as soon as there is information (in the full sense of the term), there is also the possibility for error. Because aboutness is an extrinsic relationship, it is necessarily fallible. Detection of error within an individual involves an additional level of information about the information, and thus an additional level of interpretive process. Autogens are too simple to register anything about the interpretation process. An interpretation of the interpretation processes is a higher logical type relationship. This is why it only arises at the level of autogen lineage selection. As we saw in the last chapter, this implies that natural selection is a form of distributed error detection; and as we will see in the next chapter, only when some analogue of natural selection is internalized within the interpretive process itself—in physiology, say, or brain function—does error detection become intrinsic to the system that does the interpreting.

ENERGETICS TO GENETICS

This use of the autogen model to explore the necessary and sufficient conditions to constitute a minimal interpretive capacity does not, however, address the most compelling issue: the nature and origin of genetic information. The above demonstration of how autogenic teleodynamics can provide the basis for an interpretive dynamic offers a recipe of sorts, showing how a molecular relationship (binding of substrate molecules to capsule molecules) can come to be interpreted as information about some relevant extrinsic state of affairs. Whereas this is a reciprocal coupling between this one molecular relationship and the extrinsic requirements of the teleodynamic system in which it has come to be incorporated, genetic information (and its precursor analogues) must in effect involve an additional level of referential relationship. It must be in relationship to the teleodynamics of the organism (or autogen) as this substrate-binding relationship is to the environment. Genetic information is about some aspect of this teleodynamic organization with respect to certain environmental factors. So our question now becomes: How can this infolding of reference arise?

Obviously, as there are innumerable molecular details in the autogen story that I have merely assumed to be plausible without actually investigating the chemistry involved, when we complicate this account, there are exponentially more to be faced. These constitute the critical science that must be undertaken before any of this can be said to actually apply to the origins of life, or protolife, much less to the origins of genetic information. My purpose, however, is not to explain the origin of life, but rather to get clear about the principles that must be understood in order to focus this research on the most relevant details. What I intend is only to provide what might be described as a proof of principle. Neither protolife nor genetic information may have arisen in the specific ways that I describe; but I believe that the principles exemplified in these scenarios also apply to whatever specific molecular processes actually took place at the dawn of life on Earth.

With this caveat in mind, let's explore a somewhat fanciful—but not in any sense magical or homuncular—scenario for how a simple form of genetic information might arise in an autogenic context.

The intuition behind this imaginative scenario is motivated by noticing a curious coincidence that appears to be common to all organisms: some of the building blocks of the information-conveying molecules of life (DNA and RNA) are also the principal energy-conveying molecules (e.g., *ATP* and *GTP*) and so-called second-messenger molecules (e.g., *cAMP* and *cGMP*) of the cell. All of these molecules and their DNA-RNA monomeric counterparts have a three-component structure. This includes a purine double-ring molecule at one end (*A* = adenine, *G* = guanine), one or more phosphate molecules (PO_4) forming a sort of tail at the other end, and a five-carbon (pentagonal ring) sugar molecule in the middle (ribose). In their non-informational roles, these three-component molecules are transferred or diffused from place to place to serve energy delivery or "switching" functions with respect to other molecular systems. In contrast, their role as bearers of genetic information is only realized in a polymeric form, in which each nucleotide is linked to another by having its phosphate linked to its neighbor's sugar, one after another, to produce a long sugar-phosphate-sugar-phosphate- . . . "backbone," with base residues linked alongside. In polymeric form, the purine-containing nucleotides (homologous to *AMP* and *GMP*) are joined by pyrimidine (single-ring)-containing nucleotides (that can for comparison be designated *CMP*, *TMP*, and *UMP*, where *C* = cytosine, *T* = thymine, and *U* = uracil). In their roles as information conveyors, it is the sequence of these dangling purines and pyrimidines that matters, aided by the preferential binding of *A* with *T*, *C* with *G*, and *U* with *A* (the interchangeability between *U* and *T* distinguishes RNA from DNA; RNA substitutes *U* for *T* in DNA and has a slightly modified ribose sugar), that makes replication and translation possible.

Why this coincidence? My hypothesis is that the monomeric functions (energy transfer and switching functions) came first, and the information-conveying functions evolved as an afterthought, so to speak. Again, this suggests that information is not primary.

To begin, I need to postulate the (unexplained) presence of nucleotide molecules serving one or more of these basic energetic functions. Though the spontaneous inorganic synthesis of some of these nucleotide monomers has been recently demonstrated, and claimed as support for an RNA-World origins scenario, all that matters for the scenario proposed here is for there to be some means for their synthesis. Important for this argument is that their synthesis by catalytic processes of a complex autogenic system must be plausible—which for the sake of this scenario I will take as unproblematic. The starting assumption, then, is that some aspect of autogen catalysis or self-assembly is potentiated by these phosphate-ferrying molecules, as is the case in living cells. This might involve picking up an additional phosphate molecule or two in the high-energy context of a volcanic vent; but again, for the sake of the principle being explored, this chemistry is unimportant. During the reconstitution and replication phases of the "life cycle" of these more complex energy-assisted autogens, the presence of captured energy in the form of triphosphates could make up for the need for energy-rich substrate molecules to fuel the catalytic reactions involved. So, where simple autogen catalysis is parasitic on specific energy-rich substrates, the availability of a more generic energy source, in the form of phosphate phosphate bonds, would both offer a sort of jump start for catalysis and a freedom from such specificity of substrates. Such augmentations of the autogenic reconstitution and replication process would give lineages that generated and incorporated nucleotides the ability to capture and deliver energy, a significant evolutionary advantage.

As I have surveyed the literature on the origins of life, I have found two other authors who have independently proposed this evolutionary direction from energetic to informational use of nucleotides: the evolutionary biologist Lynn Margulis and the theoretical physicist Freeman Dyson.[10] Dyson's argument is that the polymerization of these molecules might be a sort of garbage-collection trick, to remove those monomers that have given up their extra phosphates and might otherwise compete for the extra phosphate residues still available to do work carried by other nucleotides. My speculation is vaguely similar, but specific to the context of autogenic "metabolism."

Recall that unlike most living organisms, autogens do not actively maintain non-equilibrium conditions. They are not continually in a dynamic state, but may spend vastly more time as inert structures. Phosphate-mediated molecular chemistry would therefore only be important during those brief, rare periods when containment has broken down and catalysis and self-assembly processes are critical for reconstituting this stable form. Polymerization of nucleotides—binding the phosphates between sugars—would make them unavailable for interacting with other molecular components while still maintaining a store of them, available for the next replication cycle. This might further be aided if autogen enclosure tended to exclude water, increasing the dehydrating conditions that facilitate polymerization. Disruption of autogen containment would thus inversely increase exposure to water, allowing rehydration to facilitate depolymerization of the nucleotides, making them again available to capture new high-energy phosphates.

So far, this scenario offers an interesting augmentation of the autogenic logic in which the addition of energy capture and management is a significant evolutionary advantage. But it also offers something in addition—a new potential source of constraint and constraint propagation. To see this possibility, we need to consider certain functionally incidental physical properties of such a polymer. First, as a means of molecular storage, the relative positions of different nucleotides along the polymeric chain are irrelevant. If many different nucleotides are all capable of serving some phosphate-carrying function, their polymerization order will tend to occur at random, only reflecting the degree of their relative prevalence. This unconstrained ordering of nucleotides is, of course, a critical property of nucleic acid information-conveying capacity (its Shannonian entropy). If the nucleic acid sequence were to strongly favor certain bonds or combinations over others, then this would introduce a bias, and thus redundancy and a reduction of the information-bearing capacity. Second, precisely because to be useful in autogen metabolism requires that phosphate-bearing nucleotides bind selectively to the catalytic and other constitutive molecules of the autogen, there will be a tendency for these other molecules to also bind to the polymer; and if there is some specificity associated with the various nucleotide sequence variants, then this binding will have some degree of specificity.

Together, these two properties can serve as the basis for the recruitment of the polymeric form to serve a function other than nucleotide collection and phosphate inactivation; it can act as a template.

There are a number of serious limitations affecting the autogenic form of evolution. One of the most significant has to do with the size of the network of catalytic interactions that is sufficient to complete autogen replication. While an increasingly complex autocatalytic set might provide autogens with some flexibility with respect to variable environments, as the number of catalysts and the complexity of the interaction patterns increase, an upper limit to evolvability will be quickly reached. For every catalyst that is

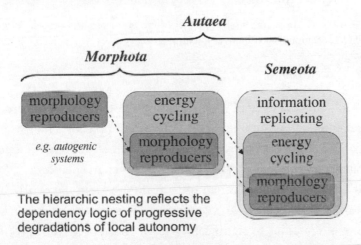

FIGURE 14.2: Depiction of an autogen thought experiment, demonstrating how a component molecule might spontaneously evolve to provide information about the production of the autogenic system of which it is a part. In this example it is assumed that variant nucleotide molecules are generated as side products of autocatalysis. In the dynamic phase of autogenesis these nucleotides could serve to capture free energetic phosphate molecules to provide generic energy for catalytic reactions (*left*). Assuming additionally that during the inert phase of the autogenic cycle free nucleotides were induced to polymerize into a randomly arranged linear molecule, different nucleotide orders would tend to provide differential substrates onto which free catalysts might tend to bind (*center*). The binding order of catalysts along the nucleotide polymer would incidentally bias interaction probabilities between catalysts by virtue of proximity and timing of release. In this way different sequences of nucleotides could come to be selected with respect to the catalytic interaction biases most conducive to autogen reproduction. The selectively favored order in this way re-presents the constraints that constitute the specific teleodynamic reaction network.

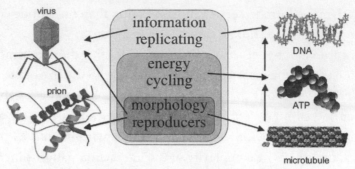

- Degenerate forms, parasitic on Semeota, reflect this logic

- Cellular exemplars with necessary hierarchic dependencies

FIGURE 14.3: The living fossil remnants of the major pre-life stages of *Morphota* evolution may still be exhibited in living systems. Some possibillities are depicted here.

added to an autocatalytic network, the number of possible non-productive molecular interactions between them and other molecules increases exponentially. The larger and more specific the interaction network necessary for autogen replication, the slower and less efficient the process. What is required is some source of constraint on the possible molecular interactions, besides that which is intrinsic to their structures. A mechanism that constrains interactions to significantly favor those that are appropriate to autogen formation, and to significantly inhibit those that are not, would be a significant aid in overcoming this explosion of possible side reactions.

The probability of interaction between molecules is, in large part, a function of relative proximity. Molecules in low concentration are thus on average seldom close enough to interact because of the vast numbers of molecules in between. An autocatalytic process is for this reason significantly sensitive to the relative proximity of any catalyst molecules capable of interacting chemically with any other catalysts. And where autocatalysis involves more than just a few interacting catalysts, and where some reactions between the catalysts in the set are relevant, the differential proximity of catalysts becomes much more important. In living cells, where molecular "machines" may involve sometimes as many as a dozen different molecules working precisely together, the different components are often bound to substrates in a specific configuration, or sequestered in different cellular partitions, or assembled by sequential availability of components to produce multipart molecular complexes (such as a ribosome), in order to avoid irrelevant interfering molecular interaction. An autogen that utilizes more than a small number of interacting components would likewise require similar constraints.

Without constraints on the relative probabilities of molecular interactions within an even modestly complex autogen, that is, with an autocatalytic set of say more than about four or five catalysts, there would be a tendency for side reactions to impede reproduction by generating inappropriate products and reducing the rates of relevant reactions. Moreover, this would rapidly grow out of control as autogenic chemistry became even modestly more complex and added new components. With respect to this limit-to-complexity problem, the availability of a randomly constructed nucleotide polymer could make a significant difference.

An important factor determining chemical reaction rate is the relative probability that the relevant molecules will interact with one another compared to the probability that competing interactions will take place. So, for example, the relative concentrations of diverse molecules will decrease the probability of any specific molecules interacting. At a molecular level, this is effectively a problem of relative proximity. One of the ways that catalysts can increase reaction rates, then, is to increase the probability that certain molecules will be near enough to interact. But in an analogous way, a large linear polymer with irregular structure can bias proximities and reaction rates of other molecules. Presuming that the various proteinlike catalysts enclosed in the inactive autogen will have slightly different affinities to bind to the linear sequence of the nucleotide polymer, this binding order will provide proximity constraints affecting the probability that any two bound catalysts will interact. Molecules bound nearby one another on the same polymer will have a higher probability of interacting than those bound far apart.

To the extent that molecules bound nearby one another are also those whose immediate interaction is conducive to autogen formation, rather than a side reaction, this reaction bias will facilitate preservation of this feature, because more of this variant will get produced as a result of their more efficient chemistry. In other words, the degree of correlation between catalyst binding order and the topology of the optimal autocatalytic reaction network will provide a source of selection favoring preservation of certain nucleotide sequences over others. Selective preservation and transmission of this correlation between nucleotide sequence and catalytic interaction sequence can thus become the crude equivalent of genetic information. It would now be accurate to describe the structural features of a given nucleotide sequence as conveying information about an advantageous chemical reaction sequence. A molecular structure would thereby inherit constraints from the more functional autocatalytic dynamics and transfer those constraints to future dynamics, as some autogen lineages out-reproduced others.

Many more details are required to complete this simple scenario. They include the preservation of advantageous nucleotide sequences, the means by which they can be replicated along with autogen replication, and the maintenance of their linear form. Most of these difficulties can be answered by assuming a double-stranded DNA-like form of these earliest nucleotide polymers, rather than a single-stranded RNA-like form. This of course contradicts RNA-World assumptions, but in part the RNA advantage is reduced in this case because catalytic functions are already assumed irrespective of nucleotide functions. The point of this sketchy scenario is not, of course, to make any claim about the relative plausibility of any particular molecular process, but simply to show how a teleodynamic process can offload certain critical constraints that it must preserve onto a separate physical substrate, and in so doing endow this substrate with semiotic functionality—in other words, aboutness.

This vision of a DNA molecule with proteins bound to it in locations determined by nucleotide sequence has a living parallel. Molecular biologists are already familiar with a number of classes of DNA-binding proteins. These play diverse roles in living cells—from merely protective and structural packing and maintenance to a variety of regulatory roles. The binding of these various proteins and protein complexes to the DNA molecule is likewise determined by various stereochemical properties, including the structural correspondence between protein structure and the distinctive "twist" of the double helix that correlates with nucleotide sequence at that point. These slight differences in twist are due to the slightly variant properties of the cross-linked bases and slightly different energetic configurations in the way the molecule "relaxes." The regulatory functions of DNA-binding proteins are also mediated by the protein-protein interactions they precipitate or block, as these will determine what other protein complexes will bind to adjacent segments. Analogously, the scenario just sketched effectively reverses what we might normally imagine the priority of genetic functions to be. What in contemporary molecular genetics we take to be a secondary supportive function of DNA-bound proteins, in service of regulating the expression of the genetic code, is in this scenario taken to be the primordial "coding" function of DNA: a substrate for organizing the interactions among various catalytically interacting proteins.

Any number of additional modifications to this scenario could be suggested to give it increased plausibility for an origins theory. For example, we might speculate that the significant temperature differences found in regions near volcanic vents could act like natural PCR machines, causing DNA-like chains to separate and re-aneal to form replicas. For the purpose of this discussion, explaining how these mechanisms evolved may improve the plausibility of this as a theory of life's origins, but it adds little to the constructive definition of biological information that results. What questions we may be predisposed to ask about the origins of these biological functions will however be quite different as a result.

It may be a disappointment to some that I have not also attempted an account of the origins of the triplet genetic code that determines protein structure. That is a far more complex problem, which I am not equipped to speculate about. So this is far from an account of the so-called code of codes that matches nucleotide sequences to the amino acid residue sequences constituting proteins. But to the extent that this scenario provides a general model demonstrating how a teleodynamic molecular processes can facilitate the transfer of critical constraints onto other substrates, then it may provide hints to help reconstruct the likely large number of incremental evolutionary steps that could lead from a reaction sequence template to a molecular code.

If we step back from the specifics of the molecular code problem and ask only how a molecule's structure could come to play a critical role in the organization of the dynamical properties of life, autogenic theory provides such an account. Interestingly, one implication is that it demotes DNA replication to a supporting role, not the defining feature of information, as in replicator theories. What must get preserved and replicated are certain constraints which are critical to the teleodynamic process that generates them, and these constraints can be embodied in many different ways. Molecular information is not, then, intrinsic to nucleic acid sequences, or to the process of replication, but to these constraints in whatever form or substrate. Which also means that there is no reason to assume, even in contemporary organisms, that the only information transmitted generation to generation is genetic. In whatever form it occurs, biological information is not an intrinsic attribute of that substrate. As this entire enterprise has been at pains to demonstrate, it is precisely this non-intrinsic character of information (and of all entential phenomena) that must be accounted for. Only when this general principle is grasped will we be able to avoid importing cryptic homuncular properties into our theories of evolutionary genetics.

This scenario demonstrates that a given component attribute of a teleodynamic system becomes informational because it comes to embody and propagate constraints relevant to the preservation of the dynamical organization that it is a part of. These constraints had to already be an implicit feature of the teleodynamics of that system, prior to becoming re-presented in this additional form. But once redundantly offloaded onto a component substrate, the maintenance of these constraints by features intrinsic to the global teleodynamic organization of the system can degrade without loss of their preservation. This will subsequently decrease the constraints on specific molecular dynamics, allowing a kind of evolutionary exploration of alternative forms that would not be possible otherwise. The highly specific dynamical and chemical features that were previously maintaining these constraints can now become multiply realized.

Thinking of biological information in these dynamical and substrate-neutral terms reframes how we think about the function of biological inheritance. In this scenario, genetic information arises from the shifting of dynamically sustained constraints onto structurally embodied constraints that have no direct dynamical role to play. The localized informational function of this structure is thus secondary to, and parasitic on, these prior, more holistically distributed constraints. In the above scenario, for example, the nucleotide sequence is initially merely redundant with certain favored dynamical constraints. However, the dissociation of this molecular pattern from the dynamics that it influences additionally protects these constraints from the thermodynamic unpredictability of the dynamical context.

Dynamically embodied constraints are probabilistic, and will vary to a greater or lesser degree by chance and with respect to such variables as the relative concentrations and distributions of catalysts and substrates in an autogenic context. The critical teleodynamic constraints will persist within an autogen lineage only because their probability of expression in this diverse milieu is just sufficient to stay slightly ahead of the increase in their thermodynamic degradation. Constraints that are, instead, redundantly embodied in a molecular structure are comparatively insulated from these sources of unpredictability, as well as being more thermodynamically stable. Moreover, both the reduction of constraints on properties that must be embodied in dynamical components and the re-embodiment of critical organizational constraints in structural form opens up new opportunities for incidental physical features to become functional features. This is because there will be an increased tolerance for interchangeable dynamical components and a structural means to "explore" distinct and reproducible variants of global dynamical organization (e.g., via structural modification of the template). In this way, molecular structure changes can give rise to organizational changes and organizational regularities can give rise to new forms of molecular information. So the capacity of a teleodynamic system to generate and transfer constraints from substrate to substrate is the key to open-ended evolution, involving ever more complicated dynamics and ever more diverse substrates.

FALLING INTO COMPLEXITY

Pursued as an origins-of-life scenario, autogenic theory leads us to ask very different sorts of questions than those typically considered by the origins-of-life research community. For example, elsewhere[11] I have argued that the large, inorganically formed polymers that are required to be available for autogen processes to emerge might be more likely to arise on colder planets in non-aqueous conditions, and consist of polymers formed from hydrogen cyanide monomers. Also, because of the generic nature of the autogenic mechanism, I suspect that this type of not-quite-living teleodynamic system would be able to form from a diverse range of molecular substrates. Thus the generic nature of the chemistry, the simplicity of the process, and the diversity of conditions that such a system could evolve within should make this sort of replicating molecular system far more prevalent in the universe than anything as complicated and delicate as life.[12]

The shift from simple autogen replication to information-based reproduction, though it might be a rare evolutionary transition in a cosmic sense, is one that would make a fundamental difference wherever and whenever it occurred. The capacity to offload, store, conserve, transmit, and manipulate information about the relationship between components in a teleodynamic system and its potential environmental contexts is the ultimate ententional revolution. It marks the beginning of semiosis as we normally conceive of it, and with it a vast virtual representational universe of possibilities, because it marks a fundamental decoupling of what is dynamically possible from immediately present dynamical probabilities—the point at which the merely probable becomes subordinate to representational possibility. This is the source of the explosive profligacy of biological evolution.

This account of the derivative origin of genetic information from teleodynamic processes does not therefore diminish its fundamental role in biological evolution. It does, however, help us to better understand the complex underpinnings of the natural selection process—"life's several powers"—and specifically the importance of morphodynamic processes in development, and the way this contributes to the evolution of complexity and the generation of new biological information. Though it is beyond the intention of this analysis to explore the implications of these ideas for evolution more generally, it is worth concluding this discussion of the nature of the evolutionary process and its role in the generation of information with a few comments about how this perspective illuminates certain recent issues in evolutionary theory.

First, consider the logic of the transfer of constraints from chemical dynamics to a nucleic acid substrate, thus imbuing the latter with information-conveying capacity. In the autogenic model presented, this originally developed from a redundancy relationship. The sequence structure of the nucleic acid polymer became correlated with the dynamical constraints conducive to efficient autogen replication by a process of natural selection because of the way this redundant source of constraint increased the probability of favorable versus unfavorable reaction networks. This redundancy decouples genetic information from metabolic dynamics, effectively re-presenting constraints on interaction probabilities as constraints on the stereochemistry of the interacting molecules. In this way, both forms of constraint can vary separately, and so novel synergies between these dynamical and structural constraints can be explored by natural selection.

An analogous role for redundancy in the creation of new genetic information occurs via the serendipitous duplication of genes. Gene duplication is an error or mutation that causes a length of DNA to be copied and spliced back into the chromosome, often just adjacent to the original. There are some instances where the extra product produced by a duplicate is beneficial and other instances where it is harmful or just neutral. Duplication, like redundancy in Shannon's information theory, is an important hedge against error. In the case of a duplicated gene, this means that the functional degradation of one copy, such as might be caused by mutational damage, might not produce any degradation of function, much less catastrophic failure. And in general, even the damaged copy will not become entirely non-functional. Point mutations that change only one amino acid residue in a large protein will usually have the effect of only modestly altering its structure. Thus its change of properties will also tend to be modest, if noticeable. By unlucky chance in a protein composed of hundreds of amino acids, it is nonetheless possible for a single substitution to affect a critical structural feature of the protein, and it is also possible for a point mutation that inserts or deletes a nucleotide to cause a "frame shift" which alters how every subsequent codon is read, and thus produces an entirely different protein, if any at all. So long as there is redundancy and no additional deleterious effect produced by the altered copy, natural selection is relaxed on this duplicate; its copies in various progeny can continue to accumulate mutations and thereby "explore" variations on the original functional theme, so to speak, without significant cost in viability.

Because even a single molecule's function is seldom simple, and is often the result of various "chemical compromises," this "exploration" can often produce synergistic effects with respect to the original. This occurs if the slightly divergent duplicate happens to contribute a useful variant version of the original function, and in so doing frees the original form from the constraints of compromises that may have been necessary to operate in multiple contexts. Randomly hitting upon such a synergistic interrelationship is made more likely in sexually reproducing organisms, because constant recombination generation-to-generation effectively samples the combinatorial options that are potentially available in a population with many different independently varying versions of the duplicate. This process of duplication, degradation variation, random recombination of variants, and selective stabilization of the most synergistic combinations can happen repeatedly in the evolution of a species' genome. The result can be a whole family of related genes, working collectively or separately to achieve a complex result.

Probably the classic example of this sort of synergy due to gene duplication is duplication of the hemoglobin genes in mammals. Bloodborne hemoglobin is a tetrameric complex in which four hemoglobin molecules—two alpha and two beta hemoglobins—fit together into a tetrahedral shape, with each hemoglobin's oxygen-binding region facing outward. A gene duplication event resulted in the separate alpha and beta forms diverging in shape over evolutionary time, allowing for the evolution of their current synergistic fit. Independent variations in the shapes of the non-oxygen-binding surfaces of the alpha and beta variants over the course of their evolution made it possible for recombination to "sample" their different tendencies to self-assemble into larger complexes, and to favor the replication of those complexes with improved self-assembly, stability, and oxygen-carrying capacity. Even more interesting is the fact that further duplications of the beta hemoglobin gene in evolution have resulted in a number of variants of beta hemoglobin, each with slightly different oxygen affinities. These appear to have been recruited by analogous recombination and selection processes to serve the changing oxygen-transfer demands of the developing mammalian fetus, whose hemoglobin needs to be able to extract oxygen from its mother's hemoglobin and deliver it to its own body. Moreover, the different variant fetal beta hemoglobins are produced in different amounts at different stages of gestation, and appear to correspond to the changing oxygenation demands of a growing fetus, thus achieving a sort of temporal synergy.

Probably the most dramatic example of the effect of gene duplication, and subsequent complementation of function, produces the theme-and-variation structure of multicellular plant and animal bodies. This turns out to be the effect of a class of genes that play regulatory roles by producing protein products which bind to other regions of the genome and act to effect the expression of large constellations of other genes. The family of genes that was found to be responsible for the segmentation of animal bodies is named the *homeobox* family of genes ("Hox genes," for short, in vertebrates). These are named for a coding region they share in common that constitutes the critical DNA binding domain of the proteins they code for. They were discovered in fruit flies because mutations of these genes produce bizarre but systematic modifications of whole body segments, such as a duplicated thoracic region or a leg where an antenna should have developed.

It was found that different variant duplicates of these genes orchestrate the expression of large numbers of "downstream" genes, but in slightly different ways; a bit like two orchestra conductors with slightly different styles. Because each variant duplicate gene is also confined to be expressed only within particular segments of the developing body, the structures they produce are both similar to one another in each segment and slightly different as well. This results in what can be described as a theme-and-variation pattern of body parts—like the slightly different legs in the different positions on an insect body, or the slightly varying pattern of digits on human hands and feet. Somewhere back in evolutionary time, prior to the point where we shared a common ancestor with flies—probably in the range of half a billion years ago—there was a single ancestral homeobox gene. It regulated the expression of many other genes. But as it was duplicated, and the duplicates came to vary in their region of expression and the ways they controlled expression of the same downstream genes, a corresponding redundancy of slightly varying body segments resulted. Analogous to the hemoglobin duplicates, combinatorial sampling allowed different combinations of these variant body structures to be "tested" within a single body, and selectively favored or disfavored by how their interaction effects affected reproduction. In this way, variant versions of duplicate genes became the means that the evolutionary process could stabilize synergistic functional relationships between different body structures. Whole repeated organs (legs or mouthparts) have thus evolved in complementary ways with respect to one another, by partially shielding each other from selection by virtue of their redundancy. So, a distant cousin to the offloading of constraints redundantly onto alternative substrates has also been the secret to the generation of new genetic information, as well as the theme-and-variation logic that is the hallmark of all complex multicellular animals.

This duplication and complementation process in evolution—like the transfer of constraints from autogen molecular dynamics to nucleotide sequence structure described in the scenario above—is a morphodynamic effect, a progressive propagation of constraints. The synergy that results at each step, under the influence of selection, introduces further constraints that differentiate each duplicate unit with respect to the other, as some fraction of their functional load is reduced reciprocally. In effect, the new higher-order synergistic relationship that results is an expression of the morphodynamic processes that are implicit in the teleodynamics of the whole. Although natural selection is a process that decreases variety, and thus maintains or increases constraint on organism form and function, the production of this redundancy provides a partial shielding from natural selection, allowing new forms of variety to emerge in parallel. In informational terms, this is an increase in the entropy of what might be called the "evolutionary channel capacity," and the means by which new information can emerge, as new higher-order combinatorial possibilities become subject to the constraint increasing influence of the natural selection process.

A whole new subfield of evolutionary research is opening up as we begin to pay attention to the contribution of the morphodynamic features of the genetic and epigenetic processes that generate organism structures and functions. On reflection, we can now see that "life's several powers" include and depend on the underlying morphodynamic processes that synergistically support and generate one another. When natural selection weakens its limiting grip on various adaptive functions—as happens in the case of gene duplication—this constraint-propagating, form-generating tendency that generates living organization can become the means by which higher-order forms of functional information can emerge. Such back-and-forth interplay between selection and morphodynamics thus opens the door to indefinite complexification and ever higher-order forms of teleodynamic organization.

15

SELF*

Now there are selves. There was a time,
thousands (or millions, or billions) of years ago,
when there were none—at least none on this
planet. So there has to be—as a matter of
logic—a true story to be told about how there
came to be creatures with selves.

—DAN DENNETT, 1989[1]

STARTING SMALL

René Descartes' now legendary claim—"I think therefore I am"—sets
the challenge for a theory of self. Who or what is this "I" of which Descartes speaks? Attempts to probe this mystery have produced a whole field
of philosophy, and have recently spawned dozens of new books each year
purporting to have solved the puzzle, or at least to have probed deeper into
its still murky depths. Unfortunately, the results of a half millennium of dissecting this conundrum from what seems to be every angle imaginable has
produced little that can count as a significant advance, much less an answer
to the riddle. I believe that this is because beginning with the challenge of
explaining Descartes' *cogito* is a serious mistake. Not because it is a case of

*Parts of this chapter are borrowed and revised from two papers that I co-authored with col-
leagues: *The Emergence of Self*, by Terrence W. Deacon, James Haag, and James Ogilvy (2010),
and *Eliminativism, Complexity and Emergence*, by Terrence W. Deacon and Tyrone Cashman
(2010).

tilting at windmills—indeed, to deny the reality of self is absurd, even if it sometimes makes interesting philosophical speculation—but because it is like trying to run before we can walk, or even crawl.

This classic way of approaching the problem begins by assuming that human subjective experience is a simple and well-characterized phenomenon. It is of course neither simple nor easy to describe, even if it is the most ubiquitous aspect of everything we know. And even though it is the one common attribute of everything experienced, self isn't always already there. It doesn't just suddenly emerge fully formed, either in evolution or in one's lifetime. Selves evolve, selves develop, selves differentiate, and selves change. This takes time. Selves mature slowly and differentiate incrementally, and more important, they are both the cause and the consequence of this process. This is another reason for assuming that there must be—as Dan Dennett points out in the epigraph to this chapter—a story to be told about how the phenomenon we call "self" came about.

The subjective self that Descartes and modern consciousness theorists have focused on is a late chapter in that story. To deal with it adequately requires more than just meditating on what one knows and is able to doubt— more than just introspecting. It requires an acquaintance with many details of animal evolution and brain function. But this isn't sufficient, either. Without the conceptual tools that we have been developing here, without being able to quite explicitly describe the dynamics that is common to what we might describe as simple organism self and also to human mental self, we will lack the means to make use of this common evolutionary and dynamical thread to reconstruct this most personal form of individuation.

So, starting with the problem of human subjective experience too narrowly focuses attention on only one highly developed and specialized variant of the self phenomenon, and it confronts us with a particular variant that is the product of vastly complex supporting processes whose byzantine details are the product of billions of years of evolution. The sort of reflective mental self that Descartes accepts as undeniable only became possible after billions of years of evolution. It is not just some general ubiquitous phenomenon that is widely represented in the world, even if it is our only window into that world. It is also the product of an astronomical number of subtle and incessant physical processes, taking place every millisecond, in the nearly unimaginably complicated and highly structured chemical-electrical network that is a human brain. And even if it weren't the product of such complexity, it would nevertheless still be surprisingly counterintuitive in its organization and properties, precisely because it is an intentional phenomenon. Why should we expect anything more than to be dazzled by this most complex variant of this most counterintuitive process?

Trying to make sense of something so impenetrably complicated and counterintuitive presents us with an intractable first step. It is almost certainly one of the main reasons that discussions of self and of subjective experience have produced little progress since Descartes' time, despite enormous advances in philosophy, psychology, and the neurosciences. I think that it's time to take a quite different approach. Descartes' question needs to be set aside until we can assess the problems of intentionality and self at much simpler levels. I suggest that we start small—as small as possible. By starting with the most minimal case where we can feel justified identifying something vaguely like self, we will find it easier to dissect apart these counterintuitive issues. Only when we have taken these first foundational steps can we safely build on them to an account of the dynamical architecture of subjective self.

Luckily for our quest to understand self, human consciousness isn't the only exemplar of this property. It is a special and highly divergent variant on a more general theme—a theme exemplified by the thousands of types of animals with brains. And this neurologically mediated form of self is in turn a special variant on an even more general theme: what it means to be an individual organism. But self is not merely *associated* with life and mind, it is what defines the very individuation that characterizes an organism and is its most fundamental organizing principle. Only living organisms are truly individual in the sense that all aspects of their constitution are organized around the maintenance and perpetuation of this form of organization. It is the circularity of this consequential architecture—teleodynamics—that both delineates and creates the individuality that is organism self. Organism functions are in this way indirectly self-referential and self-projecting. In the inanimate world around us, we find no trace of this circularity of generative processes. Though bounded and unified, neither stones, nor drops of water, nor automobiles, nor computers, nor any other non-living artifact is reflexively individuated in this way.

Understood in this more general sense, self is not a property limited to organisms with brains like humans, or to other animals likely to have some tiny semblance of subjective experience. Subjectivity is almost certainly a specially developed mode of self that is probably limited to creatures with complex brains. This special aspect of neurologically mediated self is in many ways derivative from (or rather, emergent from) features exemplified in different ways in the more basic form of self that constitutes organism individuality. It is taken for granted in biology that the development, maintenance, and reproduction of even the simplest organism are entirely reflexive. An organism's functions and adaptations are unproblematically assumed to be defined as existing "for" the persistence of this individuation or else its reproduction.

So, although these different ways we use the concept of self certainly have important qualitative and quantitative distinctions, they also share many core features in common. Self is, in all cases, the origin, target, and beneficiary of functional organization, whether molecular or mental. Thus there is good reason to believe that by first exploring self at its most basic level, before trying to tackle the problem in its most complex form, we may be able to discern some fundamental general principles that will apply as we build our analysis upward toward the most complex self-phenomenon: human consciousness.

Recognizing that even organisms as simple as bacteria have properties that qualify them as selves, in at least a minimal sense, also indicates that subjectivity is not a critical defining property of self, but probably a property only relevant to the concept as it is applied to creatures with complex brains. This allows us to at least temporarily bracket this problematic attribute from consideration, while exploring certain more basic attributes. In setting this issue aside, however, we have not evaded the core conceptual challenge that needs to be addressed. Indeed, issues of teleology, agency, representation, and value (to mention a few topics already explored) are critical elements of self that need to be considered. Unpacking these challenging concepts by exploring simpler homologues has been the major strategy throughout this text. This strategy has proved valuable for clearing away many of the confusions that have previously impeded progress toward understanding how ententional processes and relationships can be made compatible with the natural sciences. And this approach hasn't reduced these problems to merely chemical and physical processes, but instead has demonstrated the importance of looking beyond just these material-energetic embodiments. Similarly, this approach can help to illuminate to what extent the phenomenon of self is not a simple physical property of bodies or brains, but rather a critical absential character, which is ultimately the locus of the reflexive individuation that creates the distinction from non-self.

Even though the concept of self is ubiquitous to biological discourse, it has been generally unwelcome in the natural sciences, and is mostly treated heuristically in biology. The lack of a non-homuncular account for the origins and efficacy of teleological phenomena has long guaranteed that the concept of self would pose irresolvable dilemmas and consequently remain ambiguous. In this respect, there is an analogy with the concept of teleonomy. Recall that teleonomy has been used to legitimize explanations employing end-directed accounts while presuming to deny that this has any intentional implications. But as I have gone to great pains to demonstrate, organisms aren't merely mechanisms that mimic teleological tendencies; they are entirely organized around a central goal-directedness, self-generation, and perpetuation. End-directedness is an intrinsic defining characteristic of an organism, not something only assessed from outside and extrinsically or accidentally imposed. Having addressed this issue in its most embryonic form—with respect to the emergence of teleodynamic processes in simple autogenic systems—we are now at least provided with a few new tools and principles with which to approach this challenge from the bottom up, so to speak. It's time we stop treating self as a euphemism.

As Dennett suggests, it is safe to say that 4 billion years ago (before the dawn of life on Earth) there was no such thing as a self in any form on this planet, and probably not anywhere in our solar system. Physical systems with this property emerged at some point in this planetary history. I have argued that this emergent transition occurred roughly coincident with the origin of life—or, to be more explicit, with the first autogenic molecular system. The emergence of the form of self that characterizes human subjectivity is in contrast a very recent, very special, and very much higher-order variant of this first transition. So, while this complex variant includes such radically different emergent properties as subjectivity and interiority (i.e., the first-person perspective that cannot be directly shared), the experiential version of self should nevertheless reflect a common logic that traces to this original teleodynamical transition. For a more tractable and basic perspective on this problem, let's step back from issues of consciousness and subjectivity to consider the reasons that we describe organisms to be maintaining, protecting, and reproducing them*selves*.

INDIVIDUATION

The central thesis of this chapter is that the core property which links the selves of even the simplest life forms with that seemingly ineffable property that characterizes the human experience of self is a special form of dynamical organization: teleodynamics. What has not been made explicit, however, is that all teleodynamic processes are implicitly individuated, that is, they are *closed* in a fundamental sense with respect to other dynamical features of the world. This special form of closure is reflected in the phrase "reflective individuation" used earlier, which I introduce to explicitly avoid using the prefix "self-" (as in self-maintaining, self-organizing, self-reproducing) in defining the fundamental organizational logic of self. It also provides a more accurate and more explicit unpacking of the form of this relationship by implicitly invoking a mirror analogy. Indeed, it might even be analogized to mirrors reflecting one another and thereby creating a way of containing a perpetually reproduced image.

Another way to look at this closure is that it is the physical analogue of a logical type violation in logic, such as that exemplified by the famous liar's paradox: "This statement is false." The concept of logical type is basically the relationship that exists between a whole and a part, or a class and a member of that class. The sentence above is interpreted as a composite whole by virtue of the interpretation of the parts (words) with respect to one another. But the component phrase "This statement" refers to the whole of which it is a part and so assigns the value of being false to the whole. The familiar result is that with each reading, one is enjoined to reread and reinterpret the sentence. This self-undermining statement can never be assigned a final interpretation because the whole is constantly changing the reference of the parts and the parts are constantly changing the reference of the whole. The critical feature is of course that *the whole is represented and reflected in the part*. This is possible in a sentence because of the nature of symbolic representation. Since the reference of this phrase is not physically assigned, but can apply to any object, there is nothing restricting its application to the sentence of which it is a part. In an organism, the relationships are analogous. Each component function contributes to the continued existence of the whole and the whole is required to generate each component function. In this sense, each functional feature embodies a trace of the whole individuated organism, reflecting the coherent influence of the whole and contributing to its future coherence. This is the essence of reflexive individuation: a compositional synergy, functioning to determine its constituents in a way that both embodies and reinforces their synergistic relationship. The whole/part hierarchy thus becomes inextricably tangled.

If this tangled dynamical logic is fundamental to organism self, then it should apply equally well to forms of self at other levels too, including human minds. However, these parallels need not be simple and direct. This is because the neural teleodynamics of mind is hierarchically nested within the vegetative forms of teleodynamics that constitute simple cells and multicelled animal bodies. Mental self is both a higher-order form of self, and at the same time subordinate to the organism self. To rephrase this using the technical distinctions we developed in our earlier discussion of emergence, neural teleodynamic processes dynamically supervene on what might be called the *vegetative* teleodynamics of the organism, and organism self dynamically supervenes on cellular self, though in different ways. Even the simplest bacterium is organized as a self, with emergence of ententional properties and possibly a primitive form of agency in the ability to propel itself (e.g., with a flagellum). However, the intentionality and subjectivity which exemplifies species with complex brains involves higher-order properties that emerge from a distinctive higher order level of reflexive dynamics constituted by the interactions among the vast numbers of vegetative teleodynamical selves that are neurons.

This vegetative form of self exhibits some very general organizational principles, which are necessary and sufficient to explain the emergence of end directed and informational relationships from simpler components lacking these attributes. But why must a system exhibiting teleodynamic organization be an individuated unit? This is not the case, for example, with thermodynamic or morphodynamic processes. They may be physically boundable, but it is only a contingent boundedness. Organisms also typically are materially bounded, but in addition their boundedness is intrinsic to their dynamics. It is implicit in the closed reciprocity of organism functions. As a result, physical boundedness may be ambiguous in many organisms or at certain stages in their development. The necessary interdependence of organism structures and processes implicitly determines an individuated system. This intrinsically generated individuation also determines a self/other distinction, which need not be an artifact of material discontinuity or containment within a membrane. So, while biological self and other are commonly distinguishable on either side of some physical interface—a membrane, cell wall, or skin—these are only convenient means of physically embodying this otherwise purely dynamical distinction.

SELVES MADE OF SELVES

The primacy of dynamical boundedness over material boundedness becomes more evident as we ascend in level of biological and mental selfhood. And when we reach the level of a human subjective self, even the concept of boundary makes no sense, despite the fact that the reflexive closure of subjectivity is unambiguous.

With the evolution of ever more complex forms of organisms, the recursive complexity of self has also necessarily become more highly differentiated. Multicellularity offers a case in point. Multicelled organisms all develop from single-cell zygotes. This initial single cell self-reproduces vast numbers of genetically identical offspring cells—self copies—that at early stages of embryogenesis still retain some degree of autonomy. By remaining in contact with one another, maintaining a molecular and topological trace of their relative positions, and sending identifying molecular signals to one another, this local society of cellular twins begins to differentiate and cellular-level autonomy decreases. Each gives up progressively more of its autonomy to the growing whole embryo. Cellular-level dynamical reciprocity is thus progressively simplified, and intercellular forms of reciprocity begin to take its place. One level of individuation is sacrificed so that a higher level of compositional individuation can form. Again, what determines this transition in level of self is not the generation of an exterior skin, but rather a change in dynamical reciprocity. In this way, local clusters of similarly differentiated populations of cells, forming organs, begin to play an analogous role to the various interdependent classes of molecular processes within a cell.

Developing a skin or bark or shell serves a different purpose. It protects the now more distributed reciprocity from extrinsic disruption.

Of course, protective encasement is a nearly ubiquitous feature of organism design—whether for viruses, bacteria, amoebas, plants, or animals—because reciprocity is always potentially disruptable by the intrusion of extrinsic influences. Dynamical reciprocity, without which there can be no teleodynamics, is therefore forced to persist in the context of mutually exclusive requirements with respect to the external non-self world. Because it is dynamical, it is dependent upon extrinsic energy and material; because it is a form of reciprocal dependency, it is dependent upon being isolated from aspects of the non-self world that might disrupt this delicate reciprocity. For this reason, the evolution of highly specialized bounding surfaces, which are also selectively permeable, is one of the most important loci of evolutionary differentiation. Whether cell membranes with their elaborate constellations of receptor and transporter molecular complexes linking inside and outside, or animal body surfaces with their multitudes of receptor types and motor organs enabling them to find food and mates or avoid predators, this interface is where the self-other distinction mostly gets negotiated. It is just not what determines self. So, although our attention tends to be drawn to these boundary interfaces when focusing on what constitutes self, this interface is better understood as an adaptation to better preserve the self already there.

The critical but contingent relationship between selves and physical boundaries complicates the identification of biological self. Individual cells in your body are each physically bounded by a membrane within which the reciprocal processes essential to self-maintenance take place. But they are also critically incomplete as well. Unlike free-living cells, somatic cells have become co-dependently parasitic. Their integration into a larger self has enabled them to forgo the adaptations to the variable environments that an autonomous cell must encounter. One of the most fundamental is reproduction. Many of the cell types in your body are terminally differentiated, which means they cannot be modified to be any other kind, and many of these are non-reproductive, that is, they will never again divide. Apparently, the molecular apparatus that must be maintained in order to support this massive metabolic transformation would be a significant impediment to the terminally differentiated function of this cell. Thus differentiated neurons and muscle cells, for example, have either redeployed some of this apparatus for other purposes or suppressed its generation altogether. They still retain one feature of organism self but have sacrificed another. The metabolic self (so to speak) of these cells is bounded in the cell membrane, but that feature of biological self which maintains this ultimate hedge against the ravages of entropy—duplication—has been shifted to the whole organism.

Of course, the locus of individuated self is also ambiguous in other biological contexts. Lichens are the result of a co-dependent symbiosis between a fungal and algal species, neither of which can exist without the other. Their reciprocal co-dependence is also guaranteed by the degradation in each species of adaptations for autonomous living. The lichen is therefore an individual defined by this reciprocal co-dependence, and not by genetic homogeneity. This is not really an exceptional case in biology: all eukaryotic organisms share the trait in a very basic form, including us. The mitochondria—the tiny bacterial-shaped loci of oxidative metabolism in every plant and animal cell—are the distant ancestors of a once free-living bacterial organism that evolved to become an endo-symbiont to another larger-celled form. This is not merely reflected in their characteristic bean-like structure, but also in the fact that they contain and reproduce within each cell using an enclosed circular bacterial chromosome, whose DNA sequences place all mitochondria (from plant or animal) well within the bacterial lineage. Thus we too are characterized by dual genomes with separate origins.[2] Our current cellular-level individuation is in this way characterized by a synergistic reciprocity that developed between these once separately closed forms of teleodynamic unities.

Such self-components (somatic cells) integrated into higher-order self units (organisms) are each internally sustained by highly complex networks of synergistically reciprocal morphodynamic processes. They are individuated selves by virtue of the level at which there is circular closure of this morphodynamic network. This closure not only creates a distinct level of individuation but also a distinct locus of teleodynamics. I have designated each of these partially autonomous, partially dependent levels of individuated teleodynamics a *teleogen*, as distinct from an *autogen*, which is fully autonomous and simple. In the complex biological and mental worlds of the present state of life on Earth, teleogenic structures are the norm, as evolution tends to generate highly entangled forms of teleodynamic processes, given sufficient time. This is not because evolution is itself a kind of final causal process, but rather because there is no limit to how teleodynamic processes can become entangled with one another. This also leaves open the possibility that the different teleogenic units constituting a complex organism can come into conflict.

Tracing the way that higher-order forms of organism evolve was the topic of the last part of the previous chapter. There we noticed that each shift to a higher-order form of individuation was typically associated with loss of lower-level autonomy and the serendipitous self-organization of higher-order synergies. What evolves in these cases are multiple levels of self-built-upon-self, with the higher-order reciprocities and synergies emerging as a result of the degradation of certain self-features of the lower-level component selves. Because of this, the remaining lower-level teleodynamic characteristics are those most consistent with the higher-order teleodynamics. Teleogens composed of teleogens in which lower-level degradation is present are thus common. In the case of mitochondria and their "host" eukaryotic cell, for example, both the nuclear genome and the mitochondrial genome are partially "reciprocally" degraded, so that neither can persist without the other, but for most functions they remain relatively modular and individuated in their functions. This tendency for modularity, implicit in the nature of teleodynamics, is what makes complex multi-leveled selves possible. Without it, a complexity catastrophe would be inevitable—too many components, needing to interact in a highly constrained manner in a finite time, despite vast possible degrees of freedom—setting an upper limit on the complexity of self.

Because self is defined by constraints, not by particular material or energetic constituents, it can in principle exist in highly distributed forms as well. Thus a termite colony may involve millions of individuals and many generations of turnover, but only one reproducing queen. In reproductive and evolutionary terms, only the whole functioning colony is a reproducing organism, and the "self-" of the colony is quite ambiguously bounded in space or material and energetic usage, especially when vast numbers of individuals are foraging independently in the surrounding territory. Thus self, too, may become highly diaphanous at higher emergent levels, with more variable degrees of individuation and correlation with physical boundedness emerging at progressively higher-order levels of self.

These many features of teleogenic hierarchies are of course relevant to mental selves. Mental self is not a composite self in the same sense as is a multicelled organism body. Brains are composites of vast numbers of highly interdependent cells—neurons and glia (the "support cells" of the nervous system). And each neuron is extensively interconnected with other neurons. Brains evolved to enable multicelled animal bodies to move from place to place and intervene in the causality of extrinsic conditions, thereby altering the body's relationship to its local environment. Brains are in this respect part of the boundary that mediates between the teleodynamics intrinsic to the organism and the dynamics of its external world. So, in certain respects, it might be appropriate to compare brains to the specialized molecular pores, signaling molecules, and actuators (like flagellar motors) that span the membranes of cells and mediate inside/outside relationships. They are part of the boundary interface that continually creates and demarcates self. But unlike these cellular-level interface mechanisms, brains mediate the self/other distinction by using the dynamics of self itself.

NEURONAL SELF

Despite the current focus on consciousness, mental self is subordinate to and nested within the more general form of self that is characteristic of all living organisms. This is made self-evident by the fact that although the unconsciousness of anesthesia can temporarily interrupt this experience, the continuity of self persists across such gaps, so long as the body remains alive and the brain is largely undamaged. Even where there has been profound memory impairment, as in victims of Alzheimer's disease or hippocampal damage, though identity may be compromised, the sense of consciousness and agency still persists. Our worries about death, and our comparative unconcern with the state of unconsciousness, or even amnesia, thus provide evidence that we intuitively judge the self of Descartes' *cogito* to be subordinate to the self of life in general. Nevertheless, the experience of this self is the result of the way that organism self pervades and organizes neural processes.

Brains are organs which evolved to support whole organism functions that are critical to persistence and reproduction. They are not arbitrary, general-purpose, information-processing devices. Everything about them grows out of, and is organized to work in, service of the organism and the teleodynamic processes that constitute it. Animal physiology is organized around the maintenance of certain core self-preservation functions on which all else depends. Critical variables—such as constant oxygenation, availability of nutrients, elimination of waste products, maintenance of body temperature within a certain range, and so forth—all must be maintained, or no other processes are possible. Sensory specializations, motor capabilities, basic drives, learning biases, emotional response patterns, and even the structure of our memories are ultimately organized with respect to how they support these critical core variables. Additionally, the reproductive capacity, which is a ubiquitous correlate of having evolved, is also part of the larger teleodynamic background that gets re-expressed as neurological self, and similarly insinuates biases into these basic neuronal processes.

So, as a critical mediator of the self/other interface, a brain must be organized around this constellation of processes that constitutes the teleodynamics of organism self. The reciprocity and interdependence of the various physiological processes that sustain the organism thus get recapitulated as core organizing principles constituting neurological function. For the most part, however, the monitoring and regulation of these core bodily functions is automatic, and maintenance of them is relegated to the physiological analogues of thermostats and guidance systems, that is, teleonomic mechanisms. Nevertheless, these processes must also be re-represented to a higher-order adaptive process in service of mediating action within and with respect to the environment. So the vegetative self of the organism must be triply embodied: first in the cellular-molecular relationships that form the most basic organism teleodynamics; second in higher-order automatic neuronal regulatory systems; and third in the way that changes in these regulatory processes or in the values of the underlying physiological variables alter the yet higher-order adaptive activities of the brain that mediate between the vegetative teleodynamics and the world of extrinsic possibilities. It is this third level of re-re-represented organism teleodynamics that is the substrate of the subjective self.

Neural self is further complicated by virtue of its role as inside/outside mediator. Brains evolved in animals to generate alternative virtual worlds and virtual futures. In order to be able to favorably change the contexts within which the organism finds itself, brain mechanisms must be able to model an organism's surrounding conditions and also to model the possible outcomes of acting to modify those conditions. To do this, an animal must be capable of simulating more than just the moment-to-moment changing relationships between internal dynamics and external conditions. It must also be able to predict the possible consequences of its own interventions. Of course, depending on the predictability of the environment and the complexity of the organism, this modeling capacity may need only a small constellation of relatively fixed alternative responses. But where both the environment and the organism are complex, this may require elaborate, open-ended means for generating conditional scenarios. In such complex cases, it is critical to also include the capacity to simulate the teleodynamic processes that produce these adaptive behaviors and generate these scenarios. In social species, this can reach quite convoluted levels of detail. But because of the intrinsically recursive nature of behaviors that change external conditions, even modestly complicated brains need to include some self-referential features.

Thus it is inevitable that having a brain should also entail the generation of a form of teleodynamic relationship that is partially organized with respect to itself as environment. This higher-order form of teleodynamic causal circularity creates an entirely novel emergent realm of self-dynamics. It helps to explain why being an organism with a complex brain inevitably includes a doubly reflexive organization with respect to itself. This self-referential elaboration of teleodynamic logic is therefore one level more convoluted than any autogenic process. This predictive and projective function creates a more open-ended form of individuation, which must be able to generate a diverse range of possible self-environment scenarios. This premium on the flexible generation of possible futures (and thus purposes) makes the term *teleogenic* particularly apt for such a level of teleodynamic process.

SELF-DIFFERENTIATION

The development of self emerges in a process of differentiation. The self that is my entire organism did not just pop into the world fully formed. It began as a minimal undifferentiated zygote: a single cell that multiplied and gave rise to a collection of cells/selves which by interacting progressively differentiated into an embryo, a fetus, an infant, a child, and eventually an adult organism. Similarly, it is difficult to imagine one's subjective self just popping into existence fully differentiated. By the very nature of its thoroughly integrated and hierarchically organized form, it would seem to demand a bottom-up differentiation in order to arise. But if so, then this also suggests that the moment-by-moment subjective sense of self, as well, is only the most differentiated phase in a sort of micro-differentiation process happening continuously and in the space of seconds.

This is consistent with introspective experience. There are times when we are only dimly aware of our memories, our intentions, or the surrounding stimuli. For example, on waking from a sound sleep, we experience a sort of ascent into differentiated awareness, and a graded "booting-up" of our more critical and goal-directed faculties. The early phases of this process involve only the incorporation of basic physiological and somatic factors into awareness, and are in this sense largely undifferentiated forms of self. Only after we are fully awake, beginning to assess our immediate surroundings, and remembering the habitual activities, demands, and plans that attend this time of day and social condition, does our mode of self become fully differentiated. But even as one experience gives way to another and one activity gives way to another in the course of awake experience, each new focus of attention and intention must differentiate anew to replace a former, more or less differentiated phase of awareness. In this sense, each change of focus is a mini-recapitulation of waking anew. There is no deep discontinuity because the least differentiated level of self changes little from moment to moment; but even so, one characteristic of severe anterograde amnesia (in which the victim cannot consolidate new memories) is a constantly recurring sense of "just now being awake for the first time."[3]

In summary, then, it is my hypothesis that the subjective self is to be identified with this locus of neurological teleodynamics, which is variably differentiated at various stages of life and alertness, and which in its most differentiated form can include itself as recursively represented and projected into a simulated virtual world. Because teleodynamic processes depend on lower-order morpho- and homeodynamic processes, these too must be taken into account in a full theory of neurologically generated subjective experience (a topic addressed in the next chapter). And because these are emergent dynamical processes, all rely on constant exuberant metabolic activity, and are always in some stage of differentiation or dedifferentiation that takes time to unfold.

Additionally, as is also the case with simple autogenic systems, because a teleodynamic system is self-generating, self-reinforcing, and self-similarity maintaining, it can serve as a reference dynamic against which all other dynamical tendencies and influences that affect it are contrasted. Their distinction as non-self is implicit in their tendency to initiate the generation of contragrade dynamical teleodynamic processes. This orthograde/contragrade teleodynamic distinction thus defines the dynamical boundary of self. The minimal form of this teleodynamic organization is also at the core of all neurological functions, where it can be as undifferentiated as a simple autogenic process, and likewise provides the most basic self/non-self distinction by virtue of the contragrade dynamics of adaptive physiological processes. However, this teleodynamic organization is progressively re-represented and differentiated within brain systems that are progressively more entangled with external receptors and effectors. Because of the intrinsically conservative self-similarity-maintaining nature of teleodynamic processes, each level of teleodynamic activity is an effective locus of self, and is that with respect to which otherness is implicitly marked. Each locus of teleodynamic neural activity is also the dynamical "substrate" which differentiates and "evolves" moment by moment with respect to a complex environment of sensory "perturbations," present and remembered.

In this sense, the unity of consciousness may be more mercurial than commonly imagined. Different loci of teleodynamic activity at the same level may develop in parallel and come into competition as they differentiate and propagate to recruit additional neurological resources. Thus the differentiation of self may also involve a Darwinian component (also discussed in the next chapter). The often quite sophisticated alter egos that we find ourselves interacting with in dreams, and who can often act in unexpected ways, suggests that in this state of consciousness, there may be no winner-take-all exigency. In dreams, all action is virtual and thus need never be finally differentiated; but when awake and enjoined to behave by real-world circumstances, action depends on a winner-take-all logic to produce a single integrated action. So the unity of waking conscious experience may in this respect be a special case of a more pluralistic self-differentiation process.

THE LOCUS OF AGENCY

Perhaps the most enigmatic feature of self is its role as agent: as the locus and initiator of non-spontaneous physical changes in the world around it. This is often confused with the age-old problem of explaining the possibility of free will in a deterministic world. However, it is different in a number of important respects. Self as agent is indeed what philosophers struggling with the so-called free will paradox should be focused on, rather than freedom from determinate constraint. Determinate causality is in fact a necessary condition for the self to become the locus of physical work. An agent is a locus of work that is able to change things in ways more concordant with internally generated ends and contrary to extrinsic tendencies.

Approaching the self-dynamics of mental agency using this same framework, we need to look to the closure of the teleodynamic constraint generation process for the locus of the capacity to do self-initiated work. For the simplest autogenic process, this closure is constituted by a complex synergy between morphodynamic processes that makes possible both the generation of constraints and also their maintenance and replication. The teleodynamics that distinguishes the agency of organisms from mere physical work is a product of this closed reciprocity of form- (i.e., constraint-) generating processes. Specific forms of work are made possible by the imposition of specific forms of constraint, and the way this channels spontaneous change, via the expenditure of energy. So this defining dynamic of organisms amounts to the incessant generation of the capacity to perform specific forms of work to alter the surrounding milieu in ways that are determined by this locus of teleodynamics, irrespective of extrinsic causal tendencies. This persistent capacity to generate and maintain self-perpetuating constraints is therefore at the same time the creation of a locus of the capacity to do self-promoting work.

Evolution can be seen as a process that has vastly complicated both the nature of these constraints and the capacity to utilize them as means of initiating specific forms of work—work that aids the persistence and further evolution of these capacities. As new forms of teleodynamics evolve, they bring new capacities to perform work into being—new options, correlated with ever more diverse extrinsic influences. In this way, as evolution has given rise to organisms that embody vast webs of constraints in their internal dynamics, and specifically with ever more diverse means of interacting with their surroundings, they have increased the ways in which they are able to impose these or complementary constraints on the dynamics of external events and relationships. So, to again describe this in the negative, the evolution of increasingly complex forms of constraints—absences— has given rise to increasingly varied ways to impose constraints on the world with respect to these internal constraints. In this sense, the source of agency can be somewhat enigmatically described as the generation of interactive constraints which do work to perpetuate the reciprocal maintenance of the constraints that maintain organism self.

This view of self-agency, defined in terms of constraints, may seem counterintuitive because of our conviction that the emergence of life and mind has increased, not decreased, our degrees of freedom (i.e., free will). Increasing levels and forms of constraint do not immediately sound like contributors to freedom. In fact, however, they are essential. What we are concerned with here is not freedom-from, but freedom-to. What matters is not some disconnection from determinate physics, but rather the flexibility to organize physical work with respect to some conserved core dynamical constraints. This is not a breakdown of causal efficacy; in fact, just the opposite. Being an agent means being a locus of causal efficacy. Agency implies the capacity to change things in ways that in some respects run counter to how things would have spontaneously proceeded without such intervention. It also implies that these influences are organized so that they support the persistence of this capacity, and specifically the persistent self-generating system that is its locus.

What we are concerned with here is not some abstract conception of causality but rather the capacity to do work, to resist or counter the spontaneous tendencies of things extrinsic to the teleodynamics that creates self. The evolution of our higher-order capacities to create and propagate ever more complex and indirect forms of constraint—from the self-organizing processes that build our bodies to the production of scientific theories that guide our technologies—has in this way progressively expanded the capacity to restrict sequences of causal processes to certain very precise options. The ability to produce highly diverse and yet precise constraints—absences— thus makes possible a nearly unlimited capacity for selves to intervene in the goings-on of the natural world.

EVOLUTION'S ANSWER TO NOMINALISM

But where does the "freedom" come from? Clearly, it is not freedom from deterministic involvement in the world, because this would preclude the capacity to do work. The answer to this question is ultimately related to the classic realism/nominalism problem. To see this, it is first necessary to recapitulate bits of the critique of the Realism/Nominalism debate in philosophy that we initially dealt with in chapter 6.

The classic nominalistic view is that being a member of a general type or a categorized phenomenon, such as a whirlpool or a member of a species, does not in itself have any causal significance over and above the unique and distinct individual physical attributes of that particular instance. Exhibiting a general form does not, according to this view, have any independent causal status. These generalities and similarity classes are merely mental constructions, due to a necessary simplification that is implicit in the nature of mental representation.

By recognizing, however, that in the physical world the analogues to general types can be determined negatively, that is, in terms of constraints, it becomes possible to understand physical causality negatively as well. This is because constraints can propagate through physical interaction. Or to state this negatively, degrees of freedom not actualized do not tend to propagate during physical interactions. Moreover, as the discussion of work has further clarified, all non-spontaneous change (efficient cause) is a function of the coupling of constraint and constraint dissipation (i.e., the release of energy). So general tendencies, understood in this negative sense, can indeed be the loci of physical causality, because work both depends on and propagates constraints. General properties understood in this negative sense are, then, the causal determinants of other general properties.

But there is a sense in which the emergence of teleodynamics has significantly augmented this efficacy of generals; and with the evolution of the teleodynamics of brains, this mode of causality has even begun to approach a quasi-Platonic abstract form of causal influence. Ironically, the structure of the nominalistic argument for the epiphenomenality of general types provides the essential clue for making sense of this augmented notion of causal realism. This is because both living and mental processes do indeed break up the physical uniqueness of physical processes into similarity classes, due to the way they ignore details that are not relevant to the teleodynamic processes they potentially impact. But this simplification *has* causal consequences.

Seeing this clustering by simplification as necessitating epiphenomenality turns out to be both right and wrong at the same time. It is right when it comes to providing a reliable predictor of causal properties that are present irrespective of organism discernment. It is wrong, however, when one includes organism agency as a causal factor.

Both organism adaptations and mental representations necessarily ignore vastly many physical properties of things, fail to generate distinctions that correspond to physical differences, and lump together phenomena into general types for what might be described as pragmatic reasons. This categorical fallibility is even true for scientific knowledge. It is, for example, an historical commonplace for scientific investigation to reveal that phenomena once treated as members of a common general type are in fact derived from radically different origins and have radically different causal properties. So nominalism is, in this basic sense, well supported by the ubiquity of human error. It is captured in a catch phrase attributed to the humanistic psychologist Abraham Maslow (which he intended as a criticism of therapists trained in only one tradition): "If the only tool you have is a hammer, you will tend to treat everything as a nail."[4]

But physical responses, perceptions, and mental categories aren't merely passive reflections on the world; they exist to structure adaptation to the world. For this reason, the mere resemblance of an object to a perceptual class can be what causes that object to be modified in a particular way by an animal or person.

Consider, for example, the case of the boy selecting beach stones to skip across the surface of the water. The various stones he chooses may only have a few superficial features in common, but these features are what matter. The general attributes of being of a certain size, shape, solidity, and weight, as crudely assessed by comparison to a remembered type, thereby determine the production of another specific general attribute: the fact of skipping across the water. And so these attributes, constituting a vague general type maintained in the memory of a child, have played a determinate role in the relative location of stones with respect to the water's edge. Although we noticed this loophole in our earlier discussion, it is now possible to dissect the causal structure of this aspect of agency more carefully.

This augmented efficacy of generals—even such an abstract general as a "skippable stone"—is a feature that emerges with teleodynamics. Consequently, it is not just limited to minds; it is an attribute of life itself. Since the dynamical constraints that constitute teleodynamic processes are themselves not specific individuals, but effectively the recursive perpetuation of specific absences, in whatever way they become insinuated into extrinsic physical processes they interact only with respect to properties that are relevant to this teleodynamic perpetuation. Other properties are incidental to the work that results. Thus plants will grow toward artificial light as they would toward sunlight, and children (and dieters) will consume artificially sweetened candy as if it contained nutritive sugars.

Brains have elaborated this causal realism to an extreme, and minds capable of symbolic references can literally bring even the most Platonic of conceptions of abstract forms into the realm of causal particulars. To list some extreme but familiar examples, a highly abstract concept like artistic beauty can be the cause of the production of vastly many chiseled marble analogues of the human female form; a concept like justice can determine the restriction of movement of diverse individuals deemed criminal because of only vaguely related behaviors each has produced; and a concept like money can mediate the organization of the vastly complex flows of materials and energy, objects and people, from place to place within a continent. These abstract generals unquestionably have both specific and general physical consequences. So human minds can literally transform arbitrarily created abstract general features into causally efficacious specific physical events.

THE EXTENTIONLESS *COGITO*

The picture of human "selfness" that emerges from this account, then, is neither eliminative nor preformationist. It is not all or none. It is graded in its level of differentiation, both in its initial development and in its moment-to-moment dynamics. There is no ghost in the organic machine and no inner intender serving as witness to a Cartesian theater. The locus of self-perspective is a circular dynamic, where ends and means, observing and observed, are incessantly transformed from one to the other. Individuation and agency are intrinsic features of the teleodynamics that brains have evolved to generate, because of the dynamical closure, constraint generation, and self-maintenance that defines teleodynamics. However, the neurologically mediated self exhibits a higher-order form of teleodynamics than is found at any other level of life. This is because the teleodynamics of brain functions that evolved to guide animals' locomotion and their capacity to physically modify their environments inevitably must model itself. The self-referential convolution of teleodynamics is the source of a special emergent form of self that not only continually creates its self-similarity and continuity, but also does so with respect to its alternative virtual forms.

Thus autonomy and agency, and their implicit teleology, and even the locus of subjectivity, can be given a concrete account. Paradoxically, however, by filling in the physical dynamics of this account, we end up with a non-material conception of organism and neurological self, and by extension, of subjective self as well: a self that is embodied by dynamical constraints. But constraints are the present signature of what is absent. So, surprisingly, this view of self shows it to be as non-material as Descartes might have imagined, and yet as physical, extended, and relevant to the causal scheme of things as is the hole at the hub of a wheel.

16

SENTIENCE

*. . . sentience—without it there are no
moral claims and no moral obligations.
But once sentience exists, a claim is made, and
morality gets "a foothold in the universe."*

—WILLIAM JAMES, 1897[1]

MISSING THE FOREST FOR THE TREES

Do you ever worry that turning off your computer, or erasing its memory, or just replacing its operating system could be an immoral act? Many serious scientists and philosophers believe that brains are just sophisticated organic computers. So maybe this should be a worry to them. Of course, maybe turning off the machine is more analogous to sleep or temporary anesthesia, because no data need be lost or analytic processes permanently disrupted by such a temporary shutdown. But erasing data or corrupting software to the point that it is unusable does seem to have more potent moral implications. If you do sometimes contemplate the moral implications of these activities, it is most likely because you recognize that doing so could affect someone else who might have produced or used the data or software. The morality has to do with the losses that these potential users might suffer, and this would be of little concern were all potential users to suddenly disappear (though "potential" users might be an open-ended class). Aside from the issue of potential harm to users, the nagging question is whether there is "someone" home, so to speak, when a computation is being performed—something intrinsically subjective about the processing of data through the CPU of the mechanism. Only if the computer or computational process can have what amounts to experiences and thus possesses *sentience* is there any intrinsic moral issue to be considered.

Indeed, could your computer's successful completion of a difficult operation also produce algorithmic joy? Does your computer suffer when one of its memory chips begins to fail? Could a robotic arm in a factory take pride in the precision with which it makes welds? If possible, would such a normative self-assessment require a separate algorithm, or would it be intrinsic to successful or unsuccessful computation of the task? One might, for example, create an additional algorithm, equipped with sensors to assess whether a given task was accomplished or not. In the case of failure, it could modify the operation to more closely approach the target result next time around. In fact, this is the way many "neural net" algorithms work. One such approach, called *back propagation*, incrementally modifies connection strengths between nodes in a network with respect to success or failure criteria in "categorizing" inputs. In such a computation, is there anything like intrinsic feeling? Or is this just an algorithm modifying an algorithm? The normative value that comes into existence with sentience is special. No other physical process has intrinsic ethical status, because no other type of physical process can suffer. Explaining the emergence of sentience thus must at the same time be an explanation of the emergence of ethical value. This is a tall order.

When it is carefully distinguished from any specific content that is the focus of mental experience, the background "feeling of being here" is sometimes just described in terms of a distinctive quality to experience—or *quale*, to use the technical jargon of philosophers. This quality has a perspective, an internal and private locus, a self-reference frame with respect to which non-self content is discriminated. Following William James, I will refer to this core feature of conscious experience as sentience. The term *sentience* derives from the Latin, and literally means "feeling."

In what follows, I will argue that the sentience of human experience is only one highly differentiated form of a far more widespread and diverse phenomenon, and that it has its roots in far simpler processes that only dimly resemble mental experience. I will argue that this applies even to the way that organisms are sensitive to influences from their environment, but not just in the way that most material objects can be modified by inter-action with other objects and forces. Because organisms are teleodynamic systems, they do not merely react mechanically and thermodynamically to perturbation, but generally are organized to initiate a change in their inter-nal dynamics to actively compensate for extrinsic modifications or inter-nal deficits. Feeling is in this most basic sense active, not passive, and is a direct consequence of teleodynamic organization because of its incessant end-directed nature. Since there can be higher-order forms of teleodynamic processes, emergent from lower-order teleodynamic processes, we should not be surprised to find that there are higher-order emergent forms of sen-tience as well, over and above those of the simpler cellular components of the body and nervous system.

The core hypothesis of this book is that all teleodynamic phenomena necessarily depend upon, and emerge from, simpler morphodynamic and homeodynamic processes. This implies that the complex intentional features that characterize our thoughts and subjective experiences must likewise emerge from a background of neurological morphodynamic and homeo-dynamic processes. Moreover, these lower-order subvenient dynamical fea-tures must also inevitably constitute significant aspects of our mental lives. I believe that it is impossible to even approach issues of sentience without taking the necessary contributions of homeodynamic and morphodynamic aspects of mental experience and brain function into account. Once we do so, however, we will discover new ways of asking old questions about the relationship between minds and brains, and perhaps even find ways to rein-tegrate issues of subjective value into the natural sciences.

Reframing the concept of sentience in emergent dynamical terms will allow us to address questions that are not often considered to be subject to empirical neuroscientific analysis. Contrary to many of my neurosci-ence colleagues, I believe that these phenomena are entirely available to scientific investigation once we discover how they emerge from lower-level teleodynamic, morphodynamic, and thermodynamic processes. Even the so-called hard problem of consciousness will turn out to be reconceptual-ized in these terms. This is because what appeared to make it hard was our predisposition to frame it in mechanistic and computational terms, pre-suming that its intentional content must be embodied in some material or energetic substrate. As a result, the vast majority of descriptions of brain function tend to be framed in terms that not only fail to make the connection between the cellular-molecular processes at one extreme and the intentional features of mental experience at the other; they effectively pretend that making sense of this relationship is irrelevant to brain function. Descriptions of psychological and neurological processes remain unequivocally on one side or the other of this classic divide—inescapably Cartesian.

Whether we imagine that we can reduce all intentional properties of mind to the "syntactic" features of neural "computations" and avoid directly addressing issues of mental content (e.g., the computational theory of mind), or describe them only in qualitative phenomenal terms (e.g., phenomenological theories) that merely assume the existence of what we hope to explain, we still end up ignoring a fundamental step in the ladder joining material and intentional properties. And although dynamical systems approaches do a better job of accounting for the non-mechanistic features of brain function, they too are at base founded on the assumption that the content of mental representations must be physically embodied by some specific neural substrate or else be epiphenomenal.

Current techniques for correlating mental experience with brain activity using *in vivo* imaging technologies (fMRI, PET, and MEG) have recently enabled us to correlate local metabolic levels with experiential phenomenology (i.e., performing a mental task). But precisely because these tools only provide an assay of the regional thermodynamics of neuronal signaling activity, they must ignore the necessary contribution of any supervening levels of dynamics. In previous discussions of life—and of information more generally—we have outlined principled reasons to think that there can be no simple mapping between ententional and mechanistic (teleodynamic and thermodynamic) properties, even though whatever is ententional has emerged from a statistical mechanistic base. Analogously for neurological processes, it will turn out that these ignored levels of emergent dynamical processes are essential for bridging between brain processes and mental experiences.

The ententional features of life cannot be directly mapped to specific physical substrates or chemical processes, and yet they nevertheless dynamically supervene on processes involving these physical correlates. The dynamical requirements for generating ententional phenomena turn out to be highly constrained by the need to both generate and preserve the constraints essential to maintain these processes. So the functional properties of life only emerge when homeodynamic and morphodynamic processes are organized in precisely complementary and completely reflexive ways with respect to one another. If, as I believe, an analogous emergent dynamic infrastructure is necessary to produce *any* ententional property, then it must also apply to the generation of mental *intentionality*. Moreover, traces of this hierarchic dynamical dependency should be reflected in the very structure of the experience of perceiving, thinking, or acting. In other words, we should be able to find the signature of these emergent dynamical levels in the details of brain process and in the very quality of subjective experience. The ultimate goal of this chapter, then, will be to trace the relationships between neurally embodied forms of homeo-, morpho-, and teleodynamic processes and various aspects of mental experience.

As in the case of organism evolution, it is only by examining the dynamics of these lower-level emergent processes that we will be able to adequately explain the sentience, representation, perspective, and agency that are the hallmarks of mental experience. By reframing the problem in these dynamical terms, I believe we will discover that rather than being the ultimate "hard problem" of philosophy and neuroscience, the subjective features of neural dynamics are the expected consequences of this emergent hierarchy. The so-called mystery of consciousness may thus turn out to be a false dilemma, created by our failure to understand the causal efficacy of emergent constraints.

Before we can attempt this reframing, however, we need to get beyond the idea that we might be able to explain mental processes in computational or simple dynamical terms. This is not a trivial demand, since the computational approach to mental processes and neuronal interactions has been enormously successful at clearing away many prejudices and confusions about the mind/brain relationship. The analytic tools it provided heralded a veritable quantum leap in experimental and theoretical sophistication in cognitive neuroscience. And dynamical systems approaches have begun to attract serious attention as well, especially as researchers struggle to deal with the enormously complex dynamics of signal processing in a richly connected network of billions of nodes. So, before we even begin a reframing of neural and mental processes in emergent dynamic terms, it is necessary to understand why the presumed direct correspondence between neural algorithms, signal dynamics, and cognitive operations inevitably fails to even address issues of intentionality, much less the mysteries posed by sentient experience.

The point of such a critique is not to argue that we should abandon the physical modeling of brain processes—indeed, careful analysis of the details of brain processes is of critical importance to an emergent dynamic approach as well—but rather to overcome the desire to directly map mental phenomena onto neuronal phenomena. As the logic of emergent dynamics has repeatedly shown, the physical constituents of living process are only relevant to ententional phenomena insofar as they contribute to the generation of teleodynamic processes. To make sense of conscious intentionality, and ultimately subjective sentience, we need to look beyond the neuronal details to explore the special forms of teleodynamic constraints they embody and perpetuate. I believe that only by working from the bottom up, tracing the ascent from thermodynamics to morphodynamics to teleodynamics and their recapitulation in the dynamics of brain function, will we be able to explain the place of our subjective experience in this otherwise largely insentient universe.

SENTIENCE VERSUS INTELLIGENCE

Do computers think? Indeed, one still sometimes hears computers being called "thinking machines." The idea that cogitating is a form of computing has a long history, stretching back to the first half of the twentieth century. In fact, the term *computer* used to designate a person whose job it was to do the calculations supporting the accounting needs of large companies. Slowly, over the course of the last century, mechanical aids were supplied to take the drudgery out of this work, including mechanical adding machines, slide rules, and eventually digital computers. The mathematician who ultimately began the revolution that replaced people with machines was Alan Turing. The so-called Turing test—which he suggested in a paper in 1950,[2] and which now bears his name—embodies the conjecture that if mechanized operations produce indistinguishable results from human operations on the same tasks, then the two are functionally identical. In the idealized test, we assess whether a computer can produce responses that are indistinguishable from those that a normal thinking person might produce under similar circumstances—for example, by posing questions for the computer to answer. If it can, then, presumably we have no justification to claim that these are fundamentally different processes.

While superficial operational identity may not be sufficient to prove that the two processes are ultimately identical in some deeper sense, it nonetheless is sufficient to question the assumption of difference. Thus a Turing test questioner might not be able to tell if she is being answered by an algorithm or by another person. It also provides a way to assess cognitive power, because there are good reasons to believe that algorithmic complexity and reasoning difficulty are correlated. If some task requires more operations to complete, it should take more time and resources to complete, whether by computer or person. However, whether or not one accepts this as a test of machine intelligence, it is inadequate for testing sentience. Intelligence and information-processing power are probably not irrelevant, but they appear to be assessing something quite distinct from sentience. This suggests that a very different causal architecture is likely involved in the production of intelligent versus sentient processes.

With the explosive growth of computing power in the past few decades, we have come to take for granted that quite complicated and intelligent behaviors can be simulated by computation. Assume, for example, that we had the support of massive computational power and had compiled a massive database of human mental and behavioral responses to all manner of adaptive challenges and cognitive tasks, and with a level of minute detail never before imagined. Using these data and some powerful predictive inferential tools, it should be possible to create algorithms capable of recreating in exquisite detail the kinds of responses characteristic of people adapting to normal life situations and solving difficult cognitive tasks. In other words, it is not an unrealistic thought experiment to imagine a computational system able to produce an output that can fool even the most sophisticated Turing tester: a Turing test analogue to Deep Blue, the IBM chess-playing computer that matched the world chess grand master (described in chapter 3). Whatever problem-solving domain such a computing system faced, it would do as well or better than any human. Of course, this context could not differ too much from the domains from which its reference data were derived. But in these conditions, this golem would act as savvy as any human counterpart; and yet, I would argue, it would be completely insentient.

This is effectively an elaboration of an argument presented by the philosopher John Searle with his well-known "Chinese room" thought experiment.[3] Both help to demonstrate the difference between merely intelligent behavior and sentient (or conscious) behavior. In Searle's imaginative scenario, a man who is unfamiliar with the Chinese language sits in a room that only gives him access to pages of Chinese characters, which are slipped in through a slot. His task is to match these to an identical string of characters in a book (or massive database) that prescribes what characters to send back out, given that specific input. To those on the outside, the characters he produces constitute a very sensible response to a question posed in Chinese by the characters fed to him; but to the man inside, it is a simple pattern-matching task. Searle argues that this shows how such a simulation can be produced in the absence of any cognizance or subjective involvement with the meaning of the task. If a man can produce such behavior "mindlessly," then when a machine does it in an analogous way, we can reasonably assume that it is not aware of what it is doing either. Of course, the man's pattern-matching behavior is accomplished consciously; but since we can more easily imagine that pattern matching can be automated without any reference to the meaning or purpose of these activities, we intuitively appreciate the force of Searle's argument.

In this sense, a machine, or any highly sophisticated algorithm plus database that captures vast quantities of useful isomorphisms between inputs and outputs required to perform a given task, might rightly be described as exhibiting intelligence, *irrespective of making any claim about whether it is sentient*. Intelligence is about making adaptively relevant responses to complex environmental contingencies, whether conscious or unconscious. In this sense, we can analogously describe the increasing complexity, flexibility, and fittedness of evolved adaptations as a kind of embodied intelligence. And, of course, many of our most reliable adaptive capacities are produced with only limited, if any, awareness and forethought. The dissociability of intelligent and sentient functions does not imply that a Turingesque assessment of sentience is impossible, however. It only suggests that it would be a very different sort of assessment than a Turing test of intelligence. So how might it differ?

Consider the possibility that we could extend the "test" to give us access to the "behavior" of the constituent operations that are used to perform the task, and not just the input-output relationship. What additional insight could be gained from access to the details of how the operation was performed? Searle's Chinese room scenario implicitly suggests that this should make a difference. It is precisely by reflecting on *how* the process is accomplished by the man in the room that Searle argues that we know that the apparent "interpretation" of the symbols is illusory, and that it is instead accomplished without an awareness of what is being done. Our knowledge of how the man in the room is performing the task clues us in to the fact that it can be done "mindlessly," so to speak—such as by a simple computer pattern-matching algorithm and look-up table linked to a large database. But if we can discern that this process has the structure of a blind mechanism, and can confidently conclude from this that it is not a sentient process, shouldn't there also be some criterion that would allow us to determine when the task *is* being accomplished with awareness of what is being done?

Possibly. Critics might argue that even if we can determine when the process is mindless, this does not guarantee that we could recognize what a sentient organization of the process looks like, or that the process could ever be done otherwise. Nevertheless, it seems reasonable to assume that if we can recognize the ways that the process can be performed mindlessly, at least we should be able to eliminate cases that are non-sentient, if they are not too complicated. It could be the case, however, that there will be ambiguous examples. Actually, I think that we need not worry about the issue of sentient processes being too complex to assess. I believe that it is not so much complexity that matters, but dynamical architecture. For the same reason that we were able to judge that a system as simple as an autogen exhibits intentional properties (and thus were not forced to deal with the complexities of living organisms), I believe that we will be able to distinguish the emergent dynamical architecture which produces the intentional properties of mental processes. Not only do I believe that we *can* discern whether a process is sentient, but once we understand the basic criteria for making such a judgment, I believe that we will also be able to make an assessment about *how sentient* this process is! The challenge is, of course, to develop a clear model of this emergent dynamical architecture.

The first step will be to outline the ways that current computational and dynamical systems models of mental processes exemplify mechanistic and morphodynamic models of living processes, respectively. As with the understanding of the deep logic of life, we will find that these ways of conceiving of cognition fall short of explaining or even acknowledging the reality of intentional properties. Ignoring the emergence of teleodynamic processes, such properties can only have epiphenomenal status in these theories. I will argue that we lack a naturalistic account of sentience because of a similar failure to understand how the teleodynamics of sentience emerges from morphodynamic and thermodynamic/homeodynamic processes of nervous system function. A final challenge of this analysis will be to show how higher-order levels of sentience emerge from, and depend on, these lower levels.

THE COMPLEMENT TO COMPUTATION

Although there is little debate about whether the operations of human thinking and feeling involve physical processes taking place in the brain, the question that divides the field is whether the physical account as framed in terms of computation, even in its widest sense, is sufficient to explain mental experience. The computational (machine) analogy has been undeniably useful. Besides providing a methodology for formulating more precise models of cognitive processes that offer testable empirical predictions, it offered a clue to the resolution of the classical paradox concerning the bifurcation of the mental and physical realms. Computation assumes that information processes correspond to the manipulation of specific physical tokens. As a model of cognition, it implies that the processes of perceiving and reasoning about the world are also constituted by algorithms "instantiated" as physical processes, whether in symbolic or subsymbolic form. And if mental processes can be fully understood in terms of the organized manipulation of physical tokens, then the operations of thought will inherit whatever physical consequences these manipulations might be capable of. Simply stated: If computation is a mechanical process and thought is computation, then thoughts can be physically efficacious. The mind/body problem disappears. Or does it?

I take this to be an invaluable insight. However, there are two problems that it leaves us with. First, it does not in itself commit us to any specific theory concerning the nature of the physical process that constitutes the representational and sentient capacities of the brain. It applies quite generally to any account of cognition that is based on a physical process, not merely what is usually meant by computation. If what we mean by computation is *any* physical process used for sensing, representing information, performing inferences, and controlling effectors, then it would seem that to deny that brains compute is to also deny the possibility of any physical account of mental processes. If all that was meant was that thought is a physically embodied process, there would be little controversy, and little substance to the claim. But the concept of computation, even in this most broadly stated form, is not as generic as this. Implicit in this account is the assumption that computation is constituted by a specific physical process (such as a volley of neuronal action potentials or the changes in connection weights in a neural net) that corresponds to a specific teleological process (such as solving a mathematical operation or categorizing a physical pattern).

This assumption is the basis for a second, even more serious problem. What constitutes correspondence? And what is the difference between just any physical correspondence (the swirling of the wind and the rotation of a merry-go-round) and one that involves representation? In all computational theories—including so-called neural net theories—representation is understood in terms of a rule-governed mapping of specific extrinsic properties or meanings to correspondingly specific machine states. Although this may at first sound like an innocuous claim, it actually begs the fundamental question: How is representation constituted? Simply asserting the existence of a mapping between properties in the word and properties of a mechanistic process presumed to take place in the brain implicitly frames the relationship in dualistic terms. There are two domains: the physical processes, and whatever constitutes the representational relationships between them. Everything depends on the nature of the correspondence, and specifically, what makes it different from either an isomorphic or an arbitrary one-to-one mapping between components or phases of two distinct processes. Computation is merely a way of describing (or prescribing) machine processes. It is not an intrinsic property of any given machine process.

There is an instructive parallel between the logic of computation and the logic of natural selection, as we described it in the previous chapter. As we've seen, although natural selection in any form requires processes that generate, reconstitute, and reproduce these selectively favored traits, the specific details of these processes are irrelevant to the logic and the efficacy of the process. Natural selection is in this sense agnostic with respect to how these adapted traits are physically generated. It is an abstraction about certain kinds of processes and their consequences. Similarly, the logic of computation is defined in a way that is agnostic to how (or whether) representation, inference, and interaction with the world are implemented. These issues are bracketed from consideration in order to define the notion of an algorithm: an abstract description of a machine process that corresponds to a *description of* or *instructions for* performing a semiotic process, such as solving a mathematical equation or organizing a list. This agnosticism is the source of the power of computation. As Alan Turing first demonstrated, a generic machine implementation can be found for any reasoning process (or semiotic process in general) that can be precisely described in terms of a token-manipulation process. This generic description, an algorithm, and its generic mechanistic assumptions, guarantee its validity in any of an infinite number of possible physically manifested forms.

As we discovered in the case of biological evolution, however, there is only a rather restricted range of dynamics that can actually fulfill the criteria required for natural selection to occur. Analogously, in the case of mental processes, we will find that only a very restricted form of dynamics is capable of constituting mental representation. So it cannot be merely assumed to exist by fiat without ignoring what will likely turn out to be important restrictions. Further, without representation, computation is just machine operation, and without human intentionality, machine operation is just physical change. So, while at one time it was possible to argue that the computational conception of the mind/brain relationship was the only viable game in town, it now appears that it never really addressed the problem at all, but only assumed that it didn't matter.

Whether described in terms of machine code, neural nets, or symbol processing, computation is an idealized physical process in the sense that the thermodynamic details of the process can be treated as irrelevant. In most cases, these physical details must be kept from interfering with the state-to-state transitions being interpreted as computations. And because it is an otherwise inanimate mechanism, there must also be a steady supply of energy to keep the computational process going. Any microscopic fluctuations that might otherwise blur the distinction between different states assigned a representational value must also be kept below some critical threshold. This insulation from thermodynamic unpredictability is a fundamental design principle for all forms of mechanism, not just computing devices. In other words, we construct our wrenches, clocks, motors, musical instruments, and computers in such a way that they can only assume a certain restricted number of macro states, and we use inflexible metals, regularized structures, bearings, oil, and numerous thresholds for interactions, in order to ensure that incidental thermodynamic effects are minimized. In this way, only those changes of state described as functional can be favored.

A comparison with what we have learned in previous chapters about living processes can begin to remedy some of the limitations of this view. Although living processes also must be maintained within quite narrow operating conditions, the role that thermodynamic factors play in this process is basically the inverse of its role in the design of tools and the mechanisms we use for computation. The constituents of organisms are largely malleable, only semi-regular, and are constantly changing, breaking down, and being replaced. More important, the regularity achieved in living processes is not so much the result of using materials that intrinsically resist modification, or using component interactions that are largely insensitive to thermodynamic fluctuation, but rather due to using thermodynamic chemical processes to generate regularities; that is, via morphodynamic organization.

Generally speaking, these two distinct design strategies might be contrasted as regularity achieved by a means of thermodynamic isolation in machines, and regularity achieved by a means of thermodynamic openness in life. In addition, there is a fundamental inversion with respect to the source of constraints that determine the critical dynamical properties. In machines, the critical constraints are imposed extrinsically, from the top down, so to speak, to defend against the influence of lower-level thermodynamic effects. In life, the critical constraints are generated intrinsically and maintained by taking advantage of the amplification of lower-level thermodynamic and morphodynamic effects. The teleological features of machine functions are imposed from outside, a product of human intentionality. The teleodynamic features of living processes emerge intrinsically and autonomously.

But there is more than just an intrinsic/extrinsic distinction between computation and cognition. Kant's distinction between a mere mechanism and an organism applies equally well to mental processes. To rephrase Kant's definition of organism as a definition of mind, we could describe mind as possessing a self-amplifying, self-differentiating, self-rectifying formative power. A computation exhibits merely motive power but a mind also exhibits formative power. Or to put this in emergent dynamical terms: *computation only transfers extrinsically imposed constraints from substrate to substrate, while cognition (semiosis) generates intrinsic constraints that have a capacity to propagate and self-organize.* The difference between computation and mind is a difference in the source of these formal properties. In computation, the critical formal properties are descriptive distinctions based on selected features of a given mechanism. In cognition, they are distinctive regularities which are generated by recursive dynamics, and which progressively amplify and propagate constraints to other regions of the nervous system.

Like the mechanistic conception of life, the computational theory of mind, even in its most general form, assumes implicitly that the physical properties that delineate an information-bearing unit are only extrinsically defined glosses of otherwise simple physical mechanics. It also assumes that the machine operations that represent semiotic operations of any given type (e.g., face recognition or weather prediction) are also merely externally imposed re-descriptions of some extrinsically assessed correspondence relationship. Thus the mapping from computer operation to some interpreted physical process is nothing more that an imposed descriptive gloss. In the same way that Shannonian information, prior to interpretation, is only potentially information about something, the operations of a computing device before they are assigned an interpretation are also just potential computational operations. What we need is an account where the relevant interpretive process is intrinsic, not extrinsic.

COMPUTING WITH MEAT

In a now famous very short story about explorers from an alien planet investigating the source of a radio broadcast from Earth, the science fiction writer Terry Bisson portrays a conversation they have about the humans they've discovered. Taking their dialogue from a midpoint in the discussion:

"They're meat all the way through."

"No brain?"

"Oh, there is a brain all right. It's just that the brain is made out of meat!"

"So . . . what does the thinking?"

"You're not understanding, are you? The brain does the thinking. The meat."

"Thinking meat! You're asking me to believe in thinking meat!"

"Yes, thinking meat! Conscious meat! Loving meat. Dreaming meat. The meat is the whole deal! Are you getting the picture?"

"Omigod. You're serious then. They're made out of meat."[4]

With exaggerated irony, Bisson points out that we humans are pretty sloppy computing devices. It may be tongue-in-cheek, but it hints at an important challenge to the computer theory of mind: the design principles of brains just don't compute! Consider the basic building blocks of brains compared to those necessary to build devices with the formal properties that are required to produce computations. Neurons are living cells. These cells have been adapted, over the course of evolution, to use some of their otherwise generic metabolic and intercellular communication capabilities for the special purpose of conveying point-to-point signals between one another over long (in cellular terms) distances. Because they were not "designed" for this function, but recruited during the course of evolution from cells that once had more generic roles to play, they tend to be a bit unruly, noisy, and only modestly reliable transducers and conveyors of signals. Additionally, they still have to perform a great many non-signal transmission tasks that are necessary for any cell to stay alive, and this too constrains their capacity to perform purely signaling functions.

But how unruly? It would probably not be too far off the mark to estimate that of all the output activity that a neuron generates, a small percentage is precisely correlated with input, while at least as much is the result of essentially unpredictable molecular-metabolic noise; and the uncorrelated fraction might be a great deal more.[5] Neurons are effectively poised at the edge of chaos, so to speak. They continually maintain an ionic potential across their surface by incessantly pumping positively charged ions (such as the sodium ions split from dissolved salt) outside their membrane. On this electrically unstable surface, hundreds of synapses from other neurons are tweaking the local function of these pumps, causing or preventing what amount to ion leaks, which destabilize the cell surface. As a result, they are constantly generating output signals generated by their intrinsic instability and modified by these many inputs.

In addition, neural networks, in even modestly large brains (such as those of small vertebrates), are quite degenerate in the sense that precise control of connectivity is not obtained by the minimal genetic specification that comes from embryogenesis, and from the remarkably limited number of molecular signals that appear to be determining cell phenotypes and connections, when compared with the astronomical number of precise connections that are in need of specification. In fact, contrary to what might be expected, large mammalian brains (such as our own) do not appear to use significantly more genetic information for their construction than the smallest mammalian brains (such as mouse brains), even though the numbers of connections must differ by many orders of magnitude. This is a humbling thought, but it is not the whole story. The specification of brain architecture is highly conserved across the entire range of animal forms, and in vertebrate brains connection patterns are also significantly dependent upon body structure and environmental information for their final specifications. The point of bringing this up in the present context is simply to make it clear that brains are not wired up in point-to point detail. Not even close. So we could hardly expect neuronal connection patterns to be the biological equivalent of logic gates or computer circuits.

In summary, it doesn't appear on the surface that brains are made of signal-processing and circuitry components that are even vaguely well suited to fill the requirements for computations in the normal sense of this word.

So what *is* minimally required to carry out computations in the way a computer does them? In a word, predictability. What about messy neural computers? What happens if the circuits in a computer happen to garble the signals—as one should expect would happen in a biocomputer? The "virtual machine" breaks down, of course. Sort of like damaging a gear in a real machine, it will at the very least fail to function as expected, fail to compute, and it may even cease to operate altogether: it might crash. Now imagine that all the connections and all the logic elements in your computer were a little unpredictable and unreliable. Certainly, computing in the typical way would be impossible under these conditions. But this infirmity doesn't by itself rule out biological computing in brains.

The first computer designers had to take this problem seriously, because at the time, in the late 1940s and early fifties, the tens of thousands of logic elements needed for a computer were composed of vacuum tubes and mechanical relays, and tended to be somewhat unreliable. To deal with this, they employed a technique long used for dealing with other forms of unreliable communication, and which had been mathematically formalized only a few years earlier by another pioneer of information technology, Claude Shannon. As we saw earlier, the answer is to include a bit of redundancy in the process—send the message many times in sequence, or send multiple messages simultaneously, or do the same computation multiple times, and then compare the results for consistency. So, even if brains are made of messy elements, it is at least conceivable that some similar logic of noise reduction might allow brains to compute, if with some inefficiency due to the need for redundant error-correcting processes. At least in principle, then, brains could be computers, provided the noise problem wasn't so bad that error correction doesn't get too far out of hand.

Unfortunately, there are a number of reasons to think that it *would* get out of hand. The first is that brains the size of average mammal brains are astronomically huge, highly interconnected, highly re-entrant networks. In such networks, noise can tend to get wildly amplified, and even very clean signal processing can produce unpredictable results; "dynamical chaos," it is often called. But additionally, many of the most relevant parts of mammal brains for "higher" cognitive functions include an overwhelmingly large number of excitatory connections—a perfect context for amplifying chaotic noisy activity.

We are forced to conclude from this list of computationally troublesome characteristics of brains that they appear to be organized according to a very different logic than one would expect of a computing device in any of the usual senses of the word. Indeed, if we turn the question around and ask what these structural features seem to suggest, we would conclude that noise must be a "design feature" of brains, not a "bug" to be dealt with and eliminated. But does this make any sense? Is it a realistic possibility that a major function of brains is to generate and communicate and amplify noise, such as is produced by the random variations of molecular metabolic processes in nerve cells? Is it even possible to reconceive of "computation" in a way that could accommodate this sort of process? Or will we need to scrap the computational model altogether for something else? And if something else, what else?

In fact, there is another approach. It is suggested by the very facts that make the computational approach appear so ill-fitted: by the significant role played by noise in living brains, and by a neural circuit architecture which should tend to amplify rather than damp this noisiness. We need to think of neural processes the way we think about emergent dynamic processes in general. Both self-organizing and evolutionary processes epitomize the way that lower-order unorganized dynamics—the dynamical equivalent of noise—can under special circumstances produce orderliness and high levels of dynamical correlations. Although unpredictable in their details, globally these processes don't produce messy results. This is the starting point for a very different way to link neuronal processes to mental processes.

Life and sentience are deeply interrelated, because they are each manifestations of teleodynamic processes, although at different levels. Sentience is not just a product of biological evolution, but in many respects a microevolutionary process in action. To put this hypothesis in very simple and unambiguous terms: the experience of being sentient is what it feels like to *be* evolution. And evolution is a process dependent on component teleodynamic processes and their relationships to the environment on which they in turn depend, and on the perturbing influence of the second law of thermodynamics—noise, and its tendency to degrade constraint and correlation.

The presumption that we can equate the information constituting mental content to the patterns produced by specific neuronal signals or even groups of signals, as suggested by the computer metaphor, is in fact an order of magnitude too low. These signals are efficacious, and indeed control body functions, though in vertebrate nervous systems the efficacy of neural signals is almost certainly a product of activity statistics rather than specific patterns. These are sentient at the level of these cells and their relations to other cells, but do not reach the level of whole brain sentience. That is a function of dynamical regularities at a more global level. The activity of any single neuron is no more critical to the regularities of neuronal activity that give rise to experience than are the movements of individual molecules responsible for Bénard cell formation. Focusing on the neuronal level treats these individual processes as though they are computational tokens, when instead they are more like individual thermodynamic fluctuations among the vast numbers of interacting elements that contribute to the formation of global morphodynamic effects.

Focusing on the way that the activity of individual neurons correlates with specific stimuli is a seductive methodological strategy. Indeed, as external observers of these correlations, neuroscientists can even sometimes predict the relevant stimulus features or behaviors that individual neurons are responding to. So the search for stimulus correlations with neuronal activity in experimental animals has become one of the major preoccupations of neurophysiology. In this respect, large brains are not as stochastic in their organization as many non-biological thermodynamic systems. Interactions between neurons are not merely local, and are far more complex and context-sensitive than are simple dynamical interactions. The highly plastic network structures constituting brains provide the ground for a vast range of global interaction constraints which can be effected by adjustments of neuronal responsiveness, and which cycles of recursive neuronal activity can amplify to the point of global morphodynamic expression.

But neurophysiologists are extrinsic observers. They are able to independently observe both the activity of neurons and the external stimulus presented to them, whether these stimuli are applied directly in the form of electrical stimulation or indirectly through sensory pathways. Yet this in no way warrants the inference that there is any intrinsic way for the nervous system to likewise interpret these regularities, and indeed, there is no independent way to compare stimuli with their neuronal influences. The computational metaphor depends on such an extrinsic homunculus to assign an interpretation to any given correlation, but this function must arise intrinsically and autonomously in the brain. This is another reason to expect that the phenomena which constitute sensory experiences and behavioral intentions are dynamically supervenient on these neuronal dynamics and not directly embodied by them. Since morphodynamic regularities emerge as the various sources of interaction constraints compound with one another, they necessarily reflect the relationship between network structural constraints and extrinsic constraints. And different morphodynamic processes brought into interaction can generate orthograde or contragrade effects, including teleogenic effects.

Another important caveat is that correlations between neuronal activity patterns in different brain regions, and between neuronal activity and extrinsic stimuli, are not necessarily reflections of simple causal linkages. They may be artifacts of simply being included in some higher-order, network-level morphodynamic process, in the same way that causally separated but globally correlated dynamics can emerge even in quite separated regions in non-biological morphodynamic processes. For example, recall that molecules separated by many millimeters in separate Bénard cells within a dish can exhibit highly correlated patterns of movement, despite being nearly completely dynamically isolated from one another. So, in accord with the classic criticisms of correlation theories of reference (such as those discussed in chapters 12 and 13), neural firing correlation alone is neither a necessary nor sufficient basis for being information about anything. Interpretation requires something more, something dynamic: a self, and a self is something teleodynamic.

A focus on correlative neuronal activity as the critical link between mental and environmental domains has thus served mostly to reinforce dualism and to justify eliminative materialism. When the neurophysiologist making these assessments is excluded from the analysis, there is no homunculus, only simple mechanistic causal influences and correlations. The road to a scientific theory of sentience, which preserves rather than denies its distinctiveness, begins with the recognition of the centrality of its dynamical physical foundation. It is precisely by reconstructing the emergence of the teleodynamics of thought from its neuronal thermodynamic and morphodynamic foundations that we will rediscover the life that computationalism has drained from the science of mind.

FROM ORGANISM TO BRAIN

Organisms with nervous systems, and particularly those with brains, have evolved to augment and elaborate a basic teleodynamic principle that is at the core of all life. Brains specifically evolved in animate multicelled creatures—animals—because being able to move about and modify the surroundings require predictive as well as reactive capacities. The evolution of this "anticipatory sentience" —nested within, constituted by, and acting on behalf of the "reactive (or vegetative) sentience" of the organism—has given rise to emergent features that have no precedent. Animal sentience is one of these. As brains have evolved to become more complex, the teleo-dynamic processes they support have become more convoluted as well, and with this the additional distinctively higher-order mode of human symboli-cally mediated sentience has emerged. These symbolic abilities provide what might be described as sentience of the abstract.

Despite its familiarity, however, starting with the task of trying to explain first-person human conscious experience, even with all the wonderful new tools of *in vivo* brain imaging, cellular neuroscience, and human introspec-tion at our fingertips, is daunting. We scarcely have a general theory of how the simplest microscopic brains operate, much less an understanding of why a human brain that is just slightly larger and minimally different in its orga-nization from most other mammal brains is capable of such different ways of thinking. Although it may not address many questions that are of inter-est to those who seek a solution to the mystery of human consciousness, I propose to take a different approach than is typical of efforts to understand consciousness. I believe we can best approach this mystery, as we have suc-cessfully approached other natural phenomena, by analyzing the physical processes involved, but doing so in terms of their emergent dynamics. As with the problem of understanding the concept of self, we must start small and work our way upward. But approaching human subjectivity as an ana-logue of the primitive ententional properties of brainless organisms offers only a very crude guide. It ignores the fact that brain-based sentience is the culmination of vastly many stages in the evolution of sentience built upon sentience. So although we must build on the analogy of the teleodynamic processes of autogens and simple organisms, as did evolution, in order to avoid assuming what we want to explain, we must also pay attention to the special emergent differences that arise when higher-order teleodynamic processes are built upon lower-order teleodynamic processes.

But can we really understand subjective experience by analyzing the emergent dynamical features of the physical processes involved? Aren't we still outside? Although many philosophers of mind have argued that the so-called third-person and first-person approaches to this problem are fun-damentally incompatible, I am not convinced. The power of the emergent dynamic approach is that it forces us to abandon simple material approaches to ententional phenomena, and to take account of both what is present and what is absent in the phenomena we investigate. Although it does not promise to dissolve the interior/exterior distinction implicit in this problem, it may provide something nearly as useful: an account of the process that helps explain why such an inside (first-person) and outside (third-person) distinction emerged in the first place, and with it many of the features we find so enigmatic.

By grounding our initial analysis of ententional phenomena on a minimal exemplar of this emergent shift in dynamics—an autogen—we have been able to demystify many phenomena that have previously seemed outside the realm of natural science. Perhaps starting with an effort to explain the most minimal form of sentience can also yield helpful insights into the dynamical basis of this most personal of all mysteries. At the very least it can bring these seemingly incompatible domains into closer proximity. So let's begin by using the autogen concept, again, this time as a guide for thinking about how we might conceive of a Chinese room–like approach for claims about sentient processes.

This suggests that as a first step we can compare conscious mindful processes to the teleodynamic organization of a molecular autogen and compare the mindless computational processes (such as Searle envisions) to the simple mechanistically organized chemical processes that autogens are supervenient upon. In other words, understanding and responding in Chinese may be different from merely matching Chinese characters, the way autogenesis is different from, say, the growth of a crystal precipitating from a solution. The analogy is admittedly vague at this point, and the Chinese room allegory also besets us with an added human factor that could get in the way.[6] But considered in light of emergent dynamics, Searle intends to illuminate a similar distinction: the difference between a mechanistic (homeodynamic) process and an ententional (teleodynamic) process. If, as we have argued, autogenic systems exemplify precisely how ententional processes supervene on and emerge from non-ententional processes, then it should be possible to augment the Chinese room analogy in similar ways to precisely identify what it is missing. To do so would provide a useful metric for determining whether a given brain process is merely computational or also sentient.

Indeed, this analogy already suggests a first step toward resolving a false dichotomy that has often impeded progress in this endeavor: the false dichotomy between computation and conscious thought. This is not to say that conscious thought can be reduced to computation, but rather that computation in some very generic sense is like the thermodynamic chemistry underlying autogenic molecular processes. It is a necessary subvenient dynamic, but it is insufficient to account for the emergent dynamic properties we are interested in. And just as it was necessary to stop trying to identify the teleodynamics of life with molecular dynamics alone, and instead begin to pay attention to the constraints thereby embodied, it will be necessary to stop thinking of the computational mechanics as mapping to the intentional properties of mentality. In this way, we may be able to develop some clear criteria for a Turing test of sentience.

Of course, we will need to do more than just identify the parallels. We need to additionally understand how these teleodynamic features emerge in nervous systems, and what additionally emerges as a result of the fact that these teleodynamic processes are themselves emergent from the lower-level teleodynamic characteristic of the neurons that brains are made of.

In the following chapter, I offer only a very generic sketch of an emergent dynamics approach to this complex problem. Once we begin to understand how this occurs, and why, there will still be enormous challenges ahead: filling in the special details of how this knowledge can be integrated with what we have learned about the anatomy, physiology, chemistry, and signal-processing dynamics of brains. If this approach is even close to correct, however, it will mean that a considerable reassessment of cognitive neuroscience may be required. Not that what we've learned about neurons and synapses will need to change, or that information about sensory, motor, or cognitive regional specialization will be rendered moot. Our growing understanding of the molecular, cellular, connectional, and signal transduction details of brain function is indeed critical. What may need to change is how we think about the relationship between brain function and mental experience, and the physical status of the contents of our thoughts and expectations. What awaits is the possibility of a hitherto unrecognized methodology for studying mind/brain issues that have, up to now, been considered beyond the reach of empirical science.

17

CONSCIOUSNESS

Nobody has the slightest idea how anything
material could be conscious. Nobody even knows
what it would be like to have the slightest idea how
anything material could be conscious.

—JERRY FODOR, 1992[1]

THE HIERARCHY OF SENTIENCE

The central claim of this analysis is that sentience is a typical emergent attribute of any teleodynamic system. But the distinct emergent higher-order form of sentience that is found in animals with brains is a form of sentience built upon sentience. So, although there is a hierarchic dependency of higher-order forms of sentience on lower-order forms of sentience, there is no possibility of reducing these higher-order forms (e.g., human consciousness) to lower-order forms (e.g., neuronal sentience, or the vegetative sentience of brainless organisms and free-living cells). This irreducibility arises for the same reason that teleodynamic processes in any form are irreducible to the thermodynamic processes that they depend on. Nevertheless, human consciousness could not exist without these lower levels of sentience serving as a foundation. To the extent that sentience is a function of teleodynamics, it is necessarily level-specific. If teleodynamic processes can emerge at levels above the molecular processes as exemplified in autogenic systems, such as simple single-cell organisms, multicelled plants and animals, and nervous systems (and possibly even at higher levels), then at each level at which

508

teleogenic closure occurs, there will be a form of sentience characteristic to that level.

In the course of evolution, a many-tiered hierarchy of ever more convoluted forms of feeling has emerged, each dependent upon but separate from the form of feeling below. So, despite the material continuity that constitutes a multicelled animal with a brain, at each level that the capacity for sentience emerges, it will be discontinuous from the sentience at lower and evolutionarily previous levels. We should therefore carefully distinguish molecular, cellular, organismal, and mental forms of sentience, even when discussing brain function. Indeed, all these forms of sentience should be operative in parallel in the functioning of complex nervous systems

A neuron is a single cell, and simpler in many ways than almost any other single-cell eukaryotic organisms, such as an amoeba. But despite its dependence on being situated within a body and within a brain, and having its metabolism constantly tweaked by signals impinging on it from hundreds of other neurons, in terms of the broad definition of sentience I have described above, neurons are sentient agents. That doesn't mean that this is the same, or even fractionally a part of the emergent sentience of mental processes. The discontinuity created by the dynamical supervenience of mental (whole brain–level) teleodynamics on neuronal (cellular-level) teleodynamics makes these entirely separate realms.

Thus the sentient experience you have while reading these words is *not* the sum of the sentient responsiveness of the tens of billions of individual neurons involved. The two levels are phenomenally discontinuous, which is to say that a neuron's sentience comprises no fraction of your sentience. This higher-order sentience, which constitutes the mental subjective experience of struggling with these ideas, is constituted by the teleodynamic features emerging from the flux of intercellular signals that neurons give rise to. Neurons contribute to this phenomenon of mental experience by virtue of the way their vegetative sentience (implicit in their individual teleodynamic organization) contributes non-mechanistic interaction characteristics to this higher-order neural network–level teleodynamics. The teleodynamics that constitutes this characteristic form of cellular-level adaptive responsiveness, contributed by each of the billions of neurons involved, is therefore separate and distinct from that of the brain. But since brain-level teleodynamics supervenes on this lower level of sentient activity, it inevitably exhibits distinctive higher-order emergent properties. In this respect, this second-order teleodynamics is analogous to the way that the teleodynamics of interacting organisms within an ecosystem can contribute to higher-order population dynamics, including equilibrating (homeodynamic) and self-organizing (morphodynamic) population effects. Indeed, as we will explore further below, the tendency for population-level morphodynamic processes to emerge in the recursive flow of signals within a vast extended network of interconnected neurons is critical to the generation of mental experience. But the fact that these component interacting units—neurons—themselves are adaptive teleodynamic individuals means that even as these higher-order population dynamics are forming, these components are adapting with respect to them, to fit them or even resist their formation. This tangled hierarchy of causality is responsible for the special higher-order sentient properties (e.g., subjective experience) that brains are capable of producing, which their components (neurons) are not.

In other words, sentience is constituted by the dynamical organization, not the stuff (signals, chemistry) or even the neuronal cellular-level sentience that constitutes the substrate of that dynamics. The teleodynamic processes occurring within each neuron are necessary for the generation of mental experience only insofar as they contribute to the development of a higher-order teleodynamics of global signal processing. The various nested levels of sentience—from molecular to neuronal to mental—are thus mutually inaccessible to one another, and can exhibit quite different properties. Sentience has an autonomous locus at a specific level of dynamics because it is constituted by the self-creative, self-bounding nature of teleogenic individuation. The dynamical reflexivity and constraint closure that characterizes a teleodynamic system, whether constituting intraneuronal processes or the global-signaling dynamics developing within an entire brain, creates an internal/external self/other distinction that is determined by this dynamical closure. Its locus is ultimately something not materially present—a self-creating system of constraints with the capacity to do work to maintain its dynamical continuity—and yet it provides a precise dynamical boundedness.

The sentience at each level is implicit in the capacity to do self-preservative work, as this constitutes the system's sensitivity to non-self influences via an intrinsic tendency to generate a self-sustaining contragrade dynamics. This tendency to generate self-preserving work with respect to such influences is a spontaneous defining characteristic of such reciprocity of constraint creation. Closure and autonomy are thus the very essence of sentience. But they are also the reason that higher-order sentient teleogenic systems can be constituted of lower-order teleogenic systems, level upon level, and yet produce level-specific emergent forms of sentience that are both irreducible and unable to be entirely merged into larger conglomerates.[2] It is teleogenic closure that produces sentience but also isolates it, creating the fundamental distinction between self and other, whether at a neuronal level or a mental level.

So, while the lower level of cellular sentience cannot be dispensed with, it is a realm apart from mental experience. There is the world of the neuron and the world of mind, and they are distinct sentient realms. Neuronal sentience provides the ground for the interactions that generate higher-order homeodynamic, morphodynamic, and teleodynamic processes of neural activity. If neurons were not teleodynamically responsive to the activities of other neurons (and thereby also to the extrinsic stimuli affecting sensory cells), there would be no possibility for these higher-order dynamical levels of interaction to emerge, and thus no higher-level sentience; no subjective experience. But with this transition from the realm of individual neuronal signal dynamics to the dynamics that emerges in a brain due to the recursive effects that billions of neuronal signals exert on one another, there is a fundamental emergent discontinuity. Mental sentience is something distinct from neuronal sentience, and yet this nested dependency means that mental sentience is constituted by the dynamics of other sentient interactions. It is a second-order sentience emergent from a base of neuronal sentience, and additionally inherits the constraints of whole organism teleodynamics (and its vegetative sentience). So subjective sentience is fundamentally more complex and convoluted in its teleodynamic organization. It therefore exemplifies emergent teleodynamic properties that are unprecedented at the lower level.

EMOTION AND ENERGY

An emergent dynamic account of the relationship between neurological function and mental experience differs from all other approaches by vir-

tue of its necessary requirement for specifying a homeodynamic and morphodynamic basis for its teleodynamic (intentional) character. This means
that every mental process will inevitably reflect the contributions of these
necessary lower-level dynamics. In other words, certain ubiquitous aspects
of mental experience should inevitably exhibit organizational features that
derive from, and assume certain dynamical properties characteristic of,
thermodynamic and morphodynamic processes. To state this more concretely: experience should have clear equilibrium-tending, dissipative, and
self-organizing characteristics, besides those that are intentional. These are
inseparable dynamical features that literally constitute experience. What do
these dynamical features correspond to in our phenomenal experience?

Broadly speaking, this dynamical infrastructure is "emotion" in the
most general sense of that word. It is what constitutes the "what it feels
like" of subjective experience. Emotion—in the broad sense that I am using
it here—is not merely confined to such highly excited states as fear, rage,
sexual arousal, love, craving, and so forth. It is present in every experience,
even if often highly attenuated, because it is the expression of the necessary
dynamic infrastructure of all mental activity. It is the tension that separates
self from non-self; the way things are and the way they could be; the very
embodiment of the intrinsic incompleteness of subjective experience that
constitutes its perpetual becoming. It is a tension that inevitably arises as the
incessant shifting course of mental teleodynamics encounters the resistance
of the body to respond, and the insistence of bodily needs and drives to
derail thought, as well as the resistance of the world to conform to expectation. As a result, it is the mark that distinguishes subjective self from other,
and is at the same time the spontaneous tendency to minimize this disequi-
librium and difference. In simple terms, it is the mental tension that is created because of the presence of a kind of inertia and momentum associated
with the process of generating and modifying mental representations. The
term *e-motion* is in this respect curiously appropriate to the "dynamical feel"
of mental experience.

This almost Newtonian nature of emotion is reflected in the way that the
metaphors of folk psychology have described this aspect of human subjectivity over the course of history in many different societies. Thus English
speakers are "moved" to tears, "driven" to behave in ways we regret, "swept
up" by the mood of the crowd, angered to the point that we feel ready to

"explode," "under pressure" to perform, "blocked" by our inability to remember, and so forth. And we often let our "pent-up" frustrations "leak out" into our casual conversations, despite our best efforts to "contain" them. Both the motive and resistive aspects of experience are thus commonly expressed in energetic terms.

In the Yogic traditions of India and Tibet, the term *kundalini* refers to a source of living and spiritual motive force. It is figuratively "coiled" in the base of the spine, like a serpent poised to strike or a spring compressed and ready to expand. In this process, it animates body and spirit. The subjective experience of bodily states has also often been attributed to physical or ephemeral forms of fluid dynamics. In ancient Chinese medicine, this fluid is *chi*; in the Ayurvedic medicine of India, there were three fluids, the *doshas*; and in Greek, Roman, and later Islamic medicine, there were four *humors* (blood, phlegm, light and dark bile) responsible for one's state of mental and physical health. In all of these traditions, the balance, pressures, and free movement of these fluids were critical to the animation of the body, and their proper balance was presumed to be important to good health and "good humor." The humor theory of Hippocrates, for example, led to a variety of medical practices designed to rebalance the humors that were disturbed by disease or disruptive mental experience. Thus bloodletting was deemed an important way to adjust relative levels of these humors to treat disease.

This fluid dynamical conception of mental and physical animation was naturally reinforced by the ubiquitous correlation of a pounding heart (a pump) with intense emotion, stress, and intense exertion. Both René Descartes and Erasmus Darwin (to mention only two among many) argued that the nervous system likewise animates the body by virtue of differentially pumping fluid into various muscles and organs through microscopic tubes (presumably the nerves). When, in the 1780s, Luigi Galvani discovered that a severed frog leg could be induced to twitch in response to contact by electricity, he considered this energy to be an "animal electricity." And the vitalist notion of a special ineffable fluid of life, or *élan vital*, persisted even into the twentieth century.

This way of conceiving of the emotions did not disappear with the replacement of vitalism and with the rise of anatomical and physiological knowledge in the nineteenth and early twentieth century. It was famously reincarnated in Freudian psychology as the theory of *libido*. Though Freud was careful not to identify it with an actual fluid of the body, or even a yet-to-be-discovered material substrate, libido was described in terms that implied that it was something like the nervous energy associated with sexuality. Thus a repressed memory might block the "flow" of libido and cause its flow to be displaced, accumulated, and released to animate inappropriate behaviors. Freud's usage of this hydrodynamic metaphor became interpreted more concretely in the Freudian-inspired theories of Wilhelm Reich, who argued that there was literally a special form of energy, which he called "orgone" energy, that constituted the libido. Although such notions have long been abandoned and discredited with the rise of the neurosciences, there is still a sense in which the pharmacological treatments for mental illness are sometimes conceived of on the analogy of a balance of fluids: that is, neurotransmitter "levels." Thus different forms of mental illness are sometimes described in terms of the relative levels of dopamine, norepinephrine, or serotonin that can be manipulated by drugs that alter their production or interfere with their effects.

This folk psychology of emotion was challenged in the 1960s and 1970s by a group of prominent theorists, responsible for ushering in the information age. Among them was Gregory Bateson, who argued that the use of these energetic analogies and metaphors in psychology made a critical error in treating information processes as energetic processes. He argued that the appropriate way to conceive of mental processes was in informational and cybernetic terms.[3] Brains are not pumps, and although axons are indeed tubular, and molecules such as neurotransmitters are actively conveyed along their length, they do not contribute to a hydrodynamic process. Nervous signals are propagated ionic potentials, mediated by molecular signals linking cells across tiny synaptic gaps. On the model of a cybernetic control system, he argued that the differences conveyed by neurological signals are organized so that they regulate the release of "collateral energy," generated by metabolism. It is this independently available energy that is responsible for animating the body. Nervous control of this was thus more accurately modeled cybernetically. This collateral metabolic energy is analogous to the energy generated in a furnace, whose level of energy release is regulated by the much weaker changes in energy of the electrical signals propagated around the control circuit of a thermostat. According to Bateson, the mental world is not constituted by energy and matter, but rather by information. And as was also pioneered by the architects of the cybernetic theory whom Bateson drew his insights from, such as Wiener and Ashby, and biologists such as Warren McCulloch and Mayr, information was conceived of in purely logical terms: in other words, Shannon information. Implicit in this view—which gave rise to the computational perspective in the decades that followed—the folk wisdom expressed in energetic metaphors was deemed to be misleading.

By more precisely articulating the ways that thermodynamic, mor-phodynamic, and teleodynamic processes emerge from, and depend on, one another, however, we have seen that it is this overly simple energy/information dichotomy that is misleading. Information cannot so easily be disentangled from its basis in the capacity to reflect the effects of work (and thus the exchange of energy), and neither can it be simply reduced to it. Energy and information are asymmetrically and hierarchically interdependent dynamical concepts, which are linked by virtue of an intervening level of morphodynamic processes. And by virtue of this dynamical ascent, the capacity to be *about* something not present also emerges; not as mere signal difference, but as something extrinsic and absent yet potentially relevant to the existence of the teleodynamic (interpretive) processes thereby produced.

It is indeed the case that mental experience cannot be identified with the ebb and flow of some vital fluid, nor can it be identified directly with the buildup and release of energy. But as we've now also discovered by critically deconstructing the computer analogy, it cannot be identified with the signal patterns conveyed from neuron to neuron, either. These signals are generated and analyzed with respect to the teleodynamics of neuronal cell maintenance. They are interpreted with respect to cellular-level sentience. Each neuron is bombarded with signals that constitute its *Umwelt*. They perturb its metabolic state and force it to adapt in order to reestablish its stable teleodynamic "resting" activity. But, as was noted in the previous chapter, the structure of these neuronal signals does not constitute *mental* information, any more than the collisions between gas molecules constitute the attractor logic of the second law of thermodynamics.

As we will explore more fully below, mental information is constituted at a higher population dynamic level of signal regularity. As opposed to neuronal information (which can superficially be analyzed in computational

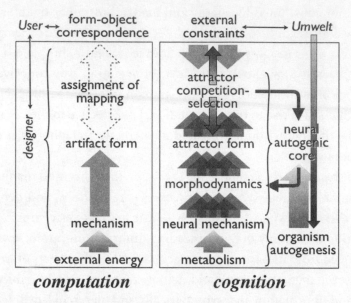

computation   **cognition**

FIGURE 17.1: The formal differences between computation and cognition (as described by this emergent dynamics approach) are shown in terms of the correspondences between the various physical components and dynamics of these processes (dependencies indicated by arrows). The multiple arrow links depicting cognitive relationships symbolize stochastically driven morphodynamic relationships rather than one-to-one correspondences between structures or states. This intervening form generation dynamic is what most distinguishes the two processes. It enables cognition to autonomously ground its referential and teleological organization, whereas computational processes must have these relationships "assigned" extrinsically, and are thus parasitic on extrinsic teleodynamics (e.g., in the form of programmers and interpreters). Computational "information" is therefore only Shannon information.

terms), mental information is embodied by distributed dynamical attractors. These higher-order, more global dynamical regularities are constituted by the incessantly recirculating and restimulating neural signals within vast networks of interconnected neurons. The attractors form as these recirculating signals damp some and amplify other intrinsic constraints implicit in the current network geometry. Looking for mental information in individual neuronal firing patterns is looking at the wrong level of scale and at the wrong kind of physical manifestation. As in other statistical dynamical regularities, there are a vast number of microstates (i.e., network activity patterns) that can constitute the same global attractor, and a vast number of trajectories of microstate-to-microstate changes that will tend to con-

verge to a common attractor. But it is the final quasi-regular network-level
dynamic, like a melody played by a million-instrument orchestra, that is the
medium of mental information. Although the contribution of each neuronal response is important, it is more with respect to how this contributes
a local micro bias to the larger dynamic. To repeat again, it is no more a
determinate of mental content than the collision between two atoms in a gas
determines the tendency of the gas to develop toward equilibrium (though
the fact that neurons are teleodynamic components rather than simply
mechanical components makes this analogy far too simple).

This shift in level makes it less clear that we can simply dismiss these folk
psychology force-motion analogies. If the medium of mental representation
is not mere signal difference, but instead is the large-scale global attractor dynamic produced by an extended interconnected population of neurons, then there may also be global-level homeodynamic properties to be
taken into account as well. As we have seen in earlier chapters, these global
dynamical regularities will exhibit many features that are also characteristic
of force-motion dynamics.

THE THERMODYNAMICS OF THOUGHT

So what would it mean to understand mental representations in terms of
global homeodynamic and morphodynamic attractors exhibited by vast
ensembles of signals circulating within vast neuronal networks? Like their
non-biological counterparts, these higher-order statistical dynamical effects
should have very distinctive properties and requirements.

Consider, for example, morphodynamic attractor formation. It is a consequence of limitations on the rate of constraint dissipation within a system
that is being incessantly destabilized. Because of this constant perturbation,
higher-order dynamical regularities form as less efficient uncorrelated tra
jectories of micro constraint dissipation are displaced by more efficient globally correlated trajectories. In neurological terms, incessant destabilization
is provided by some surplus of uncorrelated neuronal activity being generated within and circulating throughout a highly reentrant network. Above
some threshold level of spontaneous signal generation, local constraints
on the distribution of this activity will compound faster than they can dissipate. Thus, pushed above this threshold of incessant intrinsic agitation,

the signals circulating within a localized network will inevitably regularize with respect to one another as constraints on signal distribution within the network tend to redistribute. This will produce dynamical attractors which reflect the distribution of biases in the connectional geometry of that network, and thus will be a means for probing and expressing such otherwise diffusely distributed biases.

But unlike the regularities exhibited by individual neural signals, or the organized volleys of signals conducted along a bundle of axons, or even the signal patterns converging from hundreds of inputs on a single neuron, morphodynamic regularities of activity patterns within a large neural network will take time to form. This is because, analogous to the simple mechanical dynamical effects of material morphodynamics, these regularities are produced by the incremental compounding of constraints as they are recirculated continuously within the network. This recirculation is driven by constant perturbation. Consequently, without persistent above-threshold excitation of a significant fraction of the population of the neurons comprising a neural network, higher-order morphodynamic regularities of activity dynamics would not tend to form.

What might this mean for mental representation, assuming that mental content is embodied as population-level dynamical attractors? First, it predicts that the generation of any mental content, whether emerging from memory or induced by sensory input, should take time to develop. Second, it should require something like a persistent metabolic boost, to provide a sufficient period of incessant perturbation to build up local signal imbalances and drive the formation of larger-scale attractor regularities. This suggests that there should consequently be something akin to inertia associated with a change of mental content and shift in attention. But this further suggests that self-initiated shifts in cognitive activity will require something analogous to work in order to generate, and that stimuli from within or without that are capable of interrupting ongoing cognitive activities are also doing work that is contragrade to current mental processes. Third, because of the time it takes for the non-linear recursive circulation of these signals to self-organize large-scale network dynamics, mental content should also not emerge all or none into awareness, but should rather differentiate slowly from vague to highly detailed structures. And the level of differentiation achieved should be correlated both with sustained high levels of activation and with the length of time this persists. Generating more precise mental content takes both more "effort" and a more sustained "focus" of attention.

Because the basis of this process is incessant spontaneous neuronal activity, a constant supply of resources—oxygen and glucose—is its most basic requirement. Without sufficient metabolic resources, neuronal activity levels will be insufficient to generate highly regular morphodynamic attractors, though with too much perturbation (as in simpler thermodynamic dissipative systems), chaotic dynamics may result. This suggests that we should expect some interesting trade-offs.

Larger-scale morphodynamic processes should be more capable of morphodynamic work (affecting other morphodynamic processes developing elsewhere in the brain) but will also be highly demanding of sustained metabolic support to develop to a high degree of differentiation. But metabolic resources are limited, and neurons themselves are incapable of sustaining high levels of activity for long periods without periodically falling into refractory states. For this reason, we should expect that mental focus should also tend to spontaneously shift, as certain morphodynamic attractors degrade and dedifferentiate with reduced homeodynamic support, and others begin to emerge in their stead. Moreover, a morphodynamic neural activity pattern that is distributed over a wide network involving diverse brain systems will be highly susceptible to the relative distribution of metabolic resources. This means that the differential distribution of metabolic support in the brain will itself play a major role in determining what morphodynamic processes are likely to develop when and where. So metabolic support itself may independently play a critical role in the generation, differentiation, modification, and degradation of mental representations. And the more differentiated the mental content and more present to mind, so to speak, the more elevated the regional network metabolism and the more organized the attractors of network activity.

How might this thermodynamics of thought be expressed? Consider first the fact that morphodynamic attractors will only tend to form when these resources are continually available and neurons are active above some resting threshold. Morphodynamic attractor formation is an orthograde tendency at that level, but it requires lower-order homeodynamic (and thus thermodynamic) work to drive it. If neural activity levels are normally fluctuating around this threshold, however, some degree of local morphodynamic attractor activity will tend to develop spontaneously. This means that minimally differentiated mental representations should constantly arise and fade from awareness as though from no specific antecedent or source of stimulation. If the morphodynamics is only incipient, and not able to develop to a fully differentiated and robust attractor stage, it will only produce what amounts to an undifferentiated embryo of thought or mental imagery. This appears to be the state of unfocused cognition at rest—as in daydreaming—suggesting that indeed the awake, unfocused brain hovers around this threshold. Only if metabolic support or spontaneous neuronal activity falls consistently below this threshold will there be no mental experience—as may be the case for the deepest levels of sleep. Normally, then, network-level neural dynamics in the alert state is not so much at the edge of chaos as at the threshold of morphodynamics.

But entering a state of mind focused on certain mnemonically generated content or sensory-motor contingencies means that some morphodynamic processes must be driven to differentiate to the point that a robust attractor forms. Since this requires persistent high metabolic support to maintain the dissipative throughput of signal activity, it poses a sort of chicken-and-egg relationship between regional brain metabolism and the development of mental experience. Well-differentiated mental content requires persistent widespread, elevated metabolic activation, and it takes elevated spontaneous neuronal activity to differentiate complex mental states. Which comes first? Of course, as the metaphor suggests, each can be precursor or consequence, depending on the context.

If the level of morphodynamic development in a given region is in part a function of the ebb and flow of metabolic resources allocated to that region, adjusting local metabolism up or down extrinsically should also increase or decrease the differentiation and robustness of the corresponding mental representation. This possibility depends on the extent to which individual neuronal activity levels can be modulated by metabolic support. If neurons can be shifted up and down in their spontaneous activity level simply by virtue of the increased or decreased availability of oxygen and glucose, then merely shifting levels of these resources locally could have an influence on regional morphodynamic differentiation.

At this point it is only conjecture for me to imagine that this occurs, since I am not aware of evidence to support such a mechanism; but the availability of such a thermodynamically responsive neuronal teleodynamics would seem to be consistent with other aspects of emergent dynamical theory. So, for the sake of pursuing the implications of the theory as far as possible into this discussion of the neural-mental relationship, I will assume that it is a reasonable possibility. Should it prove to be the case, either directly or indirectly, that the extrinsic availability of local metabolic resources can both drive and impede levels of neuronal activity, it would offer further support for this approach. How might this work?

If an extrinsic increase in metabolic support will tend to increase the level of spontaneous neuronal activity, it will also tend to increase the probability of local attractor formation. This means that if there is a means to globally regulate the distribution of the brain's metabolic resources –that is, by up- and down-regulating rates of regional cerebral blood flow—this could also be a means of initiating and developing, or impeding and degrading, certain local morphodynamic tendencies. Simply adjusting regional metabolism could thus be a way to regulate attention and initiate operations involving specific sensory, motor, or mnemonic content. Conversely, this means that neurons subjected to high levels of excitatory stimulation may be driven beyond their metabolic capacity to react consistently. Even highly organized extrinsic input (e.g., from sensory systems or the morphodynamically organized outputs from other regions of the brain) may be insufficient to constrain attractor differentiation if there is insufficient metabolic support. But if there are also local means for neurons and glia to "solicit" an increase in metabolic support if depleted, then extrinsically driving neural activity levels beyond what is supported by the metabolic equilibrium of that region might also be a way to recruit metabolic support for morphodynamic differentiation.

Both effects will have inertial-like features. Thus in some cases it will be possible to "force" differentiation of thought through metabolic means directly, or morphodynamic work from an extrinsically metabolically active region; and in other cases it may be quite difficult, even with effort, to develop sufficiently differentiated thoughts or responses. The effect, in both cases, is that there will inevitably be what amounts to a kind of "tension" between cognition and its metabolic support.

A relevant attribute of this tension is that neuronal activity rates take place on a millisecond time scale while the hydrodynamic and diffusion changes that must support this activity take place on a multisecond time scale. Interestingly, because of the time it takes for morphodynamic activity to differentiate to a regular attractor stage, mental processes likely take place at rates that are commensurate with metabolic time scales; and in the case of highly complex and highly robust morphodynamic attractors, this may take longer—perhaps many seconds to reach a stage of full differentiation. These different temporal domains also contribute to the structure of experience.

The differential between the time that neuron to neuronal excitation can take place and the time it takes to mount a metabolic compensation for this change in activity means that there will be something analogous to metabolic inertia involved. This delay will slightly impede the ability of heightened but short-lived morphodynamic activity in one region to spread its influence into connected regions, which are not already primed with increased metabolism. This sort of dynamical recruitment will consequently require stable and robust attractor formation and maintenance, and thus stably heightened metabolic support.

What controls local metabolic support? There appear to be both extrinsic and intrinsic mechanisms able to up-regulate and down-regulate cerebral blood flow to local regions. Intrinsic mechanisms are becoming better understood because of the importance of *in vivo* imaging techniques that are based on hydrodynamic changes, such as fMRI. If, as I have argued, extrinsically regulated local increases and decreases of metabolic support can themselves induce significant changes in neuronal activity levels, with associated alterations of signal dropout and random spontaneous activity, then such a regulatory mechanism could play a significant role in directing attention, differentiating mnemonic content, activating or inhibiting behaviors, and shifting modality specific processing. How might this be effected?

Extrinsic control of regional cerebral blood flow is less well understood than intrinsic effects, and my knowledge of these mechanisms is minimal, so what I will describe here is again highly speculative. But some such mechanism is strongly implicated when considering mind-brain relationships dynamically. The answer, I believe, is something like this. Certain stimuli (or intrinsically generated representations) that have significant normative relevance (perhaps because they are associated with powerful innate drives or with highly arousing past experiences) are predisposed to easily induce characteristic patterns of activity in forebrain limbic structures (such as the amygdala, nucleus accumbens, hypothalamus). These limbic activity patterns distinguish conditions for which a significant shift of attention and mental effort is likely required, for example, because of danger or bodily need. These limbic structures in turn project to midbrain and brainstem structures that control regional differences in cerebral blood flow and regional levels of neuronal plasticity. These in turn project axons throughout the forebrain and serve to modulate regional levels of neuronal activity by adjusting blood flow and intrinsic neuronal variables, which make them more or less plastic to input patterns.

In this way, highly survival-relevant stimuli or powerful drives can be drivers of mental experience to the extent that they promote selected differentiation of local morphodynamic attractors. So, for example, life-threatening or reproductively important stimuli can regulate the differentiation of specific analytic, mnemonic, and behavioral capacities, and shut down other ongoing mental activities by simply modulating metabolic resource distribution. Because this is an extrinsic influence with respect to the formal constraints that are amplified to generate the resulting morphodynamic processes, and not the result of direct interactions between regionally distinct networks, there can be a relatively powerful inertial component in such transitions, especially if they need to be rapid, and the subsequent attractor process needs to be highly differentiated and robust (as is likely in life-threatening situations).

Rapidly shutting down an ongoing dynamic in one area and just as rapidly generating another requires considerable work, both thermodynamic (metabolic) and morphodynamic. But consider the analogy to simple physical morphodynamic processes like whirlpools and Bénard convection cells. These cannot be generated in an instant nor can they be dissipated in an instant, once stable. The same must be true of the mental experiences in such cases of highly aroused shifts of attention. Prior dynamics will resist dissolution and may require structural interference with their attractor patterns (morphodynamic work imposed from other brain regions) to shut them down rapidly. They will in this sense resist an imposed change. This tension between dynamical influences at these two levels—the homeodynamics of metabolic processes and the morphodynamics of network dynamics—is inevitable in any forced change of morphodynamic activity.

It is my hypothesis, then, that this resistance and work created by these

dependencies between levels of dynamics of brain processes constitutes
the experience of emotion. In most moment-to-moment waking activities,
shifts between large-scale attractor states are likely to be minimally forced,
and so will engender minimal and relatively undifferentiated emotions. But
there should be a gradation of both differentiation and intensity. The orthograde nature of morphodynamic differentiation does not in itself require
morphodynamic work, and so there is not necessarily extrinsic "mental
effort" required for thoughts to evoke one another or to spontaneously "rise"
from unconsciousness. This emergent dynamic account thus effectively
distinguishes the conscious from unconscious generation of thoughts and
attentional foci in terms of work. The more work at more levels, the more
sentient experience. And where work is most intense, we are most present
and actively sentient. Self and other, including the otherness created by the
inertia of our own neural dynamics, are thus brought into stark relief by the
contragrade "tensions" that arise because we are constituted dynamically.

There is, of course, still something central missing from this account. If
the contents of mental experiences are instantiated by the attractor dynamics of the vast flows of signals coursing through a neural network, where
in this process are these interpreted? What dynamical features of brain
processes are these morphodynamic features juxtaposed with in order that
they are information *about* something *for* something toward some end? The
superficial answer is the same as that given for what constitutes the locus
of self in general: a teleodynamic process. But as we have already come to
realize, the sort of teleodynamics that arises from brain process is at least a
second-order form of teleodynamics when compared to that which constitutes life. This convoluted and hierarchically tangled form of teleodynamics
includes some distinct emergent differences.

WHENCE SUFFERING?

The last chapter began by questioning whether there might be intrinsic moral implications to corrupting or shutting down a computation in
process. I agree with William James' conclusion that only if this involves
sentience do moral and ethical considerations come into play, but that if
sentience is present, then indeed such values must be considered. This is
because sentience inevitably has a valence. Things *feel* good and bad, they

aren't merely good for or bad for the organism. Because of this, the world of sentience is a world of should and shouldn't, kindness and abuse, love and hate, joy and suffering. Is this really necessary? Could there be mental sentience without its being framed between wonderful and horrible?

As we have now seen, computation is ultimately just a descriptive gloss applied to a simple linear mechanistic process. So the intrinsic intentional status of the computation, apart from this mechanism, is entirely epiphe-nomenal. This beautifully exemplifies the patency of the nominalist critique of generals. Computation is in the mind of the beholder, not in the physi-cal process. There is nothing additionally produced when a computation is completed, other than the physical rearrangement of matter and energy in the device that is performing this operation. A computer operation is there fore no more sentient than is the operation of an automobile engine. But, as we saw in the Self chapter, a teleodynamic process does in fact transform generals into particulars; and the constraints that constitute and are in turn constituted by such a process do have ententional status, independent of the physical particulars that embody them. This self-creation of constraints is what constitutes the dynamical locus of sentience, not merely some physical change of state.

In light of the hierarchic conception of sentience developed above, how-ever, I think that this general assessment now needs to be further refined. There are emergent sentient properties produced by the teleodynamics of brains that are not produced by simpler, lower-order forms of sentience. Crucially, these are special normative properties made possible because the sentience generated by brain processes is, in effect, a second-order sentience: a sentience of sentience. And with this comes a sentience of the normative features of sentience. In colloquial terms, this sentience of normativity is the experience of pleasure and pain, joy and suffering. And it is with respect to these higher-order sentient properties that we enter into the ethical realm.

To understand how this higher-order tangle has created a sentience of the normative relationships that create this sentience itself, and ultimately iden-tify the self-dynamic that is the locus of subjective experience, we need to once more revisit how the logic of teleodynamics, at whatever level, creates an individuated locus of self-creation and a dynamic of self-differentiation from the world.

In chapter 9, teleodynamics was defined as a dynamical organization that exists because of the consequences of its continuance, and therefore can be described as being self-generating over time. But now consider what it would mean for a teleodynamic process to include within itself a representation of its own dynamical final causal tendencies. The component dynamics of a teleodynamic process have ententional properties precisely because they are critical to the creation of the whole dynamic, which in turn is critical to the continued creation of these component dynamics. Were the reciprocal synergy of the whole dynamic to break down, these component dynamics would also eventually disappear. The whole produces the parts and the parts produce the whole. But then a teleogenic process in which one critical dynamical component is a representational process that interprets its own teleodynamic tendency extends this convoluted causal circularity one level further.

For animals with brains, the organism and its distinctive teleodynamic characteristics will likely fail to persist (both in terms of resisting death and reproducing) if its higher-order teleodynamics of self-prediction fails in some respect. Failure is likely if the projected self-environment consequence of some action is significantly in error. For example, an animal whose innate predator-escape strategy fails to prevent capture will be unlikely to pass on this tendency to future generations. Such a tendency implicitly includes a projected virtual relationship between its teleodynamic basis and a projected self/other condition. Generation of a projected future self-in-context thus can become a critical source of constraints organizing the whole system. But generating these virtual selves requires both a means to model the causality of the environment and also a means for modeling the causality of the teleodynamic processes that generate these models and act with respect to them. This is a higher-order teleodynamical relationship because one critical dynamical component of the whole is its own projected future existence in context. This implicitly includes a normative assessment of this possible condition with respect to current teleodynamic tendencies, including the possibility of catastrophic failure.

The vegetative teleodynamics of single-celled organisms and organisms not including brains must be organized to produce contragrade reactions to any conditions that tend to disrupt teleodynamic integrity. The various component structures, mechanisms, and morphodynamic processes that constitute this integrity must therefore be organized to collectively compensate for any component process that is impeded or otherwise compromised from without. There does not need to be any component that assesses the general state of overall integrity. But in an animal with a brain that was evolved to project alternative future selves-in-context, such an assessment becomes a relevant factor. A separate dynamical component of its teleodynamic organization must continually generate a model of both its overall vegetative integrity and the degree to which this is (or might be) compromised with respect to other contingent factors. A dynamical subprocess evolved to analyze whatever might impact persistence of the whole organism, and determine an appropriate organism level response, must play a primary role in structuring its overall teleodynamic organization.

In the previous section, we identified the experience of emotion with the tension and work associated with employing metabolic means to modify neural morphodynamics. It was noted that particularly in cases of life-and-death contexts, morphodynamic change must be instituted rapidly and extensively, and this requires extensive work at both the homeodynamic (metabolic) and morphodynamic (mental) levels. The extent of this work and the intensity of the tension created by the resistance of dynamical processes to rapid change is, I submit, experienced as the intensity of emotion. But such special needs for reliable rapid dynamical reorganization arise because the teleodynamics of organism persistence is easily disturbed and inevitably subject to catastrophic breakdown. Life and health are fragile. So the generation of certain defensive responses by organisms, whether immune response or predator-escape behaviors, must be able to usurp less critical ongoing activities whenever the relevant circumstances arise.

Teleodynamic processes have characteristic dynamical tendencies, and when these are impeded or interfered with, the entire integrated individual is at risk. But catastrophic breakdown is not just possible, it is essentially inevitable for any teleodynamically organized individual system, whether autogen or human being. The ephemeral nature of teleodynamics guarantees that it faces an incessant war against intrinsic and extrinsic influences that would tend to disrupt it—and the more disruptive of its self-similarity maintenance, the greater the work that must be performed to resist this influence.

Influences that are so disruptive that they will ultimately destroy the teleodynamic integrity of a single cell, or even a plant, will in the process induce the production of the most intense and elaborate contragrade processes possible within the repertoire of that organism's systems. But because there is no separate dynamical embodiment of the integrity of the whole, no locus of individuated-self representation, these dynamical extremes do not constitute suffering. There is a self, but there is no one home reflecting back on this process at the same time as enduring it. This is not the case for most animals.

The capacity to suffer requires the higher-order teleodynamic loop that brain processes make possible. It requires a self that creates within itself a teleodynamic reproduction of itself. This emergent dynamical homunculus is constituted by a central, teleodynamically organized, global pattern of network activity. By definition, this must be constituted by reciprocally synergistic morphodynamic processes. The component morphodynamic processes are, as we have discussed above, generated by the self-organizing tendencies of the vast numbers of signals circulating around and incessantly being introduced into the neural networks of the various brain systems. However, not all morphodynamic attractors produced in a brain contribute to this teleodynamic core, though there is a continual assimilation of newly generated morphodynamic processes into the synergy that constitutes this ultimate locus of mental self-continuity and self/non-self distinction.

A simple individual autogenic system embodies the constraints of this necessary synergy implicitly in the complementarities and symmetries of the constraints determining and generating the various morphodynamic processes that constitute it. This is also true for an animal body, though more complex and hierarchically organized. But it is not true of the teleodynamics of brains. Brains have evolved to regulate whole organism relationships with the world. Their teleodynamics is therefore necessarily parasitic on the teleodynamics of the body that they serve. Thus, for example, hypothalamic, midbrain, and brainstem circuits in vertebrate bodies play a critical role in regulating such global body functions as heart rate, digestion, metabolic rate, and the monitoring and maintenance of the levels of a wide variety of bloodborne chemical signals, such as hormones.

The local processes that maintain these systems are highly robust, and in many cases are quite nearly cybernetically organized processes.[4] In this sense, they are least like higher-order morphodynamic neural processes, and thus their functioning does not directly enter mental experience. Nevertheless, the status of these functions is redundantly monitored by other forebrain brain systems, and these deep brain regulatory systems provide the forebrain systems with signals reflecting their operation. The result is that vegetative functions of the organism are multiply represented at various levels of remove from their direct regulation. The nearly mechanistically regular constraints of their operation are thus inherited by higher-order brain processes.

So what constitutes the core teleodynamic locus of brain function? The answer is that it too is multiply hierarchically generated at different levels of functional differentiation. Even at the level of brainstem circuits, there is almost certainly a synergistic dynamic linking the separate regulatory systems for vegetative functions; but it is only as we ascend into forebrain systems that these processes become integrated with the variety of sensory and motor systems that constitute regulation of whole organism function. The constraints constituting the synergy and stability of core vegetative systems provide a global organizing influence that more differentiated levels of the process must also maintain. At the level of brain function where sensory and motor functions must be integrated with these core functions, the differentiation of morphodynamic processes that involve these additional body systems is inevitably constrained to respect these essential core regularities. In this way, the many simultaneously developing morphodynamic processes produced within the various forebrain subsystems specialized for one or the other modality of function begin their differentiation processes already teleodynamically integrated with one another. The self-locus that is constituted by this synergy of component morphodynamic processes is thus also a dynamic that is subject to varying degrees of differentiation. It can be relatively amorphous and comprised of poorly differentiated morphodynamic processes, or highly differentiated and involve a large constellation of nested and interdependent morphodynamic processes. Subjective self is as differentiated and unified as the component morphodynamic processes developing in various subregions of the brain are differentiated and mutually reinforcing. And this level of self-differentiation is constantly shifting.

The teleodynamic synergy of this brain process is ultimately inherited from these more fundamental vegetative teleodynamic relationships, which contribute core constraints influencing the differentiation of this self-locus. These core constraints provide what might be considered the envelope of variation within which diverse morphodynamic processes are differentiated in response to sensory information and the many internally generated influences. Since the locus of this self-perspective is dynamically determined with respect to the "boundaries" of morphodynamic reciprocities, and these are changing and differentiating constantly, there can be no unambiguous anatomical correlate of this homunculus within the brain. Nevertheless, the teleodynamic loop of causality that integrates sensory and motor processes with the projected self-in-possible-context must at least involve the brain systems these processes depend upon, such as the thalamic, cerebral cortical, and basal ganglionic structures of the mammalian (and human) forebrain. But even this may be variable. Comparatively undifferentiated self-dynamics may only involve perilimbic cortical areas and their linked forebrain nuclei, whereas a self-dynamics involved in complex predictive behavior may involve significant fractions of the entire cerebral cortex and the forebrain nuclei these regions are coupled with.

So why is there a "what it feels like" associated with this neural teleodynamics? And why does this "being here" have an intrinsic good and bad feel about it? The answer is that this self-similarity-maintaining dynamic provides a constant invariant reference with respect to which all other dynamical regularities and disturbances are organized. Like the teleodynamics of autogenic self, it is what organizes all local dynamics around an invariant *telos*: the self-creating constraints that make the work of this self-creation possible. In a brain, this teleodynamic core set of constraints serves as both a center of dynamical inertia that other neural activates cannot displace and a locus of dynamical self-sufficiency that is a constant platform from which distributed neural dynamics must begin their differentiation. But the teleodynamic integrity of this core neurological self is a direct reflection of the vegetative teleodynamics that is critical to its own persistence. To the extent that the vegetative teleodynamics is compromised, so too will neurological self be compromised. But vegetative teleodynamic integrity and neurological teleodynamical integrity are only linked; they are not identical.

As the possibility of anesthesia makes obvious, the mental representation of bodily damage can be decoupled from the experiential self. Under these circumstances, thankfully, sensory information from the body is not registered centrally and cannot thereby alter neural teleodynamics. Mental processes can therefore continue, oblivious to even significant physiological damage. So pain is not "out there" in the world. It is how neural teleodynamics reorganizes in response to its sensory assessment of vegetative damage. Its effect is to interrupt less critical neural dynamics and activate specific processes to stop this sensory signal. To do this, it must block the differentiation of most morphodynamic processes that are inessential to this end, and rapidly recruit significant metabolic and neural resources to generate action to avoid continuation of this stimulus. So, whereas any non-spontaneous shift of neural dynamics requires metabolic work to accomplish, the reaction to pain is preset to maximize this mobilization.

In many respects, pain is a special form of minimal perception, which helps to exemplify the relationship between what might be described as the analytic and emotional aspects of sentience. Perception is not merely the registration of extrinsically imposed changes of neural signals. It involves the generation of local morphodynamic processes that remain integrated into the larger teleodynamic integrity and yet are at the same time modulated by these extrinsically imposed constraints. Any slight dissonance that results initiates neural work to further differentiate the core teleodynamic organization to minimize this deviation. This drives progressive differentiation of the relevant morphodynamic processes in directions that at the same time are adapted to these imposed constraints and minimally dissonant with global teleodynamics. So perception is, in effect, the differentiation of self to better fit extrinsically imposed regularities. This can be looked upon as a sort of principle of work minimization that is always assessed with respect to this core teleodynamics.

In contrast to other sensory experiences, pain requires minimal morphodynamic differentiation to be assessed: only what's necessary to localize it in the body and determine what kind of pain (with only a very few modalities to sample). Unlike other percepts, however, there is no differentiation of morphodynamic perceptual processes that is able to increase pain's integration with core teleodynamics. Instead, pain blocks the differentiation of other modes of sensory analysis and rapidly deploys resources to possible motor responses to stop this sensation (such as the rapid withdrawal of one's finger from contact with a hot stove). This simple non-assimilable stimulus continues to drive the differentiation of actual and potential motor responses until the painful stimulus ceases. Of course, once damage is done, the pain stimulus will often continue despite actions to limit it. In these cases, a continual "demand" for recruitment of a motor response, and a correlated mobilization of metabolism to achieve it, will persist despite the ineffectiveness of any action. One consequence is that further damage may be averted. Another is that the continual maintenance of this heightened need to act to end the pain is experienced as suffering. Pain that can't be relieved is thus continually perturbing the core teleodynamics; contorting self-dynamics to necessarily include this extrinsic influence and the representation of a goal state that remains unachieved; and constantly mobilizing and focusing metabolic resources for processes that fail to achieve adaptation.

The capacity to suffer is therefore an inseparable aspect of the deep coupling between the neurally represented and viscerally instantiated teleodynamics of the body. Specifically, it is a consequence of the evolution of means for mobility and rapid behavior able to alter extrinsic conditions. Organisms that have not evolved a capacity to act in this way have no need for pain, and no need to represent their whole bodily relationship to extrinsic conditions. Their teleodynamic integrity is maintained by distributed processes without the need for an independent instantiation as experience. Pain is the extreme epitome of the general phenomenology we call emotion because of the way it radically utilizes the mobilization of metabolic resources to powerfully constrain signal differentiation processes, and thereby extrinsically drive and inhibit specific spontaneous morphodynamic tendencies. More than anything else, it exemplifies the essence of neurological sentience.

BEING HERE

In the last chapter, we began to explore the distinctive higher-order form of teleodynamics that emerges from a teleodynamic process that must include itself as a component: a teleodynamic circularity in which the very locus of teleodynamic closure becomes virtual. This is a tangle in the dynamical hierarchy that is superficially analogous to the part/whole tangle that defines teleodynamics more generally. In a simple autogenic system, each part (itself a morphodynamic product) is involved in an ultimately closed set of reciprocal interaction relationships with each other to create the whole, but it takes the whole reciprocally synergistic complex to generate each part. In logic, this would amount to a logical-type violation (in which a class can also be a member of itself). By including the capacity to model itself in relation to extrinsic features of the world, a neurally generated teleo-dynamic system similarly introduces a higher-order tangle to this dynamical hierarchy.

In this chapter, we discovered a further tangle in the hierarchy of neural teleodynamics. Because neurons are themselves teleodynamic and thus sensitive to and adaptive to the changes in their local signal-processing *Umwelt*, the higher-order dynamics of the networks they inhabit can also be a source for changes in their internal teleodynamic tendencies. Thus neurons "learn" by changing their relative responsivity to the patterns of activity that they are subject to. In the process, the biases in the network that are responsible for the various morphodynamic attractors that tend to form will also change. This can happen at various timescales. As neurons temporarily modify their immediate responsivity over the course of seconds or minutes, the relative lability of certain morphodynamic attractor options can change.

Finally, with the exploration of perception, and specifically the perception of pain, we have brought together these various threads, integrating all with the concept of emotion. Emotion in this generic sense is not some special feature of brain function that is opposed to cognition. It is the necessary expression of the complicated hierarchic dependence of morphodynamics on homeodynamics (specifically, the thermodynamics of metabolism), and the way that the second-order teleodynamics that integrates brain function is organized to use this dependence to regulate the self/other interface that its dynamical closure (and that of the body) creates.

Although each is discontinuous from the other by virtue of dynamical closure, neuronal-level sentience is nevertheless causally entangled with brain-level sentience, which is entangled in a virtual-self-level of sentience. And human symbolic abilities add a further, yet-higher-order variant on this logical type-violating entanglement. This latter involves the incorporation of an abstract representation of self into the teleodynamic loop of sentience. Thus we humans can even suffer from existential despair. No wonder the analysis of human consciousness tends to easily lead into a labyrinth of self-referential confusions.

There remains an immense task ahead to correlate these dynamical processes with specific brain structures and neural processes. But despite remaining quite vague about such details, I believe that the general principles outlined in these pages can offer some useful pointers, leading neuroscientists to pay attention to features of brain function that they might otherwise have overlooked as irrelevant. And they may bring attention to neural dynamics that have so far gone unnoticed and suggest ways to develop new tools for analyzing mental processes considered outside the purview of cognitive neuroscience. So, while I don't believe that neuroscience will be pursued differently as a result, or that this will lead to any revolutionary new discoveries about neurons, their signaling dynamics, or the overall anatomy of brains, it may prompt many researchers to rethink the assumptions they bring to these studies.

While we have only just begun to sketch the outlines of an emergent dynamics account of this one most enigmatic phenomenon—human consciousness—the results point us in very different directions than previously considered. With the autogenic creation of self as our model, we have broken the spell of dualism by focusing attention on the contributions of both what is present and what is absent. Surprisingly, this even points the way to a non-mystical account of the apparent non-materiality of consciousness. The apparent riddle of its non-materiality turns out not to be a riddle after all, but an accurate reflection of the fact that the locus of subjective sentience is not, in fact, a material substrate. The riddle was not the result of any problem with the concept of consciousness, but of our failure to understand the causal relevance of constraint. With the realization that specific absent tendencies—dynamical constraints—are critically relevant to the causal fabric of the world, and are the crucial mediators of non-spontaneous change, we are able to stop searching for consciousness "in" the brain or "made of" neural signals.

I believe that human subjectivity has turned out not to be the ultimate "hard problem" of science. Or rather, it turns out to have been hard for unexpected reasons. It was not hard because we lacked sufficiently complex research instruments, nor because the details of the process were so many and so intricately entangled with one another that our analytic tools could not cope, nor because our brains were inadequate to the task for evolutionary reasons, nor even because the problem is inaccessible using the scientific method. It was hard because it was counterintuitive, and because we have stubbornly insisted on looking for it where it could not be, in the stuff of the world. When viewed through the perspective of the special circular logic of constraint generation that we have called teleodynamics, this problem simply dissolves. The complex and convoluted dynamical processes that are the defining features of self, at any given level, are not embodied in molecules, or neurons, or even neural signals, but in the teleodynamics of processes generated in the vast networks of brains. The molecular interactions, propagating neuronal signals, and incessant energy metabolism that provide the substrate for this higher-order dynamical process are necessary substrates; but it is because of what these do not actualize, because of how their interactions are constrained, that there is agency, sentience, and valuation implicit in their patterns of interaction. We are what we are not: continually, intrinsically, necessarily incomplete in our very nature. Our sense of self, our experience of being the originative locus of agency, our interior subjective isolation, and the sense of emerging out of nothing and being our own prime mover—all these core characteristics of conscious experience—are accurate reflections of the fact that self is literally *sui generis*, emerging each moment from what is not there.

There can be no simple and direct neural embodiment of subjective experience in this sense. This is not because subjectivity is somehow other-worldly or non-physical, but rather because neural activity patterns convey both the interpretation and the contents of experiences in the negative, so to speak; a bit like the way that the space in a mold represents a potential statue. The subjectivity is not located in what is there, but emerges quite precisely from what is *not* there. Sentience is negatively "embodied" in the constraints emerging from teleodynamic processes, irrespective of their physical embodiment, and therefore does not directly correlate with any of the material substrates constituting those processes. Intrinsically emergent constraints are neither material nor dynamical—they are something missing—and yet as we have seen, they are not mere descriptive attributions of material processes, either. The intentional properties that we attribute to conscious experience are generated by the emergence of these constraints—constraints that emerge from constraints, absences that arise from, and create, new absences. You are in this quite literal sense something coming out of nothing, and thus newly embodied at each instant.

But this negative existence, so to speak, of the conscious self doesn't mean that consciousness is in any way ineffable or non-empirical. Indeed, if the account given here is in any way correct, it suggests that consciousness may even be precisely quantifiable and comparable, for example, between states of awareness, between species, and even possibly in non-organic processes, as in social processes or in some future sentient artifact. This is because teleodynamic processes, which provide the locus for sentience in any of its forms, are precisely analyzable processes, with definite measurable properties, in whatever substrates they arise. Because a teleodynamic process is dynamically closed by virtue of its thoroughly reciprocal organization, it is clearly individuated from its surroundings, even if these are merely other neural dynamics. Because of this individuation, it should be possible to gain a quantitative assessment of the thermodynamic and morphodynamic work generated moment by moment in maintaining its integrity. It should also be possible at any one moment to determine what physical and energetic substrates constitute its current locus of embodiment.

These should not be surprising conjectures. As it is, we already use many crudely related intuitive rules of thumb to make such assessments when it comes to assessing a patient's state of anesthesia or level of awareness after brain damage, and even when comparing different animals. We generally assume that a metabolically active brain is essential, and that as metabolism and neuronal activity decrease below some threshold, so does consciousness. We assume that animals with very small brains (such as gnats) can have only the dimmest if any conscious experience, while large-brained mammals are quite capable of intense subjective experiences and likely suffer as much as would a person if injured. So some measure of dynamical work and substrate complexity already seems to provide us with an intuition about the relative degree of consciousness we are dealing with.

The present analysis not only supports these intuitions but provides further complexity and subtlety as well. It suggests that we can distinguish between the kind of brain dynamics that is associated with consciousness and what kind is not. Indeed, this is implicit in the critique of computational theories. Computations and cybernetic processes are insentient because they are not teleodynamic in their organization. In fact, we intuitively also take this into account when we introspect about our own state of conscious awareness. For example, when acquiring a new skill—such as learning to play a piece of music on an instrument like the piano—the early stages are very demanding of constant attention to sensory and motor details. It takes effort and work of all kinds. But as learning progresses and you become skilled at this performance, these various details become less and less present to awareness. And by the time it is performed like an expert, you are able to almost "do it in your sleep," as the saying goes. Highly skilled behaviors are performed with a minimum of conscious awareness. It is as though they are being performed by an algorithmic process. Indeed, *in vivo* imaging studies demonstrate that as we become more and more skilled at almost any cognitive task, the differential level of metabolism and the extent of neural tissue involved decreases, until for highly automatic skills there is almost no metabolic differential. Computationlike processes can involve precise connections and specific signals. They need not depend on the statistics of mass-level homeodynamic and morphodynamic processes. So, if automated functions are those that have become more computationlike, we should expect that they will have a rather diminutive metabolic signature. Indeed, it makes sense that one of the functions of learning would be to minimize the neural resources that must be dedicated to a given task. Consciousness is in this respect in the business of eliminating itself by producing the equivalent of virtual neural computers.

Serendipitously, then, fMRI, PET, and other techniques for visualizing and measuring regional differentials and changes in neural metabolism may provide a useful preliminary tool for tracing the changing levels and loci of brain processes correlated with consciousness. If the three-level emergent dynamic accounts of the differentiation of mental content and emotion are on the right track, then the dynamical changes in this signature of changing brain metabolism are providing important clues about these mental states. Indeed, this intuition is provisionally assumed when studying brain function with *in vivo* imagery.

So, even though this is a theory which defends the thesis that intentional relationships and sentient experiences are not material phenomena in the usual sense of that concept, it nonetheless provides us with a thoroughly empirical set of predictions and testable hypotheses about these enigmatic relationships.

CONCLUSION OF THE BEGINNING

Although much of my professional training has been in the neurosciences, in this book I have almost entirely avoided any attempt to translate the emergent dynamic approach to mental experience and agency into detailed neurobiological terms. This is not because I think it cannot be done. In fact, I've hinted that my purpose is in part to lay the groundwork for doing exactly that. I believe that an extended effort to articulate an emergent dynamical account of brain function is necessary to overcome the Cartesian no-man's-land separating the study of the brain from the study of the mind. But the conceptual problems that remain to be overcome are immense.

I have at most sketched the outlines here of an approach that might overcome them. Despite the number of pages that I felt were required to even frame the problem correctly, I don't claim to have accomplished much more than to have described a hitherto unexplored alternative framing of these enigmatic problems. I believe, however, that once this figure/background logic of analysis becomes assimilated into one's thinking about biological, psychological, and semiotics problems, the path toward solutions in each of these domains will become evident. These paths have not been followed previously simply because they were not even visible within current paradigms. Such alternatives didn't exist in the flat materialistic perspective that has dominated thinking for much of the last few centuries. It is my hope that this glimpse of another scientifically rigorous, but not simplistically materialistic, way to view these issues will inspire others to explore some of the many domains now made visible.

I believe that despite its counterintuitive negative framing, this figure/background reversal of the way we conceive of living and mental causality promises to reinstate subjective experience as a legitimate participant in the web of physical causes and effects, and to ultimately reintroduce intentional phenomena back into the natural sciences. It also suggests that the subtitle of this book is slightly misleading. Mind didn't exactly emerge from matter, but from constraints on matter.

EPILOGUE

All sciences are now under the obligation to prepare
the ground for the future task of the philosopher,
which is to solve the problem of value . . .

— FRIEDRICH NIETZSCHE[1]

NOTHING MATTERS

Have we now arrived at where we started? Is this where we thought we were when we began? Let's take stock.

The laws of physics have remained unchallenged. The sense that I have of being a sentient and efficacious agent in the world, of being able to change things in ways that resemble my imagined ends, of recognizing beauty upon hearing a Chopin nocturne or sensing the tragedy of being part of a civilization unable to turn away from a lifestyle destroying its own future; all these have not changed. But something I now know about these experiences is different. I know something more and am something less as a result. At the very least, my experiences must be understood differently. This "I" from which I start, and from the perspective of which the whole physical world often seems alien, now appears in a different light. I am not the same I. On the one hand, I have somehow lost the solidity that I once took for granted, me-the-physical-body is no longer so certain; and yet on the other hand my uncertainty about my place in the world, the place of meaning and value in the scheme of things, seems more assured with the realization that I may be more like the hole at the wheel's hub than the rim of the wheel itself.

We began this exploration with an analogy between the challenges posed by the mathematics of zero and the challenges posed by the entential properties of living and mental processes. We then explored the many ways that modern science has, like the mathematics of the Middle Ages, attempted to exclude a role for the mark of absence in the fabric of legitimate explanation. Then, accepting the challenge of explaining how it could be that absent phenomena might be causally relevant, we began to reconceptualize some of the most basic physical processes in terms of the concept of constraint: properties and degrees of freedom not actualized. This figure/ background reversal didn't undermine any known physical principles, nor did it introduce novel, unprecedented physical principles or special fundamental forces into contemporary science. It didn't even require us to invoke any superficially strange and poorly understood quantum effects in our macroscopic explanations in order to account for what prior physical intuition seemed unable to explain about meaning, purpose, or consciousness. Rather, it merely required tracing the way that two levels of self-organizing, constraint-creating processes could become so entangled as to result in a dynamical unit—an autogen or teleogen—that enables specific constraints to create, preserve, and replicate themselves with respect to the given constraints in their physical context. But being able to trace in detail each step that is required to cross from the realm of simple mechanical processes into the realm of entential relationships changes everything. Even such basic concepts as work and information have taken on new meaning, and previously esoteric notions like self and sentience can be given fairly precise physical definitions.

When Western scholars finally understood how operations involving zero could be woven into the fabric of mathematics, they gained access to unprecedented and powerful new tools for modeling the structure and dynamics of the physical world. By analogy, developing a scientific methodology that enables us to incorporate a fundamental role for possibilities not actualized—constraints—in explaining physical events could provide a powerful new tool for precisely analyzing a part of the world that has previously been shrouded in paradox and mystery. The mathematical revolution that followed an understanding of the null quantity in this way may presage a similarly radical expansion of the sciences that are most intimately associated with human existence. It is time that we overcame our confused Zeno's paradox of mind, which makes it appear that represented purposes can never reach the finish line of causal consequences. It's time to recognize that there is room for meaning, purpose, and value in the fabric of physical explanations, because these phenomena effectively occupy the absences that differentiate and interrelate the world that is physically present.

THE CALCULUS OF INTENTIONALITY

It is with teleodynamic organization, I have argued, that for the first time one physical system is capable of influencing other physical systems via something that is merely virtual—that which is specifically absent, missing, displaced, potential, or merely abstract. In the simple thought experiment that exemplifies the core architecture of this argument—the emergence of an autogenic process—it is the premature halting of the component morphodynamic processes, a tendency that is spontaneous and would otherwise run to self-extinction, that makes self, information, and life possible. This preserves the preconditions necessary to iterate this process again and again. These preconditions are self-reconstituting and thus self-referential constraints. This failure to continue makes possible the capture, preservation, and potential propagation of the constraints that are thereby created— a process that I have also called an entropy ratchet because it prevents the decay of secondary constraints generated as a side product of an otherwise entropy-increasing, constraint-destroying process. It is the possibility of briefly building up constraints by morphodynamic action, and then halting the process before there is any dissipation of those constraints, that is the secret of this distinctive form of causality. In principle, it allows the generation of constraints to continually reconstitute this intrinsic potential endlessly. It also provides the capacity to remember and reproduce information, because self-rectifying constraint preservation is the defining criterion of referential information. This property is the foundation for all higher-order intentional processes.

To put this in somewhat enigmatic terms, teleodynamics enables the potentially indefinite to enable something intrinsically incomplete to bring itself into existence. Consider again the analogy between entential phenomena, on the one hand, and zero and infinity in mathematics, on the other. Purposes and functions have what amount to infinitesimal vectors. If, as the philosopher Ruth Millikan asks us to imagine, a lion suddenly and miraculously came into existence due to some amazing quantum accident, with all the living physiological detail of any living lion, its heart could still be said to function to pump its blood, its eyes could still be described as functioning to guide its movements, and its sexual urges could still be described as existing for the purpose of reproducing.[2] Even though none of these phenomena arose by natural selection, at the very moment this lion popped into existence, before even one beat of its heart, at that instant these tendencies were present and the entention was present as well, if these processes are poised to begin and proceed in a way that preserves the whole lion.

Similarly, though my fingers never evolved for linguistic communication, the moment I began to use them to type words on a keyboard, they came to function for this end. This is because the function was not implicit in fingers or computer keys but in how the constraints of linguistic communication by computer fit with the constraints of finger movement control. The potential of my fingers to assume this function was simply not excluded by the constraints they acquired due to natural selection. In this respect, even their grasping function can't be attributed to natural selection. Any acquired constraints that were valuable to grasping were simply maintained preferentially down the generations. It didn't require work to bring this function into existence for the simple reason that this convergence of constraints wasn't excluded, though in the course of evolution many other possibilities were. In this respect, function is effectively a geometric or formal relationship, not a material efficient one, a dynamical alignment or symmetry of some structure or process with respect to the teleodynamic system of which it is a part. Because of this, functionality can arise the instant that this potential becomes an actualized tendency, and even if in its implementation it is impeded from achieving this end.

This is the analogue to an infinitesimal velocity. Being able to ascribe a velocity to a projectile at a specific point along its trajectory, even though actual velocities are defined by finite distances and durations, was one of the powerful capabilities provided by the invention of calculus. So being able to specify an analogous basis for the assessment of entential properties provides a way to solve the Zeno's paradox of the mind that has held up our understanding of these phenomena for millennia. Specifying a function or representation is, in this respect, like the operation of differentiation in calculus: specifying an intrinsic (instantaneous) *telos*.

Similarly, the value of these physical attributes to the overall teleodynamics of the accidental lion can also be estimated, as can the value to it of objects in his surroundings. Food, water, appropriate levels of oxygen in the air, ambient temperature—all can be assigned some potential value with respect to their ability to support the ends served by the tendencies of this teleodynamics. Each behavioral option with respect to each environmental attribute can now be assigned some relative value in terms of its correspondence or not with physiological requirements for this incipient persistent tendency. And means to estimate these qualities will be recapitulated cognitively, translated into tendencies to mobilize neural work to obtain or avoid them with respect to this evaluation. This valuation, likely weighted with respect to many interdependent functional factors, is the analogue of the operation of integration in calculus.

So, by analogy, one might be justified in claiming that it is with the emergence of teleodynamics that nature finally discovered how to operate with the dynamical equivalent of zero. None of the dynamical properties associated with life and mind—such as function, purpose, representation, and value—existed until the universe had matured sufficiently to include complex molecules capable of forming into autogenic configurations. The explosive growth in dynamical complexity and causal possibility that arose with the emergence of teleodynamic processes on Earth, beginning with the origin of life, was a revolution of physical processes far more extravagant than the revolution of mathematics that followed the taming of zero. But these teleodynamic properties, whether embodied in the constraints affecting molecular dynamics or the constraints organizing neuronal signal dynamics, are the analogues of zero in what might be called the formal operations of matter—the dynamics of physical change. Life and mind are in this sense the embodied calculus of these physical processes; and with each leap from one teleodynamic level to another—from life to brain processes to the symbolic integration of millions of human minds extending over millennia—that physical calculus has now expanded in expressive power to the point it is able to fully represent itself.

VALUE

Perhaps the most tragic feature of our age is that just when we have developed a truly universal perspective from which to appreciate the vastness of the cosmos, the causal complexity of material processes, and the chemical machinery of life, we have at the same time conceived the realm of value as radically alienated from this seemingly complete understanding of the fabric of existence. In the natural sciences there appears to be no place for right/wrong, meaningful/meaningless, beauty/ugliness, good/evil, love/hate, and so forth. The success of contemporary science appears to have dethroned the gods and left no foundation upon which unimpeachable values can rest. Philosophers have further supported this nihilistic conception of scientific knowledge by proclaiming that no assessment of the way things are can provide a basis for assessing how things should be. This is the ultimate heritage of the Cartesian wound that severed mind from body at the birth of modern science. The removal of any approach to value from a scientific perspective is the ultimate expression of having accepted the presumed necessity of that elective surgery.

As I lamented in the opening chapter of this book, the cost of obtaining this dominance over material nature has had repercussions worldwide. Indeed, I don't think that it is too crazy to imagine that the current crisis of faith and the rise in fundamentalism that seems to be gripping the modern world is in large part a reaction to the unignorable pragmatic success of a vision of reality that has no place for subjectivity or value. The specter of nihilism is, to many, more threatening than death.

By rethinking the frame of the natural sciences in a way that has the metaphysical sophistication to integrate the realm of absential phenomena as we experience them, I believe that we can chart an alternative route out of the current existential crisis of the age—a route that neither requires believing in magic nor engaging in the subterfuge of ultimate self-doubt. The universe *is* larger than just that which we can see, and touch, or manipulate with our hands or our cyclotrons. There is more here than stuff. There is how this stuff is organized and related to other stuff. And there is more than what is actual. There is what could be, what should be, what can't be, what is possible, and what is impossible. If quantum physicists can learn to become comfortable with the material causal consequences of the superposition of alternate, as-yet-unrealized states of matter, it shouldn't be too great a leap to begin to get comfortable with the superposition of the present and the absent in our functions, meanings, experiences, and values.

In the title to one of his recent books, Stuart Kauffman succinctly identifies what has been missing from our current blinkered metaphysical worldview. Despite the power and insights that we have gained from this powerful way of conceiving of the world, it has not helped us to feel "at home in the universe." Even as our scientific tools have given us mastery over so much of the physical world around and within us, they have at the same time alienated us from these same realms. It is time to find our way home.

GLOSSARY

Absential: The paradoxical intrinsic property of existing with respect to something missing, separate, and possibly nonexistent. Although this property is irrelevant when it comes to inanimate things, it is a defining property of life and mind; elsewhere (Deacon 2005) described as a constitutive absence

Attractor: An attractor is a "region" within the range of possible states that a dynamical system is most likely to be found within. The behavior of a dynamical system is commonly modeled as a complex "trajectory of states leading to states" within a **phase space** (typically depicted as a complex curve in a multidimensional graph). The term is used here to describe one or more of the quasi-stable regions of dynamics that a dynamical system will asymmetrically tend toward. Dynamical attractors include state of equilibrium of a thermodynamic system, the self-organized global regularity converged upon by a morphodynamic process, or the metabolic maintenance and developmental trajectory of an organism (a teleodynamic system). An attractor does not "attract" in the sense of a field of force; rather it is the expression of an asymmetric statistical tendency

Autocatalysis: A set of chemical reactions can be said to be "collectively autocatalytic" if a number of those reactions produce, as reaction products, catalysts for enough of the other reactions that the entire set of chemical reactions is self-sustaining, given an input of energy and substrate molecules. This has the effect of producing a runaway increase in the molecules of the autocatalytic set at the expense of other molecular forms, until all substrates are exhausted

Autocell: A minimal molecular teleodynamic system (termed an *autogen* in this book), consisting of mutually reinforcing autocatalytic process and a molecular self-assembly process, first described in Deacon 2006a

Autogen: A self-generating system at the phase transition between morphodynamics and teleodynamics; any form of self-generating, self-repairing, self-replicating system that is constituted by reciprocal morphodynamic processes

Autogenic: Adjective describing any process involving reciprocally reinforcing morphodynamic processes that thereby has the potential to self-reconstitute and/or reproduce

Autogeneses: The combination of self-generation, self-repair, self-replication capacities that is made possible by teleodynamic organization; the process by which reciprocally reinforcing morphodynamic processes become a self-generating autogen

Boltzmann entropy: A term used in this work to indicate the traditional entropy of thermodynamic processes. It is distiguished from "entropy" as defined by Claude Shannon for use in information theory

Casimir effect: When two metallic plates are placed facing each other a small distance apart in a vacuum, an extremely tiny attractive force can be measured between them. Quantum field theory interprets this as the effect of fluctuating electromagnetic waves that are present even in empty space

Chaos theory: A field of study in applied mathematics that studies the behavior of dynamical systems that tend to be highly sensitive to initial conditions; a popular phrase for this sensitivity is the "butterfly effect." Although such systems can be completely deterministic, they become increasingly unpredictable over time. This is often described as deterministic chaos. Though unpredictable in detail, such systems may nevertheless exhibit considerable constraint in their trajectories of change. These constrained trajectories are often described as attractors

Complexity theory: A field of study in applied mathematics concerned with systems of high-dimensionality in structure or dynamics, such as those generated by non-linear processes and recursive algorithms, and including systems exhibiting deterministic chaos. The intention is to find ways to model physical and biological systems that have otherwise been difficult to analyze and model

Constitutive absence: A particular and precise missing something that is a critical defining attribute of "ententional" phenomena, such as functions, thoughts, adaptations, purposes, and subjective experiences.

Constraint: The state of being restricted or confined within prescribed bounds. Constraints are what is not there but could have been. The concept of constraint is, in effect, a complementary concept to order, habit, and organization because something that is ordered or organized is restricted in its range and/or dimensions of variation, and consequently tends to exhibit redundant features or regularities. A dynamical system is constrained to the extent that it is restricted in degrees of freedom to change and exhibits attractor tendencies. Constraints can originate intrinsic or extrinsic to the system that is thereby constrained

Contragrade: Changes in the state of a system that must be extrinsically forced because they run counter to **orthograde** (aka spontaneous) tendencies

Cybernetics: A discipline that studies circular causal systems, where part of the effect of a chain of causal events returns to influence causal processes further back up the chain. Typically, a cybernetic system moves from action, to sensing, to comparison with a desired goal, and again to action

Eliminative materialism: The assumption that all reference to ententional phenomena can and must be eliminated from our scientific theories and replaced by accounts of material mechanisms

Emergence: A term used to designate an apparently discontinuous transition from one mode of causal properties to another of a higher rank, typically associated with an increase in scale in which lower-order component interactions contribute global properties that appear irreducible to the lower-order interactions. The term has a long and diverse history, but throughout this history it has been used to describe the way that living and mental processes depend upon chemical and physical processes, yet exhibit collective properties not exhibited by non-living and non-mental processes, and in many cases appear to violate the ubiquitous tendencies exhibited by these component interactions

Emergent dynamics: A theory developed in this book which explains how homeodynamic (e.g., thermodynamic) processes can give rise to morphodynamic (e.g., self-organizing) processes, which can give rise to teleodynamic (e.g., living and mental) processes. Intended to legitimize scientific uses of ententional (intentional, purposeful, normative) concepts by demonstrating the way that processes at a higher level in this hierarchy emerge from, and are grounded in, simpler physical processes, but exhibit reversals of the otherwise ubiquitous tendencies of these lower-level processes

Entelechy: A term Aristotle coined for a non-perceptible principle in organisms leading to full actualization of what was merely potential. It is responsible for the growth of the embryo into an adult of its species, and for the maintenance of the organism's species-specific activities as an adult

Ententional: A generic adjective coined in this book for describing all phenomena that are intrinsically incomplete in the sense of being in relationship to, constituted by, or organized to achieve something non-intrinsic. This includes function, information, meaning, reference, representation, agency, purpose, sentience, and value

Epiphenomenal: Something is epiphenomenal if it is causally irrelevant and therefore just a redescription of more fundamental physical phenomena that are responsible for all that the causal powers mistakenly attribute to the epiphenomenal feature

Functionalism: The idea that the organization of a process can have real causal efficacy in the world, independent of the specific material components that constitute it. Thus a computer algorithm can exhibit the same global causal consequences despite being run on quite different computer architectures

Fusion: A conception of emergence proposed by the philosopher Paul Humphreys, which argues that lower-level components and dynamics merge in indecomposable ways in the emergence of higher-order phenomena. It is especially relevant to the transition from quantum to classical processes. A related concept is discussed in terms of the reciprocal co-creation of biomolecules that compose an organism body

Golem: In Jewish folklore a golem is an animated, anthropomorphic being, created entirely from inanimate matter but lacking a soul

Homeodynamics: Any dynamic process that spontaneously reduces a system's constraints to their minimum and thus more evenly distributes system properties across space and time. The second law of thermodynamics describes the paradigm case

Homunculus: Any tiny or cryptic humanlike form or creature, something slightly less than human, though exhibiting certain human attributes. In recent scientific literature, "homunculus" has also come to mean the misuse of teleological assumptions: the unacknowledged gap-fillers that stand behind, outside, or within processes involving apparent teleological processes, such as many features of life and mind, and pretend to be explanations of their function

Intentional: In common usage, an adjective describing an act that is performed on purpose. Technically, in twentieth-century philosophy of mind, it is a term deriving from the medieval Scholastics, reintroduced by the German philosopher Brentano, to designate a characteristic common to all sensations, ideas, thoughts, and desires: the fact that they are "about" something other than themselves

Mereology: Literally, the "study of partness"; in practice, the study of compositionality relationships and their related hierarchic properties

Morphodynamics: Dynamical organization exhibiting the tendency to become spontaneously more organized and orderly over time due to constant perturbation, but without the extrinsic imposition of influences that specifically impose that regularity

Multiple realizability: When the same property can be produced by diverse means; independence of certain phenomena from any of their specific constitutive material details (see also **Functionalism**)

Nominalism: The assumption that generalizations are merely conveniences of thought, abstracted from observation, and otherwise epiphenomenal in the world of physical cause and effect; thus a denial of the efficacy of types, classes, species, ideal forms, and general properties over and above that of the individuals they describe

Orthograde: Changes in the state of a system that are consistent with the spontaneous, "natural" tendency to change without external interference

Panpsychism: The assumption that a vestige of mental phenomenology is present in every physical event, and therefore suffused throughout the cosmos. Although panpsychism is not as influential today, and effectively plays no role in modern cognitive neuroscience, it still attracts a wide following, mostly because of a serendipitous compatibility with certain interpretations of quantum physics

Phase space: In mathematics and physics, a phase space is a space in which all possible states of a system are represented. Each possible state of the system corresponds to one unique point in the phase space. For mechanical systems, a phase space usually consists of all possible values of position and momentum

Preformationism: Narrowly, the assumption that the human physique was preformed from conception. More broadly as used here, the assumption that entential phenomena were performed in antecedent phenomena—that, for example, language is preformed in a universal grammar module, information is preformed in DNA, or that consciousness is preformed in the mind of God

Protected states: Insulation between levels of dynamics, in effect, micro differences that don't make a macro difference because of statistical smoothing and attractor dynamics. Introduced by the physicist Robert Laughlin to describe the causal insulation of physical processes at different levels of scale

Protocell: Any of a number of theoretical, membrane-bound multimolecular units conceived by molecular biologists as experimental or theoretical simplest possible living units, usually consisting of replicating polynucleotides within a lipid "bubble," used as possible exemplars of the precursors of life

Realism: The assumption that general properties, laws, and physical dispositions to change are fundamental facts about reality, irrespective of subjective experience, and are causally efficacious

Self-organization (Self-simplification): W. Ross Ashby (1957) defined a self-organizing system as one that spontaneously reduces its statistical entropy, but not necessarily its thermodynamic entropy, by reducing the number of its potential states. Ashby equated self-organization with self-simplification. In parallel, Ilya Prigogine explored how such phenomena can be generated by constantly changing physical and chemical conditions, thereby continually perturbing them away from equilibrium. This work augmented the notion of self-organization by demonstrating that it is a property common to many far-from-equilibrium processes; systems that Prigogine described as *dissipative structures*

Shannon entropy: A measure of the variety of possible signal configurations of a communication medium determined as proportional to the logarithm of the number of possible states of the medium. This is an entirely general quantity that can be applied to almost any phenomenon. Designating it as **entropy**, though initially due to its mathematical parallel with thermodynamic entropy, is now generally thought to be describing the same thing in informational terms

Shannon information: A way of measuring the information-carrying capacity of a medium in terms of the uncertainty that a received signal removes

Strong emergentism: The argument that emergent transitions involve a fundamental discontinuity of physical laws—cf. **Weak emergentism**

Supervenience: The relationship that emergent properties have to the base properties that give rise to them

Teleodynamics: A form of dynamical organization exhibiting end-directedness and consequence-organized features that is constituted by the co-creation, complementary constraint, and reciprocal synergy of two or more strongly coupled morphodynamic processes

Teleogen: A non-autonomous autogenically organized system that is a component within a larger autogenic system, such as a somatic cell within a multicellular organism or an endosymbiotic organism within an organism. Although such subordinate or lower-order nearly autogenic subsystems are not fully reciprocally closed in their dynamics, they nevertheless exhibit end-directed tendencies and normative relationships with respect to extrinsic factors

Teleogenic: A systemic property (or individuated dynamical system) constituted by a higher-order form of teleodynamic process, specifically where that teleodynamic process includes a self-referential loop of causality such that the causal properties of the individuated teleodynamic unit are re-presented in some form in the generation of teleodynamic adaptive processes

Teleomatic: Automatically achieving an end, as when a thermodynamic system develops toward equilibrium or gravity provides the end state for a falling rock

Teleonomy (Teleonomic): Teleological in name only. A terminological distinction that would exemplify a middle ground between mere mechanism and purpose, behavior predictably oriented toward a particular target state even in systems where there was no explicit representation of that state or intention to achieve it

Teleological (Teleology): Purposive, or end-directed (the study of such relationships). Philosophically related to Aristotle's concept of a "final cause"

Top-down causality: The notion that higher-order emergent phenomena can alter phenomena that they supervene upon (i.e., the components and interactions that collectively have given rise to the emergent phenomena). Usually proposed as a countervailing causal claim to the reductionist assumption that macro events and properties are entirely determined by the micro events and properties of components that compose them. For example, some have argued that whole brain functions, which are the product of billions of neural interactions, can alter the way individual neurons behave, and thus generate causal consequences at the neuronal level. As a temporally understood causal relationship, this is not problematic; but understood synchronically, it appears to lead to vicious regress. Anther way of understanding top-down causality, due to Roger Sperry (see chapter 5), is as global constraint. Thus atoms in a wheel are constrained to only move with respect to neighboring atoms; but if the whole wheel rolls, all the atoms are caused to follow cycloid paths of movement

Tychism: The metaphysical assumption that, at base, change is spontaneous and singular, and thus intrinsically uncorrelated

Vitalism: A theory in natural philosophy claiming that physical and chemical processes alone are insufficient to explain living organisms. An additional non-perceptible factor is necessary which Hans Driesch (1929) called *entelechy*, to honor Aristotle, and Henri Bergson (1907) called *élan vital*. For Driesch, in its earliest stage an embryo is not manifold in an *extensive* sense, but there is present in it an entelechy which is "an *intensive* manifoldness"

Weak emergentism: The argument that although in emergent transitions there may be a superficially radical reorganization, the properties of the higher and lower levels form a continuum, with no new laws of causality emerging. Often associated with epistemological emergentism because it is attributed to incomplete knowledge of the critical causality

NOTES

CHAPTER 0: ABSENCE

1. With thanks to Charles Seife, the author of *Zero: The Biography of a Dangerous Idea*, from which I take this chapter number idea and the quote from Tobias Dantzig's *Number* at p. 35.
2. This term was suggested to me by my longtime colleague Tyrone Cashman. Though the term takes liberties with the flexibility of the English language, I think that its somewhat cavalier form is compensated by the way it focuses attention on this one most central feature of these troublesome phenomena. I will resist the obvious tendency to call this metaphysical paradigm *absentialism*, because as will soon be evident, it is an explanation of the emergent and dynamical character of the processes generating these phenomena that is my primary goal.
3. David Chalmers (1995), "Facing up to the problem of consciousness," *Journal of Consciousness Studies* 2:208.
4. A calculating device made with movable beads arranged on strings, originally devised in the Middle East, but spread eastward as far as China, where it is still in use as a calculation aid.
5. For a trivial example, consider that since $0 \times 1 = 0$ and $0 \times 2 = 0$, it follows that $0 \times 1 = 0 \times 2$. We can then divide each side by zero (which is the disallowed operation) and cancel $0/0$ in each (which for any other number equals 1, but is undefined for 0), and simplify to get $1 = 2$.
6. Prior to the work of the eighteenth-century French chemist Antoine Lavoisier, a special substance given the name "phlogiston" was presumed to be present in all combustible materials and responsible for their susceptibility to catch fire.

CHAPTER 1: (W)HOLES

1. *Tao Te Ching #11:* This "translation" is my own effort to clarify the meaning of this enigmatic entry, based on comparisons between a number of different translations from the original ancient Chinese text.
2. In a famous thought experiment, the brilliant eighteenth-century philosopher-mathematician Pierre-Simon Laplace suggested that to an infinitely knowledgeable mind the world would be an entirely predictable clockwork process, right down to the movement of each atom. He says: "We may regard the present state of the universe as the effect

of its past and the cause of its future. An intellect which at a certain moment would know all forces that set nature in motion, and all positions of all items of which nature is composed, if this intellect were also vast enough to submit these data to analysis . . . it would embrace in a single formula the movements of the greatest bodies of the universe and those of the tiniest atom; for such an intellect nothing would be uncertain and the future just like the past would be present before its eyes."—Simon Laplace, *A Philosophical Essay on Probabilities* (1814/1951), p. 4.

3. The function of an organ was once presumed to be inherited from its divine designer, and thus parasitic in the same sense as artifact function. This view still motivates religious critics of evolution. Since Darwin, this analogy has been abandoned in favor of what is often described as a teleonomic view (see chapter 4).

4. The realism/nominalism problem will be faced head-on in chapter 6.

5. A Wikipedia entry (http://en.wikipedia.org/wiki/Little_green_men) suggests that the depiction of aliens as "little green men" may date to an Italian report from as early as 1910, claiming to have captured aliens, and it was also popularized by a newspaper article satirizing Orson Welles' 1938 broadcast of *War of the Worlds*. It notes that goblins and gremlins have sometimes been depicted as small green humanoids. The linkage to copper-based blood is unclear, but the blue-greenish blood of some arthropods and mollusks has probably been folk knowledge for centuries.

6. Interestingly, the legal system has come down on the side of nominalism when it comes to patentability and ownership of software. In a famous precedent-setting lawsuit, the Apple computer company brought suit against Microsoft for copying the "look and feel" of its object-based Macintosh interface. Apple argued that many features of the Macintosh interface appeared in the Windows operating system and thus violated its copyright. These included icons like file folders, click and drag effects, the use of a mouse, and a trash bin for erasing files. Microsoft won the lawsuit by arguing that although there were indeed obvious similarities of "look and feel," they were created by entirely different software instructions, and thus there was no use of the actual code that Apple had patented.

7. Richard Dawkins (1996), p. 133.

8. Steven Weinberg (1993), p. 154.

9. Francis Crick (1994), p. 3.

10. Jerry Fodor (1990), p. 156.

11. The title implicitly parodies Bishop Paley's notion that organisms, like watches, could only have been created by an intelligent designer with a purpose in mind. By describing the watchmaker as unable to see, Dawkins is effectively arguing that there is neither forethought nor purposely guided selection of components and assembly involved. See Dawkins' *The Blind Watchmaker* (1996).

12. Except for that one tiny exception, the Big Bang, which created the whole of the visible universe. But that's another story.

CHAPTER 2: HOMUNCULI

1. Daniel Dennett, *Brainstorms* (1978), p. 12.

2. *Osmosis Jones*, dir. Tom Sito and Piet Kroon with the Farrelly brothers, Warner Bros., 2001.

3. B. F. Skinner, "Behaviorism at Fifty," in T. W. Wann, ed., *Behaviorism and Phenomenology* (1964), p. 80.

4. Francis Crick and Christof Koch (2003), *Nature Neuroscience* 6:120.
5. This theory is often paraphrased by the epigraph "ontogeny recapitulates phylogeny," which means that developmental stages effectively retrace the stages of phylogenetic evolution leading up to the present individual. This was presumed to occur because prior phylogenetic stages represented less elaborated developmental sequences.
6. Steven Pinker, *The Language Instinct* (1995), pp. 81–82.
7. Stephen Hawking, *A Brief History of Time* (1988), p. 1.
8. Seth Lloyd, *Programming the Universe* (2006), p. 147.
9. Ibid.
10. See Alfred North Whitehead's *Process and Reality* (1928/1979).

CHAPTER 3: GOLEMS

1. B. F. Skinner (1971), p. 200.
2. Daniel Dennett (1978), p. 12.
3. For the classic statements of these versions of eliminativism, see, e.g., Rorty (1970), Stich (1983), Paul Churchland (1989, 1995), Patricia Churchland (1986), and Dennett (1987, 1991).
4. I am indebted to theological graduate student Adam Pryor for providing the Hebrew text that demonstrates this typographical pun.
5. Gregory Bateson (1979), p. 58.
6. Although Noam Chomsky's article criticizing Skinner's "Verbal Behavior" and a series of learning experiments by John Garcia are often credited with striking major blows, they are probably more like symptoms rather than causes of the abandonment of behaviorism for a more cognitive approach, which was also influenced by people like Jean Piaget and Jerome Bruner.
7. In contemporary digital computers this is accomplished by compiler software, which is usually machine-specific, and mapped to global machine operations.
8. Though it didn't succeed altogether, since as a young man I spent one summer on a rail crew as a "gandy dancer" using a hammer to straighten out spikes bent by a modern version of a motorized spike driver.
9. From IBM Research Lecture RC-115, given in 1958. See Irving J. Good (1959).
10. Jerry Fodor (1980), pp. 148–49.

CHAPTER 4: TELEONOMY

1. This quote from E. Von Bruecke, a nineteenth-century physiologist, was quoted in Cannon (1945), p. 108. Thanks to Don Favareau for suggesting this quote.
2. See Ernst Mayr (2001), "The philosophical foundations of Darwinism," *Proceedings of the American Philosophical Society*, vol. 145, no. 4 (December), pp. 492–93.
3. Ernst Mayr (1974), chap. 3; italics in the original.
4. Ibid.
5. Ibid., p. 98.
6. Jesper Hoffmeyer (1996), p. 56.
7. Quoted in ibid.
8. Paul Weiss (1967), p. 821.

CHAPTER 5: EMERGENCE

1. The author's translation of a line from Ilya Prigogine and Isabelle Stengers, *La Nouvelle Alliance* (1979), p. 278.
2. George Henry Lewes (1874), Vol. II, p. 412.
3. John Stuart Mill (1843), Bk III, chap. 6, sect. 1.
4. Gregory Bateson (1979), pp. 455–56.
5. See Robert Wilson's *Genes and the Agents of Life* (2005).
6. This quote from Stephen Hawking appeared in a 2000 newspaper interview. See Hawking: "I think the next century . . ."
7. Mill (1843), Bk III, chap. 6, sect. 1.
8. Strict Darwinians of the time followed August Weismann's dictum that acquired somatic changes could not directly influence what was inherited via the germ line in the next generation.
9. See Conwy Lloyd Morgan's *Emergent Evolution* (1923).
10. See, e.g., Leo Buss, *The Evolution of Individuality* (1987), or Eörs Szathmáry and John Maynard Smith, *The Major Transitions in Evolution* (1995).
11. See C. D. Broad, *The Mind and Its Place in Nature* (1925).
12. Samuel Alexander, *Space, Time, and Deity* (1920).
13. For more detailed philosophical taxonomies of emergence, see, e.g., Philip Clayton, (2003), Michael Silberstein and Anthony Chemero (2009), or the Stanford Encyclopedia of Philosophy at http://plato.stanford.edu/entries/properties-emergent/.
14. See Paul Humphreys (1997 a & b). Humphreys would probably object to the much broader analogical extension of his concept of fusion that I employ in this comparison. However, the heuristic usefulness of the general idea implicit in his more technical usage is that it exemplifies the system dependence of the concept of a component part.
15. "Synchrony" refers to processes and events occurring simultaneously, whereas "diachrony" refers to processes and events that succeed and follow one another in time.
16. Jerry Fodor (1998), p. 61.
17. See Conwy Lloyd Morgan (1923).
18 Donald Davidson, "Mental Events," in L. Foster and J. W. Swanson, eds., *Experience and Theory* (1970), p. 214.
19. See, e.g., Pepper (1926).
20. Mark Bickhard, "Process and Emergence" (2003). To support this claim from quantum physics, Bickhard cites numerous distinguished physicists including, notably, Steven Weinberg (1995, 1996, 2000).
21. The phrase *phase space* originates from the use of graphs to represent the way that substances (like water) change phase from solid to liquid to gas with respect to different values of temperature and pressure. Graphs of other forms of dynamical processes can involve many dimensions, and so may not be easily depicted. The term *state space* is in many respects more accurate, since each point in such a graph represents a specific state of the system; however, we can also think of the points along a trajectory of change as being phases of change. In real-world systems, and even complex mathematically represented systems, such trajectories never intersect themselves. But in finite automata this will ultimately occur, at which point the trajectory will form a complex loop following an identical path over and over from that time forward.
22. The phrase is borrowed from the philosopher Daniel Dennett, who used it to character-

ize an argument that while not demonstrative, nevertheless aids in thinking more clearly about a difficult problem.

23. Mark Bickhard (2005).

24. To continue the analogy, fMRI effectively produces a snapshot that is averaged across distances many orders of magnitude larger than neurons, averaged across many seconds or minutes, and often even averaged across multiple trials and multiple subjects.

25. See Timothy O'Connor (2000).

26. Douglas Hofstadter coins this expression in his *Gödel, Escher, Bach* (1979).

CHAPTER 6: CONSTRAINT

1. W. Ross Ashby (1962), p. 257.

2. A notion that can be compared with the indeterminacy of modern quantum physics.

3. C. S. Peirce, *Collected Papers*, Vols. 1–6 (1931–35), 1:16, 4:1.

4. More subtly, in chapter 11 when we return to consider the concept of work—i.e., that which is necessary to change things in non-spontaneous ways—we will discover that it takes constraint to produce work; and in chapters 12 and 13 we will discover that constraint is also the defining attribute of information.

5. A non-linear equation is one in which one or more variables in that equation take on the value of the result of performing the calculation that the equation specifies. Thus, given an initial value, the same calculation is repeatedly performed, with each successive calculation incorporating the resultant value of the previous calculation as the updated value of that variable. A classic example is the calculation of compounded interest, in which the interest on a principal is added to the remaining principal to yield the new principal on which the next calculated interest value is based, and so forth.

6. Robert Laughlin would even extend this logic to make it a basic factor distinguishing quantum from classical dynamics (though I am not qualified to assess this claim). As has been understood for nearly a century, although quantum fluctuations are ubiquitous and inject irreducible indeterminateness into all quantum-level interactions, these noiselike effects typically wash out above the atomic scale to produce material interactions that are nearly indistingushable from those predicted by Newtonian dynamics.

CHAPTER 7: HOMEODYNAMICS

1. Curiously, there were efforts to explain gravitation in efficient causal terms following Newton and before Einstein. One theory hypothesized that the gravitational "force" might occur due to a kind of shielding effect. Laplace imagined that there were vast numbers of particles careening around the universe, constantly bumping into solid matter and moving it about, a bit like atoms in Brownian motion. An object isolated in space would be evenly bumped on all sides and would therefore remain at rest. Two objects brought into close proximity would, in contrast, partially shield one another from collisions on the sides facing each other, and this would produce an asymmetrical force on each, propelling them together.

2. You might also say that gravitational-inertial mass is itself just a local warp in space-time, though this is not the standard interpretation.

3. This partial isolation of properties and the possibility of linking previously isolated dimensions of change, like transforming heat into linear motion in a heat engine, will be explored in extensive detail in chapter 11, as it pertains to the concept of *work*.

4. The analogy between Newtonian and thermodynamic interactions is actually closer than might first be apparent. Real billiard balls also have microscopic atomic structure, and the transfer of momentum from one to the other is mediated by a transient distortion of the relationships between atoms that is quickly propagated through each ball during the exchange of momentum. In this respect, this transient period of atomic-level deformation leading up to elastic rebound is analogous to the period of heat transfer between contiguous media, and the resultant state of whole ball movement is analogous to the new equilibria that the two media settle into. And although the heat (i.e., average molecular momentum) in each medium may be equivalent after interaction, if they are composed of different numbers of molecules, the total amount of molecular momentum in each will be appropriate to this difference.

5. The term *phase space* initially referred to graphs that depict the way variables like temperature and pressure interact to determine the change in phase (e.g., from liquid to gas).

6. The metaphor of a warped phase space is actually not the way attractors are usually graphed. More commonly, the space is Euclidean and attractors show up as areas of more dense trajectories. Indeed, this analogy is only heuristic because the dimensionality of a thermodynamic system of even modest size is high and thus quite beyond depiction. This way of thinking about the attractor bias in trajectory orientation is intended to highlight the geometric nature of the causality.

7. It can be objected that the so-called modern scientific paradigm being invoked here is actually a hundred-year-old pre-relativistic and pre-quantum understanding of causality. This is in fact the case. Does this mean that these concepts and this way of fractionating the notion of causality are irrelevant to modern physics, or that these distinctions are likely to be undermined by considering them in light of more contemporary physical theory? I don't think so. I suspect that pursuing a similar logic in an analysis of quantum-level and relativistic processes might indeed help explain some of the more paradoxical aspects of these processes. This is a matter that would at least require another book, and for the purpose of making sense of the teleological features of life and mind, I believe that these extremes of scale in space, time, and energy are not relevant. We need go no further than classical statistical mechanics to make sense of these phenomena.

8. C. S. Emmeche, S. Køppe, and F. Stjernfelt, *Levels* (2000), p. 31.

9. Whether we can identify homeodynamic processes at even deeper quantum levels is unclear to me, though I suspect that some variant of these reciprocal orthograde-contragrade dynamical relationships will even apply in this strange domain as well.

CHAPTER 8: MORPHODYNAMICS

1. Peirce's 1898 series of Harvard lectures to which he was invited by William James. Reprinted in *Reasoning and the Logic of Things: The Cambridge Conferences Lectures of 1898* (1992), Cambridge, MA: Harvard University Press, p. 258.

2. Paul Weiss (1967).

3. Literally, "in glass," meaning in a laboratory experiment performed within a dish, not within an organism.

4. Note that dependence on what occurred previously in the sequence also describes the additive logic of the Fibonacci series.

5. For a recent demonstration of the spontaneous formation Fibonacci spirals in a non-organic material, see Li et al. (2005, 2007).
6. Ashby later cautioned against a too literal interpretation of self-organization for reasons analogous to those discussed here.
7. Bénard (1900).
8. For example, hexagonal close packing occurs spontaneously when same-size balls are packed together tightly on a planar surface.
9. For the sake of narrative flow, after having painstakingly dissected Bénard cell dynamics, I will simplify each of the following accounts, ignoring the many subtle differences that make their emergence slightly disanalogous to Bénard cell emergence, and focusing mainly on the major features that exemplify their logic of constraint transfer and amplification.

CHAPTER 9: TELEODYNAMICS

1. David Bohm, "Some Remarks on the Notion of Order" (1968), p. 34.
2. Ulanowicz (2009) argues that the dynamics of living systems contradict the fundamental axioms of Newtonian dynamics.
3. See Ilya Prigogine, *Introduction to Thermodynamics of Irreversible Processes* (1955).
4. Dewar in Whitfield (2005), p. 907.
5. Kleidon, in ibid.
6. Swenson (1989), p. 46.
7. E. D. Schneider and J. J. Kay (1997), p. 165.
8. See Stuart Kauffman (2000).
9. Maturana and Varela (1980).
10. Bickhard (2003).
11. See John von Neumann (1966), p. 67.
12. See E. V. Koonin (2000).
13. See R. Gil et al. (2004).
14. See recent reviews in Rassmussen et al. (2004) and also Szathmáry et al. (2005).
15. T. R. Cech (1986).
16. But see Frietas and Merkle (2004) for a review of the state of this research.
17. See, e.g., Anet (2004); Andras and Andras (2005); de Duve (1996); Kauffman (1995); and Morowitz (1992).

CHAPTER 10: AUTOGENESIS

1. Immanuel Kant, *The Critique of Judgment* (1790), Sect. 65, p. 558.
2. Kauffman (2000).
3. Eigen and Schuster (1979).
4. Kant (1790), Sect. 65, p. 557.
5. See Deacon (2004, 2005, 2007).
6. Though neither are literally *self* replication. Both require reciprocal relationships between molecules; and even if RNA molecules are able to assume both template and catalytic functions, they can't do both at once, even in the simplest cases envisioned. See the more detailed critical discussion of this concept in chapter 14.

7. Von Uexküll (1957).
8. Francisco J. Varela (1992), p. 11.

CHAPTER 11: WORK

1. From a textbook on thermodynamics: H. C. Van Ness, *Understanding Thermodynamics* (1969).
2. Gergory Bateson, *Steps to an Ecology of Mind* (1972), p. 453.
3. I.e., there is total energy symmetry across time in accordance with the first law of thermodynamics.
4. The curious fact that it takes work to change or produce constraints, and that constraints are required to do work, is discussed as an important clue to what Stuart Kauffman describes as the missing "theory of organization," in his book *Investigations* (2000).

CHAPTER 12: INFORMATION

1. John Collier (2003), p. 102.
2. Franz Brentano (1874), *Psychology from an Empirical Standpoint*, pp. 88–89.
3. Indeed, this is what makes self-organizing dynamics so intriguing and makes living dynamics often appear planned and executed by an external or invisible agency.
4. Shannon based his analysis on the model of a transmission channel such as a telephone line with a fixed limit to the variety and rate of signals that it could carry. This model system will be used throughout the discussion to be consistent with Shannon's terminology; but the analysis equally applies to anything able to convey information, from text on a page to physical traces used as clues in a criminal investigation.
5. Thus the probability of appearance of each character at each position is maximally uncorrelated with all others, as is the movement of each molecule in a gas at equilibrium. This justified calling both conditions maximum entropy.
6. See Warren Weaver and Claude Shannon (1949).
7. See Deacon (2007, 2008).
8. For example, negentropy plays a central role in Erwin Schrödinger's famous essay *What Is Life?*

CHAPTER 13: SIGNIFICANCE

1. Personal communication from Stuart Kauffman. See also Kauffman et al. (2008).
2. This does not necessarily mean that there is a loss of functionality accompanying this reduction. In fact, in biology just the opposite is true. The functional features of protein molecules are mostly a function of their three-dimensional structure. This is a secondary consequence of the different properties of their constituent amino acids (coded in the DNA) and how these properties in this order interact with one another and with the surrounding chemical milieu. Because of this, a protein embodies a considerably higher potential Shannon entropy than does the DNA sequence that codes for it. This contributes to an enormous amplification of information from genome to cellular organization.
3. See, e.g., Karl Popper (2003) and Donald T. Campbell (1974).

4. E.g., Fred Dretske (1988) and Ruth Millikan (1984).
5. See, e.g., critiques by Mark H. Bickhard (1998, 2000, 2003, and 2005).
6. Although other organisms sharing the same ecosystem may comprise a significant source of relevant selection "pressures," they do not generally directly modify the information passed from generation to generation, but simply impose complex changing boundary conditions on one another. Parasites, predators, and opposite-sex conspecifics may however exert a more direct constraining influence.
7. See Charles Sanders Peirce (1931–35).

CHAPTER 14: EVOLUTION

1. See D. Batten, Stanley Salthe, and F. Boschetti (2009), p. 27.
2. Peter A. Corning (1995), p. 112.
3. James Mark Baldwin (1896), Sect. V.
4. Less well known is the fact that for much of his career, Pasteur himself was convinced that life could be spontaneously generated. He just believed that others working on the problem were less careful about contamination and had missed what he considered to be the crucial factor: something he had discovered in his effort to understand the cause of problems in the winemaking industry—the asymmetrical left-handed twist of organic molecules, like the sugars produced by grapes. He had discovered this by virtue of the way that sugar crystals from healthy wine barrels rotated light in one direction as seen through polarization filters. After exploring this phenomenon in other contexts as well, he came to the realization that despite the lack of an obvious inorganic preference for this way of contorting molecules, those which were formed by organisms invariably exhibited crystalline forms producing this same one-way twist. The cause of this universal asymmetric "chirality" (i.e., handedness, like the mirror difference between left and right hands) of biologically generated molecules remains somewhat of a mystery, and is not irrelevant to understanding the origins and nature of life. To Pasteur, this looked like the key to the "secret of life." His notebooks turn out to be filled with elaborate techniques for introducing twists into chemical processes, hoping to produce the left-handed chemistry of life from otherwise sterilized and sealed chemical soups.

Today, most biologists assume that the left- as opposed to right-handed twist of biologically produced sugars is not an essential feature of life. Rather, because of the necessity to pass molecules around a circuit of synthetic and catalytic metabolic steps, mixed chiralities would be far more likely to produce incompatibilities and competitive interactions that would in effect get in each other's way. With such a mix of nearly similar but incompatible small building blocks, complex webs of metabolic processes would be far less likely to evolve. Why left and not right twists? Perhaps just an accident of evolution— a consequence of a slight asymmetry of percentages of molecular forms present in the initial conditions in which life formed; perhaps because of a slight bias in nature that makes the left forms slightly more probable in prebiotic synthesis; or maybe, as Pasteur thought, there is something else that is special about life that requires molecular left-handedness. For our purposes, however, coming to a conclusion about this mystery will not be necessary, since it is likely contingent on a strictly chemical explanation, not one that should affect an account of the emergence of intentional properties in general.
5. The republication of many Bastian's books and papers, along with commentaries by his contemporaries, has rekindled interest in many aspects of his work, especially in areas

that have only recently again become of scientific interest as representing certain over-simple dogmas concerning the nature of life, the causes of disease, and the role of inorganic processes in the origins of life.

6. In fact, I will argue that conditions on the primordial Earth may not have been conducive even to autogenic emergence. The problem is also the solution: water. Water is the ideal solvent for biochemical reactions, but it poses a barrier to one of the crucial requirements for protolife to emerge, whether in autogen form or any other form. The problem is that large organic polymers tend to break down in water, and it is hard to get them to form, because they require dehydration reactions. Ultimately, I think this forces us off the early Earth to explain the first steps toward life.

7. Although, due to the remarkable conservatism of many genes, the result is surprisingly often *not* noise; however, it may be a very different sort of consequence than would be produced in the donor organism.

8. There is a trivial exception. Alternating sequences of *ATAT* or *GCGC*, etc., will produce replicas, but note that the information-carrying capacity is nil, except for length difference and alternation redundancies.

9. I am less satisfied with my effort to map Peirce's object terms onto this process.

10. See Margulis and Sagan (1995), and Dyson (1999).

11. See Deacon (2006a).

12. Some of the implications of this argument are that we should expect to find diverse forms of extraterrestrial protolife in planetary and even extraplanetary environments quite different from those on Earth. Consequently, we will need to significantly expand the scope of what we consider astrobiology (or exobiology). I have elsewhere suggested that autogenic, non-template-based replication will be widespread and diverse, whereas template-based, informational replication and control of organism development will be rare because of the much more limiting requirements. I have given the name *Autaea* to all protolife and life forms, and *Morphota* to those like autogens that replicate via holistic molecular dynamic constraints, and have designated those, like Earth life, that use information-based reproduction and morphogenesis *Semeota*. Modern fossils of ancient autogenic mechanisms are not likely present except as aspects of still more complex organisms, or else parasitic on them. Thus viruses and structures like microtubules still utilize processes characteristic of autogenic chemistry.

CHAPTER 15: SELF

1. Daniel Dennett, "The Origin of Selves," *Cogito* 3 (Autumn 1989), pp. 163–73.

2. The story is even more complicated for plants, since besides mitochondria, they house organelles called chloroplasts (which carry out photosynthesis) that have yet a different bacterial origin and bacterial genomic structure.

3. This is a quote from the well-known, very sad case of an anterograde amnesic patient named Clive Waring, about whom a number of video documentaries have been made.

4. This catch phrase has been attributed to Maslow by many authors in slightly different forms, though I have searched in vain for a published citation. I first came across a mention of it decades ago in the introduction to Robert Ornstein's *The Psychology of Consciousness* (1977).

CHAPTER 16: SENTIENCE

1. From William James' essay "The Will to Believe," in *The Will to Believe and Other Essays in Popular Philosophy* (1897), Cambridge, MA, & London: Harvard University Press, 1979, p. 198.
2. See Alan Turing's 1950 paper on "Computing Machinery and Intelligence."
3. First presented in John Searle's 1980 article "Minds, brains and programs."
4. Terry Bisson, "They're Made Out of Meat," *Omni* (April 1991).
5. These are no more than guesses. It is not obvious how one would actually go about assessing such a problem. The idea that a good deal of neural signal transduction is highly stochastic is strongly implicated by the spontaneous resting firing rates of many neurons, the highly unstable nature of the resting potential, and the molecular variability intrinsic to cellular metabolism.
6. For example, it involves the comprehension of symbols—a capacity that is uniquely complex in itself, and distinctively human.

CHAPTER 17: CONSCIOUSNESS

1. Jerry Fodor in the *Times Literary Supplement*, July 3, 1992, p. 5.
2. I would even venture to speculate that there could be emergent levels of sentience above the human subjective level, in the higher-order dynamics of collective human communications—sentience that we large-brained, symbolically savvy individuals would never be able to experience, even though our sentient conscious interactions happened to be its necessary constituents. Of course such a sentience could only arise if these human interactions constituted a higher-order teleodynamic individual; a reciprocally organized, self-perpetuating complex of morphodynamic processes.
3. See Gregory Bateson (1972), "Form, Substance, and Difference," in *Steps to an Ecology of Mind: Collected Essays in Anthropology, Psychiatry, Evolution, and Epistemology*, pp. 457–61.
4. In vertebrate brains, these neural regulatory centers for vegetative processes function nearly as deterministically as simple cybernetic control circuits, and yet they still typically tend to involve homeo- and morphodynamic processes generated in local neural networks. But they have evolved such strict constraints that they essentially simulate the deterministic predictability of cybernetic dynamics.

EPILOGUE

1. From Friedrich Nietzsche, "Good and Evil, Good and Bad," in *On the Genealogy of Morals* (1887/1967), Vol. 1, Part 1, Sect. 17, p. 798. The original four-volume Schlechta edition in German includes this text as a note (Anmerkung) in small print. The original German sentence reads: "Alle Wissenshaften haben nunmehr der Zukunfts-Aufgabe des Philosophen vorzuarbeiten: diese Aufgabe dahin verstanden dass der Philosoph das Problem vom Werte zu loesen hat, dass er die Rangordnung der Werte *zu bestimmen hat."
2. The philosopher Ruth Millikan's discussion (1984) of the thought experiment in which an exact living replica of a lion instantaneously materializes was intended to explore the role that prior natural selection history plays in our understanding of biological function.

Because she, along with Fred Dretske (1988), explains functionality in terms of prior selection history, the duplicate lion's various organs are not considered to have functions (except by reference to being copies of evolved lion organs). Here I take a view, also expressed by Mark Bickhard (2003), that function is not created by natural selection but is rather preserved and tested by it. Function emerges instantaneously as an alignment with the dynamical tendency of some process (e.g., the beating of the heart) with the teleodynamic tendencies of the whole system of which it is a component.

REFERENCES

Alexander, Samuel (1920). *Space, Time, and Deity.* 2 vols. London: Macmillan.

Anderson, Philip W. (1972). "More is different," *Science,* vol. 177, no. 4047 (Aug. 4, 1972): 393–96.

Andras, P., and C. Andras (2005). "The origins of life—the 'protein interaction world' hypothesis: Protein interactions were the first form of self-reproducing life and nucleic acids evolved later as memory molecules," *Medical Hypotheses* 64: 678–88.

Anet, F. (2004). "The place of metabolism in the origin of life," *Current Opinion in Chemical Biology* 8: 654–59.

Ashby, W. Ross (1952). *Design for a Brain.* New York: John Wiley & Sons.

—— (1956). *An Introduction to Cybernetics.* London: Chapman & Hall.

—— (1962). "Principles of the Self-Organizing System," in H. Von Foerster and G. W. Zopf, Jr., eds., *Principles of Self-Organization: Transactions of the University of Illinois Symposium.* London: Pergamon Press.

Baldwin, James Mark (1896). "A new factor in evolution," *American Naturalist* 30: 441–51, 536–53.

Bastian, Henry C. (1870). "Facts and reasonings concerning the heterogeneous evolution of living things," *Nature* 2: 170.

Bateson, Gregory (1972). *Steps to an Ecology of Mind: Collected Essays in Anthropology, Psychiatry, Evolution, and Epistemology.* New York: Random House/Ballantine.

Batten, D., Stanley Salthe, and F. Boschetti (2008). "Visions of evolution: Self-organization proposes what natural selection disposes," *Biological Theory* 3: 17–29.

Baum, L. Frank (1900). *The Wonderful Wizard of Oz.* New York: George M. Hill Co. Film adaptation by Metro-Goldwyn-Mayer, *The Wizard of Oz,* 1939.

Bénard, Henri (1900). "Les tourbillons cellulaires dans une nappe liquide," *Sci. Pure Appl.* 11: 679–86 [also *de Phys.,* 9: 513].

Bentley, W. A., and W. J. Humphreys (1931). *Snow Crystals.* New York: McGraw-Hill.

Bergson, Henri (1908). *Evolution Créatice.* Paris: Félix Alcan.

Bertalanffy, Ludwig von (1952). *Problems of Life.* London: Watts.

Beveridge, W. I. B. (1950). *The Art of Scientific Investigation.* London: William Heinemann.

Bickhard, Mark H. (1998). "Levels of representationality," *Journal of Experimental and Theoretical Artificial Intelligence* 38.

—— (2000). "Autonomy, function and representation," *Communication and Cognition—Artificial Intelligence* 17(3–4): 111–31. Special issue on the contribution of artificial life and the sciences of complexity to the understanding of autonomous systems. Guest editors: Arantza Exteberria, Alvaro Moreno, and Jon Umerez.

—— (2000). "Emergence," in Peter Bøgh Andersen et al., eds., *Downward Causation*. Aarhus, Denmark: University of Aarhus Press.

—— (2003). "Process and Emergence: Normative Function and Representation," in J. Seibt, ed., *Process Theories: Crossdisciplinary Studies in Dynamic Categories*. Dordrecht, Netherlands: Kluwer Academic.

—— (2005). "Consciousness and reflexive consciousness," *Philosophical Psychology* 18(2): 205–18.

Bohm, David (1968). "Some Remarks on the Notion of Order," in C. H. Waddington, ed., *Towards a Theoretical Biology*. Vol. 2. Chicago: Aldine.

Boltzmann, Ludwig (1866). "Über die Mechanische Bedeutung des Zweiten Hauptsatzes der Wärmetheorie," *Wiener Berichte* 53: 195–220.

Brentano, Franz (1874). *Psychology from an Empirical Standpoint*. London: Routledge & Kegan Paul.

Broad, C. D. (1925). *The Mind and Its Place in Nature*. London: Routledge & Kegan Paul.

Brooks, D. R., and E. O. Wiley (1988). *Evolution as Entropy: Toward a Unified Theory of Biology*. Chicago: University of Chicago Press.

Buridan, John (1959). *Questions on the Eight Books of the Physics of Aristotle*, Book VIII. English translation in Marshall Clagett (1959), *Science of Mechanics in the Middle Ages*. Madison: University of Wisconsin Press.

Burks, A. W. (1970). *Essays on Cellular Automata*. Urbana: University of Illinois Press.

Buss, Leo (1987). *The Evolution of Individuality*. Princeton: Princeton University Press.

Campbell, Donald T. (1960). "Blind variation and selective retention in creative thought as in other knowledge processes," *Psychological Review* 67: 380–400.

—— (1974). "Evolutionary Epistemology," in P. A. Schilpp, ed., *The Philosophy of Karl R. Popper*. LaSalle, IL: Open Court.

Cannon, Walter (1932). *The Wisdom of the Body*. New York: W. W. Norton & Company (rev. and enlarged 1963).

—— (1945). *The Way of an Investigator*. New York: W. W. Norton & Company.

Carnot, Sadi (1824/1977). "Reflections on the Motive Power of Fire," in Eric Mendoza, ed., *Reflections on the Motive Power of Fire and Other Papers on the Second Law of Thermodynamics*. Gloucester, MA: Peter Smith.

Carroll, Lewis (1917). *Through the Looking-Glass, and What Alice Found There*. London: Macmillan & Company.

Cech, T. R. (1986). "A model for the RNA-catalyzed replication of RNA," *Proceedings of the National Academy of Sciences* 83: 4360–63.

Chalmers, David (1995). "Facing up to the problem of consciousness," *Journal of Consciousness Studies* 2: 200–219.

—— (1999). "Materialism and the metaphysics of modality," *Philosophy and Phenomenological Research* 59: 473–93.

Chomsky, Noam (1959). Review of B. F. Skinner's *Verbal Behavior* in *Language* 35: 26–58.

—— (1956). *Syntactic Structures*. New York & Berlin: Mouton de Gruyter.

—— (1965). *Aspects of the Theory of Syntax*. Cambridge, MA: MIT Press.

—— (2006). *Language and Mind*. 3rd edn. Cambridge, UK, & New York: Cambridge University Press.

Churchland, Patricia (1986). *Neurophilosophy: Toward a Unified Science of the Mind-Brain*. Cambridge, MA: MIT Press.

Churchland, Paul (1989). *A Neurocomputational Perspective: The Nature of Mind and the Structure of Science.* Cambridge, MA: MIT Press.

—— (1995). *The Engine of Reason, the Seat of the Soul: A Philosophical Journey into the Brain.* Cambridge, MA: MIT Press.

Clausius, Rudolf (1850/1977). "Ueber die bewegende Kraft der Wärme und die Gesetze die sich daraus für die Wärmelehre selbst ableiten lassen," in Mendoza, ed., *Reflections on the Motive Power of Fire and Other Papers on the Second Law of Thermodynamics.*

Clayton, Philip (2004). *Mind and Emergence: From Quantum to Consciousness.* New York: Oxford University Press.

Collier, John (2003). "Hierarchical dynamical information systems with a focus on biology," *Entropy* 5: 102.

Conway, John (1970). *The Game of Life*—cf. Martin Gardner.

Copeland, B. J., ed. (2004). *The Essential Turing.* Oxford: Clarendon Press.

Corning, Peter A. (1995). "Synergy and self-organization in the evolution of complex systems," *Systems Research* 12: 89–121.

Crane, H. R. (1950). "Principles and problems of biological growth," *Scientific Monthly* 70: 376–89.

Crick, Francis (1995). *The Astonishing Hypothesis: The Scientific Search for the Soul.* New York: Scribner's.

——, and Christof Koch (2003). "A framework for consciousness," *Nature Neuroscience* 6: 119–26.

Dantzig, Tobias (1930). *Number: The Language of Science. A Critical Survey Written for the Cultured Non-Mathematician.* New York: The Macmillan Company.

Darwin, Charles (1859). *On the Origin of Species by Means of Natural Selection or the Preservation of Favored Races in the Struggle for Life.* London: John Murray.

—— (1872). *The Descent of Man and Selection with Relation to Sex.* London: John Murray.

Davidson, Donald (1970). "Mental Events," in L. Foster and J. W. Swanson, eds., *Experience and Theory.* London: Duckworth.

Dawkins, Richard (1976). *The Selfish Gene.* New York: Oxford University Press.

—— (1996). *The Blind Watchmaker: Why the Evidence of Evolution Reveals a Universe Without Design.* New York: W. W. Norton & Company.

—— (1996). *River Out of Eden: A Darwinian View of Life.* New York: Basic Books.

Deacon, T. (2003). "The Hierarchic Logic of Emergence: Untangling the Interdependence of Evolution and Self-Organization," in B. Weber and D. Depew, eds., *Evolution and Learning: The Baldwin Effect Reconsidered.* Cambridge, MA: MIT Press.

—— (2006a). "Reciprocal linkage between self-organizing processes is sufficient for self-reproduction and evolvability," *Biological Theory* 1(2): 136–49.

—— (2006b). "Emergence: The Hole at the Wheel's Hub," in P. Clayton and P. Davies, eds., *The Re-Emergence of Emergence.* Cambridge, MA: MIT Press.

—— (2007). "Shannon–Boltzmann–Darwin: Redefining information. Part 1," *Cognitive Semiotics* 1: 123–48.

—— (2008). "Shannon–Boltzmann–Darwin: Redefining information. Part 2," *Cognitive Semiotics* 2: 167–94.

—— (2010). "What Is Missing from Theories of Information?" in Paul Davies and Niels Henrik Gregersen, eds., *Information and the Nature of Reality: From Physics to Metaphysics.* Cambridge, UK, & New York: Cambridge University Press.

de Duve, Christian (1996). *Vital Dust: The Origin and Evolution of Life on Earth.* New York: Basic Books.

Dennett, Daniel (1978). *Brainstorms: Philosophical Essays on Mind and Psychology*. Montgomery, VT: Bradford Books.

——— (1987). *The Intentional Stance*. Cambridge, MA: MIT Press.

——— (1989). "The Origin of Selves," *Cogito* 3 (Autumn): 163–73.

——— (1991). *Consciousness Explained*. New York: Little, Brown.

——— (1995). *Darwin's Dangerous Idea: Evolution and the Meanings of Life*. New York: Simon & Schuster.

Depew, David, and Bruce Weber (1995). *Darwinism Evolving: Systems Dynamics and the Genealogy of Natural Selection*. Cambridge, MA: MIT Press.

Dewar, Roderick C. (2003). "Information theory explanation of the fluctuation theorem, maximum entropy production, and self-organized criticality in non-equilibrium stationary states," *Journal of Physics, A, Mathematics and General* 36: L631–L641.

Dretske, Fred (1988). *Explaining Behavior. Reasons in a World of Causes*. Cambridge, MA: MIT Press.

Driesch, Hans (1929/2010). *The Science and Philosophy of the Organism*. London: A. & C. Black.

Dyson, Freeman (1999). *Origins of Life*. New York: Cambridge University Press.

Edelman, Gerald (1987). *Neural Darwinism. The Theory of Neuronal Group Selection*. New York: Basic Books.

Eigen, Manfred, and P. Schuster (1979). *The Hypercycle—A Principle of Natural Self-Organization*. Heidelberg: Springer.

Eigen, Manfred, and R. Winkler-Oswatitsch (1992). *Steps Towards Life: A Perspective on Evolution*. New York: Oxford University Press.

Einstein, Albert (1916). "Die Grundlage der allgemeinen Relativitatstheorie," *Annalen der Physik* 49.

——— (1920). *Relativity: The Special and General Theory*, trans. R. W. Lawson, New York: Henry Holt.

Emmeche, Claus, Simo Køppe, and Frederik Stjernfelt (1997). "Explaining emergence: Towards an ontology of levels," *Journal for the General Philosophy of Science* 28: 83–119.

——— (2000). "Levels, Emergence, and Three Versions of Downward Causation," in Andersen et al., eds., *Downward Causation: Minds, Bodies and Matter*.

Empedocles, *On Nature*, in G. S. Kirk, J. E. Raven, and M. Schofield, eds. (1990), *The Presocratic Philosophers*. 2nd edn. Cambridge, UK: Cambridge University Press.

Farmer J. D., S. A. Kauffman, and N. H. Packard (1987). "Autocatalytic replication of polymers," *Physica D* 22: 50–67.

Fodor, Jerry (1990). "Making Mind Matter More," in Fodor, *A Theory of Content and Other Essays*. Cambridge, MA: MIT Press.

——— (1992). "The Big Idea: Can There Be a Science of Mind?" *Times Literary Supplement*, July 3, 1992, p. 5.

——— (1998). *Concepts: Where Cognitive Science Went Wrong*. New York: Oxford University Press.

Foerster, Heinz von (1984). *Observing Systems*. Systems Inquiry series. Salinas, CA: Intersystems Publications.

Franklin, K. J. (1951). "Aspects of the circulations economy," *British Medical Journal* 1: 1347n.

Freitas, R. A., Jr., and R. C. Merkle (2004). *Kinematic Self-Replicating Machines*. Georgetown, TX: Landes Bioscience.

Gardner, Martin (1970). "Mathematical games: The fantastic combinations of John Conway's new solitaire game of 'Life,' " *Scientific American* 223 (October): 120–23.

Gil, Rosario, Francisco Silva, Juli Pareto, and Andres Moya (2004). "Determination of the core of a minimal bacterial gene set," *Microbiology and Molecular Biology Reviews* 68: 518.

Gödel, Kurt (1931/2001). *Collected Works*. Vol. 1: *Publications 1929–1936*. New York: Oxford University Press.

Good, Irving J. (1959). *Speculations on Perceptrons and Other Automata*. IBM Research Lecture RC-115 (Yorktown Heights, NY, 1959), based on a lecture to the Machine Organization Department, IBM, Dec. 17, 1958.

Haeckel, Ernst (1876). *The History of Creation*. New York: Appleton.

Haldane, J. B. S. (1929). "The Origin of Life," in *Rationalist Annual*. Reprinted in J. D. Bernal, ed. (1967), *The Origin of Life*. Cleveland: World Publishing Co.

Hawking, Stephen (1988). *A Brief History of Time*. New York: Bantam Books.

—— (2000). "I think the next century will be the century of complexity," quoted in *San Jose Mercury News*, Jan. 23, 2000. See also *Complexity Digest* 5 (March): 10.

Hoelzer, Guy A., John Pepper, and Eric Smith. "On the Logical Relationship Between Natural Selection and Self-Organization." Unpublished MS.

Hoffmeyer, Jesper (1996). *Signs of Meaning in the Universe*. Bloomington: Indiana University Press.

Hofstadter, Douglas (1979). *Gödel, Escher, Bach: An Eternal Golden Braid*. New York: Basic Books.

Holland, John (1998). *Emergence: From Chaos to Order*. Reading, MA: Helix Books.

Hull, David (1982). "Biology and Philosophy," in G. Floistad, ed., *Contemporary Philosophy: A New Survey*. The Hague: Martinus Nijhoff.

Hume, David (1739–40/2000). *A Treatise of Human Nature*. New York: Oxford University Press.

Humphreys, Paul (1997a). "How properties emerge," *Philosophy of Science* 64: 1–17.

—— (1997b). "Emergence, not supervenience," *Philosophy of Science* 64: S337–S345.

James, William (1897/1979). "The Will to Believe," in *The Will to Believe and Other Essays in Popular Philosophy*. Cambridge, MA: Harvard University Press.

Joule, James (1843). *On the Heat Evolved During the Electrolysis of Water*. Manchester, UK: Joseph Gillett.

Kant, Immanuel (1790/1952). *The Critique of Judgment: II. Teleological Judgment*, trans. James Creed Meredith. *The Great Books* 42:550–613. Chicago: University of Chicago Press.

Kauffman, Stuart (1986). "Autocatalytic sets of proteins," *Journal of Theoretical Biology* 119: 1–24.

—— (1993). *The Origins of Order: Self-Organization and Selection in Evolution*. New York: Oxford University Press.

—— (2000). *Investigations*. New York: Oxford University Press.

——, R. Logan, R. Este, R. Goebel, D. Hobill, and I. Shmulevich (2008). "Propagating organization: An enquiry," *Biology and Philosophy* 23: 27–45.

Kelso, J. A. Scott (1995). *Dynamic Patterns*. Cambridge, MA: MIT Press.

Kessler, M. A., and B. T. Werner (2003). "Self-organization of sorted patterned ground," *Science* 299: 380–83.

Kim, Jaegwon (1993). *Supervenience and Mind*. Cambridge: Cambridge University Press.

—— (1999). "Making sense of emergence," *Philosophical Studies* 95: 3–36.

Kleidon, A. (2004). "Beyond Gaia: Thermodynamics of life and earth system functioning," *Climatic Change* 66: 271–319.

Koffka, Kurt (1924/1980). *Growth of the Mind. An Introduction to Child Psychology*. New York: Harcourt, Brace. New Brunswick & London: Transaction Books.

Koonin, E. V. (2000). "How many genes can make a cell: The minimal-gene-set concept," *Annual Review of Genomics and Human Genetics* 1: 99–116.

Krebs, H. A. (1954). "Excursion into the borderland of biochemistry and philosophy," *Bulletin of the Johns Hopkins Hospital* 95: 45–51.

Langton, Chris, ed. (1989). *Artificial Life*. New York: Addison-Wesley.

Lao Tsu. *Tao Te Ching*. See R. Henrick (1989), *Lao Tzu Te-Tao Ching*. New York: Ballantine Books.

Laplace, Pierre-Simon (1825/1951). *A Philosophical Essay on Probabilities*, trans. from 6th French edn. by Frederick Wilson Truscott and Frederick Lincoln Emory. New York: Dover Publications.

Laughlin, Robert B., and David Pines (2000). "The Theory of Everything," *Proceedings of the National Academy of Sciences* 97(1): 28–31.

——, D. Pines, J. Schalian, B. P. Stojkovic, and P. Woyles (2000). "The Middle Way," *Proceedings of the National Academy of Sciences*, 97(1): 32–37.

Lavoisier, Antoine (1783). *Essays, on the Effects Produced by Various Processes on Atmospheric Air; With a Particular View to an Investigation of the Constitution of Acids*, trans. Thomas Henry. London: Warrington.

Leff, H. S., and A. F. Rex, eds. (1990). *Maxwell's Demon: Entropy, Information, Computing*. Bristol: Adam Hilger.

Lewes, G. H. (1875). *Problems of Life and Mind*. Vol. II. London: Treubner & Co.

Li, Chaorong, Xiaona Zhang, and Zexian Cao (2005). "Triangular and Fibonacci number patterns driven by stress on core/shell microstructures," *Science* 309: 909.

Li, Chaorong, Ailing Ji, and Zexian Cao (2007). "Stressed Fibonacci spiral patterns of definite chirality," *Applied Physics Letters* 90: 164102.

Lloyd, Seth (2006). *Programming the Universe: A Quantum Computer Scientist Takes On the Cosmos*. New York: Alfred A. Knopf.

Locke, John (1690/1997). *An Essay Concerning Human Understanding*, ed. Roger Woolhouse. New York: Penguin Books.

Lorenz, Edward (1963). "Deterministic nonperiodic flow," *Journal of Atmospheric Sciences* 20: 130–41.

Lucretius (n.d./1994). *On the Nature of Things (De rerum natura)*, trans. R. E. Latham. London: Penguin Books.

—— (n.d./1995). *On the Nature of Things: De rerum natura*, trans. Anthony M. Esolen. Baltimore: Johns Hopkins University Press.

Malthus, Thomas Robert (1826). *An Essay on the Principle of Population, or a View of its Past and Present Effects on Human Happiness; with an Inquiry into our Prospects respecting the Future Removal or Mitigation of the Evils which it Occasions*. London: John Murray.

Margulis, Lynn, and Dorian Sagan (1995). *What Is Life?* New York: Simon & Schuster.

Maturana, H., and F. Varela (1980). *Autopoiesis and Cognition: The Realization of the Living*. Dordrecht, Netherlands, & Boston: D. Reidel.

Maxwell, James Clerk (1867). Letter to P. G. Tait, reprinted in C. G. Knott (1911), *Life and Scientific Work of Peter Guthrie Tait*. Cambridge, UK: Cambridge University Press.

—— (1871/2001). *Theory of Heat*. New York: Dover Publications.

Maynard Smith, J., and E. Szathmáry (1995). *The Major Transitions in Evolution*. Oxford: Oxford University Press.

Mayr, Ernst (1974). "Teleological and teleonomic: A new analysis," *Boston Studies in the Philosophy of Science* 14: 91–117.

—— (1985). "How Biology Differs from the Physical Sciences," in D. Depew and B. Weber, eds., *Evolution at a Crossroads.* Cambridge, MA: Harvard University Press.

—— (2001). "The philosophical foundations of Darwinism," *Proceedings of the American Philosophical Society* 145(4) (December): 492–93.

McCulloch, Warren (1965). *Embodiments of Mind.* Cambridge, MA: MIT Press.

Mill, John Stuart (1843). *A System of Logic.* London: Longmans, Green, Reader, & Dyer.

Millikan, Ruth G. (1984). *Language, Thought, and Other Biological Categories.* Cambridge, MA: MIT Press.

Minsky, Marvin (1996). *The Society of Mind.* New York: Simon & Schuster.

Monod, Jacques (1971). *Chance and Necessity: An Essay on the Natural Philosophy of Modern Biology.* New York: Alfred A. Knopf.

Morgan, Conwy Lloyd (1923). *Emergent Evolution.* London: Williams & Norgate.

Morowitz, Harold J. (2002). *The Emergence of Everything: How the World Became Complex.* New York: Oxford University Press.

Neumann, John von (1966), *The Theory of Self-Reproducing Automata.* Urbana-Champaign: University of Illinois Press.

Nicolis, G., and I. Prigogine (1977). *Self-Organization in Nonequilibrium Systems.* New York: John Wiley & Sons.

Nietzsche, Friedrich (1887/1967). *On the Genealogy of Morals*, trans. Walter Kaufmann and R. J. Hollingdale. New York: Vintage.

O'Connor, Timothy (1994). "Emergent properties," *American Philosophical Quarterly*, vol. 31, no. 2: 91–104.

—— (2000). *The Metaphysics of Free Will.* Oxford: Oxford University Press.

Odum, Howard T. (1971). *Environment, Power and Society.* New York: John Wiley & Sons.

Ornstein, Robert (1977). *The Psychology of Consciousness.* New York: Harcourt Brace Jovanovich.

Pattee, Howard (1996). "Simulation, Realization and Theory of Life," in Margaret. A. Boden, ed., *The Philosophy of Artificial Life.* Oxford: Oxford University Press.

Peirce, Charles Sanders (1898/1992). *Reasoning and the Logic of Things: The Cambridge Conferences Lectures of 1898*, ed. Kenneth Laine Ketner. Cambridge, MA: Harvard University Press.

—— (1931–35). *Collected Papers of Charles Sanders Peirce.* Vols. 1–6, ed. Charles Hartshorne and Paul Weiss; Vols. 7–8 (1958) ed. Arthur W. Burks. Cambridge, MA: Harvard University Press.

Penzias, Arno, and Robert Wilson (1965). "A measurement of excess antenna temperature at 4080 Mc/s," *Astrophysical Journal* 142: 419–21.

Pepper, Stephen (1926). "Emergence," *Journal of Philosophy* 23: 241–45.

Pinker, Steven (1995). *The Language Instinct: How the Mind Creates Language.* New York: HarperCollins Perennial.

Pittendrigh, Colin S. (1958). "Adaptation, Natural Selection, and Behavior," in A. Roe and George Gaylord Simpson, eds., *Behavior and Evolution.* New Haven: Yale University Press.

Polanyi, Michael (1968). "Life's irreducible structure," *Science*, vol. 160, no. 3834: 1308–12.

Popper, Karl R. (2003). *Conjectures and Refutations: The Growth of Scientific Knowledge.* London: Routledge.

Prigogine, Ilya (1955). *Introduction to the Thermodynamics of Irreversible Processes.* Chicago: Thournes.

——, and Isabelle Stengers (1984). *Order Out of Chaos: Man's New Dialogue with Nature.* New York: Bantam Books.

Putnam, Hillary (1975). *Mind, Language, and Reality*. New York: Cambridge University Press.

Rasmussen, S., L. Chen, D. Deamer, D. C. Krakaur, N. H. Packard, P. F. Stadler, and M. Bedau (2004). "Transitions from nonliving to living matter," *Science* 303: 963–65.

Robbins, Tom (1984). *Jitterbug Perfume*. New York: Bantam/Dell.

Rorty, Richard (1970). "In defense of eliminative materialism," *Review of Metaphysics*, vol. 24, no. 1: 112–21.

Schneider, E. D., and J. J. Kay (1994). "Life as a manifestation of the second law of thermodynamics," *Mathematical and Computer Modeling*, vol. 19, nos. 6–8: 25–48.

Schopf, J. W., ed. (2002). *Life's Origin: The Beginnings of Biological Organization*. Los Angeles: University of California Press.

Schrödinger, Erwin (1944). *What Is Life? The Physical Aspect of the Living Cell*. Cambridge, UK: Cambridge University Press.

Searle, John (1980). "Minds, brains and programs," *Behavioral and Brain Sciences* 3: 417–57.

Seife, Charles (2000). *Zero: The Biography of a Dangerous Idea*. London: Penguin.

Shannon, Claude (1948). "A mathematical theory of communication," *Bell System Technical Journal* 27 (July & October): 379–423, 623–56.

Shelley, Mary (1818/2009). *Frankenstein or The Modern Prometheus: The 1818 Text*. Oxford: Oxford University Press.

Silberstein, Michael, and Anthony Chemero (2009). "After the philosophy of mind: Replacing Scholasticism with science," *Philosophy of Science* 75: 1–27.

Skarda, C. A., and Walter J. Freeman (1987). "How brains make chaos in order to make sense of the world," *Behavioral and Brain Sciences* 10: 161–95.

Skinner, B. F. (1964). "Behaviorism at Fifty," in T. W. Wann, ed., *Behaviorism and Phenomenology*. Chicago: University of Chicago Press.

——— (1971). *Beyond Freedom and Dignity*. New York: Bantam/Vintage.

Spencer, Herbert (1862). *A System of Synthetic Philosophy: I. First Principles*. Available online at http://praxeology.net/HS-SP.htm#firstprinciples.

Sperry, Roger (1980). "Mind-brain interaction: Mentalism, yes; dualism, no," *Neuroscience* 5: 195–206.

Stanford Encyclopedia of Philosophy at http://plato.stanford.edu/entries/properties-emergent/.

Stich, Stephen (1983). *From Folk Psychology to Cognitive Science: The Case Against Belief*. Cambridge, MA: MIT Press.

——— (1996). *Deconstructing the Mind*. New York: Oxford University Press.

———, and Michael Bishop (1998). "The flight to reference, or how not to make progress in the philosophy of science," *Philosophy of Science*, vol. 65, no. 1: 33–49.

Strick, James E. (2000). *Sparks of Life: Darwinism and the Victorian Debates Over Spontaneous Generation*. Cambridge, MA: Harvard University Press.

Swenson, Rod (1988). "Emergent attractors and the law of maximum entropy production: Foundations to a general theory of evolution," *Systems Research* 6: 187–97.

——— (1989). "Emergent evolution and the global attractor: The evolutionary epistemology of entropy production maximization," *Proceedings of the 33rd Annual Meeting of the International Society for the Systems Sciences*, ed. P. Leddington, vol. 33, no. 3: 46–53.

Szathmáry, Eors (2005). "Life: In search of the simplest cell," *Nature* 433: 469–70.

Thompson, Benjamin Count Rumford (1968). *Collected Works of Count Rumford*. Vol. I: *The Nature of Heat*, ed. Sanborn C. Brown. Cambridge, MA: Harvard University Press.

Thorndike, Edward (1905). *The Elements of Psychology*. New York: A. G. Seiler.

Turing, Alan M. (1939). "Systems of logic defined by ordinals," *Proceedings of the London Mathematical Society*, ser. 2, 45: 161–228 (Turing's PhD thesis, Princeton University, 1938).

—— (1950). "Computing machinery and intelligence," *Mind* 50: 433–60.

Uexküll, Jakob von (1957). "A Stroll Through the Worlds of Animals and Men: A Picture Book of Invisible Worlds," in Claire H. Schiller, ed. and trans., *Instinctive Behavior: The Development of a Modern Concept*. New York: International Universities Press.

Ulanowicz, Robert (2009). *A Third Window: Natural Life beyond Newton and Darwin*. West Conshohocken, PA: Templeton Foundation Press.

Van Ness, H. C. (1969). *Understanding Thermodynamics*. New York: Dover Publications.

Varela, Francisco (1992). "Autopoiesis and a Biology of Intentionality," in Barry McMullin and Noel Murphy, eds., *Autopoiesis and Perception*. A Workshop with ESPRIT BRA 3352. Addendum to the print proceedings distributed during the workshop. Dublin: Dublin City University.

Watson, James D. (2003). *DNA: The Secret of Life*. New York: Alfred A. Knopf.

——, and Francis Crick (1953). "Molecular structure of nucleic acids: A structure for deoxyribose nucleic acid," *Nature* 171: 737–38.

Weaver, Warren, and Claude Elwood Shannon (1963). *The Mathematical Theory of Communication*. Urbana-Champaign: University of Illinois Press.

Weinberg, S. (1977). "The search for unity, notes for a history of quantum field theory," *Daedalus*, vol. 106, no. 4: 17–35.

—— (1993). *The First Three Minutes: A Modern View of the Origin of the Universe*. New York: Basic Books.

—— (1995). *The Quantum Theory of Fields*. Vol. I: *Foundations*. New York: Cambridge University Press.

—— (1996). *The Quantum Theory of Fields*. Vol. II: *Modern Applications*. New York: Cambridge University Press.

—— (2000). *The Quantum Theory of Fields*. Vol. III: *Supersymmetry*. New York: Cambridge University Press.

Weiss, Paul (1926). "Morphodynamik," *Abhandl. Z. Theoret. Biol. H.* 23: 1–46.

—— (1967). "One Plus One Does Not Equal Two," in G. C. Quarton, T. Menlenchuk, and F. O. Schmitt, eds., *Neurosciences. A Study Program*, New York: Rockefeller University Press.

Whitehead, Alfred North (1929). *Process and Reality: An Essay in Cosmology*. London: The Macmillan Company.

Whitfield, John (2005). "Complex systems: Order out of chaos," *Nature* 436: 905–7.

Wiener, Norbert (1948). *Cybernetics, or Control and Communication in the Animal and Machine*. Cambridge, MA: MIT Press.

Wilson, Robert (2005). *Genes and the Agents of Life: The Individual in the Fragile Sciences Biology*. New York: Cambridge University Press.

Wolfram, Stephen (2002). *A New Kind of Science*. Champaign: IL: Wolfram Media, Inc.

INDEX

Page numbers in *italics* refer to illustrations.